Humans versus Nature

Humans versus Nature

A Global Environmental History

DANIEL R. HEADRICK

OXFORD
UNIVERSITY PRESS

OXFORD
UNIVERSITY PRESS

Oxford University Press is a department of the University of Oxford. It furthers
the University's objective of excellence in research, scholarship, and education
by publishing worldwide. Oxford is a registered trade mark of Oxford University
Press in the UK and certain other countries.

Published in the United States of America by Oxford University Press
198 Madison Avenue, New York, NY 10016, United States of America.

Library of Congress Cataloging-in-Publication Data
Names: Headrick, Daniel R., author
Title: Humans versus nature : a global environmental history / Daniel R. Headrick.
Description: New York, NY : Oxford University Press, [2020] |
Includes bibliographical references and index.
Identifiers: LCCN 2019021017 | ISBN 9780190864712 (hardback : acid-free paper) |
ISBN 9780190864729 (pbk : acid-free paper) | ISBN 9780190864736 (updf) |
ISBN 9780190864743 (epub)
Subjects: LCSH: Nature—Effect of human beings on. |
Global environmental change—History.
Classification: LCC GF75 .H4347 2020 | DDC 304.2/8—dc23
LC record available at https://lccn.loc.gov/2019021017

Contents

Acknowledgments

A work of this nature is necessarily the product of many minds. I am grateful to all those who have blessed me with their knowledge, wisdom, and advice, leaving to me alone the responsibility for whatever errors remain.

Among the many people who have read and commented on parts of this book in manuscript, I offer special thanks to John R. McNeill on chapters 1 through 6, Alan Mikhail on the Middle East, Robert Marks on China, Geoffrey Parker on the early modern period, Alex Roland on twentieth-century warfare, Sulmaan Khan on several early chapters, Bruce Campbell on the Black Death, and Michael Bryson on climate change.

I also grateful to Yale University and its magnificent libraries, without which I could not have written this book. I also thank Susan Ferber at Oxford University Press and Jeremy Toynbee for editing this work and guiding it through the publication process.

But most of all, I thank my wife Kate for carefully and patiently reading my manuscript and making it so much better than it would have been without her. It is to her that I dedicate this book.

Humans versus Nature

Introduction

Global Environmental History

In 1928, John Widtsoe, director of the US Federal Bureau of Reclamation, wrote: "The destiny of man is to possess the whole earth; and the destiny of the earth is to be subject to man. There can be no full conquest of the earth, and no real satisfaction to humanity, if large portions of the earth remain beyond his highest control."[1]

Since Widtsoe wrote these words, the conquest of the Earth has proceeded apace. World population has more than tripled. Consumer goods are more plentiful, living standards have risen, and new technologies have transformed the lives of millions. Roads, railroads, airports, cities, and mines cover far more of the surface of the planet. Land once forested has been turned into pastures and farmland. Large parts of the oceans have been depleted of fish and polluted with trash and oil. The atmosphere contains more methane and carbon dioxide. And more species of wild animals and plants have become extinct or reduced to tiny numbers. The changes in the world since 1928 have largely been man-made, at the expense of wild plants and animals and natural landscapes.

Until recently, most people in the West applauded this achievement. After all, didn't it have divine sanction? As the Bible said, "And God blessed them, and God said unto them, Be fruitful, and multiply, and replenish the earth, and subdue it: and have dominion over the fish of the sea, and over the fowl of the air, and over every living thing that moveth upon the earth."[2]

Yet in recent decades, an increasing number of voices have decried the "possession of the earth" as an environmental crisis, marked by pollution, depletion of resources, deforestation, and the extinction of species. This raises the question: When did humans begin to damage the Earth; in other words, when did the "crisis" begin? Historians and scientists have answered this question in a variety of ways.

In 2000, Nobel Prize–winning atmospheric chemist Paul Crutzen and ecologist Eugene Stoermer coined the term "Anthropocene" to refer to the era when humans began to have a major impact on the environment.[3] He argued that this era began with the Industrial Revolution of the late eighteenth century, "when analyses of air trapped in polar ice showed the beginning of growing global concentrations of carbon dioxide and methane." He even gave a specific date—1784—to coincide with James Watt's design of the steam engine.

Historian Stephen Mosely has argued that the root causes of the current crisis can be found back in the sixteenth century, when "the world of nature was reconceptualised as machine-like to meet the needs of an emergent capitalism. . . . [T]his new scientific worldview saw nature as dead matter for human utilisation."[4]

Going further back in time, medievalist Lynn White Jr. in a famous article entitled "The Historical Roots of Our Ecologic Crisis" argued that indifference to nature originated in early medieval Europe when "the distribution of land was no longer based on the needs of a family but, rather, on the capacity of a power machine [the eight-oxen plow] to till the Earth. Man's relation to the soil was profoundly changed. Formerly man had been part of nature; now he was the exploiter of nature." Behind the machine, White continued, was a new religious view of the world: "By destroying pagan animism, Christianity made it possible to exploit nature in a mood of indifference to the feelings of natural objects."[5]

Long before the Middle Ages, environmental scientist Earle Ellis and his colleagues argued, "Relatively small human populations likely caused widespread and profound ecological changes more than 3,000 years ago."[6]

Taking an even wider chronological view, environmental historian Sing Chew claimed that "the history of civilizations, kingdoms, empires, and states is also the history of ecological degradation and crisis. Such a historical trajectory of human 'macro-parasitic' activity has occurred at the systemwide structural level for at least the last five thousand years."[7]

Finally, climatologist William Ruddiman traced the beginning of the human impact on the planet to the first farmers and herders of the Neolithic Age, some 10,000 years ago.[8]

What this debate shows is the very human, yet futile, custom of attaching dates to a long-term process. The human impact on the environment began in a very small way very long ago, but grew gradually—in fits and starts— until the eighteenth century, then began to rise ever more sharply at an accelerating pace until the present. What has changed over time was not the desire of humans to exploit their environments, but the technological and organizational means they developed and employed against the rest of nature—and their consequences.

Humans and Nature

All living beings survive by extracting resources from their environments. In the process, they compete with other living beings: some flourish, some barely survive, and some become extinct. Yet nature has ways of preventing any one species from overwhelming and destroying the others. One way is by limiting the supply of resources available in a given environment. For each species, the environment's carrying capacity fluctuates widely, due to predation, diseases, climate change, and other factors. Lynxes survive by eating rabbits, so once they have decimated the rabbits in their area, many lynxes will starve, allowing the rabbit population to recover, followed by a recovery of the lynx population, and so on. Another way is by limiting the amount of resources any given creature can absorb at one time; a lion, having lunched on a gazelle, will take a nap in the shade of a tree, digesting his meal before undertaking another hunt.

Humans, however, continue to take resources from their environment long after their needs are met. In nineteenth-century America, hunters armed with rifles killed as many bison as they could, taking only their tongues, sometimes shooting them for the sheer pleasure of killing. Long before Europeans came with their rifles, American Indians stampeded herds of bisons over cliffs in order to take the meat from a few of them. Everywhere they went, Stone Age hunters exterminated many of the largest and fiercest animals that roamed the Earth. Since the appearance of *Homo sapiens*, the motivation to take as many resources from the environment as technology allows has been a human characteristic.

Much of history consists of humans' gradual acquisition of the means to realize their desires at the expense of the rest of nature. Humans have time and again found ways to overcome the limitations of their physical abilities through ingenious technological and organizational innovations, allowing them to thrive in almost every environment. Already in prehistoric times, they set fire to forests, drained wetlands, plowed the soil, domesticated plants and animals, and built elaborate irrigation systems. Most of these activities allowed more people to survive and reproduce, and later to create states and high cultures.

Despite many setbacks, human victories over nature have outnumbered their defeats. In modern times, as technological changes have accelerated, humans have acquired ever greater powers to transform the natural environment for their own benefit. Today, humans have the means to clear-cut entire forests, appropriate all accessible fresh water, change the climate, and "possess the whole earth."

Yet it would be wrong to think of the human impact on the environment in solely negative terms. Domesticated animals, as well as hangers-on such as roaches and sparrows, are far more numerous than they would be without us; the same is true of domesticated plants and weeds.

At the same time, the natural world still has agency and periodically disrupts the human urge to subdue, dominate, and possess. Droughts, floods, hurricanes, and other weather anomalies are reminders of this power. Volcanoes, earthquakes, and tsunamis wreak havoc. And periodically epidemics put entire populations at risk.

Some natural events that harm humans are themselves the unintended consequences of human actions. Burning fossil fuels emits greenhouse gases that cause global warming, and global warming in turn causes weather anomalies that affect human societies. New means of transportation allow diseases like cholera, SARS, and Zika to spread quickly. Antibiotics fed to farm animals encourage the emergence of antibiotic-resistant bacteria that endanger human health.

Finally, humans are highly dependent on the natural world—not just those parts of nature we have domesticated, but also those we have not: the forests that absorb carbon and emit oxygen; the wildernesses that offer respite from the stresses of civilization; and the myriad plants, animals, and microorganisms that promise scientific breakthroughs to benefit humanity. In other words, as ecologists stress, everything in nature, including humans, is interconnected and interacts in complex feedback loops.

This book examines these complex interactions between humans and the rest of nature. Two aspects particularly stand out. One is the human impacts on the rest of nature and how they have changed over time. The other is how nature, in turn, has affected humans and human civilizations, especially the "natural disasters" that have disrupted (and sometimes reversed) the advancing power of humans over the natural world. Examples of both kinds of interactions abound in the chapters that follow.

Chapters 1 through 3 are largely chronological, following the evolution of human societies from hunters and gatherers to farmers and herders and to the first civilizations. Not all regions of the world went through all of these phases, and those that did followed the sequence at different times.

The first chapter relates the difficult and contentious relations between early humans and the environments in which they lived. Early humans were extremely vulnerable to the forces of nature and at one point almost went extinct. Yet their descendants migrated to every continent except Antarctica and learned to survive in environments for which their bodies were totally unsuited. Once arrived in a new environment, using only fire and simple handheld weapons, they exterminated numerous species of large animals as no creature had done before.

Chapter 2 describes the domestication of plants and animals that allowed the rise of two kinds of communities: settled agricultural villages and no-madic herding bands. In the process, humans changed landscapes by transforming forests and natural grasslands into farms and pastures. Though the human population increased, their health and stature declined, and new diseases appeared.

Chapter 3 recounts the appearance of complex hierarchical societies around the world. Remarkably, many early civilizations were organized around the control of water: irrigation, drainage, and the struggle against floods. Some societies were almost destroyed by droughts while others proved more resilient.

Starting in Chapter 4, the narrative takes a geographical turn as the history of the two hemispheres diverges. Chapter 4 describes how humans increas-ingly farmed on rain-watered lands and vastly expanded the area in Eurasia and Africa they occupied from the mid-second millennium BCE to the sev-enth century CE. But the increased contacts between large-scale societies made them vulnerable to pandemics, such as the plague.

The following chapter recounts the consequences of the Medieval Climate Anomaly (a particularly warm period from the eighth to the fourteenth cen-turies) for the development of large-scale societies throughout Eurasia and sub-Saharan Africa. In many parts of the Eastern Hemisphere, however, human encroachments on nature were reversed by the beginnings of the Little Ice Age and the calamity of the Black Death.

Chapter 6 describes the enormous biological changes that accompanied the opening of contacts between the Eastern and Western Hemispheres from the sixteenth to the eighteenth centuries. Diseases imported from Europe and Africa reduced the indigenous population of the Americas by nine-tenths or more, allowing the recovery of forests and wildlife. At the same time, the

Americas were invaded by Old World plants and animals, some of which went feral and dramatically changed the environments of the New World.

Chapter 7 looks at this period in the Eastern Hemisphere. Climatically, this was the Little Ice Age that precipitated major economic and political crises. The peoples of the Eastern Hemisphere survived the crisis (and in a few places even flourished) thanks to the bounty provided by New World crops.

Chapters 8 and 9 focus on industrialization and the sudden increase in the power of humans over nature that it provided. Its effects diverged sharply, with the Western nations as centers of power, and non-Western regions as objects of transformations imposed from outside.

Chapter 8 introduces the Industrial Revolution and its impact on the environments of two major industrializing countries. In both Great Britain and the United States, cities and industries expanded dramatically, polluting local air and water. The impact of American industrialization was especially severe and widespread, leading to the plunder of natural resources, the destruction of arable soils and forests, and the decimation or the extinction of several species of wildlife. At the same time, industrialization encouraged the population to grow fast and reduced people's vulnerability to natural shocks.

Chapter 9 looks at the non-industrializing part of the world, especially monsoon Asia in the same period. India, China, and Southeast Asia (but also Egypt and Brazil) were profoundly affected by Western industrialization, especially by the demand for tropical crops that led to a vast expansion of arable land at the expense of forests and their fauna and flora. Although the population of the affected regions increased, standards of living did not and people remained as vulnerable as ever to floods, droughts, and epidemics.

Two thematic chapters cover the twentieth century. Chapter 10 deals with the environmental impact of both world wars and conflicts such as the Vietnam War. It also discusses major development schemes, as in the Soviet Union, the United States, China under Mao, and Brazil, and the effect of these schemes and projects on forests, wildlife, and other environments. Chapter 11 looks at peacetime economies and the rise of mass consumerism and its environmental costs, especially the impact of automobiles, petroleum, and industrial agriculture. The areas covered include the United States, Western Europe, Japan, and China after Mao.

The next three chapters take up current environmental issues in their historical contexts. Chapter 12 describes the recent climate change and its causes, as well as scientific predictions and possible future scenarios. It also discusses the politics of global warming, both nationally and internationally, and the public reaction to the issue and to the debates.

Chapter 13 goes underwater to reveal the impact of hunting on whale populations and the collapse of cod stocks through overfishing. It discusses the sustainability of salmon populations through farming and the control of wild salmon fishing. It also describes the impact of humans on the oceans in the form of dead zones, coral bleaching, and the accumulation of garbage.

Chapter 14 addresses the extinction of terrestrial species as a natural phenomenon, and with the five extraordinary mass extinctions in the history of the Earth, such as the one that wiped out the dinosaurs. We are currently witnessing a sixth mass extinction, this one caused by human beings through habitat destruction, hunting, and global warming, especially in the tropical rainforests and in the Arctic. The chapter also deals with the survival and multiplication of plants and animals selected and encouraged by humans, but also weeds, pests, and pathogens that benefit from unintentional human actions.

The last two chapters address responses to the current environmental crisis. Chapter 15 considers human attitudes toward nature. Traditional societies devised rules and taboos to protect aspects of the natural world that they valued, such as sacred woods or hunting preserves. While deeply held, these beliefs and behaviors slowed down, but never reversed, people's desire to use nature for their own benefit.[9] Since the nineteenth century, environmentalists have been decrying the damage that untrammeled development has inflicted on the natural environment. In recent decades, their voices have entered the political discourse. As a result, modern states have tried to mitigate the impact on the natural world through restrictions on pollution, laws to protect endangered species, national parks and wilderness areas, international agreements on fishing and whaling, and nuclear arms limitation treaties.

In the twenty-first century, humanity faces difficult decisions, as discussed in the Epilogue. The need to protect what is left of the natural environment is clear. At the same time, the pressure to continue along the path of expansion and development is more powerful than ever. We humans have now outsmarted all other living beings and taken over much of the planet. Yet we march backward into the future, blind to what is to come. What can the story of the past teach us about ourselves and how we should interact with the planet we live on? Will technological breakthroughs allow us to continue enjoying the benefits of past innovations while mitigating the harm they have inflicted on the environment? As we take ever greater control over the Earth as though it were the Planet Machine, will we operate it with wisdom, restraint, and balance, qualities humans have so seldom displayed in the past?

1

The Foragers

It is tempting, looking back at the history of humanity, to believe that it was our destiny—by divine right or because of our superior intelligence—to become the dominant species on Earth. Yet, 200,000 years ago, when the first *Homo sapiens* appeared, there were other members of the genus *Homo*— *Homo erectus*, Denisovans, Neanderthals, to name a few. One by one, they went extinct, and only *Homo sapiens* survived. It is by chance that we survived while other species died out.

Yet humans differed from other species of the genus *Homo* from very early times. Even as they lived at the mercy of natural forces, they also began transforming their environments in ways no other species did. At a site in the Czech Republic, archaeologists found the bones of a thousand mammoths killed by humans. At sites in Colorado and Wyoming, they found the remains of herds of bison stampeded over cliffs. And in New Zealand, hundreds of thousands of skeletons of moas—gigantic flightless birds—were found near Maori hunting sites.[1] In many parts of the world, humans killed large numbers of great animals, some of whom went extinct soon after humans arrived on the scene. Humans also transformed plant life, either directly by setting fire to forests to encourage open grasslands, or indirectly by killing the animals that kept certain plants in check. In short, *Homo sapiens* caused major changes in the environment almost as soon as they appeared on Earth.

Homo Erectus

The designation *Homo* is an honorific bestowed upon a genus of bipedal animals by those of us who call ourselves *Homo sapiens*, or knowing man. Between australopithecines and ourselves were many hominid species. Most went extinct, but one lineage survived to become the ancestors of *Homo sapiens*. The most recent of our now-extinct ancestors were creatures we call *Homo erectus* ("standing man"), a species that appeared in Africa between 1.9 and 1.6 million years ago.

Unlike earlier hominins who spent part of their time in trees, *Homo erectus* were full-time ground dwellers. Their teeth and jaw muscles were smaller than those of their predecessors, too small to chew raw meat efficiently. They had a smaller gut than earlier hominins, indicating that they did not need to spend as much energy and time digesting as other carnivores. They may also have lost the fur that covered other primates and instead developed sweat glands that allowed them to spend long stretches under the hot sun. Their brain volume measured up to 950 cubic centimeters, two-thirds the size of ours. Such a brain required a lot of energy, which they obtained from a diet that included meat.[2]

With a larger brain came better tools. Because the only tools that have survived were those made of stone, archaeologists call this period of prehistory the Paleolithic, or Old Stone Age. Earlier hominins, as far back as the australopithecines 3 to 4 million years before, had made rough choppers by breaking a piece off a cobble, leaving a sharp edge. *Homo erectus* were able to obtain 60 centimeters of cutting edge from a kilogram of flint, four times more than previous hominins. They created a tool kit consisting of stones carved to form hammers, cleavers, or choppers, along with the associated flakes.[3] Such tools remained in use, with some refinements, for a million years. In addition, *Homo erectus* almost certainly used wooden sticks or, in East Asia, bamboo, though evidence thereof has long since perished.

Homo erectus were the first hominins to migrate out of Africa, probably following herds of herbivores seeking better grazing lands during a dry period some 1.8 million years ago.[4] In many places, they joined the ranks of the dominant predators. In Africa, they displaced sabertooth cats and the remaining australopithecines.[5] Archaeologists have found remains of *Homo erectus* and the animals they killed in China, the Caucasus, Hungary, Java, Spain, and France, as well as Africa. At two sites in Spain, Torralba and Ambrona, they found the bones of thirty elephants, as well as deer, aurochs (wild cattle), and rhinoceroses. That does not mean that *erectus* attacked such large and dangerous prey in the open; they killed and butchered animals mired in a swamp.[6]

Fire and Food

What most sharply distinguished *Homo erectus* from all other creatures was their systematic use of fire. According to some anthropologists, direct physical remains of their fires, such as charcoal, burned bones, and hearths made of stones, occur at Bouche de l'Escale in France, dated between 350,000 and 250,000 years ago, followed by Vertesszölös in Hungary and Terra Amata in France around 250,000 years ago and by 150,000-year-old ash deposits in Hayonim Cave in Israel. Traces of fire become more numerous after that, both in western Europe and in China; after 110,000 years ago, there are abundant remains.

Earlier hominins may have made use of natural fires started by lightning. Archaeologists have found burned seeds, wood, and flint at Gesher Benot Ya'akov dating back at least 790,000 years.[7] In addition, according to anthropologist Richard Wrangham, there are "provocative hints" of fire control at two sites in Kenya—Chesowanja dating back 1.42 million years and Koobi Fora from 1.5 million years ago—and more definitive evidence of human control of fire at Swartkrans, South Africa, 1 million years ago.[8] These findings remain controversial, however.

This does not mean that Homo erectus knew how to start a fire. As recently as the early twentieth century, some of the indigenous people of Tasmania and the Andaman Islands knew how to keep a fire going but not how to start one. More likely, Homo erectus found natural fires caused by lightning and used a burning branch to set fire to other wood. These activities—seizing a burning branch, carrying it to a safe place, collecting firewood, and feeding the fire—had to be learned over thousands of years.[9]

Wrangham argues that besides the physical remains of fires, there is other evidence that Homo erectus used fire. Kill sites prove that Homo erectus ate meat. Yet their small teeth, jaw muscles, and guts could not efficiently chew or digest raw meat. Therefore they must have roasted it. Tubers and roots also needed to be cooked to make them edible by softening them or removing toxins. Hence they must have used fire systematically.[10]

Control of fire had other consequences. It scared away other animals, making it possible for Homo erectus to spend the night on the ground, safe from nocturnal predators and snakes. They may have used fire to harden the points of wooden spears or objects of bone. Having lost their body hair, they needed fire to keep warm at night, especially in cooler regions like Europe and China.[11] Keeping a fire going and sitting around it at night may also have helped them socialize and communicate, maybe even take the first step toward language.[12]

Possession of fire provides a clue as to the evolution of Homo erectus. Climate change probably played a role, but it is more likely that fire was the determining factor. While the ancestors of *Homo erectus* ate their food raw, a few may have chanced upon meat or tubers roasted in a natural fire, or may accidentally have dropped a piece of meat or a tuber into a fire and found the results delectable enough to try again. Armed with this new method of preparing food, those willing to approach fires may have become more successful hunters, been less often the victims of predators, and seen more of their children survive. Eventually, the fire users predominated over the others, and their descendants developed jaws and guts adapted to their new diets.

Fire control also challenged their brains, for the use of fire gave an advantage to those with bigger brains, whose offspring then survived in larger numbers than their smaller-brained counterparts. If so, it was the first instance of cultural evolution preceding, even causing, biological evolution. In Wrangham's words, "The reduction in tooth size, the signs of increased energy availability in larger brains and bodies, the indication of smaller guts, and the ability to exploit new habitats all support the idea that cooking was responsible for the evolution of *Homo erectus*."[13]

Did the appearance of *Homo erectus* also transform the environments in which they lived? Forest historian Michael Williams thinks so: "With fire humans accomplished the first great ecological transformation of the earth." But that transformation came later, for other scholars have found no evidence of environmental changes caused by *Homo erectus*. They may have been human, but they were not human enough to transform the natural world in which they lived.[14]

From *Erectus* to Neanderthals

Some 600,000 years ago or more, a new creature, *Homo heidelbergensis*, appeared in Africa, probably descended from African *erectus*. Its brain, measuring about 1,200 cubic centimeters, was close to that of *Homo sapiens*. It too migrated out of Africa, reaching Europe half a million years ago. Most anthropologists believe that it was the ancestor of *Homo neanderthalensis* (in Europe and the Middle East) and *Homo sapiens* (in Africa).[15]

Until they vanished 28,000 years ago, Neanderthals inhabited Europe and much of the Middle East. They were short (1.67 meters on average), stocky, and barrel-chested. They weighed, on average, 76 kilograms, considerably more than most humans before the twentieth century.[16] Their bones, heavier than those of modern humans, supported powerful muscles. At 1,300 cubic centimeters, their brains were larger than those of modern humans. Their genes and vocal tract indicate that they were able to articulate more sounds than any earlier hominin, although we do not know whether they had a vocabulary. These were not the simpletons depicted in cartoons but humans like us, albeit built like weightlifters.

The reason for their squat, muscular bodies is that they evolved in Eurasia during the Ice Age, when temperatures were much lower than today. Theirs was a case of successful biological evolution, though they probably also draped themselves in animal pelts to keep warm. They made a variety of stone flake-tools such as spear points and curved scrapers, which remained much the same for over 5,000 generations.

Where they differed from later *Homo sapiens* is that Neanderthals were big-game hunters. Maintaining their muscular bodies in frigid weather required a diet of 5,000 calories a day. Analysis of their teeth shows that this diet consisted largely of meat. To obtain enough, they hunted relentlessly. Wielding spears and clubs, they attacked large animals at close quarters, such as woodland elephants, red deer, wild boar, and bears. This was a dangerous way of life, as shown by the fractures found on many of their remains. But they took care of the injured, the handicapped, and the elderly. They buried their dead. All their campsites show signs of fire, even stone firepits. In winter they retreated into caves and rock shelters. In short, they were specialists in an age of ice.[17]

What Makes Us Human

Meanwhile in Africa other descendants of *Homo heidelbergensis* had become recognizably the ancestors of *Homo sapiens*. These new creatures were creative and imaginative; used language and symbolic thought; shared their knowledge and passed it on to younger generations; relied on tools and other devices in almost everything they did, and changed them with increasing frequency; and, rather than adapting to their environments as Neanderthals did, began to change the environment to meet their needs. With an intelligence far greater than they needed to survive, they often overshot the mark, hunting other animals to extinction, pushing beyond the carrying capacity of their environment, eventually enslaving and murdering each other in huge numbers.

Homo sapiens

The earliest known remains of *Homo sapiens*, recently discovered in Morocco, date back between 300,000 and 350,000 years ago.[18] Other remains, discovered in Ethiopia, date back some 195,000 years. Analysis of mitochondrial DNA shows that all humans can trace their ancestry back to a single woman who lived between 200,000 and 95,000 years ago. Similarly, recent analyses of Y-chromosomes, transmitted from fathers to sons, show an early male ancestor at about 200,000 years ago. These estimates will certainly be revised as the technology of genetics improves.[19]

A question that has been much debated is whether the earliest *Homo sapiens* were only anatomically, or also behaviorally and intellectually, like humans today and, if so, when they began to exhibit modern behavior and symbolic thought. Until recently, archaeologists associated modern behavior with the cave paintings and other symbolic artifacts found in Europe dating to 40,000 years ago. These findings led scientists to believe that a dramatic increase in intelligence, creativity, and culture took place in Europe at that time, but they have now been discredited, thanks to recent archaeological evidence that the transition to modern behavior and symbolic thinking began in Africa long before anywhere else.[20]

During a cold and arid period that lasted from 195,000 to 90,000 years ago, a small number of *Homo sapiens*, perhaps only a few hundred, found shelter in caves on the coast of South Africa. Excavations at Pinnacle Point show evidence of human habitation between 164,000 and 35,000 years ago. The inhabitants of these caves survived by eating shellfish as well as whales and seals washed up on the shore. They probably also harvested tubers. They attached stone points onto spears with tree sap and other sticky substances; if so, these are the oldest composite tools known. To prepare stones, they heated them in a fire, a technique used sporadically at Pinnacle Point as early as 164,000 years ago, and consistently from 72,000 years ago.[21]

People who lived at Blombos Cave 100,000 years ago used to pound and grind red ochre and abalone shells, then mixed the powder with bone marrow and charcoal, perhaps to decorate their bodies. These objects, of no immediate practical use, are evidence of a higher level of symbolic thinking.[22]

Further evidence of modern behavior, such as perforated seashells probably used as beads, have been found in South Africa, Algeria, Morocco, and Israel perhaps as far back as 135,000 years ago, but certainly 70,000 to 80,000 years ago. More practical objects, such as spearhead points meant to be attached to wooden handles and intentionally marked ostrich shells, have also been found in South Africa.[23] Barbed harpoon points found in the Democratic Republic of the Congo date back 80,000 years. *Homo sapiens* also killed and butchered zebra and Cape warthog, animals previously too dangerous to hunt. These objects and behaviors show a sophistication unknown among earlier hominins.[24]

Nonetheless these advances did not lead right away to a flowering of new cultures because humans at the time were so few in number and so widely scattered. As archaeologist Chris Stringer explains: "It is as if the candle glow of modernity was intermittent, repeatedly flickering on and off again. Most of the suite of modern features does not really take root strongly and consistently until much later, close to the time when humans began their final emergence from Africa about 55,000 years ago."[25]

Bottleneck and Exodus

Like *Homo erectus, Homo heidelbergensis,* and Neanderthals before them, *Homo sapiens* migrated from Africa in search of resources as soon as environmental conditions permitted.[26] At Jebel Faya on the east coast of Arabia, archaeologists have found stone tools dated about 125,000 years ago that are similar to those used by *Homo sapiens* in Africa around the same time. At the time, with the most recent ice age at its peak and the oceans at their lowest level, the Straits of Bab el-Mandeb that separate Eritrea from Arabia were at their narrowest and the climate of Arabia was wetter than it is today.[27] After that, the next *Homo sapiens* remains outside of Africa were found at Skhul and Qafzeh in Israel, dated between 80,000 and 110,000 years ago, near Neanderthal artifacts using the same Mousterian technology.[28]

No sooner had *Homo sapiens* shown the ability to think in symbols and travel the world than they almost vanished. Their population, which geneticists estimate at some tens of thousands of members 100,000 years ago, crashed. How low it fell is a matter of much speculation: "a few thousand, perhaps even a few hundred, members," according to Ian Tattersall; "a few thousand breeding pairs," according to Kate Ravilious; "as low as ten thousand individuals," according to Jonathan Wells and Jay Stock; "no more than a mere two thousand people," according to Spencer Wells.[29]

The most likely cause of the crash was the eruption of the volcano Toba in Indonesia sometime between 75,000 and 71,000 years ago. This was the largest natural disaster since the Chicxulub asteroid hit the Earth 65.5 million years ago and exterminated the dinosaurs.[30] Toba expelled 2,800 cubic kilometers of magma and sent 800 cubic kilometers of ash into the atmosphere.[31] It deforested most of Southeast Asia and covered much of India with a layer of ash. It caused six years of winter throughout the Northern Hemisphere, followed by a thousand years in which the temperatures remained lower than during the worst of the Ice Age, decimating the flora and fauna and causing widespread famine. *Homo sapiens* living in Arabia and the Middle East died. Others found refuge in equatorial Africa, southern India, and the Malay Peninsula, but in greatly reduced numbers; the Neanderthals migrated to Europe.[32]

Once the volcanic cold abated, the remaining *Homo sapiens* in Africa began to multiply again and seek new lands. With the return of warmth and vegetation, descendants of the survivors increased in numbers. Those who survived the catastrophe were the toughest, most energetic, and most adaptable of their species. Between 65,000 and 45,000 years ago, some of them ventured out of Africa, carrying with them more complex and effective tools than their predecessors had, including composite weapons and sharper spearheads.[33] From analyzing the DNA of different peoples around the world, geneticists have determined that the number of emigrants was tiny: "at most 550 women of childbearing age, and probably considerably fewer" according to Vincent Macaulay; "at most a few hundred colonists" according to Paul Mellars.[34]

For a long time, it was believed that they could have left Africa only by walking down the Nile Valley, across the Sinai Peninsula, and into western Asia. Increasing evidence now points to a more likely route across the Straits of Bab el-Mandeb to Arabia. Even during the Ice Age, when the sea was at its lowest and the straits at their narrowest, this would have required a boat or a raft. Once having mastered that technology, the pioneers and their descendants could have followed the coasts of Arabia, India, and Southeast Asia. This route had the advantage of a warmer climate than the interior of Eurasia during the Ice Age. Unfortunately for archaeologists, the sea level has since risen, obliterating their campsites.[35]

From the small number of survivors grew the humanity we know today. Until about 12,000 years ago, it grew exceedingly slowly, for foragers deliberately kept their numbers low. They did so by avoiding sexual intercourse as long as a woman was nursing her child, which took up to three years; as a result, the average time between births was four years. Foragers also

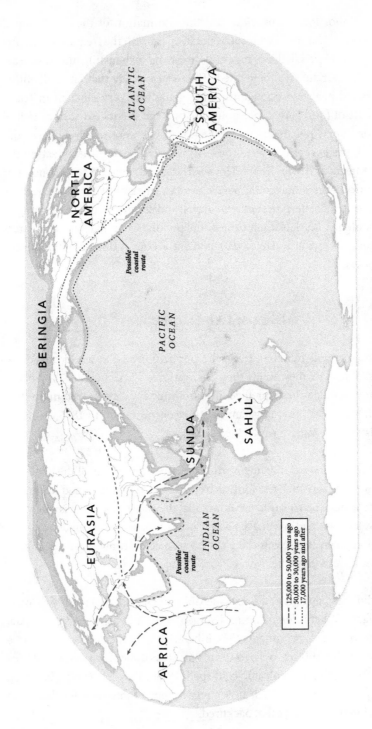

Figure 1.1. Map of migrations of *Homo sapiens* from Africa to Eurasia, Sahul, and the Americas.

practiced abortion, infanticide, and the elimination of the handicapped and the severely injured who could not keep up with the frequently moving band. They took all these measures to avoid exhausting the resources on which their survival depended. This is not to say that late Paleolithic foragers led healthy lives. Hunting big game was very efficient in terms of calories of food gained per hour of labor, but it entailed serious risk of death or injury. By constantly moving, foragers were relatively free from parasites; the exceptions were those who risked trichinosis by eating the meat of bears and wild boars. However, they all suffered from a number of chronic diseases that could persist even in small hunting bands, such as yaws, salmonella, herpes, staphylococcus, and streptococcus. According to anthropologist Mark Nathan Cohen, the population of Paleolithic foragers grew very slowly; at a rate of 0.01 percent a year, it doubled about every 7,000 years.[36]

Diaspora And Extinctions

From these small numbers came the population of the world outside of Africa. Wherever they went, they changed the environments they occupied. In particular, these intelligent, well-armed humans proved more than a match for all the other large animals they encountered. In 1876 the noted biologist Alfred Wallace wrote:

> We live in a zoologically impoverished world, from which all the hugest, and fiercest, and strangest forms have recently disappeared; and it is, no doubt, a much better world for us now that they have gone. Yet it is surely a marvelous fact, and one that has hardly been dwelt upon, this sudden dying out of so many large Mammalia, not in one place only but in over half the land surface of the globe.[37]

How did this happen, and what role did *Homo sapiens* play in this "marvelous fact"? Where *Homo sapiens* diverged from earlier species is the impact they had on the environments they inhabited. All other creatures found a niche to inhabit, subject to the forces of nature, as did *Homo sapiens* in Africa before and during the volcanic cold spell. When they left Africa, however, they began to transform their new environments, sometimes for their own benefit, but often in ways that backfired.

The first immigrants into a formerly human-free environment often found what must have seemed like infinite resources there for the taking. Unrestrained by resource scarcities or predators, they lived richly and multiplied—for a while at least. Once they had depleted the landscape, a period of disillusionment and recrimination must have followed. After this, some, as in Australia, reached a sustainable relationship with the now-depleted environment. Others, elsewhere, developed new technologies that allowed them to exploit new resources.

We now know what Wallace could only speculate about. Australia lost 86 percent of its megafauna (animals with an adult weight of over 44 kilograms); South America lost 80 percent; North America lost 73 percent; and Europe lost 14 percent. Among mammals, the extinction rates are even starker: 94 percent in Australia, 73 percent in North America, 29 percent in Europe, and 5 percent in Africa.[38] The enormous contrasts between these rates of extinction reflect the kinds of animal *Homo sapiens* encountered, but, more important, the reactions of these animals to humans. Where animals had long been in contact with hunter-gatherers of the genus *Homo*, as in Africa and (to a lesser degree) in Europe, they were wary, hence better able to survive. Where they encountered humans for the first time, as in Australia and the Americas, they were naive and trusting, and more easily killed.

What caused these extinctions is the subject of intense controversy. Two hypotheses—climate change and human predation—have been proposed, while some scholars suggest that it was a combination of the two that drove so many large animals to extinction. The combination of human predation and climate change applies best to North America, where the extinctions took place at a time of rapidly fluctuating climate. Elsewhere, the extinctions occurred at times—from 35,000 years ago or more in Australia to less than 800 years ago in New Zealand—that rarely coincided with a major change in the climate. Furthermore, the animals in question had survived repeated climate changes over thousands of millennia. Unlike earlier mass extinctions that affected animals of all sizes, the more recent extinctions affected only large animals.[39]

Finally, there is the coincidence that these large animals died at about the same time that humans entered their environments. So we cannot exonerate *Homo sapiens*. Yet hunter-gatherers were thinly scattered on the ground; geographer Ian Simmons estimates that in Upper Paleolithic Europe there may have been one for every 10 square kilometers, in Australia as few as one per

250 square kilometers, and globally one per 25 square kilometers.[40] How could so few hunter-gatherers have exterminated entire species of animals? To understand this, we must look at the human migrations out of Africa after Toba and their impact on the megafauna of the regions they invaded.

Australia

Until recently, archaeologists believed that the earliest humans reached Australia between 46,000 and 42,000 years ago, long before they reached Europe, despite its far greater distance from Africa. Newer research, however, showed that humans inhabited the arid interior of the continent as early as 49,000 years ago.[41] Among the migrants who left Africa, those who crossed the Red Sea at the straits of Bab el-Mandeb were able to continue along the Indian Ocean coast, surviving on fish, mollusks, seabirds, turtles, and seashore vegetation until they reached Indonesia. Archaeologists have found evidence of humans in the Malay Peninsula before the Toba eruption.[42] Even at the height of the Ice Age, there was still a 100-kilometer stretch of open water between Sunda (a continent that included Southeast Asia and most of Indonesia) and Sahul (a continent that included New Guinea and Australia). Having learned to build seaworthy rafts or boats, *Homo sapiens* finally reached and colonized Sahul. This was evidently a brief episode, for the indigenous peoples of these lands share genetic markers found nowhere else, showing that they did not return to Sunda. All evidence of navigation and coastal life disappeared under water when the Ice Age ended and the seas rose.[43]

Before humans arrived, Australia was a different world from the other continents. It had been separated by seas from Eurasia and the Americas for so many millions of years that the plants and animals that evolved there were quite unique; for instance, in place of placental mammals, there evolved marsupials like kangaroos that continued their gestation in an external pouch, or monotremes like platypuses that laid eggs but nursed their babies after they were hatched.

Many Australian animals were also very large. Three species of marsupials, including *Diprotodon*, a wombat the size of a rhinoceros, weighed up to 3 tons. *Procoptodon* was a 250-kilogram, 3-meter-high kangaroo. There were enormous flightless birds, such as *Dromornis* that stood 4 meters tall and weighed up to 500 kilograms and *Genyornis*, over 2 meters tall and weighing

220–240 kilograms. A reptile named *Megalania* was over 7 meters long and weighed up to 1,950 kilograms.[44]

Australian Extinctions and Environmental Changes

To the first humans who landed in Australia, these animals must have seemed not only strange, but tame, for they had no reason to fear humans. As late as 1802, there were still emus, wallabies, wombats, and elephant seals on islands off the southern coast of Australia that Aborigines had evidently never visited, for the first European visitors had no trouble catching (and exterminating) them.[45]

After humans arrived, fifteen out of sixteen species of large mammals, or nineteen out of twenty species including large birds and reptiles, vanished. Numerous smaller animals also went extinct around the time that humans arrived or soon after. All in all, sixty species of vertebrate animals went extinct. Whether humans caused their extinction, or merely witnessed it, is still controversial.[46] Ecologist Tim Flannery is convinced that "the weight of evidence is now clearly in favour of a very rapid, human-caused extinction for the Australian megafauna."[47] However, archaeologist Donald Grayson calls the idea that humans hunted these animals to extinction "even more speculative than it is in North America."[48] So far, archaeologists have found no butchering sites, bones with butchering marks, or other evidence of human killing. However, it is not necessary to believe that humans killed all the large animals they encountered. Almost all large animals breed slowly and have very few offspring. By killing a few breeding females, hunters lowered their birth rate below their natural death rate, until there were none left. Furthermore, there is proxy evidence of their extinction in the form of spores of the fungus *Sporormiella* that reproduces by being ingested and excreted by large herbivores. Around 41,000 years ago, these spores disappeared— evidence that their hosts had also disappeared.[49]

Not all large animals became extinct, of course. Those that are left— kangaroos, wallabies, walleroos, and quolls (small carnivorous marsupials)— are smaller than their pre-human ancestors because hunters killed off the larger ones. Small animals (up to 5 kilograms) did not shrink. Humans migrated to Australia several times before Europeans arrived. On one such migration, some 3,000 or more years ago, they introduced the dingo, or wild Australian dog, causing the extinction of the thylacine, a carnivorous marsupial in Australia, although a small number survived in Tasmania until the twentieth century.[50]

The disappearance of the largest animals had ripple effects throughout the Australian biota. Big herbivorous animals kept the soil fertile by recycling the nutrients in the vegetation they ate. Fires burned more fiercely because, in the absence of large herbivores, the vegetation they would have eaten was left to accumulate and, when lightning struck, it burned uncontrollably. Without herbivores to recycle the nutrients in plants, fires impoverished the soil, turning the carbon in plants into carbon dioxide in the atmosphere and leaving the ground open to erosion. As they burned, fire-sensitive plants were replaced by fire-promoting ones like eucalyptus.[51] In Flannery's view, "fire has made Australia—originally the most resources-poor land—an even poorer one."[52]

Eventually, after the largest animals were gone, the ancestral Aborigines adapted to their new homeland. Abel Tasman, the first European to see Tasmania in 1642, noted fire and smoke all along the coast. What he saw was "firestick farming," fires deliberately set by the inhabitants before so much dead vegetation accumulated that natural firestorms would burn out of control. This burning also encouraged the growth of new grass and encouraged the population of kangaroos, bandicoots, and wallabies that were the Aborigines' main food source. On the Cape York Peninsula, their fires stimulated the growth of fire-resistant cycad trees that produced edible kernels.[53] The inhabitants also set aside "story places," off-limits to humans, where favored game animals like tree-kangaroos could reproduce and then repopulate nearby hunting grounds.

Thus did the Aborigines of Australia learn to survive for tens of thousands of years in an unpredictable environment, living in small numbers close to the carrying capacity of their arid land and devoting much time to cultural activities and to maintaining social relationships over long distances. To Nicholas Wade, they "settled into a time warp of perpetual stagnation." To Tim Flannery, this demonstrates the sustainability of their way of life.[54]

Homo sapiens in Eurasia

From the Middle East, some Homo sapiens made their way to Central Asia, Europe, and Siberia between 40,000 and 35,000 years ago, to East Asia some 35,000 years ago, and by land to India around 30,000 years ago. Whether any

Homo erectus were there to receive them or had long since vanished is not known. At the same time, others migrated to Europe, becoming the ancestors of today's European population.[55]

Eurasia at the time was in the midst of the Ice Age. Ice sheets covered the northern half of the continent, tundra and steppe covered the southern half. This vast open grassland was home to herds of reindeer, bison, antelope, and wild horses, but also to solitary animals like wooly mammoths and wooly rhinos. Though cold and windy, it was a hunter's paradise.

Then the climate turned even colder, reaching the Last Glacial Maximum between 26,000 and 19,000 years ago. Ice sheets covered Scandinavia and Scotland and the northern half of Germany and France, while alpine glaciers covered Switzerland, Austria, northern Italy, and parts of southern France. South of the ice, arctic conditions prevailed. With a sixty-day growing season, only the hardiest grasses, mosses, and lichens could survive. Dry dusty winds whipped across the treeless land. Then, 18,800 years ago, the climate began to warm, fluctuating quickly between warm and cold until 12,700 years ago.[56] In this rapidly changing environment, *Homo sapiens* flourished.

The first *Homo sapiens* in Europe, called Cro-Magnon, are famous for their cave paintings representing the mammoths, reindeer, bison, horses, lions, and other dangerous animals that they hunted, for their relationship with the animals in their world was filled with symbolism and spirituality. Just as astonishing were their technology and their customs. Instead of using the same hand axes that earlier *Homo sapiens* and Neanderthals had used for hundreds of thousands of years, the Cro-Magnon devised new tool kits, changing them periodically. Their earliest tool kit, called Aurignacian, lasted from 40,000 to 28,000 years ago and included carved bone and antler and oil lamps to light the way in the caves in which they painted; they also used throwing spears rather than the handheld lances of their predecessors. The next, known as Gravettian (28,000 to 22,000 years ago), involved carved "Venus" figurines as well as elaborate burials. During the Solutrean period from 22,000 to 17,000 years ago, craftsmen heated rocks and used a pressure-flaking technique to produce bifacial points, tanged spearheads, and flint knives and saws. Finally, the Magdalenian period from 18,000 to 10,000 years ago saw the introduction of spear-throwers and bows and arrows, as well as elaborate carvings on bone and antler. Out of a core of flint, these inhabitants made blades, chisels, scrapers, and microliths, tiny sharp shards embedded in spear or arrow shafts. They also carved flutes and made harpoons and fish traps. In short, the Cro-Magnon were creative and sophisticated people.[57]

Since they flourished during the Ice Age, their most important invention may well have been the needle. The earliest one found, in Russia, is dated at 40,000 years ago. With needles and sinew the Cro-Magnon—or more likely Cro-Magnon women—could sew fitted garments of animal pelts that protected wearers from the cold. Geneticists have uncovered other evidence for the appearance of fitted clothing: body lice (*Pediculus humanus corporis*) that live in clothing.[58]

With fitted garments, *Homo sapiens* ventured out onto the steppes of eastern Europe and central Asia, a land without trees or caves and with few edible plants, places so cold that even Neanderthals avoided them. *Homo sapiens* reached Lake Baikal in Siberia 30,000 to 35,000 years ago. Their diet was mostly meat, some of which they stored frozen in underground pits during the winter. For shelter, they dug substantial houses into the ground and covered them with mammoth bones and animal skins or sod, as archaeological excavations in Ukraine have shown. For fuel, they burned animal bones. Thanks to their culture and ingenuity, they flourished in environments for which their bodies were totally unsuited. During the Last Glacial Maximum, when conditions got too harsh even for them, they retreated to the lands closer to the Mediterranean. But after it passed, they moved back north and east, reaching Beringia, the land bridge that connected Siberia and Alaska when the seas were lower, 28,000 years ago.[59]

How can the extraordinary efflorescence of human culture and technology in Upper Paleolithic Eurasia be explained? Thousands of years earlier, *Homo sapiens* in Africa had demonstrated the capacity for symbolic thought, language, and the ability to adapt their technology to changing conditions. If neither brains nor language were the crucial factors, the efflorescence may have been a response to the climate. Africa also underwent climatic changes—sometimes cooler and drier, sometimes warmer and wetter—but humans there were able to adapt to these shifts without radically changing their behavior. The natural conservatism of hominins, so amply demonstrated over the previous millions of years, still sufficed for survival and reproduction, and humans could occasionally experiment with new behaviors and cultural expressions without introducing major changes in their way of life. No doubt, further archaeological research will narrow the gap between our knowledge of Middle Stone Age Africa and that of Upper Paleolithic Eurasia.

In Eurasia, Neanderthals had adapted to the glacial climate of the Ice Age, physically by becoming stockier and more powerful, and culturally by building fires and wrapping themselves in animal skins. *Homo sapiens*, in contrast, remained biologically animals of the tropics, but they adapted to the cold by sewing fitted clothing and experimenting with new objects and behaviors. The challenge of the Ice Age unleashed the potential for rapid innovation that had been dormant for thousands of years.

Anthropologist Rachel Caspari has recently investigated the ratio of older (defined as twice the reproductive age) to younger individuals (up to reproductive age) among the remains of Neanderthals and Cro-Magnon. She found that ratio was five times higher among the Cro-Magnon than among Neanderthals. What caused the higher longevity among the Cro-Magnon is not known. But the consequences were dramatic, for cultural evolution depended on the transmission of knowledge between generations. Older people, grandparents in particular, could pass on their knowledge and experience to the younger generations. By contributing their social and intellectual resources, they could increase the number of surviving grandchildren. Thus, Caspari points out, "the hallmark features of the Upper Paleolithic—the explosive increase in the use of symbols, for instance, or the incorporation of exotic materials in tool manufacture—look as though they might well have been the consequence of swelling population size."[60]

Eurasian Extinctions

Not only were humans ingenious in adapting to a new environment, but they were also intent on adapting their environment to their own desires. Painting pictures of large animals on cave walls is one sign of this eagerness. Another is what they did to the fauna of Eurasia.

They hunted the largest animals first. These animals had had half a million years to adapt to hunters, first *Homo erectus*, then Neanderthals, and finally *Homo sapiens*. In the process, they had become wary. The impact of the newcomers was therefore not nearly as severe as it was among the naive animals of places that humans had never visited. Nonetheless, many large animals became extinct during the first 20,000 years of human occupation, among them the wooly mammoth, the wooly rhinoceros, the Irish elk or giant deer, and the cave bear and cave lion. The last mammoth died in northern Siberia between 11,000 and 9,600 years ago. Dwarf elephants, the size of circus ponies, survived on Wrangel Island in the Arctic Ocean until 3,700 years ago. Others, such as the musk ox and the bison, disappeared from Eurasia but found a refuge in the Americas, while the wild horse (*Equus przewalskii*) survived only in the most remote area of Mongolia.[61]

Just as interesting is what happened to the Neanderthals. From the time *Homo sapiens* returned to Eurasia after the eruption of Toba, the two species coexisted for close to 15,000 years. *Sapiens* and Neanderthals occasionally mated, at least outside of Africa; between 2 and 5 percent of the genes of Europeans are from Neanderthals.[62]

Why then did the Neanderthals vanish? There is no evidence that they and *Homo sapiens* ever fought; *Homo sapiens* occasionally fought one another, so the possibility cannot be ruled out.[63] Perhaps it was competition for prey by the newcomers, who could improve their weapons and hunting techniques faster than could the Neanderthals. On the steppe-tundra of Ice Age Eurasia, animals adapted to the frigid climate—wooly mammoths, wooly rhinos, musk oxen, reindeer, and antelopes—replaced the woodland animals that Neanderthals had previously hunted. Out in the open, such animals could not be ambushed, and evidently the Neanderthals did not learn the cooperative hunting techniques required to stampede them.

Homo sapiens had three other advantages. One was fitted clothing, which allowed them to survive in temperatures that Neanderthals could not withstand. Another was the willingness of *Homo sapiens* to eat things that the carnivorous Neanderthals seldom did, such as plant foods, small game, and fish.[64] A third possible advantage is the partnership between human hunters and wild dogs, a stage in the evolution of wolves into domesticated dogs.[65] In the competition for scarce resources during the Last Glacial Maximum, *Homo sapiens*' cultural adaptations proved more effective than the Neanderthals' biological adaptations. As *Homo sapiens* multiplied, the Neanderthals retreated. Trapped between the cold treeless land to the north and the Mediterranean Sea and deserts and mountains to the south, they took refuge in the Balkans, Iberia, and the Caucasus, surviving in ever declining numbers. The last known trace of them, found in Gorham's Cave in Gibraltar, dates back 28,000 years.[66]

Humans Reach the Americas

Because of the climate barrier, the Americas were among the last places to be settled by human beings. Until about 12,000 years ago, ice sheets covered northern Siberia and blocked access to Beringia and the route to America.

In North America, the Cordilleran ice sheet extended along the West Coast, while the Laurentide ice sheet covered the rest of the continent. At its maximum extent from 21,000 to 17,000 years ago, the Laurentide sheet reached Ohio and southern Illinois. As the Ice Age tapered off, it retreated to the northern Great Lakes 14,000 years ago and to northern Canada 11,000 years ago. The warming climate opened an ice-free corridor between the two ice sheets that allowed animals and humans to wander south from Alaska into a new continent.

Genetic analyses of Native Americans and indigenous inhabitants of northeast Asia confirm that the newcomers came from Siberia.[67] Archaeologists have only begun to investigate northeastern Siberia, but it seems that until about 1,000 years ago, this vast region was too cold and game too scarce to support much human life. Perhaps what made it possible for humans to survive there was the domestication of the dog, which could pull a sled, help them hunt game, protect them from predators, keep them warm at night, and, if necessary, be eaten.[68] From Siberia, hunting bands migrated to Beringia, a land that included northeastern Siberia, Alaska, and the nearby continental shelves then covered with steppe-tundra and inhabited by wooly mammoths, giant sloths, steppe bison, musk oxen, and caribou. After several thousand years, as the climate warmed, their descendants began drifting south.[69]

When and how humans arrived in the Americas has been a contentious question for many years. The long-held belief that hunters equipped with Clovis points (a unique shape of spearheads dated between 13,500 and 12,900 years ago) were the first humans in the Americas has been discredited by more recent findings. Most remarkable is the discovery at two sites in Texas of thousands of stone tools and other man-made objects dating back 15,500 years, long before the first Clovis points were made. A site at Monte Verde, in southern Chile, shows evidence of human habitation dating back at least 12,500 and possibly 16,000 years. Other still disputed pre-Clovis sites include the Meadowcroft Rockshelter in Pennsylvania, which may be 19,000 years old; Pedro Furtado in Brazil, dated to 17,000 years ago; and Paisley Five Mile Point Cave in Oregon of 14,400 years ago. In other words, there were humans in the Americas before an ice-free corridor opened up, but their numbers were small and the evidence they left behind is subject to controversy.[70]

How small hunting bands could have made the trip from Beringia to northwestern North America before the opening of the ice-free corridor is a mystery. Archaeologist Jon Erlandson and others have argued that these people traveled by boat or raft along the west coast of the continent, just as others made their way by boat to Australia. All along the North Pacific coast, from Japan to Baja California, the shallow coastal waters were rich in kelp, which attracted sea otters, fish, and birds. This rich ecosystem would have provided more nutrition for coastal navigators than the harsher and more dangerous interior of the continent. The hypothesis has been disputed, for the northern Pacific is a cold and violent ocean, and during the Ice Age it would have been filled with icebergs. Besides, it was argued, what traces these early navigators might have left behind have long since been covered by the rising sea. But recent findings have revealed that 12,200 to 11,200 years ago, humans periodically visited an island off the coast of Southern California to hunt Canada geese, cormorants, seals, and sea lions and to collect mollusks.[71]

American Extinctions

Interesting from an ecological point of view are the changes in the natural environments that coincided with the multiplication of humans in the Americas. Around the time of their arrival, 73 percent of all mammal species over 44 kilograms and all species over 1,000 kilograms vanished from the continent. Among them were mammoths, mastodons, *Eremotherium* (giant ground sloths that stood 2 meters tall and weighed 3 tons), *Smilodon* (saber-toothed cats the size of lions), *Castoroides* (beavers the size of black bears), *Glyptodon* (giant armadillos), straight-horned American bison, five species of horses, short-faced bears, dire wolves, and many others.[72] With their extinctions, the fauna of the Americas were dramatically diminished.

These extinctions did not happen overnight. Analysis of lake sediments in Indiana show that giant animals began to disappear between 14,800 and 13,700 years ago, long before the first Clovis points. Others died out between 13,800 and 11,400 years ago, coinciding with the Clovis period. And in Alaska, mammoths and horses survived until at least 10,500 years ago. In short, the extinctions took a long time.[73]

What could have caused them? One answer might be climate changes. The climate began warming 18,000 years ago. Then it cooled abruptly 12,800 years ago, possibly caused by the impact of an extra-terrestrial object.[74] Then, 1,300 years later, it began warming again. Although such climate changes may have stressed some animals, the native American megafauna had survived many climate shocks before. In a few thousand years, the Americas lost more genera of large land mammals than in the preceding 1.8 million years. In the 2 million years before humans arrived, the Americas lost fifty species of large mammals. Then, in just 2,000 years after humans came, fifty-seven species went extinct. If climate change could cause extinctions, it would have affected smaller animals as well, but it hardly affected smaller land mammals or marine mammals.[75] Furthermore, there is evidence that the megafauna collapse began a thousand years before the cooling phase and continued after it. In short, the climate hypothesis is unconvincing.[76]

That leaves humans. In 1967, anthropologists Paul Martin and H. E. Wright Jr. advanced the so-called blitzkrieg or overkill hypothesis that when humans first entered the Americas, they advanced like a wave front at a rate of 16 kilometers a year, killing every large animal they encountered.[77] It is not necessary to accept such a dramatic scenario to identify the role of humans in the extinction of large animals. In place of the "blitzkrieg" analogy, a more complex picture is emerging.

The animals of the Americas had never met humans before and had no instinct to flee from them. As in Australia, the larger animals reproduced so slowly that culling a few reproductively active females each year could have driven their population into an irreversible decline. In biologist Edward Wilson's words, "As a rule, inbreeding starts to lower population growth when the number of breeding adults falls below five hundred. It becomes severe as the number dips below fifty and can easily deliver the coup de grâce to a species when the number reaches ten."[78] Humans did not need a blitzkrieg to exterminate the American megafauna.

The megafauna did not go without a fight. The small number of sites of human occupation before the Clovis "explosion" is probable evidence that few of the earlier human migrants survived. Naturalist and grizzly bear expert Doug Peacock offers an intriguing hypothesis to explain the failure of the first migrants to thrive and multiply, namely, short-faced bears (*Arctodus pristinus* and *A. simus*). Weighing almost a ton and up to 3.7 meters tall on their hind legs, these were the largest of all carnivorous land mammals and could have kept human numbers very low until the introduction of Clovis points led to the extinction of their prey, the large herbivores.[79]

Not all large animals became extinct, of course. Bison, elk, moose, and grizzly bears still roam, because they were recent arrivals that crossed over the land bridge of Beringia from Siberia. Having survived millennia of encounters with humans in the Old World, they knew to keep their distance. Thus the grizzly, descendant from the Eurasian brown bear, replaced the American short-faced bear; the moose replaced the American stag-moose; the gray wolf replaced the dire wolf. The original bison, related to the Eurasian wisent, inhabited the Americas for 3 million years; its descendant, a large long-horned herbivore, evolved under the pressure of human hunting into the familiar *Bison bison* or "American buffalo," a smaller animal with short horns that travels in herds for protection. Among the larger native American animals, only the black bear and the mule deer survived the arrival of *Homo sapiens*.[80]

Unlike Neanderthals, Paleo-Indians were not exclusively hunters. They were opportunists who hunted and gathered all sorts of foods—fish, mollusks, small animals, seeds, and nuts—even before the easy hunting had petered out. They used fire to open up forests and encourage herbivores to multiply.[81] After about 11,000 years ago, they learned to stampede herds of bison into narrow canyons or over cliffs, probably using fire.[82] In the varied environments of the New World, the simple Clovis hunting culture was replaced with a great diversity of cultures, some of which grew food and created complex societies.

Melting and Mesolithic

Ice cores taken from Greenland glaciers tell the story of the earth's climate over the past hundred thousand years, as do tree rings and pollen and shells brought up from the bottoms of lakes. After about 18,000 years ago, the climate began to warm, melting the great northern ice sheets. As the oceans rose, Beringia, the land bridge between Siberia and Alaska, was flooded. Sunda became the archipelagos of Indonesia and the Philippines. The British Isles were severed from Europe, as was New Guinea from Australia. Many islands that dotted the sea shrank or vanished under the waters. In the northern parts of Eurasia and North America, tundra gave way to forests. As reindeer, horses, and other herd animals moved north with the retreating steppe, they were replaced by forest dwellers like elk and moose and, farther south, by deer and boars. In the Middle East, as the climate grew warm and moist, grasses with edible seeds spread widely, while the Sahara became a grassland dotted with lakes and wetlands.[83]

The Mesolithic

The changes in climate, flora, and fauna posed a challenge to all living beings. Many succumbed; others survived in reduced numbers; yet others multiplied. Among the winners in this changed environment were *Homo sapiens*. As large animals became rare, humans turned to catching fish that abounded in the rivers as well as mollusks, birds, eels, and sea-mammals along the coasts; they also collected nuts, acorns, tubers, berries, and other plant foods.[84] With more abundant foods, their population rose. One demographer estimated that the number of *Homo sapiens* in Europe increased from about 6,000 during the Last Glacial Maximum (25,000–19,000 years ago) to almost 29,000 by 13,000 years ago.[85]

To kill the smaller or more skittish animals that inhabited the forests, such as rabbits, beavers, boars, and deer and also fur-bearing carnivores like foxes and wolves, hunters had to learn new techniques and create new weapons. Throwing spears replaced the heavy thrusting spears of their ancestors. The bow and arrow was especially well adapted to the new way of life. Instead of a single spear, a hunter could carry a quiver full of arrows, each one with tiny blades embedded in the shaft. A 1.5-meter-long bow could propel an arrow at up to 100 kilometers per hour accurately up to 50 meters. Even at that distance, it could pass through a bear.[86]

For hundreds of thousands of years, humans and their predecessors had used fire for heating, cooking, and protection. The first evidence that fire was used in hunting was at the end of the Ice Age, when *Homo sapiens* deliberately set fire to forests. In North America, hunters set fire to forests to clear the underbrush and open up pastures to attract grazing animals like deer and elk and to create habitats for beavers, turkeys, and quail. The prairie that stretched from Wisconsin to Texas was created and maintained by Indians to encourage the herds of bison that were their main prey. A similar use of fire was true of foraging people in the Middle East and in much of tropical America.[87] Humans were no longer living off the bounty of nature but deliberately manipulating nature for their own benefit. Anthropologists refer to them as Mesolithic (Middle Stone Age) peoples.[88]

The luckiest of the Mesolithic foragers lived in environments so rich that they could settle down and build permanent dwellings. Five thousand years ago some Indians of California lived in villages of a thousand or more inhabitants. Along the northwest coast of North America, the Haida built long houses of planks that they could disassemble and transport from one location to another, living through the winter off dried salmon and herring and the berries that abounded in the nearby forests. Along the shores of some Mexican lakes, foragers found enough fish and wildfowl to sustain them year-round. Foragers and fishermen built the first permanent (or at least seasonal) settlement in Europe about 8,400 years ago. In short, humans were eager to settle down and did so whenever they could, even before they learned to produce their own food. In order to settle down, however, they needed to store food, for few environments provided food year-round. In the steppe-tundra, hunters dug pits to keep frozen meat in the permafrost; in grasslands, they stored seeds in baskets or pots. In Japan, a foraging and fishing people called Jomon began making pottery 16,000 years ago.[89]

After the Ice Age, the climate of the Fertile Crescent—a part of the Middle East that stretches from the eastern Mediterranean through Syria and into northern Iraq—turned mild and rainy in the winter, with long, hot, dry summers. Foragers living in the area found an abundance of edible plants and animals. The region, at the crossroads of North Africa and western Asia, had a very rich biota, with several hundred species of native trees and plants with edible seeds or fruits, many of which could be harvested and stored for future use. Some hillsides were covered with forests of pistachio trees. Herds of gazelles roamed the open grasslands.[90]

In this rich environment, a nomadic foraging people called Kebarans hunted gazelles, sheep, and goats and collected pistachios, almonds, and acorns that they pounded into meal with mortars and pestles. To harvest wild grains, they used bone-handled sickles with flint blades. Their successors (or perhaps their descendants) the Natufians fished and hunted waterfowl and collected enough wild wheat and barley during a few weeks in the fall to eat during the rest of the year, or as insurance during times when foraging failed to satisfy their hunger. They settled into permanent villages of ten to twenty houses with hearths and grindstones. They gathered wild grains and acorns, hunted gazelles, and caught fish, turtles, and birds in nearby lakes. With enough food at hand, they had more children and their population grew.[91]

The Younger Dryas

Then the climate turned against them. In far-away North America, as the huge Laurentide ice sheet melted, the remaining ice dam held back an enormous body of water in central Canada. Some time around 12,900 years ago, the waters of this lake broke through the ice barrier and cascaded down the St. Lawrence Valley into the Atlantic Ocean or northward into the Arctic Ocean. Once there, it formed a layer over the denser saltwater of the ocean and froze. This shut down the flow of the Gulf Stream that normally warmed western Europe. Within a decade, the average temperature of the North Atlantic fell by 10 to 15 degrees Celsius (18° to 27°F) and that of the world fell by 5 to 7 degrees Celsius (9° to 13°F). This return of the Ice Age, which lasted for approximately 1,300 years, is known as the Younger Dryas.[92]

During that time, the climate of North America and Europe turned as cold as that of Siberia today, with severe winter storms. Ice covered Scandinavia and Scotland while tundra replaced forests south of the ice. Drought descended upon the Middle East. Gazelles, once abundant, became rare. Grasslands replaced the oak, almond, and pistachio forests. As wild nuts became scarce, Natufians relied more on gathering the seeds of wild grasses such as wheat, barley, and rye. Even wild grasses became less abundant as a decline in atmospheric carbon dioxide stunted their growth, and scrub replaced grasslands. As the fertility of the land declined, the Natufians abandoned their villages and became nomads again. However, there were too many people and too few resources to return to a hunting and gathering way of life.[93]

Animals, when faced with a natural calamity, migrate if they can, or starve. So had humans many times before. This time, however, the inhabitants of the Middle East reacted in a new way. Here and there, some of them—probably women, who had always gathered while men hunted—experimented with planting the seeds of edible grasses in a few moist spots in river valleys, returning a few months later to gather the grain.

Conclusion

Human evolution has been the subject of scientific investigation for over a century, with remarkable results. Though there is still much to discover, we now know when and where hominins originated, how they changed over time, and how *Homo sapiens* evolved from earlier hominins. The evidence of paleontology shows a shift from biological to cultural evolution. Over the millions of years of evolution, members of the genus *Homo* that were best adapted to particular environments were replaced by those that could learn and adapt more readily to new or changing environments. They did so by possessing larger brains, hence greater intelligence, and by creating a communal knowledge base that could be transmitted from individual to individual and from generation to generation.

Humans have long thought of themselves as different but always superior to all other animals. The study of ancient humans confirms this difference, but in a very different way. While all living beings extract resources from their environments, humans regularly go beyond the needs of survival and reproduction. Before *Homo sapiens,* hominins scavenged, gathered, and hunted as best they could but were limited in their ability to change their environments by the difficulty and cost in energy of obtaining food. Even big predators stop when they have killed enough to satisfy their hunger and that of their kin, after which they spend long hours digesting.[94] Of all the animals, only *Homo sapiens* kills to the point of exterminating other species of animals.

Our intelligence and ability to communicate have made it easy. The technological innovations of Paleolithic humans—better stone weapons, the use of fire, spears and bows and arrows, and the domestication of dogs—allowed them to kill more animals with less effort and danger. Hunting was not just a means of providing meat; when it was easy and safe enough, it became a sport. Throughout the world, wherever humans appeared, many other species of large animals vanished. Other species of the genus *Homo* went extinct, and the eruption of Toba came close to extinguishing *Homo sapiens* as well. For humans in the Old Stone Age, survival was neither smooth nor certain.

Yet humans are also adaptable to changing circumstances. They had evolved in Africa, yet they were able to flourish in Ice Age Eurasia. Then, when the Ice Age ended, they adapted their skills and way of life to the new climate. Their creative ability to extract new resources from their environments led humankind in an entirely new direction: from depending on the bounty of nature to producing food and multiplying in numbers, bringing about the next major transformation of the planet.

2

Farmers and Herders

Nineteenth-century archaeologists coined the term "Neolithic" or "New Stone" Age to refer to the polished stone tools that succeeded the chipped stone tools of the earlier Paleolithic or Old Stone Age. Though later anthropologists realized that polished stone tools were part of a much more important change, the beginnings of food production, the name has stuck.

Despite its name, the Neolithic Revolution was not a sharp break with the past but a continuation of the more intense exploitation of resources that began in the Mesolithic. For the people who lived through this period, the change would have been almost imperceptible, not only because it took centuries but also because they retained their old way of life even as they experimented with new ways.

Just how they moved from gathering to cultivating plants is unclear. There must have been intermediate stages, such as weeding competing plants or watering desirable plants in dry weather. In a few instances, people manipulated plants but stopped just short of cultivating them, as in the following story that environmental historian Neil Roberts tells:

> The fine dividing line between farming and foraging is well illustrated by aboriginal Australia, where women customarily exhort plants to be generous and yield a big tuber as they dig them up. Once out of the ground, no matter how large the tuber, tradition decrees that the woman should now complain and berate the plant, "Oh you worthless plant, you lazy thing, You stingy plant. Go back and do better." Saying this, she would chop off the top of the plant, put it back in the hole from which it came, and urinate on it.[1]

Archaeologists have long discussed the motives that compelled people to begin growing crops and raising animals: Was it overpopulation or climate change that made them do so? Or was it a desire to settle down? It is likely that in many parts of the world, Mesolithic foragers had filled up the landscape, so when a hunting-gathering band had depleted its local resources, it could not simply move into a new unoccupied territory. If, in addition, the carrying capacity of the land was reduced, the need for new resources was even more urgent. The search for security might have been a strong motive as well. In many places, sharp seasonal variations provided an incentive for people to put aside food for the lean months of the year by storing grains and nuts or by capturing and confining animals to be eaten later. Besides, even the richest natural environments went through cycles of abundance and scarcity and seasons of plenty and want. Certain food sources, such as nut-bearing trees, varied their yields from year to year, even in a steady climate. Finding sources of food that could tide people over through hard times must have been a powerful motivation to experiment.

Knowledge and Technologies

Since the Neolithic people left no writings, it is impossible to know what their motives were. But we have good evidence of the means they used in the transition from foraging to producing food. Those who tried to produce their own food rather than rely on nature needed to understand wild plants and animals. Knowledge of nature was not personal but collective. In many places, during seasons when wild foods were abundant, people gathered in large numbers to celebrate, exchange ideas, find mates, and maintain ties with friends and kin. Celebrations often involved feasting, and bands competed to be seen as the most successful and the most generous. *Homo sapiens* were naturally curious and eager to learn from others and to share their considerable knowledge of wild foods.

Mesolithic technologies—flint-bladed sickles, baskets, digging sticks, and mortars and pestles—were just as useful in dealing with cultivated as with wild plants. In this assembly of tools, those used to process foods by soaking, grinding, or boiling were as important as those used to catch or harvest foods. The technological innovations of the Neolithic—polished stone tools and, later, pottery—were elaborations of tools used in earlier times. In short, the Mesolithic broad spectrum diet and the intensification of hunting and gathering it required were prerequisites to producing food.[2]

The Domestication of Plants

What mattered as much as motives and means were the opportunities that allowed people to begin producing food, such as locally available plants and animals suitable for domestication and the area's climate and water supply. Great disparities between different parts of the world influenced the development of farming and herding, in some cases determining how feasible and easy food production was.

Domestication means accelerating the evolution of plants and animals through human selection, partly to develop certain desirable characteristics and partly to make it difficult or impossible for them to return to the wild. According to biologist Jared Diamond, among the 200,000 known species of wild plants in the world, only a hundred were amenable to domestication.[3] Those plants had to have large seeds or fruits with thin coats, a good taste, or some other feature, such as fibers, that humans found attractive. Useful plants had to be easy to harvest—for instance, by ripening at the same time. And within such a species, some individuals had to contain a mutation useful to humans, such as seeds that did not fall to the ground when ripe but stayed attached to their stalks "waiting for the harvester." To prevent a favorable mutation from being diluted in the next generation by hybridizing with plants that lacked it, domesticable plants had to be either self-pollinating or deliberately planted at a distance from their wild relatives.

Domestication was a mutual arrangement between humans and particular plants and animals. Michael Pollan, exaggerating slightly, explains the role of plants and animals in their own domestication:

> Though we insist on speaking of the "invention" of agriculture as if it were our idea, like double-entry bookkeeping or the lightbulb, in fact it makes just as much sense to regard agriculture as a brilliant (if unconscious) evolutionary strategy on the part of the plants and animals involved to get us to advance their interests. By evolving certain traits we happen to regard as desirable, these species got themselves noticed by the one mammal in a position not only to spread their genes around the world, but to remake vast swaths of that world in the image of the plants' preferred habitat.[4]

The Domestication of Animals

To be candidates for domestication, animals had to tolerate being crowded together in a small space without panicking and be willing to reproduce in captivity. Only a few animals possessed these qualities.[5] Their domestication usually involved a mutation called neoteny, or retaining youthful physiology and behavior into adulthood. Young mammals are full of curiosity, unafraid

of animals of other species, and eager to learn new tricks; house cats, for instance, act like the kittens of wild cats. Such qualities were particularly useful
when animals migrated to a new territory, as happened frequently during
and after the Ice Age. Humans took advantage of these behaviors by capturing young animals, keeping those who retained their juvenile behaviors
longest, and letting them mate. After a few generations, they had animals that
were permanently juvenile in behavior and kept their juvenile appearance,
such as shorter muzzles, rounded heads, crowded teeth, and smaller brains.
After many generations, their captors found that they had bred out the un-
desirable traits. Not all animals lent themselves to this. Steers became bulls,
dangerous and violent animals. Male boars easily returned to the wild. Even
rams and billy goats could be violent. For aggressive male animals, humans
found another means of control, namely, castration.[6]

In addition to the recognizable domesticated animals, others occupy a
niche between wild and domesticated. Asian elephants, if captured young,
can be trained to carry logs. The Lapps of northern Scandinavia follow reindeer herds as they migrate across the tundra, occasionally culling one. Most
reindeer are skittish, but they tolerate humans in their proximity. Some even
let themselves be milked or harnessed to a sled.

Then there are the commensals, literally animals that "share the table" with
humans. Mice, rats, and sparrows like to live close to humans because of the
garbage they produce and the foods (especially grains) they store. The same
is true of flies, fleas, roaches, and other human-loving insects. Then there
are those that follow humans at a slightly greater distance: seagulls, barn
swallows, raccoons, deer, rabbits, and pigeons, among others.[7]

Cats occupy a position somewhere between commensal and domesticated. The earliest evidence of domesticated cats is found in ancient Egyptian
tomb paintings and in the temples of the sun-god Ra, represented as a male
cat, and of the fertility goddess Bast, a female cat. But long before that, wild
cats (*Felis sylvestris libyca*) were attracted to the rats, mice, and sparrows that
proliferated wherever grain was stored in Neolithic villages. Eventually, their
descendants became social among humans.[8]

Herbivorous herd animals like wild sheep, goats, cattle, and horses found
it useful to associate with humans, as did omnivorous animals like pigs.[9]
Humans and their dogs provided protection from other predators, especially
for vulnerable young. Individual animals might be eaten by the humans they
lived with, but for the species, domestication was clearly beneficial. Today
while domesticated animals are in the millions, few of their wild relatives,

such as wolves, bobcats, wild boars, and wild horses, have survived. Some, like aurochs, the ancestors of cattle, have vanished.

How dogs became domesticated is controversial. Some scholars maintain that Paleolithic hunters adopted wolf-pups that were the most cooperative and the least aggressive; after many generations, their descendants evolved into hunting companions.[10] Others argue that the ancestors of dogs were not hunting wolves but commensal scavengers that hung around Mesolithic villages.[11] Having settled among Mesolithic foragers, dogs later adapted to the world of Neolithic herders. When people began to herd sheep and goats, dogs raised among these animals imprinted on them and became their guardians. Sheepdogs herd sheep the same way wolves attack herds of wild herbivore, circling the herd and cutting off stragglers, except they do not kill.

The Spread of Farming and Herding

For food production, climate and water mattered as much as available plants and animals. Some regions were too cold. Others, like the Sahara that was once covered in grass and supported herds of cattle, had become too arid.[12] Food production therefore originated in only a few places on Earth: China, the Fertile Crescent, Mesoamerica, the Andes, Amazonia, New Guinea, West Africa, and Ethiopia. All the world's domesticated plants and animals come from these eight regions. Because each of these regions had its own set of domesticable plants and animals, the forms of agriculture and animal husbandry that arose varied enormously, as did their history, both human and environmental, for millennia thereafter.

For hundreds of thousands of years during the Ice Age, humans had survived by hunting, fishing, and gathering. Why then did people in those eight regions turn to farming and herding within a few thousand years of each other? The most likely explanation is that by 10,000 years ago, the growing population of Mesolithic farmers had occupied most of the good hunting, fishing, and gathering areas. After that, climate changes were sufficient to challenge foragers to seek new sources of food but not so violent as to frustrate their efforts.[13]

The locus of domestication was less important than the diffusion of plants and animals to other regions. Geography favored some regions over others. Eurasia, oriented east-west, made it easy for plants domesticated in one region to be transplanted to other regions, like wheat from the Middle East to northern China and Europe or rice from southern China to South Asia and the Middle East. Sub-Saharan Africa, however, is oriented north-south, and its tropical climate made it impossible to grow Middle Eastern crops. The Americas, too, were oriented north-south, so it took thousands of years before maize (American corn) was bred to thrive in what is now the United States.[14]

Since the process of domestication, and its consequences for humans and for the environment, varied so much from one region of the world to another, the experience of the regions where agriculture began must be considered separately.

China and Southeast Asia

China lies between two climate zones: monsoon Asia to the south and Siberia to the north. These two zones, and their climatic fluctuations, have determined the agrarian history of China from the early Neolithic. According to recent research, the earliest domestication of plants occurred in central and southern China, with tropical warmth and abundant rains during the summer months. As far back as 12,000 years ago, foragers began collecting the grains of wild rice (*Oryza rufipogen*) growing in seasonally flooded areas along riverbanks and in low-lying wetlands in the lower Yangzi Valley and around Hangzhou Bay. By 9000 BCE, they began to gather and cultivate a mutant form, *Oryza sativa*, that has non-brittle ears and therefore could not reproduce on its own. By 6400 BCE, farmers had learned to build low dams to trap the rainy season runoff in order to reproduce the conditions preferred by the wild plants. From then on, they were able to rely more on domesticated rice than on wild plants. Archaeological sites in the lower Yangzi Valley include pottery in which rice was boiled and stored. In the prosperous village of Hemudu, built around 5000 BCE, archaeologists have found pottery, bone, stone, and wooden tools, and the bones of water buffalo, pigs, chickens, and dogs.[15]

From southeastern China, rice-based agriculture spread throughout eastern and southern Eurasia. It reached Thailand in the fifth millennium; southern China, India, and Vietnam in the third; Indonesia in the second; and Japan in the first millennium. People who took up rice farming soon outnumbered those who remained foragers.

The original inhabitants of southeastern China spoke proto-Austronesian, the ancestor of the languages spoken today in Malaysia and in the archipelagos of the western Pacific. Everywhere they settled, Austronesian-speaking people brought with them a vocabulary that clearly proves that they were farmers. In the sub-tropical regions north of Indonesia, they used words for field, paddy, garden, rice, sugarcane, plow, cattle, water buffalo, ax, and canoe. When they reached the tropics in Indonesia and beyond, they acquired words for taro, breadfruit, yam, banana, sago, and coconut. Corroborating the linguistic evidence is archaeological evidence of pottery, polished stone adzes and knives, and spindle-whorls, all of them used by farmers but not foragers. Finally, pollen evidence shows that the farmers set fire to large areas of forest. Throughout much of Southeast Asia and the Pacific, the practices and languages of farmers spread together.[16]

Unlike the south, northern China has a harsh continental climate; summers are hot, winters are bitterly cold. Rainfall varies enormously from droughts to floods; half or more falls between June and August, much of it as sudden downpours. The North China Plain is covered with over 150 meters of loess, a soft and fertile soil deposited by dust storms from Mongolia. The Yellow River, which flows through it, carries a tiny fraction of the amount of water carried by the Amazon or the Mississippi, yet it carries as much silt as these much larger rivers. Of that silt, 40 percent to 50 percent is deposited before it reaches the sea, causing the riverbed to rise above the floodplain and periodically change its course.[17]

Along the Yellow River, the first farming villages appeared around 6500 BCE, after the inhabitants had learned to grow drought-resistant foxtail and broomcorn millet. Later they added bottle gourds, sorghum, hemp, and mulberries and domesticated water buffaloes, chickens, and pigs. They obtained wheat seeds from the Middle East, perhaps brought by the same people who domesticated horses and used them to cross the great Eurasian steppe. They built substantial houses of clay and made distinctive pottery.[18] In many ways, their experience parallels that of the first farmers in the Fertile Crescent, with one difference: since the soils of China were easily cultivated with digging sticks and hoes, farmers had little need for draft animals.

The Middle East

The inhabitants of the Middle East had numerous domesticable plants and animals from which to choose. Among the wild plants that lent themselves to domestication were emmer and einkorn wheat, rye, and barley, and later chickpeas, lentils, and flax. Domesticating them took decades or even centuries.[19] Sheep and goats, the most amenable animals, were domesticated in the Zagros highlands of Iran in the tenth millennium, as were pigs soon thereafter. Domesticating cattle occurred a millennium later because their wild ancestors, aurochs or *Bos primigenius*, were very large (approximately 2 meters at the shoulder) and fierce.[20]

Sometime between 9700 and 9500 BCE, temperatures rose by 7° Celsius (13°Fahrenheit) in less than fifty years, in some places in less than ten. The new climate, with cool rainy winters and hot dry summers, favored annual plants like grasses that completed their life cycle by late spring and left dried seeds that could survive until winter rains made them germinate. This encouraged people to gather and store enough seeds to carry them through the rest of the year and to settle in villages where they could protect their food stores.

The place where the climate change was most acutely felt was the valley of the Jordan River, which had once been dotted with lakes. When it became drier, only three lakes remained: Huleh, Galilee, and the Dead Sea, the last of which was too salty to sustain life. People and animals congregated near the remaining sources of fresh water. It is here, around 8000 BCE, that the Natufians intensified their gathering of wild seeds, and then, to supplement them, began to sow seeds of emmer wheat, barley, lentils, and peas. The practice of sowing seeds then spread to other parts of the Fertile Crescent. People began to store their harvests in jars, clay-lined baskets, or bins kept off the ground. They learned to roast grains without scorching them, to prevent them from sprouting in storage. Small hamlets turned into substantial villages, some with protective walls. Though their inhabitants still engaged in hunting and gathering, they relied increasingly on domesticated crops. Soon thereafter, hundreds of villages lived entirely from agriculture.[21]

The village of Abu Hureya near the Euphrates River in northern Syria epitomizes the long and complicated dance between humans and plants and animals known as the Neolithic Revolution. Sometime before 11,000 BCE, foragers built twenty small mud-brick houses with thatch roofs that they inhabited at least part of the year. Though blessed with abundant wild foods, they also experimented with planting rye and wheat. They continued to do so until they abandoned the village around 9500 BCE. When people returned to the site a thousand years later, they built a much larger village for up to 6,000 inhabitants.

Until around 6400 BCE, 80 percent of the animal food they consumed consisted of the meat of gazelles. So abundant were the gazelles that the inhabitants continued to rely on them for meat for a thousand years after they had begun to grow grains. Then, between 6400 and 6100 BCE, among the animal bones excavated at Abu Hureya, the proportions shifted to 80 percent sheep and goats and only 20 percent gazelles. The reason is probably that other hunters along the gazelle migration paths were decimating the herds. As the gazelle herds vanished, the people of Abu Hureya turned to domesticated sheep and goats for their meat supply. With the grain they grew and the sheep and goats they raised, the people of Abu Hureya flourished off and on until about 5000 BCE, when they finally abandoned the village.[22]

Europe

Compared to other sub-continents such as India or China, Europe has an unusually high ratio of coastlines to land area; it is, in fact, an archipelago of peninsulas. Europe has three very different climatic and geographic zones with very different environmental responses to human actions: a southern zone, with a climate similar to that of the eastern Mediterranean; a northeastern zone, with a continental climate similar to that of northern China; and a northwestern zone, with a temperate oceanic climate. The Alps and Carpathian mountains form a sharp boundary between the Mediterranean and the two northern zones. Between the two northern zones, however, there is only a continuum along the great northern European plain.

Around the Mediterranean, farmers planted the same crops and raised the same animals as in the Middle East. Soon after 6400 BCE, farmers at Nea Nikomedeia in Greece built mud-walled houses; grew wheat, barley, peas, and lentils; and raised sheep and goats, while still foraging. By 6000 BCE, there were many small farming communities in Greece. Later, farmers began growing olive trees and vineyards and building terraces. By 5000 BCE, this kind of mixed farming had spread throughout the Mediterranean region, from Egypt to Tunisia and from Greece to Spain.[23]

North and west of the Alps, farmers found a very different climate from that of the Mediterranean. Western European summers were warm, winters ranged from cool to cold, and rains fell year-round. The soils were heavy and often waterlogged. A colder climate with frequent rains delayed the spread of farming until 6000 BCE or later. Farmers from Anatolia may have brought

their way of life to Europe after being forced to move by a natural catastrophe. The Mediterranean Sea, having risen 17 meters as a result of global
warming, stood almost 200 meters above the level of the Euxine Lake, a small
shallow inland sea located where the Black Sea is now. According to a controversial hypothesis, in 5600 BCE, the waters of the Mediterranean suddenly
burst through the Bosphorus, filling up the lake until it became the Black
Sea. As the water rose, it moved inland at a rate of nearly 2 kilometers a day.
The people who lived and farmed near the lake had to flee suddenly, losing
their land and homes and many of their animals. Some may have escaped
up the Danube River into the Hungarian plain. When they took up farming
again, their skills and way of life persuaded the native foragers to follow their
example.[24]

Once in the interior of Europe, farming people settled in areas of rich
soils, such as river bottoms that they could cultivate with hoes and digging
sticks, and left the forested uplands to hunters and gatherers with whom they
coexisted for thousands of years.[25] Like Middle Eastern and Mediterranean
farmers, they cultivated wheat, barley, rye, peas, lentils, and flax. Rather than
goats, however, they raised sheep, cattle, and pigs. These animals foraged in
the nearby woods and their manure fertilized the fields. Woods also provided
game and edible wild plants to supplement the domesticated foods.

After about 4400 BCE farmers started to move beyond the river valleys.
The spread of agriculture to the forests required a powerful but docile animal
that could pull a plow. When farmers learned to castrate young steers, they
obtained such an animal. The first pictures of a pair of oxen yoked to a plow
appeared in Mesopotamian art around 3500 BCE. The plow and the animals
to pull it mark the transition from gardening to agriculture and opened up
the possibility of farming the forested uplands. Farmers also began to cultivate rye, which grew well in the climate of central and western Europe.

By 3500 BCE, those parts of Europe were filling up with hamlets and villages
of wooden houses surrounded by fields, meadows, and wood lots. Cows that
grazed in the meadows gave milk. They and oxen not only plowed the land but
also produced manure that renewed the fertility of the soil. Sheep ate stubble
and weeds and furnished wool. Pigs foraged in nearby forests, returning
to the farms at night. Hunter-gatherers living in nearby forests and coastal
wetlands traded their catch for farm products. Mixed farming, with its symbiosis between plants, animals, and humans, was an almost closed system
that required only modest amounts of energy from outside, such as cutting
firewood, letting pigs forage in the forest, or supplementing home-grown

foods with some hunting and gathering. It proved to be a sustainable way of life that lasted, in places, until well into the twentieth century.[26]

Nomads of the Steppe

While China and the Middle East developed agriculture, an entirely different way of life arose in the grasslands of southern Russia where the climate was too dry and too extreme for farming, yet the natural vegetation could support herbivores. In the region lying between the Black and Caspian seas, called the Pontic-Caspian Steppe, residents began herding cattle and sheep in the river valleys in the sixth millennium. Sometime after 4800 BCE, they succeeded in domesticating the horse, not to ride but as a source of meat, for horses can forage for grass through the snow during the winter, when cattle will stand and starve. By 4000 BCE, they had bred more docile horses and learned to ride them. This gave them unprecedented mobility and the ability to herd far more cattle than they could on foot.[27]

Meanwhile in the mid-fourth millennium, the inhabitants of lower Mesopotamia had created the first wheeled vehicle, a cart with four solid wheels pulled by oxen. This invention spread to the Pontic-Caspian Steppe between 3500 and 3300 BCE. The combination of ox-drawn carts and horses that could be ridden opened up the grasslands between the river valleys that had previously been too difficult for pedestrian herders. The people who possessed these tools, called Yamnaya, could herd far more animals and move them from place to place to find fresh pastures. On horseback, they could scout, raid, and trade over long distances, while their ox-carts slowly carried water, food, shelter, and other necessities across the plains. Here and there, they cultivated barley or millet, mined ores, and made metal tools and weapons.[28]

So wealthy and powerful did the Yamnaya people become that they began to migrate with their herds into the lower Danube Valley, overcoming the resident farmers.[29] To the causes of their success—their wealth in horses and cattle and their prowess as horseback-riding warriors—anthropologists Gregory Cochran and Henry Harpending have added another, a genetic mutation that allowed them to drink milk after weaning. Lactose tolerance is common among people who raise cattle, such as the Maasai of East Africa, but is rare in the rest of the world. This mutation first appeared among the Yamnaya and gave them far more food per animal than was available to lactose-intolerant herding peoples, who could digest only the meat of their animals. It also made them mobile and self-sufficient in their encounters with farmers. When the Yamnaya entered the Balkans, the farmers abandoned their villages and fled, or died. Their graves show them to be four inches shorter than the invaders, signaling their inferior health and strength.[30]

To the north and west of eastern China, the land is covered with steppe, giving way to evergreen forests in Siberia and to the Gobi Desert in Mongolia. The steppe, too dry to farm, was the native habitat of wild cattle and horses. Steppe dwellers coming from the west in the late fourth millennium brought domesticated horses that helped them herd cattle, horses, and sheep. Living in areas too marginal for any settlements and with a climate that varied enormously from year to year, herding peoples had to keep moving to find fresh grass for their animals. Tribes of nomadic herders often engaged in warfare with one another and with farmers to the south and east. These were the ancestors of the Huns and Mongols who would cast a shadow on Chinese history for centuries.[31]

New Guinea

At about the same time that people in China and the Middle East were producing food, so too were the inhabitants of New Guinea. That island has a rugged topography, a hot and rainy climate, and thick rainforest vegetation. In this unique environment, farming developed in ways that were very different from both the open-field and the rice-paddy farming of other lands.

Hunting and gathering never flourished in New Guinea as it did elsewhere because of the scarcity of game and of edible plants in the rainforest and because of the mountainous terrain. Instead, it seems the inhabitants were manipulating their environment as early as the seventh millennium BCE. At a place called Kuk Swamp, archaeologists have found artificial mounds dating to the late seventh millennium and drainage ditches from the early fourth millennium BCE. Foragers may have started by selecting sago palms or mountain pandanus and clearing away competing trees, keeping channels open in sago swamps, and felling mature trees in order to promote new shoots.

When the inhabitants began farming, they cleared the land by felling the trees, then setting fire to the resulting deadwood, a method of farming called swidden or slash-and-burn. Not only did this open up land and kill weeds, but the ashes also fertilized the soil. Farmers who cleared land this way, then moved on when its fertility was exhausted, could not continue doing so in-definitely. When untouched land ran out, they returned to fields they had used once before and then left fallow. After a field had lain fallow for a time, wild plants that had invaded it could be burned to replenish its nutrients for the next crop. The length of the fallow—from a year to several decades—depended in part on how quickly vegetation regenerated on it and in part on whether farmers were able to provide other fertilizers besides ashes, such as organic wastes. Fallowing was therefore not one system but a continuum.

Some plants could be reproduced by taking a cutting from an existing plant and sticking it in the ground. In this way, New Guineans encouraged the growth of plants with edible parts: sago, pandanus, yams, sugarcane, ba-nana, and taro. Unlike the grains that nourished farmers in other parts of the world, the edible roots and fruits of the New Guinea rainforest could not be stored for long, nor was there any need to store food because the warm wet climate produced food year-round.[32]

No animals are known to have been domesticated in New Guinea; the animals that people eventually kept—pigs, chickens, and dogs—were all introduced from Southeast Asia. In this food production system, pigs were almost the only source of meat in a protein-deficient diet. The inhabitants raised sows and let them forage in the forest and mate with wild boars during the day but return to the village at night; these pigs were feral rather than fully domesticated. Since the forest provided too little food even for pigs, the farmers grew crops to fatten them up for special occasions.[33]

Tropical Africa

Africa, the homeland of the human race, has always been a particularly dif-ficult region in which to grow food. The climate, mostly tropical, varies from extremely wet rainforests to the driest of deserts. Furthermore, it has fluc-tuated enormously between wet and dry. In many places, the soil is poor in nutrients. The disease ecology of sub-Saharan Africa—homeland to many pathogens that prey upon humans—kept the population density much lower than in the inhabited parts of Eurasia.

Between circa 10,000 and 4000 BCE, northern and eastern Africa received considerably more rain than today. The Sahara was covered with grasslands and lakes. Forests of oak, pistachio, lime, elm, and pine grew in the moister areas. Foragers hunted gazelles, hares, and Barbary sheep; caught fish, turtles, snails, and mollusks; and painted pictures of giraffes, crocodiles, and other animals on rock walls at Tassili n'Ajjer, now in the desert.

During these lush times, the people of the Sahara began to manipulate nature. They started with animals. Sometime after about 8000 BCE, they captured and penned Barbary sheep in a cave in the eastern Sahara, the first step toward domestication. Around 7000 BCE, migrants brought herds of cattle to the Sahara. This breed of longhorn cattle was probably domesticated independently of the western Eurasian cattle of the time. As nomadic pastoralism spread across the Sahara, new paintings at Tassili n'Ajjer show these longhorn cows with their herders.

Then the monsoon winds that had brought rain to the northern half of the continent shifted northward toward Europe, causing a permanent desiccation of the Sahara and unreliable rains in the Sahel to the south of it. Herders abandoned the Sahara, some migrating to the Nile Valley. Between 5000 and 4000 BCE, as the flood levels of the Nile declined, Neolithic farmers in Egypt began cultivating crops like wheat, barley, and flax and raising sheep and goats, all of them adopted from the people of the nearby Fertile Crescent.[34]

Elsewhere in Africa, the domestication of plants occurred independently in several different places. The idea of farming may have come from the Middle East, but as Middle Eastern crops would not grow south of the Sahara, the crops that Africans domesticated were entirely different: African red rice (*Oryza glaberrima*) in West Africa, yams (*Dioscorea cayennensis rotundata*) and oil palms (*Elaeis guineensis*) in the savanna south of the Sahel, drought-resistant finger millet (*Eleusina coracana*) and sorghum (*Sorghum bicolor*) in the Sahel, and coffee (*Coffea*) and teff (*Eragrostis tef*) in Ethiopia.[35]

During the mid-second millennium BCE, a people who spoke a language we call proto-Bantu began migrating out of their homeland in northeastern Nigeria and northern Cameroon, perhaps pushed out by a drying climate.[36] Some migrated south along the Atlantic coast into the equatorial rainforest of the Congo basin, while others made their way into the savanna belt of East Africa. As farmers, they outnumbered and either absorbed or overwhelmed the indigenous foragers. After settling for a few years, the farmers moved on, for tropical soil was quickly depleted of nutrients. Canoes aided their travels on the rivers of equatorial Africa. By 1500 BCE they had populated the Great Lakes region and by 500 CE they had reached South Africa. In some areas, Bantu-speaking herders preceded farmers.

Whatever drove their early migrations, another factor—iron—helps explain the success of the migrants. By the mid-first millennium BCE,

Bantu-speaking ironsmiths were producing farming tools such as axe and hoe blades and weapons such as cutlasses and spears. Iron was used not only to fell trees, cultivate land, and smite enemies but also as a form of wealth. As a result, smiths attained high social status in Bantu-speaking societies.

During the first millennium BCE, Bantu-speakers settled along rivers and in open forests. What allowed them to penetrate the deeper forests of equatorial Africa was the introduction of the banana and plantain from Southeast Asia by Malay mariners in the early first millennium CE. These crops were ideally suited to the rainforest, where they produced ten times the yield of yams. Furthermore, they did not require full clearing like yam fields and hence created fewer habitats for *Anopheles* mosquitoes, carriers of malaria. With farming, pottery, iron tools, and bananas, Bantu-speaking peoples occupied every environment in central, eastern, and southern Africa, except a few pockets of dense rainforests in the Congo basin and the Kalahari deserts of the south, gradually replacing the foragers who had occupied sub-Saharan Africa before them.[37]

Diseases affected both humans and domesticated animals. In the moister parts of Africa, along the West African coast and in the equatorial rainforest, nagana, the animal analogue of human sleeping sickness, made it impossible to keep cattle. In the Sahel and the savannas of East Africa, cattle could survive in many places, but not others. Horses, however, could not survive even where cattle could. Thus, cattle herders like the Maasai in East Africa herded their animals on foot, which limited their mobility and their wealth compared with that of the herders of Central Asia and Mongolia. Wild animals were evidently not affected by nagana, which is why there are still herds of wild herbivores in Kenya, Tanzania, and South Africa.[38] In Somalia, an area too dry for cattle, camels imported from Arabia served Somali herders as a source of milk and meat, but not as a means of transportation.[39]

The Americas

In the Americas, as in Eurasia, population growth and climatic instability led people to try domesticating plants and animals. After about 6000 BCE, as the climate turned cooler and drier, people began to experiment with cultivating edible plants.[40] Yet the differences between the Old World and the New are stark. Native Americans found many plants that lent themselves to domestication, and many of the ones they domesticated have since become staples of diets around the world: potatoes, maize, tomatoes, avocados, manioc (or cassava), squash, sweet potatoes, many varieties of beans, and many peppers—and also tobacco. Plant domestication began about the same time as in China and the Middle East but, due to the nature of these plants, it took much longer than it had in Eurasia before Native Americans could rely on farming alone for their subsistence.

Though rich in domesticable plants compared to the Old World, they were at a disadvantage in animals. After the megafauna extinctions of the Pleistocene, the Americas had few animals amenable to domestication: dogs, which had accompanied the first human immigrants across Beringia; guanacos, vicuñas, and guinea pigs in South America; and turkeys and Muscovy ducks in Mesoamerica. None of them were strong enough to carry a person or pull a cart or a plow.[41]

Plants were first domesticated in four distinct parts of the Americas: Mesoamerica, Peru and the Andes, eastern North America, and Amazonia. The first three have been carefully investigated by archaeologists; the fourth is still largely unknown. Because the Americas lie on a north-south axis, distances and climate barriers delayed the diffusion of domesticated plants and animals between the regions. As there was no trade and few human contacts between South and North America, llamas, alpacas, guinea pigs, quinoa, and potatoes were never transferred from the Andes to Mesoamerica or to North America, nor were turkeys brought to South America. Maize, manioc, and sweet potatoes, found on both continents, were the exception. Even between Mesoamerica and North America, plant transfers were difficult; it took several thousand years for maize, a sub-tropical plant, to be bred into a variety that could tolerate the short summers of the temperate zone.[42]

Mexico

The inhabitants of Mexico faced a situation similar to that of the eastern Mediterranean, namely, good soil and abundant sunshine, but highly seasonal rainfall. As large game became scarce after 7000 BCE, they increasingly came to rely on collecting the seeds of wild plants. The first cultivated plant in the Americas was the bottle gourd (*Lagenaria siceraria*), a native of Africa that had reached the Americas about 10,000 years earlier. It probably came by drifting across the Atlantic on ocean currents and then being dispersed by animals before it was cultivated by Paleo-Indians. After it came squash, culti-vated in Mexico from the seventh millennium on.[43]

The origin of maize (*Zea mays*) has been the subject of research and con-troversy for decades because there is no comparable wild plant, and the do-mesticated variety cannot reproduce without human help. Archaeologists have found cobs of a grass called *Zea tripsacum*, a possible ancestor of maize, in the Tehuacán and Balsas river valleys south of Mexico City.[44] Another pos-sible ancestor of maize is the weedy grass teosinte that grows in Mexico and Guatemala. Whichever it was, hundreds of years went by before farmers pro-duced a domesticated maize with inch-long cobs.

Excavations at Guilá Naquitz, a rock shelter near Oaxaca, showed that the inhabitants grew maize, beans, and squash. But they continued to rely heavily on hunting and gathering, returning to the site to harvest what they had planted. Not until about 2000 BCE did they grow cobs large enough and in sufficient quantity to support permanent villages. Because it took so long to create a productive form of maize and because domesticable animals were so few, the transition from pure foraging to fully developed agriculture took over four millennia, much longer than in the Middle East.[45]

In addition to the difficulty of finding suitable plants, Mexican farmers had to cope with rainy winters and dry summers. Some early farmers made the most of their environment by tapping springs. Starting around 800 BCE, they created a network of canals and aqueducts 1,200 kilometers long that irrigated 330 square kilometers of farmland. In the Tehuacán Valley, as early as 750 BCE, they constructed the Purrón Dam, the largest in the Americas before the eighteenth century. On hillsides above the Oaxaca Valley in south-central Mexico, early farmers built long narrow terraces out of stones and filled them with soil brought up from the valley floor mixed with organic wastes and potsherds. In these they planted their crops. To water the plants during the dry season, they constructed 6.5 kilometers of canals from a spring to their fields. All of this work was done by the farmers without the benefit of draft animals, wheeled vehicles, or metal.[46]

Western South America

The west coast of South America, from Ecuador to central Chile, contains more diverse environments in a small area than anywhere else on Earth. Just off the coast, the Humboldt Current that brings cold water up from

Antarctica teems with fish that in turn nourish huge numbers of sea birds. The predominant winds, coming from the Atlantic, lose all their moisture while crossing the Andes. As a result, the coast is among the driest places in the world, where the only water is in the rivers that flow down from the mountains. In the midst of the Andes lies the Altiplano, a high plateau that was home to herds of guanacos and vicuñas, distant relatives of the camels of Eurasia. Finally, on the eastern slopes of the Andes begins the rainforest of Amazonia.

Humans foraged in all three environments. On the Altiplano, hunters domesticated llamas from wild guanacos for their meat and as pack animals, and alpacas from the smaller vicuñas for their wool. By the time maize appeared in Peru around 3200 BCE, the inhabitants of the Altiplano had already begun domesticating quinoa and, soon thereafter, potatoes and lima beans.

Of the three environments, the most challenging was the coast. Around 2500 BCE, its inhabitants turned from hunting and gathering on land to exploiting marine resources full time. After 1700 BCE, some gave up fishing for farming, building irrigation channels to water the fields closest to the rivers. The combination of rich alluvial soils, irrigation water, and guano (the excrement of sea birds that roosted along the coast) as fertilizer produced astonishing yields, hence attracting dense populations.[47]

Amazonia

The ecology and prehistory of Amazonia, like that of other tropical rainforests, is far less well known than that of the dry tropics or the temperate zone. To outsiders, the rainforest seems to offer very little edible vegetation or game. Today, those parts of the forest not yet invaded and "developed" by people of European or African descent are inhabited by a few widely scattered tribes of hunter-gatherers who practice a limited form of slash-and-burn agriculture. Much of the forest seems uninhabited, hence the common belief that before Europeans arrived Amazonia was a virgin forest, an illusion produced by this strange and hostile environment.

It was not always the case. The first Europeans to visit Amazonia had astonishing tales to tell. In 1542, Francisco de Orellana, one of the Spanish conquistadors who accompanied Pizarro, led an expedition down the Amazon River from Peru to the Atlantic Ocean. The chaplain on the expedition, Gaspar de Carvajal, wrote a report of the expedition's adventures in which he described "cities that glistened white" inhabited by thousands of people. When he returned to Spain, his report was received with disbelief, for it contained many tall tales, such as women warriors he called "amazons." As no gold or silver was found in Amazonia, Europeans lost interest in it until the nineteenth century.[48]

Yet Carvajal had it right. In fact, before 1492 Amazonia supported far more inhabitants than it does today. Recent investigations have revealed a landscape, now covered with forest, that had once contained clusters of villages and towns of up to a thousand inhabitants, with central plazas and wide straight roads, surrounded by palisades beyond which lay gardens, fields, and orchards. The lack of stone in this vast alluvial plain meant that all human constructions were made of organic materials; once they were abandoned, they were destroyed by the constant humidity and by voracious insects and microorganisms. Any land left fallow for more than a few years was reclaimed by the forest.

Instead of the cities and monumental temples associated with long-lost civilizations, the ancient inhabitants of Amazonia left behind soils, plants, and ceramics. Here and there in the forest one finds patches of *terra preta* or "dark soil," layers of rich black compost up to a meter deep that includes charcoal, fish and animal bones, human excrement, and potsherds. These soils, rich in carbon, nitrogen, phosphorus, and other nutrients, contrast with the generally acidic and sterile soils of the rest of the rainforest. The *terra preta* soils of the Amazonia have been dated to between 450 BCE and 950 CE or later. On these soils, the Indians did not clear land to create open fields of grains but cultivated trees and tubers like the people of New Guinea. In some places, almost half the trees, such as cashews and peach palms, carry fruits or nuts. The inhabitants also raised manioc, a tuber that grows wild in the Amazon basin, as well as peanuts, pineapples, papaya, cacao, tobacco, and beans. The ceramics that the Amazon Indians made are the oldest in the Americas and among the oldest in the world, some dating back to 2000 BCE.[49]

North America

North of the Rio Grande, agriculture spread slowly. Some Indians obtained seeds and farming techniques through contacts with Mexico; others domesticated a completely new set of plants; and yet others did not take up agriculture at all.

The ones who went the furthest toward full dependence on agriculture lived in the Southwest, a challenging environment with low and unpredictable rainfall, thin fragile soils, and delicate vegetation. There, foragers hunted mountain sheep, mule deer, and pronghorn and collected mesquite, agave, and pinyon nuts. During the late first millennium BCE, when the climate was unstable, the inhabitants turned increasingly to growing maize, squash, sunflowers, and beans as a means of supplementing erratic supplies of wild plants and game. In the first millennium CE, many came to depend almost entirely on foods that they cultivated in canyon bottoms where the water table lay close to the surface or where they could irrigate their crops from springs and seasonal rainwater pouring over the canyon rims.[50]

In contrast to the Southwest, the climate of the northeastern part of the United States is rainy, with warm summers and cold winters. The land was covered with oak, chestnut, hickory, and other deciduous trees, and the forests teemed with game. Despite the richness of the natural environment, the Indians turned to agriculture to supplement wild foods during the lean seasons. However, their attempts to begin cultivation were hampered by the lack of plants amenable to domestication. Beginning in the fifth millennium BCE, they domesticated goosefoot (or lamb's quarter), sumpweed, ragweed, chenopod, knotweed, maygrass, sunflowers, and squash; only the last two have survived as crops. So unproductive was this agriculture that they continued to rely heavily on hunting, fishing, and gathering, and they could not build permanent settlements.

It was maize that made farming a viable substitute for foraging in the Northeast. Maize, long cultivated in Mexico, had reached the Southwest by 1200 BCE, but it did not reach the Northeast until the first two centuries CE. Even then, it was not until the ninth century that varieties of maize were developed that would grow well in the short summers so far north of their native habitat. Once beans reached the region in the twelfth century, the Indians of the Northeast finally had the full complement of domesticated plants that had long supported a dense population in Mexico.[51] They practiced slash-and-burn agriculture. It took eight to ten years before new growth was ready to be burned and crops planted anew. The result was a patchwork of meadows and open woods.[52]

In southern New England, Indians harvested maize, beans, squash, and other crops in the fall and stored some of the harvest in underground pits. Their houses were made of wood frames covered with bark or grass that could be dismantled and moved in a few hours. In the fall, they moved to scattered sites where they could hunt deer and bears for meat and hides. People who were settled farmers in the summer and early fall turned into nomadic foragers in the late fall, winter, and spring. Women did most of the farming and processed the meat and hides that men brought back from their hunts. Thus did the Indians make the best use of the great diversity of the environment in which they lived.[53]

The Midwest was one of the last parts of North America in which farming began. There, a people referred to as the Hopewell Culture farmed the fertile river bottoms growing knotweed, maygrass, a little barley, and, later, maize. From Wisconsin to Texas, as fires set by Plains Indians turned the forests into prairie, herds of bison expanded into the new grasslands, followed by Indian hunters who used grass fires to drive them into ravines and over cliffs.[54] Between 600 and 1400 CE, the population had grown and agriculture had developed to the extent of supporting a city on the Mississippi River; Cahokia, as this site is now known, may have had up to 40,000 inhabitants.

The Domestication of Humans

Among the animals domesticated during the Neolithic era were, of course, humans. In foraging bands, members who did not get along with the others or who simply wanted other opportunities could leave and try their luck elsewhere. Farmers, however, were tied to their land, and clearing and cultivating it required enormous efforts from them. Even swidden farmers, who had to start a new plot of land when the previous one was exhausted, returned to earlier plots after a certain number of years. Moving to a completely new environment was difficult, risky, and undertaken only under duress. Farming was also much more work than foraging. During the planting, weeding, and harvesting seasons, farming demanded back-breaking labor with little or no free time. The biblical story of the expulsion from the Garden of Eden—"So the Lord God banished him from the Garden of Eden to work the ground from which he had been taken"—and the hard life of farmers—"in the sweat of thy face thou shalt eat bread"—clearly reflects this transformation in the lives of humans.[55] By the time Neolithic peoples realized how bad the bargain was that they had struck, there was no turning back.

Population

Food production allowed the population to grow, which both required and permitted more crops to be grown and animals to be raised, leading to more population growth. Women's consumption of cereals high in carbohydrates removed the hormonal check on ovulation during lactation that had ensured a long gap between the children of foragers.[56] Demographer Mark Nathan Cohen estimates that the Neolithic population grew at a rate of 0.1 percent a year, doubling every 700 years, ten times faster than among hunter-gatherers.[57]

Just how many people there were in prehistoric times is a matter of educated guesswork. Scholars have based their figures on estimates of the carrying capacity of the land, hence the population density of humans, before the Neolithic era and at some time thereafter. Though the estimates vary, they fall between 4 and 10 million people at the beginning of the Neolithic era and between 40 and 200 million by the start of the Common Era, a tenfold or larger increase. As for the population density, farming may have supported 250 times more people (or more in some places) per unit of area than foraging did.[58] From the perspective of evolutionary biology in which the success of a species is measured by how well its members survive and reproduce, this increase represents a huge victory for *Homo sapiens*. However, as historian William McNeill reminds us: "Looked at from the point of view of other organisms, humankind therefore resembles an acute epidemic disease, whose occasional lapses into less virulent forms of behavior have never yet sufficed to permit any really stable, chronic relationship to establish itself."[59]

Evolution

That the Neolithic Revolution ushered in a dramatic increase in the number of people is incontrovertible. But were they better off? It was once believed that they must have been, since farming and herding represented a major step toward higher levels of civilization. But the evidence proves the contrary. Analyses of skeletons from the Eastern Mediterranean area show that between 5000 and 3000 BCE, Neolithic men were, on average, 161.3 centimeters tall, 15.8 centimeters shorter than their Paleolithic male ancestors. Women lost 12.2 centimeters, from 166.5 centimeters in the Paleolithic to 154.3 in the Neolithic. Likewise, the median life span of men dropped from 35.4 to

33.1 years and that of women from 30 to 29.2 years.[60] Even as their numbers increased, farmers and herders were shorter and less healthy and died younger than their foraging ancestors.

Anthropologist Henry Harpending and physicist Gregory Cochran go one step further and accuse domestication of causing psychological and even physiological changes in humans: "In fact, there are parallels between the process of domestication in animals and the changes that have occurred in humans during the Holocene period. In both humans and domesticated animals, we see a reduction in brain size, broader skulls, changes in hair color or coat color, and smaller teeth."[61]

Brain size changed after the end of the Ice Age. All works on human evolution stress the increase in brain size from australopithecines to *Homo erectus* to *Homo sapiens*, implying that brain size correlates well with intelligence. What happened to *Homo sapiens* brains after the Ice Age is either downplayed or completely ignored. Yet measurements of skulls show a decline of 9.9 percent among men and 17.4 percent among women between the Mesolithic and today. What caused it and what it means is not clear. A decline in bone thickness and body size may be part of the cause. Other possible causes may be an internal reorganization of the brain that superseded further increases in size. But so far, causes and consequences remain unclear, for brain size does not correlate well with intelligence.[62]

Two factors caused a decline in the health of Neolithic farmers: diet and diseases. Unlike foragers, who dined for the most part on meat and fresh fruits and vegetables, farmers relied on cereals and tubers rich in carbohydrates but low in proteins and vitamins. These starchy foods caused tooth decay and diabetes. Nutritional deficiency diseases included scurvy (caused by lack of vitamin C), pellagra (vitamin B3), beriberi (vitamin B1), and goiter (iodine). Native Americans who relied on maize for most of their calories were shorter than their ancestors; they were also prone to anemia, for maize is low in bio-available iron. Bad harvests brought years of hunger, stunting the growth of children. Nomadic herders, however, enjoyed a high-protein, low-starch diet.[63]

The consequences of the new way of life were never fixed but varied for genetic reasons. As the human population grew, so did the number of genetic mutations, some of them beneficial. Eight thousand years ago, Europeans and Central Asians developed lactose tolerance, and cattle-herding Africans did so some centuries later. Early food-producers who learned to make fermented beverages also developed a tolerance for alcohol. The lack of

Vitamin D from sunlight caused farmers living in the northern latitudes to develop light skin that would admit more sunlight, allowing their bodies to synthesize the Vitamin D needed to prevent rickets.[64]

Diseases

Farming caused diseases that were rare or non-existent among nomadic foragers. Clearing large areas of land produced ideal conditions for mosquitoes. In sub-Saharan Africa, as farmers moved into the rainforest to grow Guinea yams and oil palms, and later bananas and plantains, they created pools of water in sunlight, ideal mosquito conditions. Because the females of these mosquitoes get almost all their nutrition from human blood rather than from other animals, they are very efficient transmitters of malaria. Of the five types of malaria that affect humans, two are widespread in Africa. Vivax malaria, a disease that dates back 2 or 3 million years, is relatively mild, and has a death rate of 1 or 2 percent; 97 percent of West and Central Africans carry a mutation of their red blood cells called Duffy antigen negativity that makes them immune to this disease, the result of the long exposure of hunter-gatherers to this disease.

The other widespread kind of malaria, caused by the *Plasmodium falciparum*, invades and destroys up to 80 percent of red blood cells and kills over half of newly infected humans. Because it has a fourteen-day reproductive cycle and female anophelines live ten to twenty-one days, it can survive only where there are high densities of both mosquitoes and humans—in other words, near farming villages. There, people were constantly reinfected. Those who survived early childhood gradually built up resistance to this disease. Because falciparum malaria was especially dangerous to hunter-gatherers who came into contact with villagers, farmers gradually replaced foragers in large parts of Africa.

However, they paid a high price. After millennia of contacts with falciparum-infected mosquitoes, Africans who lived in infested areas developed a genetic defense called hemoglobin S mutation or sickle-cell trait. In some regions, 25 to 30 percent of the population carry this mutation, which reduces the death rate among children by 90 percent. Unlike the Duffy antigen negativity that is harmless to humans, there is a heavy price to pay for this defense: when both parents carry this mutation, their children will suffer—and die young—from sickle-cell anemia.[65]

Living in villages, often under crowded conditions, also caused other diseases. The *Aedes aegypti* mosquito, carrier of yellow fever, could breed in the smallest water container. Flies, attracted by garbage and feces, spread bacteria and intestinal parasites; so did rats and mice that ate the grain that farmers stored. Domesticated animals carried many diseases that spread to humans. Dogs carried rabies, sheep and pigs carried whooping cough, horses carried tetanus, cats carried toxoplasmosis, pigs and ducks carried influenza, and cattle carried diphtheria, measles, and cowpox, a disease that may have mutated into smallpox. All in all, it is estimated that over half the infectious diseases that afflicted humans came from animals, especially the domesticated ones with which humans came into close contact. Of course, humans probably also gave tuberculosis to cows, yellow fever to monkeys, measles to mountain gorillas, and polio to chimpanzees.[66]

Like plants and animals, diseases were not spread uniformly around the world. They were most common and dangerous in tropical Africa, where human beings originated. Falciparum malaria and trypanosomiasis (sleeping sickness) probably explain why until recently human population densities were much lower in sub-Saharan Africa than elsewhere. Humans who migrated from the tropics to the temperate zone left behind the diseases that depended on tropical animals or insects.

The New World had a very different set of diseases. The first small bands of hunters who migrated to the Americas had spent generations surviving in Siberia and Alaska before setting out for the temperate zone. Trekking on land or navigating down the Pacific coast must have weeded out the sick and limited the number of illnesses the migrants could have brought over from the Old World, eliminating insect-borne infectious diseases with complex life cycles such as malaria. In the Americas, there were few domesticated animals that could transmit diseases to humans. Furthermore, there was little or no contact between the Andes, Mesoamerica, the Southwest, and the Northeast, hence fewer opportunities for diseases to spread.

Yet Amerindians were by no means free of diseases. According to environmental historian Alfred Crosby, they were familiar with pinta, yaws, syphilis, hepatitis, encephalitis, polio, tuberculosis, pneumonia, and intestinal parasites. Overall, however, they were healthier and stronger and lived longer than their counterparts in the Old World, for they had no experience with the acute infectious diseases that ravaged the Eastern Hemisphere. So unimportant were infectious diseases among American Indians that their immune systems had none of the genetic resistances that Eastern Hemisphere peoples had developed in their long experience with infectious diseases. When the two hemispheres were joined after 1492, what had been their good fortune was to prove their undoing.[67]

TRANSFORMING NATURE

In order to produce food, Neolithic farmers and herders favored one small set of plants (their crops) and animals (their livestock and pets) at the expense of the rest of nature. In the process, they transformed the land.

At one end of the spectrum of transformations was the forest gardening as practiced by the peoples of New Guinea and Amazonia that mimicked natural growth and left minimal traces on the land. At the other end was monoculture: cultivating only one species of plant or raising only one species of animal. The beginnings of monoculture can be seen in the wheat fields of the Middle East, the rice paddies of China, and the herds of sheep and goats on the Eurasian steppe. Biologically speaking, these species suddenly became very successful, measured by their rates of survival and reproduction. So did other, unwanted species. Crops that ripened or were stored after harvesting attracted rats, mice, sparrows, and roaches. Puddles provided habitats for mosquitoes. Garbage and human or animal excreta attracted flies. Thanks to humans, weeds, pests, and vermin were also biological winners.

Where there were winners, there were losers, chief among them the natural vegetation. Clearing the land for crops denuded it for part of the year of its protective cover, leading to soil erosion; so did letting animals overgraze. In many places, finding good land for farming meant clearing trees, for the crops that humans favored grew best in places where there was enough moisture, warmth, and soil fertility for deciduous trees. Not only did farmers need the land that trees occupied, but they also needed much more wood than their foraging ancestors. Wherever possible, they built their houses and their farming tools out of wood. They often surrounded their villages with stout palisades of wood and their flocks with fences. Once settled, they made pottery, and firing it required a lot of firewood, as did cooking grains and tubers. Villagers in the Middle East made plaster by burning limestone. As the demand for both land and wood rose, so did the destruction of forests.

Examples abound. On the island of Crete, where land was cleared 6,000 years ago to make way for olive groves, a culture based on the export of olive oil flourished for almost 2,000 years until soil erosion undermined its prosperity. In Taiwan, pollen records reveal deforestation 6,000 years ago and, in Java and Sumatra, 3,000 years ago or earlier. In western Europe, deforestation began in earnest when farmers acquired ox-drawn plows in the fourth millennium BCE. Yet in the cool rainy climate of western Europe, forest growth kept up with timber felling by humans until the introduction of iron axes. In the British Isles, originally 95 percent covered with forests, the impact of Neolithic farming on the forests was slight before the first millennium BCE.[68]

Before iron axes, the most powerful tool that farmers had was fire. In many places, slash-and-burn farming fertilized the land by turning existing vegetation into ashes that provided nutrients for crops. This method worked best where the natural vegetation grew fast and the human population density was very low, as in tropical rainforests. In dry areas or when the population rose or for some other reason farmers returned to a once-burned plot before it could regenerate enough vegetation, its fertility declined, sometimes irreversibly.[69]

Next to fire, livestock caused the most damage to natural environments. In dry regions such as around the Mediterranean, goats were especially destructive. The climate was mild enough that they could forage year-round. Unlike cattle and sheep, which must be supervised, goats are intelligent, agile, and thirst-resistant animals that can find food on their own and return to their village at night. Furthermore, goats can eat woody material such as twigs, bushes, and tree saplings, preventing the regeneration of these plants. Overgrazing was common, leaving bare and eroded soils in some places, and thin vegetation in others. Sometimes herders burned forests to promote the growth of new grasses. As a result, land that had once been covered in forests of oak, pistachio, cedar, and other trees was reduced to scrub plants best able to resist fire and goats.[70]

The Neolithic village of Ain Ghazal in Jordan illustrates the impact of Neolithic people on a dry environment. Ain Ghazal was occupied from 8000 until about 6000 BCE. At first the inhabitants built substantial houses with walls that they covered with lime plaster, as often as every year. Burning limestone to make a ton of lime plaster required four tons of wood. To obtain enough wood for construction, plaster, and other uses, they felled all the trees within 3 kilometers of the village. Meanwhile, they let their goats feed on the stubble after the harvest, leaving the fields exposed to the fall rains that eroded the soil. Beyond the fields, the animals prevented the regrowth of the forests by eating the tree seedlings. Toward the end, as their resources were depleted, the people of Ain Ghazal built houses with smaller rooms that required fewer timbers, burned smaller branches in their hearths, replaced lime plaster with crushed limestone or mud, and raised fewer animals. After 6000 BCE, a slight deterioration of the climate forced them to abandon the village entirely and become wandering shepherds.[71]

Humans and Climate Change

The Earth's climate is subject to three factors: the amount of sunshine that reaches different parts of the planet, the ocean currents that distribute heat around the world, and the greenhouse gases that trap the heat.

The first factor that determines the climate is the amount of sunshine the Earth receives. The Earth normally gets enough sunshine in the summer to melt the snow that accumulated during the previous winter. Snow affects the climate, for it reflects sunshine much better than land, water, or vegetation. If the summer sunshine does not melt the snow, then it accumulates from year to year, turning into ice sheets and starting an ice age.[72]

These variations in solar radiation change the average climate of the Earth by only 1.5 or 2 degrees Celsius (2.7–3.6 degrees Fahrenheit), but the effect on the higher latitudes is as much as 5°C (9°F), with consequences for the whole globe. Thus, during the most recent period of low solar radiation, the Last Glacial Maximum (LGM) of 20,000 years ago, so much water was locked up in the great ice sheets that covered Eurasia and North America that the sea level was 130 meters lower than it is today. Then the solar radiation in the Northern Hemisphere increased, causing the ice to melt and the seas to rise to their current level.

The second phenomenon that affects the climate of different parts of the Earth is the flow of water in the oceans. Ocean currents, moved partly by the wind and partly by the rotation of the Earth, are blocked or diverted by continents and islands. Over the geological ages, plate tectonics moved the continents around a great deal, while islands appeared, merged with others, or disappeared, depending on rising and falling sea levels and other forces. Saltwater flows from the tropics into the North Atlantic, but as it approaches Greenland, it cools and sinks and flows back south along the bottom of the sea, creating what oceanographers call the "Great Conveyor Belt." Until recently, the Arctic Ocean, deprived of warm water and receiving little sunshine, froze over, while snow and ice accumulated on the nearby continents.

The third natural phenomenon that affects the climate is greenhouse gases, especially methane (CH_4) and carbon dioxide (CO_2). The amount of methane in the atmosphere is a function of the intensity of solar radiation: the more sunshine, the more vegetation and the more decay that releases methane. That amount followed a 22,000-year cycle until 5,000 years ago. The carbon dioxide cycle is more complex: CO_2 is released by erupting volcanoes and is absorbed by the oceans and by the chemical reaction of rainwater with exposed rocks. Glaciation and ocean currents also affect atmospheric CO_2. As a result, over the past 100 million years the amount in the atmosphere has slowly declined, causing a gradual global cooling. That change has not been smooth but has varied on a 100,000-year cycle, with amounts increasing during ice ages until there was sufficient CO_2 in the atmosphere to contribute, along with methane and sunshine, to interglacials.[73]

A Human Impact?

It is widely believed that before the Industrial Revolution and the burning of fossil fuels two and a half centuries ago, the Earth's climate was determined only by natural forces. In fact, the relatively steady climate that has reigned in much of the world over the past 12,000 years, known as the Holocene, is quite abnormal.[74] Because the cycles of solar radiation are regular and predictable, the climate of the past 10,000 years should long since have been getting colder and headed toward a new ice age.

As climatologist William Ruddiman has argued, "Had nature remained in full control, Earth's climate would naturally have grown substantially cooler. Instead, greenhouse gases produced by humans caused a warming effect that counteracted most of the natural cooling. Humans had come to rival nature as a force in the climate system."[75]

The two greenhouse gases that most influence the climate are methane (CH_4) and carbon dioxide (CO_2). A record of their concentration in the past, preserved in bubbles of air trapped in the ice in Greenland and Antarctica, has recently been revealed in ice cores extracted from these ice sheets. The concentration of these gases in the atmosphere varies for natural causes. That of methane follows the cycle of solar radiation, because more sunshine in the lower and mid-latitudes causes more vegetation to grow, hence more to die and decay, releasing more methane into the atmosphere. In the higher latitudes, decaying peat bogs also release methane. As a result, the methane concentration in the atmosphere normally follows a 22,000-year cycle. Had it followed the pattern of previous interglacials, its concentration would have declined from 725 parts per billion 10,000 years ago to around 475 today, a drop of 250 parts per billion. It did decline to 575, but then, 5,000 years ago, instead of declining further it began to rise to over 700 parts per billion just before the Industrial Revolution. Natural sources of methane, namely, wetlands, cannot be the cause of this rise, since tropical wetlands have shrunk over the last 10,000 years and the boreal wetlands of Siberia have been stable.

As for carbon dioxide, its concentration in the atmosphere over the past 350,000 years shows rapid increases every 100,000 years, followed by slow and erratic declines. Following the normal cycles that operated during ice ages and warming periods in the past, the concentration of carbon dioxide, having peaked at 265 parts per million 11,000 years ago, should have declined to 240 parts per million. Instead, it declined only to 260 parts per million. Then, 8,000 years ago, instead of continuing to decline as it would normally have done, it began to rise to 280 or 285 parts per million, even before the Industrial Revolution.[76]

What caused these deviations from the norm? The only new factor that could explain them, Ruddiman argues, is farming. The deforestation of the temperate parts of Eurasia began with the introduction of farming and accelerated when farmers began using plows and oxen, each year adding a small amount of carbon dioxide to the atmosphere. In northern and central Europe, Neolithic farmers needed to clear 3 hectares of forest per person every year; in a few places, they burned peat. In China, the first large region to be deforested, people began burning coal 3,000 years ago. According to Ruddiman, "Somehow, humans had seemingly added 300 billion tons or more of carbon to the atmosphere between eight thousand years ago and the start of the industrial era."[77]

As for the extra methane, it too derived from farming, but in a different way. The most important contributor was the cultivation of rice in paddies in China, India, and Southeast Asia, releasing methane as the vegetation, especially weeds, decayed in the water. This method of farming increased substantially 5,000 years ago. Second only to rice farming was the growth in the herds of livestock, especially cattle that release methane in their manure and in belches from their rumen, the second stomach in which they digest grass. Beyond these two sources, methane was released by the burning of biomass and the decay of human excrement, the result of an increasing population.

The amount of carbon dioxide and methane added to the atmosphere every year was very small; but multiplied by the thousands of years from the beginning of farming to the Industrial Revolution, the totals added up to as much as has been added in the past 250 years. Had it not been for this human intervention in the climate, the Earth would have been much cooler than it was, even before industrialization, and an ice sheet would have started to grow in northeastern Canada. Instead, as Ruddiman explains, "this warm and stable climate of the last 8,000 years may have been an accident. It may actually reflect a coincidental near-balance between a natural cooling that should have begun and an offsetting warming effect caused by humans."[78]

Needless to say, Ruddiman's explanation for the climate of the Holocene is controversial. Whether the additional amount of carbon dioxide and methane released by humans before the Industrial Revolution was sufficient to compensate for what would have been a natural cooling of the climate or whether the steady climate of the Holocene had other causes is still being debated.[79]

Conclusion

The transformations that took place during the Neolithic began slowly in a few places and initially had a very small impact. But farming and herding had the potential to develop in several significant ways. By increasing the capacity of land to produce food, they lifted the constraint on the number of humans who could survive and reproduce. Domestication transformed a small number of species of plants and animals into new varieties that depended on humans for their survival and reproduction, some more completely than others. In order to farm and herd, humans felled forests, cleared land of unwanted plants, and decimated wild animals, especially those that were predators or competitors of the domesticated ones. Besides transforming plants, animals, landscapes, and humans, farming and herding seem also to have changed the climate. The entire Earth was beginning to feel the impact of human actions.

Important as the changes imposed on the natural world by human actions were, they did not represent an unalloyed blessing for humankind.

As plants and animals evolved toward dependence on humans, humans evolved as well, psychologically and physiologically, toward dependence on their flocks and crops. Meanwhile, the increase in human population was accompanied by a worsening of human health and a decline in life spans due to crowding and changes in diet. What some have called "progress" was a double-edged sword.

3

Early Civilizations

Gradually, over hundreds or even thousands of years, where the soil was good and the climate suited domesticated plants, the landscape became dotted with Neolithic villages and fields. In many areas, rain was the most important factor that affected farming. Too much or too little rain or rain at the wrong time of year could ruin farmers' crops. Not surprisingly, farmers were attracted to places, such as rivers, lakes, and wetlands, where they were less beholden to the vagaries of the weather but could control the amount and timing of the water their crops received. Once they had learned to control water, they were rewarded with bigger yields than on rain-watered lands, often more than they needed for their own consumption. Surplus food that could be stored tided them over during hard times or allowed them to feed larger families.

But the surplus could also be taken from the farmers by people who did not farm but instead provided services such as protecting their communities from attack, performing religious rituals, directing the construction of monuments, or making goods the farmers could not make for themselves. Ultimately, complex societies—traditionally called "civilizations"—depended on farmers producing more food than they needed to survive, especially food that could be stored for long periods of time. Such civilizations arose in places where controlling water was the key to producing surpluses of food that could support an elite class of warriors and priests and those who served them as traders, artisans, and servants.

The first complex societies have been called "river valley civilizations," developing in places where rain was scarce but river water was abundant. In Mesopotamia, Egypt, the Indus Valley, and the west coast of South America, complex societies arose where farmers practiced irrigation. In other areas, such as Central Mexico or the Andes, shallow lakes or wetlands could be turned into croplands during dry times. In yet others, such as the Southwest of the United States, springs provided water for crops in dry seasons.

Water control took different forms in different environments, not just in river valleys. In relatively flat plains where a river inundated the land before the growing season, farmers built low dikes to hold the water until the soil was thoroughly soaked. Where a river flowed at or above the level of the land, they dug canals to bring water to their fields. Where water was too abundant, as in swamps, they dug drainage ditches or raised their fields above the water level. On hillsides, they built terraces to prevent run-off and soil erosion. Where floods threatened their crops, they constructed levees to contain flood-prone rivers. And in regions with frequent wet and dry seasons that did not correspond to the needs of the crops, they built reservoirs to hold water until it was needed. In all such cases, farmers were no longer beholden to the rain but had learned to have much better yields by manipulating nature. Despite their labors and their success, however, such farmers were still vulnerable to the vagaries of nature. Rivers, lakes, and springs were not reliable; they could still afflict farmers, and the civilizations that depended on them, with droughts and floods.

The appearance of hydraulic civilizations in different parts of the world was far from simultaneous as it depended on the plants that were domesticated in each region, the crops farmers chose to grow, and the environmental conditions they faced. In Mesopotamia, that happened 6,000 years ago; in parts of the Americas, it was less than 2,000 years ago. The difference in timing reflected differences in climate and topography but also the differing nature of the plants that people found and the much longer time needed to domesticate the plants of the Americas compared to those of the Eastern Hemisphere.

First Civilizations of the Eastern Hemisphere

The first complex societies with cities, kingdoms, and organized armies arose in three regions of southwestern Eurasia and northeastern Africa—southern Mesopotamia, Egypt, and the Indus Valley—where conditions were remarkably similar. All had a very dry climate but were crossed by large rivers that brought water down from distant mountain ranges. They were also close to the first centers of plant and animal domestication. Finally, they were in contact through trade, allowing ideas about farming and irrigation to spread among them.

Mesopotamia

The first place where a complex society arose was Mesopotamia (now Iraq). This great valley receives hot dry winds from the Sahara and Arabia in the summer, and cold dry winds from Central Asia in the winter. Northern Mesopotamia receives enough rainfall in the winter to make rain-watered agriculture possible. To the north and west, the Taurus Mountains get rain in the fall and snow in the winter, feeding the Tigris and Euphrates rivers. The south of Mesopotamia, like the Arabian and Syrian deserts to the east, gets little rain and would be a desert too if it were not for the two rivers. As their water comes from winter rains and spring snow melt, they crest in April and May, just as plants are ripening.

Mesopotamia is a flood plain formed as the Tigris and Euphrates have deposited their silt over the eons. The flood plain is extremely flat with a very shallow gradient. In ancient times, floods were often devastating, causing the rivers to shift their beds. Once the floods abated, the rivers meandered sluggishly through marshes and swamps until they reached the Persian Gulf.[1]

In this environment, all forms of life depended on the rivers. The waters teemed with fish and waterfowl. Before farmers transformed the land, reeds grew in the marshes, date palms on the natural levees, and grasses on land watered during the floods. For several thousand years after 9500 BCE, the climate was rainier than it is today. As the seas rose, the Indian Ocean invaded the lower valley, enlarging the Persian Gulf. After about 5000 BCE, as foragers who migrated into the valley took up farming and herding, their communities gradually filled the floodplain. In southern Mesopotamia, where the rains were insufficient for crops, the rivers deposited their silt in the lower valley, raising their beds above the level of the land. To bring water to their fields, farmers cut gaps in these natural levees and dug small channels. Every fall and winter, parties of farmers had to dig new channels and clear the silt from the small feeder canals. Fields nearest the rivers produced fruits and vegetables. Farther away, farmers sowed half the land with wheat, barley, or flax, leaving the other half fallow until the next year so that the soil could recover and animals could graze on the stubble. In the nearby deserts and in the Zagros Mountains of western Persia where farming was not possible, pastoral nomads raised herds of sheep and goats.[2]

After circa 4000 BCE the climate became more arid. The grasslands near Mesopotamia became so dry that herding peoples began to move into the valley in search of water and grass. Those already there congregated in towns, perhaps for protection, as disputes arose over valuable land. The concentration of people in a land of rich soils and abundant river water produced bigger harvests than were needed to feed the farmers. The surplus supported a complex society with religious and political leaders, merchants, clerks, artisans, and servants. By the early fourth millennium, towns such as Uruk,

Nippur, Eridu, and Ur had grown into cities with imposing walls, temples, and ziggurats that dominated the plain. Trade between cities and with outlying areas flourished.

In *Oriental Despotism* (1957), sociologist Karl Wittfogel argued that early irrigation systems required top-down management, thereby leading to a despotic "hydraulic-bureaucratic state."[3] Later research by archaeologists disproved his idea that only despotic states could manage irrigation systems. Instead, they found that most irrigation works were built and maintained by local communities, for communal effort was required to dig new canals, clear old ones of silt and vegetation, and maintain dikes and levees. Agriculture required year-round labor, as irrigation in the hot season permitted double-cropping. Farmers also kept sheep, goats, and pigs, and later cattle and donkeys.

In the fourth millennium, once they had learned to yoke oxen, they could cultivate much larger fields than with hoes and digging sticks. Oxen were harnessed to a simple wooden scratch-plow, or *ard*, well suited to the soft alluvial soils of Mesopotamia and other river valleys. A share of the resulting harvests was collected by political or religious administrators, stored in temple granaries, and used to pay farmers recruited to maintain the canals and levees. In the process, the priests and administrators began to keep records of collections, payments, debts and property; the first writing evolved from such record-keeping.[4] At this point, when the names of cities, rulers, gods, and wars were written down for posterity, prehistory became history. But for most of the inhabitants, life continued to depend on the land, the weather, and their plants and animals—in short, on their environment.

Mesopotamian agriculture was productive but unsustainable, for it was vulnerable to prolonged droughts. Toward the end of the third millennium BCE, the climate of the Middle East and North Africa turned arid for more than 300 years, possibly due to a volcanic eruption. Northern Mesopotamia was covered with a thick layer of windblown dust from the nearby deserts. Wheat and barley crops failed and sheep and goats died. People abandoned the region and migrated to the south, where they overwhelmed the stores of grain and caused political chaos. As the environment deteriorated, the kingdom of Akkad, the first state that had ruled all of Mesopotamia from the headwaters of the Tigris and Euphrates to the Persian Gulf, collapsed.[5]

The Mesopotamian environment was also vulnerable to human abuse. The Tigris and Euphrates carried not only water and silt but also salts dissolved from the rocks in the mountains where they originated. For thousands of years before the rise of cities, farmers had learned, by trial and error, how to prevent salt from damaging their crops. They did so by leaving half of their fields fallow every year. Shok and agul, two plants that grew on the fallow land, dried out the soil, leached salt away from the root zone of crops, and deposited nitrogen, making the land more productive for crops the following year.

Once cities arose, their leaders, eager for more revenue, demanded that farmers grow crops every year. Farmers could no longer keep the salt at bay by alternating sown and fallow. By irrigating the land, farmers raised the water table in southern Mesopotamia to within 50 centimeters of the surface. Capillary action then raised sub-surface water, which was slightly brackish, up to the level of the roots of plants. During the period of the Ur III Dynasty (2400–1700 BCE), a new canal linking the Tigris and the Euphrates brought copious water to southern Mesopotamia, leading to over-irrigation and raising the water table. As fields became saline, farmers had to replace salt-sensitive wheat with more salt-tolerant barley. Yields declined over time, from an average of 2,537 liters per hectare in 2400 BCE to 897 liters per hectare in 1700 BCE. As salinity caused the productivity of the land in southern Mesopotamia to decline, Sumer and Akkad were eclipsed by Babylonia to the north, and later Babylonia was eclipsed by the Assyrians and the Persians.[6]

Egypt

Today Egypt, like Mesopotamia, is a river valley surrounded by deserts. But the Sahara was not always a desert. For several millennia after 9000 BCE, it received enough rain to support a savanna dotted with lakes and water holes and inhabited by hippopotami, elephants, crocodiles, giraffes, and myriad other animals. After about 4000 BCE, as the climate began to dry, cattle herders appeared where there had once been only hunter-gatherers. When the climate became even drier and vegetation died, herders and foragers made their way to a more inviting environment, the Nile Valley.

The Nile had years of low and years of high water, yet overall its fluctuations were less frequent and extreme than those of the Tigris and Euphrates. It received its water from the monsoon rains that fell on the highlands of East Africa in the spring and early summer. At Aswan in upper Egypt, the Nile usually began to rise in June. The flood peaked in September and reached the north of the valley by late November. As the flood waters slowly made their way northward down the valley, they covered the land with a fresh coat of mud. Alluvial deposits also created natural levees between which the river rose above the flood plain. When the waters broke through the levees, they settled in natural basins. Before reaching the Mediterranean Sea, they deposited the rest of their silt to form the Nile Delta.

Farmers first settled the Nile Valley a little before 5000 BCE. They found it almost ideal for agriculture.[7] The silt deposits renewed the fertility of the fields every year, and the flow of water was sufficient to wash away any salts. By 3600 BCE, farmers had cleared the valley of its natural vegetation and built dikes, irrigation channels, and drainage ditches. They did not need to bring water from the river to the land in canals; instead, they built low weirs to retain the flood water for a few days before releasing it downstream to the next field. This method, called basin irrigation, ensured that the soil was thoroughly soaked and that sufficient silt was deposited on it. The Egyptians named the seasons after the cycles of the river and of agriculture: *Akhet*, the flood season from August to November; *Peret*, the growing season in the fall and winter; and *Shemu*, the season of harvest and drought from May through early August.

Unlike Mesopotamia, agriculture in Egypt did not erode the fertility of the land. The silt was so rich and the flow of the Nile so regular that they profoundly marked Egyptian culture.[8] Ancient Egyptian civilization lasted from 3200 BCE to the conquest by Alexander in 332 BCE, longer than any other, save China. As a balance between humans, the land, and the Nile, it lasted much longer. Only in the 1970s, with the opening of the Aswan Dam, did Egypt sever its ties to a 5,000-year-old symbiosis between humans and nature.

Egypt moved more quickly than any other part of the world from Neolithic villages to a full-blown civilization known for its elites and monumental architecture. That is because the fertile Nile Valley was enclosed between two deserts, so farmers could not migrate elsewhere if the burdens of tribute and taxation became too heavy. This entrapment has characterized Egyptian civilization from the fourth millennium BCE to the present.

In the early third millennium BCE, after many small kingdoms were united by Menes, the legendary founder of the first dynasty, water control was developed and extended. The pharaohs and landowners of ancient Egypt regarded the farmers' crops as a source of taxes and rents and carefully noted the level of the flood each year. A few pharaohs had canals constructed. Pepi

I (2390–2360 BCE) boasted: "I made upland into marsh, I let the Nile flood the fallow land. . . . I brought the Nile to the upland in your fields so that plots were watered that had never known water before."[9] Most, however, showed little interest in the methods of water control that the farmers used, leaving these matters to local communities.[10]

Despite its regularity, the flow of the Nile, dependent as it was on the Indian Ocean monsoon, was not completely reliable. In a wet year, the flood raised the level of the water by 2 meters and covered two-thirds of the flood plain; weak floods, however, could leave three-quarters of the valley dry. Low-water periods and the resulting famines, though rare, determined the dynastic history of Egypt and the well-being of its people. The first five dynasties (ca. 3100–2345 BCE) were blessed with abundant water. When the floods were low after about 2200 BCE, the kingdom of Egypt disintegrated into civil wars and famine, and cannibalism stalked the land. This era, which Egyptologists call the First Intermediate Period (2181–2133 BCE), was followed by a time of more abundant floods, when Egypt was prosperous and even engaged in wars against the neighboring Nubians.[11]

Sometime during or soon after the Middle Kingdom (2040–1786 BCE), farmers began to sow a second crop during the summer months, thanks to the invention of the *shaduf* or swape, a device invented in Mesopotamia that partially compensated for the deficiencies of the river. This was a long pole pivoting on an upright beam, with a bucket at one end and a counterweight at the other. With it, farmers could lift up to 2,500 liters of water a day, much more than with a bucket and a rope. Though it allowed them to grow a second or even a third crop of vegetables in their gardens during the dry season or when the flood was low, it required too much labor to irrigate staples like wheat, barley, or flax.[12]

Once again, starting in 1797 BCE, Egypt suffered from low Nile floods. The result was the Second Intermediate Period (1786–1552 BCE), a time of famines, wars, and an invasion by people called the Hyksos. The pattern was repeated when the high waters returned after 1550 BCE, bringing a time of prosperity called the New Kingdom (1552–1069 BCE), best known for the reign of Rameses II (1290–1224 BCE) and for his foreign expeditions and massive stone monuments. Then, once again, low floods from 1069 to 715 BCE—the Third Intermediate Period—was a time of warlords and invasions by Libyans, Nubians, and Assyrians.[13] Herodotus famously called Egypt "the gift of the Nile," for when the Nile was generous with its waters, the kingdom and its people flourished. When it failed, they suffered.

Figure 3.1. Egyptian swapes used to water gardens when the Nile River was low (Chadoufs de la Haute Egypt/Zangaki). Prints and Photographs Division, Library of Congress, LC-DIG-ppmsca-03921.

The Indus Valley

The Indus Valley, covering most of today's Pakistan, is the flood plain of several rivers that come down from the Tibetan Plateau and merge to form the Indus, one of the great rivers of Asia. Like other rivers that depend on the monsoon for their water, these rivers periodically but irregularly flood the land and shift their beds. Furthermore, the region is tectonically active; as the Indian plate pushes against the Eurasian plate, mud barriers occasionally form in the valley. It is in this region that the third of the classic river valley civilizations arose in the fourth millennium BCE.

Far less is known about this civilization than about Mesopotamia, Egypt, and China. Archaeologists have found it difficult to investigate because many likely sites are barely above the water table and excavations quickly fill with water. The writing of the Indus Valley people—what little of it has been preserved—has still not been deciphered. We do not know what language the inhabitants spoke, where they came from, how they organized their polities, or what happened to them. Yet the general picture that emerges is similar to that of their neighbor, Mesopotamia, allowing comparisons to be made.

The crops that the first farmers cultivated in the fifth millennium BCE were the same as those of Mesopotamia: wheat, barley, peas, and lentils. Likewise, the animals that the first herders raised were also of Middle Eastern origin, namely goats, sheep, and cattle. Clearly there was contact from the beginning. The soil and climate were also similar to those of Mesopotamia, and so was the response of the farmers, namely, to dig small channels to their fields and small dams to hold back water, and to replace them every year after the floods had washed away the previous year's water controls.

So productive was the agriculture in the valley that it supported some of the largest cities of the ancient world: Mohenjo-Daro, Harappa, and many others. These cities were laid out in a grid, with solid houses built of identically shaped bricks. Thousands of years before the Romans, the Indus Valley cities had efficient drainage systems and pools for bathing. They had large granaries but evidently no temples or palaces. The skeletons found in their cemeteries show no nutritional stresses or differences between the social classes, nor are there signs of struggle or warfare. It is believed that the cities were ruled by merchants rather than priests or warriors, for they conducted an active trade with the inhabitants of Baluchistan, Afghanistan, the Hindu Kush, and Mesopotamia, importing metal, stone, precious stones, and craft goods.[14]

The end of the Indus Valley civilization is as mysterious as its origin and efflorescence. After several hundred years, once prosperous cities were abandoned. Squatters failed to maintain the houses or built inferior houses with recycled bricks, then with mud and thatch; pottery kilns and other noxious activities arose in once residential areas; and drains were neglected. There is evidence of an increase in infectious disease and in interpersonal violence, and corpses were buried in a perfunctory way.

What could have brought about the decline and fall of this civilization? Many causes have been proposed. Some are environmental. As in Mesopotamia and Egypt, the climate became more arid in the late third millennium, undermining the agriculture that supported the cities.[15] Earthquakes may have diverted rivers, flooding some areas and leaving others too dry to farm. The region around Harappa became desiccated when the tributaries of the Saraswati River were captured by the Sutlej and Indus rivers to the west and the Yamuna to the east. Epidemics of malaria, and possibly cholera, have been suggested.[16]

Also plausible is a botanical explanation. The crops the farmers brought from the Middle East—wheat, barley, and pulses—grew well in the Indus Valley if they were sown in the fall to grow in the winter. But they did not grow farther south or east, where the rains came in late summer. When farmers domesticated or acquired crops that grew well in the summer, such as millet, sorghum, and rice, they were no longer circumscribed by the environment but could break out of the Indus Valley and move south into Gujarat and southeast into the Yamuna-Ganges Valley and from there into the Deccan. Likewise, the animals they domesticated—zebu (*Bos indicus*) and water buffalo (*Bubalus bubalis*)—were native to the subcontinent. In short, the cities lost their hold on the farmers, and when the farmers moved away, the cities imploded.[17]

At least one explanation has been put to rest, namely, the Aryan invaders who were once thought to have swept through the Indus Valley like a horde of barbarians, burning and pillaging. Aryans did exist; they were nomadic pastoralists who spoke an Indo-European language and drifted into the subcontinent with their herds. But by the time they did so, the cities of the Indus were already in decline.[18]

Early Civilizations of the Americas

The civilizations of the Americas differed from those of the Eastern Hemisphere in several important ways, largely due to the plants and animals amenable to domestication. Squash, beans, and other vegetables were valuable additions to the diet but could not provide full nutrition. Only maize could fully replace foraged vegetation, but unlike Eurasian grains, maize took many centuries of patient development before it could be relied upon year-round. The Americas also had few domesticable animals. Only llamas and alpacas could provide the secondary products that were so important in the Eastern Hemisphere, namely, wool and the ability to carry loads, but they were not strong enough to pull carts or plows. Furthermore, they were found only in the Andes.

Despite these disadvantages, the peoples of the Americas developed several brilliant civilizations. Some complex societies, like those in central Mexico and Mesoamerica, arose where water was the critical variable and water control produced agricultural surpluses sufficient to support large non-farming elites and their associated classes. Others appeared where irrigation, rain-watered agriculture, and other activities complemented each other. This was the case of the civilizations of western South America.

The West Coast of South America

The Andes lie along the western edge of South America, separating the region into four distinct environments. To the east of the mountains stretches Amazonia, a huge and relatively flat terrain that receives 90 percent of the rain brought by the winds from the Atlantic, creating one of the world's most extensive rainforests. The Andes form two mountain ranges—the Cordillera Blanca to the east and the Cordillera Negra to the west—between which lies a third environment, the Altiplano, an arid plateau with poor soils. The high mountains cast a rain shadow over their western slopes. The meager amount of rain that crosses the two cordilleras feeds sixty short rivers that cascade into the Pacific. Along that ocean lies a fourth environment, the Atacama Desert, a narrow shelf of land that receives an average of 30 millimeters of rain a year and in many years none at all, making it one of the most arid regions in the world.

Not only is the terrain rugged and the climate harsh, but the region also suffers from frequent earthquakes, droughts, and floods. As the South American plate pushes against the Nazca plate under the Pacific, the Andes experience at least one magnitude seven earthquake per decade. Ice cores drawn from glaciers in the Andes show that droughts hit the region in 534–540, 563–594, 636–645, and a most severe one between 1245 and 1310.[19]

For early settlers, the region contained one source of great wealth, however. Most years, the nutrient-rich Humboldt Current brings cold water northward from Antarctica to the coasts of Chile and Peru, water teeming with anchovies that provide food for other fish, marine mammals, sea birds, and humans. The first settlers were fishermen who built villages and small ceremonial centers near the coast between 2500 and 1800 BCE. They ground up anchovies into fish meal that could be stored for long periods of time. These fishing peoples were followed by farmers who settled along the rivers that tumbled down from the Andes, digging short irrigation canals and fertilizing their fields with ashes, manure, and guano. They grew maize, beans, cotton, manioc, and sweet potatoes. The two types of communities complemented each other, exchanging fish protein, salt, and seaweed for vegetables, cotton cloth, and fishing nets.

Their livelihood was at the mercy of a dangerous weather phenomenon, however. Several times a century, the region receives the full brunt of El Niño, a climatic inversion in which warm waters and westerly winds replace the cold waters of the Humboldt Current, bringing torrential rains and floods to the coastal deserts. As the waters turn warm, the schools of fish depart for deeper, colder waters. Flash floods cause landslides that destroy fields and irrigation works. Winds blow sand dunes inland from the coast, covering coastal communities and forcing their inhabitants to flee to higher ground.[20]

Into this difficult and risk-filled environment, the first civilizations of South America arose. The earliest evidence of agriculture and of irrigation anywhere in the New World dates from the mid-fifth century BCE. Here, a Neolithic culture gave rise to the first hierarchical social organization, the Norte Chico culture that flourished between 3000 and 1800 BCE, followed by the Cupinisque (1500–1000 BCE) and Paracas (600–175 BCE), cultures that began growing maize and making ceramics.[21] Around 100 CE, a powerful state in the Moche Valley extended its power over the nearby Lambayeque, Jequetepeque, and Santa river valleys. The people of Moche were experts at irrigation who constructed stone-lined canals, tunnels, and aqueducts, some connecting neighboring rivers. For their lords, they built palaces atop adobe pyramids. Their rulers were evidently powerful men who controlled irrigation waters and the food supply.

Their complex society, however, was vulnerable to natural disasters. The flow of the rivers varied with the seasons and from year to year; only a few carried water year-round. When rivers dried up, farmers survived by digging "sunken gardens" closer to the water table. Between 563 and 594, at a time when the interior was experiencing a prolonged drought, the Moche and Jequetepeque valleys, two of the richest and most populated, were hit by El Niños that caused major floods. Meanwhile, a powerful earthquake altered the course of the Moche River, destroying parts of cities, farmlands, and irrigations channels and causing the inhabitants to flee. The Moche civilization never recovered.[22]

After 850, a new culture, the Chimu, replaced the Moche. Its capital city, Chan Chan, with its quarter million inhabitants, was much larger than any previous one in South America. The Chimu realm was also much more powerful than its predecessor, controlling twelve river valleys and 50,000 hectares of arable land on which farmers grew two or three crops a year, using hoes and digging sticks. Their leaders were even more enthusiastic engineers than the Moche, building over 400 kilometers of canals, all with forced labor. What allowed them to survive natural disasters better than previous cultures was their superior engineering skills.[23]

Andean Civilizations

The people who lived in the southern Altiplano around Lake Titicaca in Bolivia faced very different conditions. Here the best-watered lands are the least fertile, and the most fertile get little rain. Furthermore, above 2,500 meters the climate is cold and the air is thin. Hypoxia, the lack of oxygen in the air, stresses animals and plants as well as humans. Finally, the region is subject to droughts that correspond to the El Niño events along the coast farther north.

By 1200 BCE, villages appeared around Lake Titicaca, where farmers grew potatoes and quinoa and herded llamas and alpacas. To cope with the lack of rain and to prevent erosion, highland farmers built terraces covering 11,000 square kilometers in Peru and Bolivia, to which they carried soil from the river bottoms. In the swampy margins of the lake, they built raised fields. Despite their efforts, they could expect only one good harvest every three to five years. To cope with the erratic climate, they stored food in the form of freeze-dried potatoes and sun-dried llama jerky. They also sought safety in diversity by farming different crops at different altitudes, such as maize, cotton, and coca on the warmer eastern slopes of the mountains. They obtained salt, seaweed, dried fish, and other marine products from the coast. For transportation, they relied on llamas, beasts of burden that could carry up to 32 kilograms.

As in the coastal lowlands, a harsh environment sustained a series of civilizations. The Chavín people built their capital Chavín de Huantar at an altitude of over 3,100 meters in the mountains of central Peru. Between 900 BCE and 250 CE they dominated the trade routes between the mountains, the coastal plain, and the eastern foothills of the Andes.

Meanwhile, from 100 BCE to 1100 CE the Tiwanaku dominated the high plateau near Lake Titicaca. There, with the disciplined labor of thousands of farmers, they drained the lakeshore wetlands, reclaiming nearly 80,000 hectares of land on which they grew potatoes and quinoa. Llamas pro-vided meat and also carried tropical fruits, maize, coca, and other products from the lowlands. Despite their efforts, the Tiwanaku were vulnerable to droughts that struck repeatedly, such as the catastrophic one in the late twelfth and early thirteenth century that caused the level of Lake Titicaca to drop between 12 and 17 meters and destroyed the farms along its shores. These droughts caused the ruin of Tiwanaku and its neighbor and dependent state, Wari.[24]

The last of the highland states, the Inca, was also the most powerful and best organized. From their predecessors the Tiwanaku, the Inca adopted organizational skills, religious ideas, labor discipline, and skills in building with large, painstakingly fitted stones. To these qualities, they added con-siderable military prowess. Their empire, which eventually stretched from northern Chile to Ecuador and from the Pacific coast to the Amazon forest, was supported by the complementarity of its varied agricultural zones. A net-work of roads and llama caravans provided economic security in a region rich in ecological challenges and prone to natural disasters. The Incas forced conquered peoples to build extensive terraces with water channels, even aqueducts and reservoirs. In addition to the potato and quinoa, the staple crops of the highlands, they greatly expanded the area of state lands in the lowlands devoted to growing maize. They also controlled state-owned herds of llamas as means of transport. By mastering several difficult environments from the ocean to the Altiplano to the tropical rainforest, the Inca were able to compensate for the vulnerability of each to the vagaries of nature.[25]

Central Mexico

In central Mexico it was not rivers but springs and lakes that provided the ecological foundations for civilization. Rain in this region is unpredictable and falls mainly on the mountains. Even the earliest farmers understood the need for water control, and the dams and terraces they built in the Tehuacán and Oaxaca valleys to capture water from springs preceded cities and governments by hundreds of years.

In the Valley of Mexico, surrounded by mountains, farmers devised a unique type of field called *chinampa*. In the summer and early fall, the runoff from the rains covered a quarter of the valley (about 1,000 square kilometers) with a shallow sheet of water. From October through May, when it seldom rained, the water was reduced to five lakes: Zumpango and Xaltocán in the north, Texcoco in the center, and Xochimilco and Chalco in the south.[26] In the shallow parts of the lakes, farmers built narrow fields up to 100 meters long and 5 to 10 meters wide surrounded by water on three or four sides. To do so, they cut the thick floating vegetation from the marshy areas of the lakes and piled it up in rectangular plots separated by canals. To the floating vegetation, they added mud dredged from the bottom of the canals as well as garbage and human manure. To hold the mud and compost in place, they surrounded their plots with posts and vines or branches, and planted willow trees along the edges. Thus they achieved almost 100 percent recycling of the valley's nutrients. They planted maize directly in the mud; they germinated other plants in seedbeds, such as amaranth, chili peppers, tomatoes, beans, and flowers, then transplanted them to the fields. Since the land was no more than a meter above the level of the surrounding canals, the crops obtained water during the wet season by capillary action; in the dry months the farmers watered them by hand. The water also moderated the temperature of the chinampas, reducing both frosts and overheating. Besides abounding in fish, turtles, and waterfowl, the canals also allowed the Indians to travel and transport goods by canoe, for they had no pack animals or wheeled carts.

Maintaining the chinampas and farming—or rather, gardening—on them required constant effort year-round. Although the extent of the chinampas was limited—120 square kilometers at their peak in 1519—their productivity per hectare was the highest in the world. Farmers could grow three or four crops a year, sometimes as many as seven, two of which were of maize. One hectare of chinampa could support up to fifteen people, compared to three people per hectare in China.[27]

The first civilization in central Mexico was that of Teotihuacán, a city famed for its many pyramids and rich residential quarters that began its life around 300 BCE and flourished between 100 and 650 CE. At its height in the mid-first millennium, Teotihuacán had as many as 100,000 inhabitants, making it one of the largest cities in the world at the time. It traded and received tribute from as far away as Mesomerica and northern Mexico. Teotihuacán obtained food from fields in the nearby alluvial plain irrigated by springs and channels. In waterlogged areas of the plain, farmers dug drainage ditches. Seasonal floodwaters were captured to irrigate the piedmont area. The San Juan River, channeled to flow through the center of the city, provided water to the inhabitants.[28]

The population of Teotihuacán and its trade with other parts of Mesoamerica began to decline after 600 CE. Skeletons of the inhabitants show they had stunted growth and high rates of infant and child mortality, signs of malnutrition.[29] Around 700, the center of the city and its wealthier neighborhoods burned down. What caused this massive destruction is not known. Perhaps nearby deforestation had led to the silting and erosion of farmland, or perhaps a drought ruined the agriculture of the area, leading to an uprising or an invasion. In any event, the collapse of this once-great city paralleled that of the Mayan cities to the south.[30]

In 1325, the Aztecs, a small tribe from the north, settled on an island in the middle of Lake Texcoco where they founded the city of Tenochtitlán. In order to feed the city dwellers and soldiers, they extended the practice of chinampa gardening to new parts of the lake. They built two double aqueducts to supply the city with fresh water. Because much of Lake Texcoco was brackish, they constructed a sophisticated system of dikes and sluices to protect the chinampas near the city, in particular the Nezahualcoyotl dike that separated the freshwater lagoon from the brackish parts of the lake to the east. As a result of their engineering and labor, the Aztec capital grew to over 150,000 inhabitants, larger and richer than any European city of its time except Ottoman Istanbul. The Aztecs soon outgrew their lakeside home. As their military power grew, so did the area they dominated. By the early sixteenth century, they held sway over central Mexico from the Atlantic to the Pacific oceans, exacting tribute from both allies and subjugated peoples. Thus, they came to depend on rain-watered as well as chinampa agriculture.[31]

The Southwest

Compared to the civilizations of central Mexico, the cultures of the Southwest of what is now the United States were small-scale affairs. Yet they are among the best known of the pre-Columbian cultures, thanks to the careful investigations of generations of American archaeologists.

The Colorado Plateau and the lands to the south of it contain a huge diversity of elevations, rainfall, and vegetation. Toward the end of the first millennium BCE, the inhabitants of the region turned increasingly to farming as a means of supplementing their erratic harvests of pinyon nuts, mule deer, and other wild foods. In the first millennium CE, many of them came to depend almost entirely on foods that they cultivated in canyon bottoms where the water table lay close to the surface or where they could irrigate their crops from springs and seasonal rainwater pouring over the canyon rims. In dry years, they moved to more favorable environments, gathering wild plants and cultivating a variety of crops in scattered plots as a form of insurance. As in Mexico and Mesomerica, maize was the main staple, and in the dry climate of the Southwest, it could be stored for two years or more.[32]

The place that became the most highly developed in the entire region, but also the most fragile, was Chaco, a canyon in central New Mexico 30 kilometers long and up to 2 kilometers wide. Though the canyon itself received relatively little rain, streams pouring over the cliffs that lined it and seeps at their foot covered the canyon floor with water after every rain.

From the seventh century on, and especially during the eleventh century, the rains were more abundant than they had been in millennia. As farming conditions improved, the Anasazi people (or ancestral Puebloans) moved into the canyon and increased in numbers. Agriculture in the canyon bottom flourished, for the water table was close enough to the surface for roots to reach it, and the growing season was longer than on the Colorado Plateau. To supplement the natural flow of water, the inhabitants dug ditches and channels and built small dams to capture runoff.[33]

At its height around 1150, the population of the canyon may have reached several thousand. How many lived in the canyon year-round and how many came only on ceremonial occasions is a matter of debate. One thing is certain, however: the people could not have survived nor could the buildings have arisen without massive imports from outside the canyon. By 1100, Chaco Canyon had become the center of a mini-empire, supported by satellite communities in the surrounding region linked to the canyon by a network of paths across the landscape. In exchange for political leadership and spiritual protection, the members of these communities brought pottery, stone tools, and food to the canyon from up to 100 kilometers away. Turquoise, which the people of the canyon made into ornaments and jewelry, was imported from over 150 kilometers away, as were copper, bird feathers, and other luxury items.

Most important was the import of lumber used to build the houses and ceremonial structures of the canyon. At first, the builders used pinyon pine and juniper logs from nearby forests. By 1000 these trees were all gone and the Anasazi began to fell ponderosa pine, spruce, and fir from the mountain slopes 300 meters higher than the canyon. To build the Great Houses, some 200,000 logs weighing up to 320 kilograms apiece were carried, rolled, or dragged by humans—for they had no carts or draft animals—from the Zuni and Chuska mountains 75 or more kilometers away. The impact was permanent; to this day, 900 years later, the trees have not grown back, and the land is a semi-desert covered with scrub.[34]

The prosperity of Chaco Canyon lasted as long as enough rain fell to grow the food needed by its people and those of its outlying satellite areas. Then came a drought that lasted from 1090 to 1096. As the elites, perhaps seeking to appease the spirits that brought rain, ordered the construction of ever larger and more elaborate buildings, farmers began leaving the area. After another series of dry years with unpredictable rains from 1130 to 1180, construction ceased. Outlying communities stopped sending food, perhaps because they lost faith in the ceremonies performed by the priests of Chaco Canyon. Archaeologists have found evidence of civil unrest, warfare, massacres, and cannibalism. The survivors abandoned the canyon, some of them in a hurry, leaving behind their cooking utensils. The great drought of 1275–1299 that affected the Southwest and northwestern Mexico reduced the population of the entire region. Some survived on the plateau in remote and more easily defensible sites such as Mesa Verde, which were also abandoned soon thereafter; others moved south where their descendants, the Pueblo Indians, still live, but in much simpler societies than their ancestors.

Figure 3.2. Chetro Ketl: Great House used during Ancestral Puebloan religious ceremonies in Chaco Canyon National Historical Park, New Mexico. iStock. com/powereofforever.

Many causes of the rise and fall of the Anasazi have been suggested, most of them involving environmental conditions. The dry spell that began in 1130 played a part, but there had been droughts a hundred years earlier without producing the same consequences, and the Pueblo Indians have survived many droughts since then. Rather, it was human-induced changes to the environment that made them vulnerable to the drought.

When heavy rains came, flash floods washed away dams and turned the channels farmers had dug into arroyos or gullies that were lower than the level of the fields, leaving the fields without water. Deforestation on the uplands above the canyon let rains wash away the nutrients in the soil, leaving barren ground. As long as the population was low, farmers could move to new land, but once all usable land was filled up, those whose land was exhausted had nowhere to go. Before the drought, the people of Chaco Canyon and surrounding areas had reached the limit of what their land could support. When drought hit the area, their society collapsed and the people fled.[35]

THE MAYA

In Central and North America, the transition from isolated foraging bands and farming communities to urban civilizations took place in the same time frame as in South America and much more slowly than in the urban centers of the Eastern Hemisphere. The most spectacular of these were the Maya, who built a civilization in the rainforests of Mesoamerica.

Mayan Civilization

When the Mayan ruins were first discovered by people of European origin, they were deemed the archetypal "lost civilization in the jungle." Later, the Maya were cited as exemplars of a peace-loving society of brilliant artists and architects. When their hieroglyphs were decrypted, they became famous for their bloodthirsty wars and exquisite tortures. Why this elaborate civilization collapsed and what happened to its people remain puzzling.

The region the Maya once occupied consists of three very different environments: the highlands of southern Guatemala, the Petén or lowlands of northern Guatemala, and the Yucatán Peninsula of eastern Mexico and Belize. The plateau that slopes down from the mountains of Guatemala toward the Yucatán consists of karst, a porous limestone that absorbs rain as soon as it hits the ground, leaving few rivers or wetlands on the surface.

Like many tropical areas, the Mayan land has two distinct seasons. The rainy season lasts from May or June to November, bringing between 44 centimeters in northernmost Yucatán to 400 centimeters in the southern Petén. From December to April or May, it rains little or not at all, and the *bajos* or swamps that collect surface water during the rainy season dry out. The rains are not reliable; some years have three or four times more rainfall than others, with occasional hurricanes wreaking havoc. In the Yucatán, the water table lies close to the surface so it can be reached year-round in *cenotes*, or circular sink-holes formed by the collapse of underground caves. Though the Petén gets more rain, there are no lakes or rivers to conserve it and the water table is inaccessible.[36]

Before humans started to farm, the Petén was a forest of mahogany, sapodilla (the chewing gum tree), breadnut, avocado, and other trees and shrubs, while the Yucatán was covered with thorny bushes and savannas. As in many tropical forests, only 10 to 20 centimeters of topsoil covered the bedrock, and most of the nutrients were contained in living or recently dead organic matter. Felling and burning the trees therefore meant losing nutrients, turning the soil into hard laterite or exposing the bare limestone.[37] As bishop Diego de Landa, one of the first Spaniards to investigate the area in the sixteenth century, wrote: "Yucatán is the country with the least earth that I have seen, since all of it is one living rock and has wonderfully little earth."[38]

People began farming this land as early as 1800 BCE. Their method, called *milpa*, was similar to the slash-and-burn agriculture practiced in New Guinea and other tropical forests. During the dry season, farmers cut and burned the vegetation, leaving the ashes to fertilize the soil for a year or two. During the rains they planted and harvested maize, beans, squash, chili peppers, manioc, and cotton. After two years, the soil lost its fertility and the farmers burned a new patch of forest, leaving the earlier fields fallow for between four and twenty years. Their plots were dispersed throughout the countryside, and their villages were often mobile.

Gradually, the population grew. At its peak in the early ninth century CE, the density reached some 400 people per square kilometer, as high as that of rural China in the twentieth century. At such densities, slash-and-burn was no longer an option. In its place, farmers tried a number of techniques to wrest more food from the soil. They shortened the fallow period. They cleared the forests and cultivated marginal lands on hillsides. Near rivers and wetlands, they dug short canals and built raised fields with the mud dredged from the canals. Elsewhere, they constructed terraces to prevent soil erosion and retain water. They mulched and fertilized with human wastes or composted vegetation and watered their crops from cenotes and underground cisterns.[39]

Despite their efforts, the productivity of their agriculture was low compared with other Amerindian civilizations. As the Maya had no beasts of burden and very few navigable lakes or rivers, almost everything had to be carried on the backs of humans. Food could not be transported much more than 30 kilometers; beyond that the porters would eat all the food they could carry. Thus a food shortage in one place could not be alleviated by a surplus elsewhere. The only domesticated animals were dogs, turkeys, and Muscovy ducks, and the diets of the poor were deficient in proteins. As maize could not be stored long in the humid climate, survival depended on every harvest being sufficient to last a year.

Not only was the productivity of agriculture low, but it also declined over time. Continuous cultivation without allowing fields to lie fallow leached the nutrients from the soil. Deforestation left the hillsides barren and eroded. As the cultivated area expanded, it left fewer wild resources to fall back on. This is reflected in the declining health of the people, as shown by stress marks on their teeth. By the eighth century, the Maya had reached the limits of the carrying capacity of their environment.[40]

Yet the Maya are known for their magnificent monumental architecture. At a site called San Bartolo in the Petén, archaeologists found an ancient Mayan tomb and murals dated to around 100 BCE.[41] Their civilization enjoyed a first efflorescence between 150 BCE and 50 CE, when the city of El Mirador grew to almost 10 square kilometers and up to 80,000 inhabitants before being abandoned after 150 CE. In the third century CE, there began what archaeologists call the Classic Maya period. By the eighth century, many cities arose, the largest of which were Tikal, Copán, Calakmul, and Palenque. They were adorned with spectacular palaces, temples, and plazas built by farmers recruited during the dry season.

At its height, 60,000 to 80,000 people lived in or near Tikal, in the central Petén. The heart of the city was located on a ridge. To survive the yearly dry seasons, the inhabitants paved and plastered the plazas and terraces at its core to catch rainwater that was then channeled into six major reservoirs that could store over a billion liters of water, enough for the 9,800 members of the elite who lived in the core. The nobles kept their reservoirs clean by growing water lilies, hyacinths, and ferns that removed pollutants, and referred to themselves as *Ah Nab* or "water lily people." Around the core were densely inhabited residential neighborhoods with several smaller reservoirs. Beyond the residential neighborhoods, large basins caught the used water from the city to water and fertilize outlying fields during the dry season, producing two or even three crops a year. These reservoirs together contained enough water for up to eighteen months. In short, the people of Tikal had built an artificial oasis that thrived as long as there was sufficient rain.[42]

The Collapse of the Maya

Mayan society exhibited a more extreme polarization than was found in other early civilizations. The kings and nobles ruled not only by their spiritual leadership and by their monopoly of the means of violence but also by controlling access to water during the dry season. In exchange for water, the nobles could demand food and labor to build their plazas, temples, and palaces. As the population grew, so did the demands of the elites. The nobles adorned their costumes with the feathers of the rare Quetzal bird, driving it almost to extinction. The kings constructed monuments at an ever growing pace, while aristocrats built palaces. All surviving Mayan inscriptions describe the kings and nobles, their conquests and diplomatic triumphs, and their relations with the gods. None mention commoners.

No Mayan city was able to dominate the others and create an empire. As the growing population filled in the area between the city-states, they fought one another in a crescendo of wars and violence, seeking in conquest and human sacrifices a compensation for the decline of their economies.[43]

In the ninth century, the cities of the Petén began to implode. In Copán, the last mention of a king dates from 822, and in 850 the royal palace burned. Elsewhere in the region, the last inscription dates from 909. From a peak of 20,000 or more, the population of the Copán valley dropped to 15,000 in 950, to less than 8,000 in 1150, and to almost zero by 1250.

Mayan society did not collapse everywhere at once. Even as the Petén became depopulated, cities like Mayapán and Chichen Itzá arose in the Yucatán Peninsula, until they too were abandoned around 1450. After that, there were no more cities, no more monumental stone buildings, no more inscriptions; in short, no more civilization. The Maya survived, but in greatly diminished numbers; archaeologists estimate that the population declined by 90 to 99 percent.[44]

What caused such devastation? Archaeologists have been debating this question since the ruins of this once impressive civilization came to light. Their many explanations can be classified as either social or environmental. The social explanations revolve around the issues of population and politics. Skeletons of people who died after 800 show signs of malnutrition and of rising female and infant mortality rates.[45] Hunger among the farmers and increasing demands on the part of the kings and nobles may have triggered social unrest. According to archaeologist Michael Coe, "The royalty and nobility, including the scribes . . . may well have been massacred by an enraged population."[46]

The alternative explanation emphasizes the relations between the Maya and their environment. By cutting down their forests and repeatedly cultivating every scrap of land, the farmers exhausted the fertility of the soil and eroded the hillsides.[47] Erosion, in turn, filled canals and cisterns with mud, while warfare impeded maintenance, leading to a decline in food production.[48] In Coe's words, "By the end of the eighth century, the Classic Maya population of the southern lowlands had probably increased beyond the carrying capacity of the land, no matter what system of agriculture was in use. . . . The Maya apocalypse, for such it was, surely had ecological roots."[49] David Webster concluded, "Many archaeologists believe that the Maya found themselves in a kind of ecological trap of their own making. Population growth spiraled out of control and the productive potential of the landscapes degraded as human carrying capacity was reached or exceeded."[50]

A third explanation emphasizes climate change, which some scholars, such as Richardson Gill, offer to the exclusion of all others: "The Collapse was not the result of bad management, administrative deficiencies, or poor agricultural techniques. It was the result of forces over which they had no control and against which there was no solution that they could implement."[51] The forces Gill mentions were the periodic droughts that afflicted the land. At various times before the collapse, the Maya had survived such droughts. A major drought from 150 to 200 CE caused El Mirador to be abandoned. After a long series of wet years from 250 to 760, especially from 550 on—the peak of the Classic Maya civilization—came two centuries of dry years, the driest in the last 8,000 years. A drought between 535 and 593 caused a sharp drop in the population and in construction, especially at Tikal, followed by revolts and civil wars. But after each of these episodes, the population recovered and urban construction began anew, for the population had not yet exceeded the carrying capacity of the land. Then came the mega-drought of 750 to 850 that depopulated the Petén and ended the Classic Period of Mayan civilization. Finally, the drought of 1451–1454 ruined the last Mayan cities and depopulated the Yucatán.[52]

Evidence backing up the drought hypothesis has recently come to light. According to the findings of climatologists Gerald Haug and his co-authors, lake bottom sediments that reflect the amount of precipitation show the intensity of the drought: "A seasonally resolved record of titanium shows that the collapse of the Maya civilization in the Terminal Classic Period occurred during an extended regional dry period, punctuated by more intense multi-year droughts centered at approximately 810, 860, and 910 A.D."[53] Nor was drought entirely natural. Deforestation by the Maya reduced the evapotranspiration of the trees, which in turn reduced secondary rainfall downwind of the deforested areas. In other words, the Maya may have exacerbated the droughts that nature inflicted upon the land.[54]

The drought hypothesis is not without its critics. As Patricia McAnany and Tomás Gallareta Negrón point out, cities most vulnerable to drought survived for a time, while others with better access to water did not. Because the collapse of the Petén cities stretched out over a century or more, they concluded: "It is thus unlikely that drought was a prime mover of societal change."[55]

To cause a total collapse required more than reaching the limits of the carrying capacity of their environment when the climate was favorable. It was the combination of environmental stresses and a rigid social hierarchy that left Mayan society vulnerable to environmental changes. To understand why the droughts of the ninth century and after led to the collapse of Mayan civilization demands examining Petén's unique reservoirs controlled by the elite. The power of the rulers rested on an implicit bargain between the commoners, the rulers, and the gods: in exchange for the sacrifices and labor of the commoners, the rulers would perform the ceremonies that ensured that the gods would provide the people with water. In a multi-year drought, when all other sources had dried up, the elite controlled the last supplies of drinking water. When the weather failed them, people lost faith in their rulers. It was not just hunger that drove commoners to attack their kings and lords, burn buildings, and flee the cities; it was thirst. That is why in the southern lowlands, the cities imploded so completely and the population shrank by up to 99 percent, while in the Yucatán, with its cenotes, cities endured for several centuries more.

Conclusion

As they were domesticating plants and animals, humans also learned to control water. In particularly favored locations that combined good soil, a warm climate, and access to freshwater, but where rain was insufficient or came at the wrong time, water control gave farmers astonishing yields. Bountiful harvests, in turn, had two kinds of consequences.

Socially, they led to increases in population, which allowed more water control and more intensive agriculture, in a self-reinforcing mechanism. Plentiful food also allowed some of the surplus to be diverted to support classes of people who did not farm but instead built cities and monuments, organized formal religions, engaged in trade or warfare, or exercised power over others. These developments were especially effective where a rich agricultural area was surrounded by deserts or mountains, preventing farmers from moving away. Thus it is not a coincidence that the first civilizations arose in areas where water control was possible, surrounded by areas where farming was not. Outside these few and relatively small areas, some peoples practiced a Neolithic form of farming, others herded animals, and yet others continued the hunting and gathering lives of their ancestors.

Water control, by its very nature, involved interacting with the environment. These interactions were not one-sided, however, and the consequences varied depending on the environment in each case. In some areas, under the pressure of intensified agriculture, the soil lost its fertility or became saline and unproductive, undermining the civilization that had created that pressure; such was the case of lower Mesopotamia. In contrast, in Egypt the Nile floods kept agriculture sustainable for millennia. In other cases—as in coastal Peru, the Andes, Mesoamerica, the Southwest of the United States, and even, every few centuries, Egypt—the reliance on water in sufficient quantities made agriculture, hence entire civilizations, vulnerable to changes in the climate.

Yet, the fate of these early water-based civilizations was not entirely at the mercy of nature. What mattered was the interactions between the variability of the climate and the resilience of the societies it affected. Some civilizations, like that of Egypt and northern Mesopotamia, proved to be remarkably resilient and reconstituted themselves after every crisis. Others, like the Maya and the Anasazi, collapsed under the combination of a volatile climate and a rigid society, and never recovered.

4

Eurasia in the Classical Age

By the second millennium BCE, as the human population of the Eastern Hemisphere increased, farming spread beyond the original river valleys into areas where crops grew without the need for irrigation. The process was neither smooth nor easy, as the borders between areas of state control and autonomous Neolithic villages were ill-defined, and the fate of farmers on rain-watered lands fluctuated with weather conditions.

The political and cultural history between the mid-second millennium BCE to the mid-first millennium CE—often called the Classical Age—underscores a long-range pattern. Small kingdoms and city-states that had been independent were incorporated (often by force) into ever-larger polities, such as the Roman, Han Chinese, and Persian empires.

In the process of spreading over ever-wider areas, states and civilizations impacted their environments in multiple ways, most importantly by replacing forests with fields. Other environmental consequences that resulted were neither expected nor desired. A rising population, the growth of densely crowded cities, and the increase in long-distance trade were the breeding grounds of diseases and, eventually, of devastating epidemics. This chapter traces these changes up to the disastrous sixth century.

Population and Urbanization

Though statistics on the population of ancient times are educated guesses at best, a few demographers have estimated that the population of the world 10,000 years BCE, before the beginnings of plant domestication, was around 6 million. By 1200 BCE, as a result of the greater quantity of food that farming produced, it had grown to 100 million. By the year 1 CE it had more than doubled, to over 200 million. It then went into a long and very unsteady decline until 700 CE and did not recover until 1000 CE. Then it began to climb again, reaching over 250 million in 1500.[1]

Put another way, the density of human population varied from 0.01 to 0.9 persons per square kilometer for foragers, to 0.8 to 2.7 for nomadic herders, to 10 to 60 for Neolithic farmers, and to 100 to 950 for traditional farmers in civilized societies.[2] The increasing population therefore reflected the spread of traditional farming at the expense of herding and Neolithic farming.

Far from being spread out evenly across the landscape, the population density of civilized societies varied enormously, from very thinly populated forests, mountains, and grasslands to cities and towns packed with people. Demographers have estimated that in 1360 BCE there was only one city with over 100,000 inhabitants: Thebes in Egypt. By 100 CE the number had risen to sixteen; then it fluctuated between eight and fifteen until 1000 CE, when it reached eighteen. After that it grew to twenty-five in 1150, then fluctuated between nineteen and twenty-three until 1500.[3] The numbers of cities with over 50,000 and over 250,000 inhabitants similarly fluctuated.

Both the growth and the fluctuation of urban populations were intimately tied to trade, for trade brought the timber and stone to build cities and the food to sustain large numbers of city dwellers. By water, such goods could be transported at low cost over long distances. So it is not surprising that most large cities were located near bodies of water or—as in the case of China— where a canal could substitute for a navigable river.

Cities needed more than just access to cheap water transportation. The food they imported had to be storable; only crops that could be stored like barley and wheat in the Middle East and the Mediterranean or rice in East Asia could sustain urban life, hence civilization itself, from one harvest to the next. (In South America, that role was played by dried maize and freeze-dried potatoes.) In New Guinea, Amazonia, equatorial Africa, and other tropical regions, farmers grew root crops such as yams and cassavas or fruit crops like bananas and plantains that could not be carried far and did not keep once picked. That is why the inhabitants of these regions, even after millennia of agriculture, did not build cities or create states and empires.[4]

In regions where storable crops grew, cities paid for their imports with artisanal products or—in the case of capital cities—with protection and administration. Athens expanded most rapidly during the fifth century BCE, when it was at the height of its power. Alexandria grew to over 300,000 in the first century BCE. Rome grew to about 1 million in the second century CE. China's cities grew and shrank with the country's political fortunes, but by 1100 CE, it had five cities of over a million inhabitants.[5]

The Middle East and the Mediterranean World

In the Middle East and around the Mediterranean Sea, as in all pre-industrial civilizations, agriculture was the foundation of the economy and the source of wealth and power of the states that ruled those areas. Two types of farming coexisted: irrigated agriculture in the river valleys of Mesopotamia and Egypt and a few other places, and rain-watered agriculture elsewhere. In addition, areas with rainfall too marginal for farming were occupied by pastoralists and their herds, often outside the control of states.

Mesopotamia and the Levant

Mesopotamia had been occupied, irrigated, and cultivated for millennia, but that did not prevent it from being transformed even more deeply by powerful states. The Sasanians, who ruled Persia, Mesopotamia, and surrounding regions from 224 to 651 CE, were especially eager to develop the agriculture of their domains. This meant expanding the irrigated areas. In the hilly areas of western Persia, water was more precious than land, and those who controlled water were more powerful than landowners. The Sasanians irrigated the valley of the Helmand and Amu Darya rivers in Central Asia. The government and wealthy men constructed tunnels called *qanats*, some of them several kilometers long, that brought water from distant mountains to dry land.[6] In the Mesopotamian plain, the Sasanian kings undertook massive projects using as labor thousands of war captives resettled from other areas. Led by a strong centralized bureaucracy, workers built five large dams and expanded the Nahrawan and Katul al-Kisrawi canal systems.[7] This allowed farmers to grow both winter crops of wheat and barley and summer crops of cotton, rice, and sugarcane. By expanding the irrigated area of the Diyala basin north of Baghdad, they transformed it into a rich agricultural region covering 8,000 square kilometers. Under Sasanian rule, the population of Mesopotamia rose to 5 million, a number not reached again until the twentieth century.[8]

Creating such a massive artificial environment had a price, however. Large and complex irrigation systems needed constant maintenance, causing the regime to impose heavy taxes and forced labor on the population. These systems were also unsustainable. The watering required for thirsty crops such as cotton raised the water table, causing the silting of canals and the waterlogging and salinization of the soil. As cotton drained the soil of nutrients, yields declined. In 608 the bubonic plague reached Mesopotamia and returned in 627, reducing both the population and the taxes needed to maintain the canals and other irrigation works. In 628 a massive flood caused the Tigris to shift its bed, turning parts of the district into swamps or salty desert. The shortage of manpower led to the neglect of canals, weirs, and dikes. Though taxes remained oppressive, tax revenues dropped by half or more. Half the settled area in the Diyala basin was abandoned. Sasanian rule became unstable, "a headless, floundering organism with neither the will nor the capacity even to maintain the agricultural system on which its wealth had been based."[9] Meanwhile, the Sasanians engaged in frequent and costly wars with their neighbors. When raiders from Arabia appeared in 632, they found an empire too weak to resist them.

Greece and Italy

Only one-fifth of the soil of Greece was arable, not enough to support a dense population. Furthermore, the rain fell in the fall and winter when temperatures were low and crops did not grow; in the spring and summer, when temperatures were best for growing plants, there was scarcely any rain. So farmers specialized in those crops that best tolerated the terrain and the weather. In southern Greece, in the few areas of relatively flat land, farmers grew barley, a hardier plant than wheat. On the hillsides, they planted olive trees and grapevines. They exported olive oil and wine in pottery jars called amphoras. In exchange, they imported wheat and barley from Sicily, Egypt, and North Africa. They also raised sheep and goats on land that could not be cultivated, such as on hillsides, in forests, or in the mountains during the summer. Necessity led the Greeks to become traders and seafarers and to establish colonies in Sicily and around the western Mediterranean.

Farming implements were relatively simple. The scratch-plow cut a furrow in the soil but did not turn it over, so fields had to be plowed twice, at right angles. By breaking up clods of earth, this method prevented evaporation, a benefit in a dry land. In places where cereals were inter-planted with grapevines and olive trees, as was common in Greece, farmers preferred hoes to plows. Hoe blades were made of iron, as were picks, mattocks, and spades; sometimes the plowshare was of iron as well.

Italy had more fertile land than Greece and was self-sufficient in grains and animals until the first century CE. It was the growth of the city of Rome that led the Romans to depend on grain imported from overseas. Until the late Roman republic in the first century BCE, most farms in Italy were small and family operated, sometimes with the help of a few slaves. Farmers left half their fields fallow each year, plowing them repeatedly. They offset soil exhaustion by applying manure and, near cities, human waste as well. They understood that certain plants—lupine, beans, vetch—could restore the fertility of the soil if they were plowed under. In a few places, they practiced irrigation. But all these methods required inputs of capital, something that small farmers lacked or was taken from them as taxes and rent. As a result, Roman agriculture was extensive but unproductive; for each grain of wheat or barley that farmers sowed, they could expect only four grains in return.[10]

The Roman conquests brought great wealth to successful politicians and military leaders but ruined small farmers. Wealthy landowners squeezed out the farmers and created vast estates called latifundia on which they turned arable land into pastures for animals. Only in milling grain was there technological advance, such as the water-powered mills that the Romans built to produce flour in industrial quantities for their armies and cities.

The Roman Empire

The Roman Empire of the first century CE consisted of four concentric circles. The inner circle was Rome, a city of up to a million or more inhabitants that imported three-quarters of its food by ship from overseas provinces. Surrounding it was central Italy with its unproductive latifundia. The outermost circle held the armies that protected the empire from the barbarians to the north and the Persians to the east. In between was a middle ring consisting of Gaul, Spain, North Africa, and Egypt, productive lands that sent much of their harvest either to Rome or to the armies on the frontier in return for protection and administration.

Egypt was especially hard hit. It had been conquered by Alexander the Great in 323 BCE and inherited by the Greek dynasty of the Ptolemies, who ruled it efficiently, expanding the use of water-lifting devices for summer crops and introducing irrigation to the Faiyum oasis west of the Nile, thereby increasing the amount of effective arable land from 2.2 to 2.7 million hectares. According to geographer Rushdi Said, when the Romans took over Egypt in 30 BCE, "organization and efficiency were restored, but the new system brought a new dimension of ruthlessness. The Ptolemies had at least lived in

Egypt, and the money they exacted had stayed in the country. The Romans, on the other hand, were absentee landlords who milked Egypt mercilessly."[11]

As old lands became exhausted or turned into unproductive latifundia, the empire needed ever more new lands to put under the plow, hence the constant urge to conquer. The empire settled veterans of the wars on newly conquered lands, where they were expected to clear trees and plant crops that could be sent to feed Rome or the armies. Constant wars within the empire, such as those that wracked the Roman world in the third century CE, either damaged farmland directly or prevented peasants from maintaining their terraces and drainage ditches. Starting in the second century CE, epidemics decimated the population.[12] When the armies grew too weak to conquer new lands and then to repel the influx of Germanic tribes on the northern frontier, the economic basis of Roman life entered a downward spiral.[13]

Deforestation in the Levant and the Mediterranean

Humans have been at war with forests for a very long time. Hunter-gatherers set fire to woods in order to open up meadows to attract prey. Swidden farmers burned trees to clear land and used the ashes to fertilize their fields. Early civilizations, with their growing populations and building projects, increased the pressure on forests. In the Epic of Gilgamesh, written in the mid-third millennium BCE, the wild man Enkidu helped Gilgamesh, the king of Uruk, slay the giant Humbaba: "He slew the monster, guardian of the forest, at whose cry the mountains of Lebanon [trembled] . . . the cedars they then cut." After that, Enkidu said to Gilgamesh: "My friend, we have felled the lofty cedar, whose crown once pierced the sky. I will make a door six times twelve cubits high, two times twelve cubits wide, one cubit shall be its thickness." [14] In the eighth century BCE, the Assyrian king Sargon II (r. 722–705 BCE) led his troops against the king of Urartu in the Zagros Mountains: "High mountains covered with all kinds of trees, whose surface was a jungle, over whose area shadows stretched as in a cedar forest. . . . Great cypress beams from the roof of his substantial palace I tore out and carried to Assyria. . . . [T]he trunks of all those trees which I had cut down I gathered them together and burnt them with fire."[15]

The most famous trees in the Middle East were the cedars of Lebanon. During the fourth dynasty of Egypt, pharaoh Sneferu (r. 2613–2589 BCE) imported forty ships filled with cedar logs from Lebanon. The Egyptians used cedar wood to build boats, coffins, and palaces and the resins from these and other conifers to preserve mummies. The Israelites used cedar to build the Temple of Jerusalem; as Solomon wrote to Hiram, the king of Tyre: "Send me cedar logs as you did for my father David when you sent him cedar logs to build a palace to live in."[16] The Phoenecian cities of Byblos, Tyre, and Sidon on the Lebanese coast grew wealthy from the export of cedar, pine, and oak.

Exports of Lebanese cedar continued throughout the Hellenistic and Roman periods. By the second century CE, the cedar forests of Lebanon were so depleted that the Roman emperor Hadrian (r. 117–138 CE) tried to protect them by declaring them an imperial domain.[17] In the Middle East and around the Mediterranean, however, dry summers, brief but torrential winter rains, and frequent droughts slowed the growth of trees. Forests grew back only if they were protected from lumberjacks and goats; over the course of history, such protections were seldom permanent. Governmental efforts at conservation and replanting mattered little in times of warfare when peasants fled to the hills to escape the fighting and cleared trees to plant crops. While their animals could not harm a mature forest, they could prevent new growth— cattle by eating leaves on low-hanging branches, goats by eating bushes and young trees, and swine by eating the acorns, chestnuts, and beechnuts from which new trees would have sprung.[18] By the third century CE, they had left only vestiges of once great forests.[19] Likewise, the mountains of Persia and the shores of the Caspian Sea, which were once heavily forested, were largely denuded in the first centuries CE.[20]

Metallurgy and Deforestation

Historian Theodore Werteim blamed the deforestation of the Mediterranean on metallurgy, especially iron, because of all the fuelwood used in smelting and forging. The first to smelt iron ore were smiths in Anatolia and in west-central Africa in the late third millennium BCE. It took several centuries, however, before they learned to make iron cheaply enough for common objects such as axe and hoe blades, saws, arrowheads, knives, and other tools. After 1000 BCE, iron-making became common from Scandinavia to sub-Saharan Africa and from western Europe to China and India.

The fuel used to smelt iron ores was charcoal, or wood that had been baked (but not burned) in order to extract the impurities, leaving only carbon. Wertime estimated that 20 to 28 million hectares of trees were required to smelt enough iron to leave behind the 50 to 90 million tons of Iron-Age slag found around the Mediterranean. But since such an enormous consumption of fuel was spread over several thousand years during which trees grew back, such consumption was sustainable. Near mining centers, however, deforestation was severe. Smelting a ton of iron consumed 72,320 tons of wood. Every year, the iron smelters of Populonia in Italy are said to have consumed, on average, 375 hectares of forest, the silver and copper mines of Rio Tinto in Spain 950 hectares, and the copper mines of Cyprus 2,000 hectares.[21]

Uses and Sources of Wood

Ancient rulers encouraged the spread of agriculture into forested lands. A Roman law of 111 BCE allowed anyone who brought into cultivation 12 hectares of public land to keep them.[22] The Romans used wood for cooking and heating buildings and baths. Roman baths included a tepidarium (or warm air room) kept at 55° Celsius (130° F) and a caldarium (or hot bathroom) at 70° C (158° F), both heated by hot air ducts from a huge furnace. In the fourth century, the city of Rome boasted eleven imperial baths and 856 smaller ones. Even in the small town of Welwyn in Roman Britain, the baths consumed 114 tons of wood a year, the equivalent of 0.4 hectares of mature hardwood trees or 9.3 hectares new-growth woodland.[23]

Construction was another major consumer of wood. Large buildings like temples, palaces, and baths needed large timbers to support their roofs. Chariots, battering rams, siege machinery, and fortifications were made of wood. In Roman Britain, building a typical fort that covered 2 hectares required felling from 6 to 12 hectares of mature forest. Eighty oak trees went into building a farmhouse. Making 1 cubic meter of bricks took 150 cubic meters of wood; burning a ton of lime for plaster, mortar, or cement required from 5 to 10 tons of wood.[24]

Shipwrights used pine to build merchant ships, but preferred fir for warships, for it was lighter than pine and did not decay as quickly. Oars were of oak and so were keels able to withstand hauling ships up onto beaches at night.[25] In the Persian War of 480 BCE, the Athenian fleet consisted of 200 to 300 triremes, warships with three rows of oarsmen on each side. In the First Punic War (264–241 BCE), the Romans lost 700 quinqueremes, ships with five rows of oarsmen. To replace them, they built 120 ships in sixty days and later another 220 ships in forty-five days.[26] Since the power and prosperity of states so often depended on their fleets, it is not surprising that they fought wars for access to forested areas.[27]

Rome imported large construction timbers from the Appenine forests and smaller wood from the nearby Alban hills. As the city grew in the first century CE and those forests became depleted, the Romans imported timber from the Alps, North Africa, northeastern Gaul, Asia Minor, and the Caucasus.

Long-Run Consequences

Deforestation exposed the soil to erosion and desiccation. In central Italy before the rise of Rome, soil eroded at a rate of 2–3 centimeters every thousand years. With the growth of cities and the expansion of agriculture, soil erosion increased tenfold to 20–40 centimeters per thousand years, in some places even up to 100 centimeters.[28] Where forests had once absorbed rain and gradually released the waters, deforestation led to flooding; Rome was flooded in 241 BCE and many times thereafter.[29] Areas of North Africa that had once supported cities with their baths, cisterns, and bridges were abandoned. When the deforested hillsides were eroded, rivers filled with silt that was then deposited on their flood plains and at their mouths, silting up harbors and creating malarial swamps.[30]

The decline of the Roman Empire accompanied the deterioration of its environments. When the Germanic invasions began, people fled from the plains into the hills, felling trees for land and fuel and pasturing their sheep and goats on once-arable land, changing the economy of many places from agriculture to herding. Environmental historians disagree on the extent of the transformation. John R. McNeill wrote: "Without a doubt a substantial measure of Mediterranean deforestation and consequent erosion happened in classical times, say between 500 B.C. and A.D. 500," but J. V. Thirgood believed that "extensive timber forests, though often depleted, still remained at the end of the classical period, and that natural regeneration was able to maintain the forests in being."[31]

These phenomena illustrate the costs of civilization in the Classical Age. In Mesopotamia, maintaining irrigation systems both demanded and supported a large population, leaving it vulnerable to both natural and political crises. Throughout the Middle East and around the Mediterranean, the demand for land and wood led to deforestation on such a scale that it transformed the landscape.

South Asia

The same causes that had contributed to the deforestation of the Middle East and around the Mediterranean affected the forests of India as well. The first cities on the sub-continent arose in the floodplain of the Indus River, which was covered with scrub and was almost as treeless as Mesopotamia.[32] For several centuries after 2600 BCE, over a thousand settlements dotted the Indus Valley. Then, around 1700 BCE, the cities were abandoned and replaced by villages of farmers and herders.

In Neolithic times, agricultural settlements also appeared along smaller watercourses in northwestern India and the Deccan, and in clearings in the Ganges Valley. The Iron Age in the early first millennium BCE made it easier for farmers using iron axes and iron-tipped plows to clear the middle Gangetic plains for agriculture, especially for wet rice paddies. Between 500 BCE and 300 CE, the valleys of the Godavari, Kaveri, Vaigai, and Krishna rivers were also cleared and brought under the plow. Early Brahmanic rituals involved burning wood as part of the worship of the fire god Agni, a reflection of the destruction of the forests. Forest clearance, agriculture, and urban settlements arrived simultaneously in much of Hindustan.[33]

The Mauryan Empire, which united most of the sub-continent between 322 and 185 BCE, colonized vacant lands by settling prisoners of war and providing land grants and tax remissions to farmers. It also initiated some large-scale irrigation projects. The resulting deforestation was so intense, however, that the emperor Ashoka (r. 269–232 BCE), a convert to Buddhism, preached restraint in the killing of animals and encouraged planting and protecting trees. Other rulers set aside some of the remaining forests as hunting and elephant-breeding preserves.

Deforestation and settlements continued after the collapse of the Mauryan Empire. To cope with the sudden and brief but heavy monsoon rains, farming communities dug wells and built tanks, even chains of tanks. During the Gupta Empire (320–550 CE), rulers gave grants of uncultivated lands to religious communities, both Buddhist and Brahmanical. It was in this era that India began to suffer from a resource crunch caused by soil exhaustion and perhaps also by climate change. Environmental historians Madhav Gadgil and Ramachandra Guha argue that this is when the caste system crystallized as a means of alleviating the competition for resources.[34] Overall, central and southern India was subject to the same pressures of civilization as the Middle

East and the Mediterranean world. Not until the nineteenth century did it suffer as much deforestation as those drier regions.

China

In China, all land amenable to farming lies in the east. In the northeast, the valley and floodplain of the Yellow River have fertile soils but a harsh and erratic climate, while the center, east, and southeast of China are very mountainous, with a wet tropical climate. What united these diverse zones was the migration of the Han people, who gradually absorbed or displaced all other peoples, and a political system that periodically broke apart into warring states but always eventually reunited. When it was united, China was the most environmentally diverse and resilient empire in the world before the nineteenth century. Thanks to the fertile soils of the northeast and the abundant rain of the southeast, China could support a larger population than any comparable area of land in the world. But agriculture required constant and labor-intensive manipulation of the natural environment, making it unusually vulnerable to disruptions, both natural and human.[35]

Northern China

In its upper reaches, the Yellow River cuts through a plateau composed of loess, a light soil carried by winds from the desert to the west. The first farmers who settled there in the second millennium BCE lived on a narrow strip of wetlands along the river, surrounded by forests of oaks, elms, maples, and other deciduous trees. The soil of the loess plateau could be cultivated with a hoe without the need for draft animals, but the lower reaches of the valley consisted of heavier alluvial soils that had to be plowed by oxen. Throughout the region, the soils were so fertile that they could produce bountiful crops without irrigation, as long as there was sufficient rain.[36]

Under the first historical dynasty, the Shang (1766–1046 BCE), North China was much warmer and wetter than today. The North China Plain may have supported 4 to 5 million people. Then the climate turned colder and drier, causing food shortages, population flight, and the fall of the Shang.

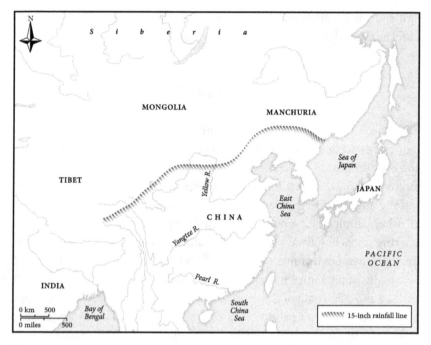

Figure 4.1. Map of China, showing the 15-inch rainfall line north and west of which rainfall is insufficient for agriculture.

After that, the climate remained both extreme and unpredictable, for the region lay at the northern edge of the monsoon belt around Asia. Heavy rains in the summer or fall could cause the river to burst its banks and flood the surrounding plain, sweeping away villages and ruining the land for years. At other times, the rains failed, causing famines. By the seventh century BCE, famines appeared in official records, and in the fourth century they reached crisis proportions. On the subject of farmers, the philosopher Mencius noted: "In good years, their lives are continually embittered, and, in bad years, they do not escape perishing."[37]

Under the Shang, a great deal of wood was consumed smelting copper and tin to make the massive quantities of bronze objects for which that dynasty is famous.[38] The Zhou dynasty (1046–221 BCE) that succeeded the Shang was, in the words of Sinologist Mark Elvin, "a civilization based on deforestation. Self-consciously, passionately so."[39] The last two centuries of the Zhou dynasty are known as the era of the "Warring States" because the introduction of iron contributed to constant warfare among the regional lords. Iron also allowed farmers to clear forests and use moldboard plows in the heavier alluvial soils of the lower Yellow River Valley.[40]

Forests were once the habitats of tigers, elephants, rhinoceroses, and other animals that have long since disappeared from China. The Shang kings organized royal hunts every winter. The court and the urban elite consumed a lot of game, mainly venison, but also bears, panthers, snow geese, turtles, and other animals, as well as honey, hazelnuts, and other forest products. The Zhou justified their overthrow of the Shang, in part, by the need to rid the region of animals. A Zhou ruler "drove the tigers, leopards, rhinoceroses, and elephants far away, and the world was greatly delighted."[41] By 1000 BCE, elephants had been pushed south of the Huai River that divided North from South China.[42]

During the Qin (221–206 BCE) and Han (206 BCE–220 CE) dynasties, the Yellow River Valley remained the core of the Chinese state, as it had been under the Shang and the Zhou. It was home to the majority of China's population, which rose to 40 to 60 million, similar to that of the Roman Empire. Warfare, city-building, and state efforts to settle farmers in order to increase food production sharply accelerated the felling of forests. Iron axes made it easier to cut down trees to clear the land for farming and to provide fuel and timber. Farmers learned new methods that required intensive labor inputs: deep tilling, breaking clods into fine particles to retain moisture, weeding and fertilizing, and alternating grains with legumes.[43] People living in coastal areas drove pine pilings into mud flats as a foundation for sea walls. The pine forests of the Taihang Mountains between Shanxi and Hebei were felled to make ink out of pine soot for the growing Chinese bureaucracy. Pottery making and metallurgy required large quantities of wood; iron smelting and forging consumed charcoal. In the cold climate of northern China, trees regenerated more slowly than they were felled.[44] The result, according to geographer Walter Mallory, was "a deforestation more complete than that of any other great nation."[45]

As the forests vanished, the rains eroded the loess soils and washed them into the rivers. As a result, the Yellow River carries an enormous amount of silt.[46] The Chinese, who had previously called it simply "the river," began calling it "yellow" from the color of the silt it carried. Han engineers estimated that the water was 60 percent silt. Unlike the silt of the Nile that fertilized the land, Yellow River silt contained a lot of gravel and sand. Once it reached the floodplain, the Yellow River deposited that silt, gradually raising its bed 3 to 12 meters above the surrounding land, creating what the

Chinese called a "hanging river." As the river rose above the floodplain, the inhabitants built ever higher levees to contain it. As early as the seventh century BCE, the government recruited laborers to reinforce the natural levees. These levees were built up to 10 kilometers from the river bank, but farmers, desperate for good land, would begin farming the land between the levees, then built their houses and villages there. Their short-term gain brought long-term risk.[47]

Despite the government's efforts, heavy rains periodically caused the river to break through its levees. When that happened, water spread over thousands of square kilometers. Because the floodplain was so flat, it drained slowly; two or three years might elapse before the land dried out. Governments and rebels sometimes deliberately caused flooding for tactical purposes. During the "Warring States" era, the armies of the small kingdoms of northern China frequently broke levees, destroying crops and ruining farmers. From 186 BCE to 153 CE, the river broke its levees on average once every sixteen years; during the worst period, it broke through every nine years on average, washing away villages and fields, killing people by the thousands and making thousands more homeless. Such breaches were frequent even during periods of peaceful development, when trees were felled, land was farmed, and silt washed into the rivers.[48] Emperor Wu, considered the greatest of the Han emperors (r. 141–87 BCE), built his reputation in part on reconstructing the Yellow River dikes after they were destroyed in a massive flood. In 11 CE, after having repeatedly burst its levees, the river changed course completely, a disaster that was to occur several times thereafter.[49]

To the north and west of the North China Plain lay the vast Eurasian steppe that stretched from Manchuria to Hungary. Once the domain of wild animals, it had become the land of horseback riding pastoral nomads who raised herds of goats, sheep, horses, and cattle. These animals turned grass into milk, meat, leather, and wool that the herders consumed or traded with the agrarian states to the south. Sometimes peaceful trade turned into border raids by peoples the Chinese called "barbarians." For many centuries, the Han fought wars against their nomadic neighbors the Xiongnu. To defeat them, the Han needed horses, and horses could only be bred on the pastures that the Xiongnu occupied. Starting in 129 BCE, Emperor Wu raised an army of 500,000 men, half of whom were cavalry troops, in order to attack the Xiongnu. After their final defeat four decades later, Wu ordered veterans of the campaign and poor farmers whose lands had been destroyed by floods to settle the newly conquered lands west and north of the Yellow River.[50]

Southern China

In contrast to the northeast, the eastern and southeastern regions of China regularly received abundant rain. Here the challenge was not to protect the land from flooding but to bring water to the hillsides and to drain the lowlands.[51] In exchange for water control, the land and climate rewarded farmers with abundant yields of rice. While the plants grew, the blue-green algae *Anabaena* that formed on the surface of the paddies transformed the nitrogen in the air into ammonia fertilizer. Other nutrients came from human and pig feces and composted vegetation. Thanks to these fertilizers, mature paddies produced up to two tons of rice per hectare per year, four times as much as rice grown on dry land, without the need to leave land fallow. Furthermore, farmers needed to set aside only one grain in fifty to a hundred as seeds for the next crop, compared to one wheat grain in three or four in Europe.

To produce abundant yields, however, rice required much more labor than other crops, first to grow seedlings in seed-beds and transplant them into paddies, then to weed the fields and fertilize them with compost and excrement, then to harvest and process the crops. Rice agriculture therefore both supported and demanded a high population density. In some places a second crop of wheat or vegetables could be grown after the paddies were drained and the rice harvested. Farmers raised pigs, chickens, ducks, and fish, but few cattle, as there was no room for pastures.

Much of southeastern China is mountainous. Once flat areas were fully occupied after 2000 BCE, farmers began building terraces and irrigation channels on the hillsides. By the first millennium, they were reclaiming land from swamps and pumping water in or out as needed by the plants.[52] Some water-control projects date back to the beginnings of China as a unified state. In the third century BCE, an engineer named Li Bing built the Capital River Dam on the Min River, a tributary of the Yangzi in Sichuan. This involved creating an artificial island in the middle of the river, a network of canals, and dikes and spillways to keep a constant year-round flow of water in the canals despite the large swings in the volume of the Min between summer and winter. This system turned the Chengdu Plain into the most fertile in China.[53]

Han Chinese from the northeast migrated slowly into the lower Yangzi Valley and around Hangzhou Bay. The delay was probably due to the prevalence of malaria, dengue fever, and schistosomiasis, a water-borne disease

Figure 4.2. Rice paddies in southern China built by farmers to retain water during the growing season. iStock.com/Kobackpacko.

transmitted by snails that lived in the rice paddies. Until the end of the Han dynasty in 220 CE, southern China was still a land of forests, described by northern literati as "a region of swamps and jungles, diseases and poisonous plants, savage animals and even more savage tattooed tribesmen." Historian Sima Qian explained: "In the area south of the Yangtse the land is low and the climate humid; adult males die young."[54] Guizhou, a province in south-central China, was known to harbor malaria, a disease the Chinese blamed on "noxious aethers of the mountains and the marshes."[55]

After the Han emperor Guangwu (r. 25–57 CE) abandoned the active defense of the northern frontier, opening the way to renewed incursions and destruction by the Xiongnu, many peasants fled to the south, leaving vast areas of the north depopulated. It took several generations for these northern farmers to adapt to the rainy climate and marshy land of the Yangzi basin. Their yields were still lower than in the north, for they planted rice seeds directly into the ground and let the fields lie fallow every other year. Not until the end of the Han dynasty were new drainage techniques introduced that opened up swampy lands to cultivation.[56]

Northern migrants brought deforestation to the Yangzi basin as they had to the North China Plain. As the needs of farming and construction gradually denuded the forests that covered southeastern China, the shortage of wood became so acute that people began using fuel-efficient stoves that burned straw, hay, grain husks, and animal dung. They even economized fuel by stir-frying rather than boiling or roasting their food.[57] As the forests shrank, wildlife disappeared. Elephants, once common throughout China, were killed because they endangered the farmers' crops and to provide tusks for the ivory trade and the meat of their trunks, considered a delicacy.[58] Deforestation, in turn, caused erosion of the hillsides and silting of the floodplains and deltas, at first with fertile humus soil but later, once the humus had washed away, with sterile subsoils.

Diseases and Civilization

The interactions between humans and the natural environment were no more one-sided in the age of agrarian states than they had been before or were to become. Not only did changes in the natural environment affect humans, but the rise of urban life, far-flung empires, and long-distance travel made humans more vulnerable to natural phenomena. Two natural phenomena—diseases and weather—had particularly serious consequences.

Humans, like all living beings, have always been vulnerable to diseases, though it is difficult to determine which diseases afflicted which peoples at which point in time. Writers in ancient times seldom described the symptoms they witnessed (or suffered from) with enough detail for scholars to definitely identify the specific disease they referred to. Furthermore, the pathogens that cause disease mutate over time, as do the defenses of the human bodies they infect, so that the same pathogen can cause very different symptoms in different periods. As a result, knowledge about diseases in ancient times is hazy. Little is known about the diseases that afflicted the ancient Mediterranean world, less about the diseases of ancient China, and even less about those of India and sub-Saharan Africa.

Many diseases that humans suffered from were mutated forms of diseases of animals, and the first people who lived with domesticated animals—Neolithic farmers and herders—caught more diseases from their animals and lived less healthy lives than their foraging neighbors or predecessors. When the pathogens that affected animals mutated into human diseases, they often killed or immunized all potential hosts and thereby burned themselves out.

Such outbreaks must have happened, sporadically, among Neolithic herders and farmers.

Just as herders and farmers were more vulnerable to diseases than hunters and gatherers, so were the inhabitants of civilized states and empires. There are several reasons for this. One is the crowding that occurred in cities; the more closely packed together people lived, the more likely they were to transmit diseases such as typhoid, dysentery, cholera, and tuberculosis. In cities, humans continued to cohabit, as they had in villages, with rats, mice, birds, dogs, cats, barnyard animals, and insects. With few exceptions, sanitation was very poor, for sewage was thrown into the streets and animal corpses and the refuse of butcher shops, tanneries, and other noxious activities were dumped just outside the city limits. Water supplies were contaminated with human and animal wastes. Furthermore, access to food was very unequal. While the elites enjoyed ample diets, caloric intake, especially of proteins, declined among common people; malnutrition was especially pronounced among infants and children.[59] Trade, migrations, troop movements, and long-distance pilgrimages affected rural as well as urban areas; as the movements of people intensified, so did the movements of the pathogens they carried.[60]

The city of Rome represents an extreme case of urban pathologies. Romans were on average four centimeters shorter than their ancestors or the peoples of central Europe. Their average life expectancy at birth was only twenty to thirty years, less than that of Paleolithic peoples. Infant mortality was 225 to 290 per thousand and 180 to 220 per thousand among children between one and three years of age. The first cause of these appalling death rates was a diet deficient in animal protein because animals were rare and expensive to feed in Italy, hence there was little milk or meat. Though the Roman world experienced food shortages, thanks to the trade in grains, it did not suffer from famines.[61]

The second cause was the appalling level of sanitation. To their credit, municipal administrators enacted many measures to make their city healthier. Rome's eleven aqueducts, built over a period of 500 years, supplied 1 billion liters of clean water a day, half of it for public baths. The Cloaca Maxima or "Great Sewer" flushed wastes out to sea. Food supplies were inspected and streets were cleaned.[62] Nonetheless, for lack of indoor toilets, the poor urinated and defecated in the streets. Animals were slaughtered in front of butcher shops. Even the famous baths were contaminated, for the water in the *tepidaria* or warm baths was rarely changed. As a result, the annual adult death rate in the city was between fifty and sixty per thousand, much higher than in the very poorest countries in the world today. Nor did conditions improve during the long Pax Romana. Like other large cities, Rome experienced much higher death than birth rates; its survival was only because of the constant influx of migrants from the countryside and of slaves forcibly brought to the city.[63]

Before the Common Era, travel in the Eastern Hemisphere was largely confined to four geographic regions—the Mediterranean and Middle East, South Asia, East Asia, and sub-Saharan Africa. Travel between them was so slow and difficult that it was rarely attempted, and only healthy travelers survived such expeditions. During the first centuries CE, however, trade increased, as more ships navigated the Indian Ocean and more merchants and pilgrims braved the rigors of crossing Central Asia or the Himalayas. In the second century, a few traders from the Roman Empire reached China, while others settled in India. The spread of diseases reflected these new contacts.[64]

Endemic Diseases

In small societies, diseases that kill or immunize their hosts soon run out of new victims and die out. Only in large populations are enough children born that diseases always find new victims. Measles is a case in point. The measles virus can be transmitted only from human to human, and it either kills or immunizes its victims quickly. In smaller populations, such as those on islands with little contact with the rest of the world, when measles appeared, it caused horrendous death tolls before burning itself out. To maintain themselves on a permanent basis, measles viruses needed to find 40,000 to 50,000 new hosts every year—either newborn children or unexposed immigrants—to replace the dead and immune. Thus they could only perpetuate themselves in populations of 300,000 to 500,000. Measles found a permanent home in the large populations of civilized societies, becoming an endemic children's disease.[65] Such societies could tolerate childhood diseases; as historian William McNeill explains, "Since children, especially small children, are comparatively easy to replace, infectious disease that affects only the young has a much lighter demographic impact on exposed communities than is the case when a disease strikes a virgin community, so that young and old die indiscriminately."[66]

Of the other diseases that proliferated in large civilizations, malaria, transmitted by mosquitoes, is well known for its impact on Rome. The city contained many small ponds and bodies of water where mosquitoes proliferated; even the *impluvia* or basins found in the courtyards of wealthy houses provided breeding grounds. The nearby Pontine Marshes were also infested with mosquitoes. Though the Romans did not understand the connection between these insects and malaria, they knew that summer and early fall were seasons when the disease was most common. Erecting three temples to Dea Febris, the goddess of fever, did not alleviate the problem. Though various strains of malaria probably existed since Neolithic times if not earlier, *falciparum* malaria, the most virulent kind, became endemic during the period of Roman domination of the Mediterranean.[67]

Epidemics before 540

Not all diseases became endemic or childhood diseases. Some, like influenza, smallpox, or bubonic plague, appeared seldom, but when they did, they caused epidemics even in large populations. These are the ones that left the greatest mark on the civilizations that they afflicted.

In histories of the Israelites and their neighbors, much had been made of the litany of plagues and pestilences, of boils and tumors and festering sores and "emerods in their secret parts" mentioned in the Hebrew Bible.[68] For centuries, scholars have argued over these strange symptoms. Unfortunately, biblical descriptions are too succinct to allow a reasonable diagnosis, because the ancients who wrote the Bible saw in them only the hand of an angry God who periodically smote the Israelites or their enemies for some trespass or other.

Not until the epidemic of 430–429 BCE that struck Athens in the midst of the Peloponnesian War is there a plausible description of a disease. Thucydides, the first historian, listed the symptoms: headaches, inflamed eyes, bleeding at the mouth, burning throat, vomiting, unquenchable thirst, skin reddened with spots, and dysentery, followed by death on the seventh or eighth day. It attacked the people of Athens at a time when the city was overcrowded with refugees escaping the fighting; in short, it was a civilized epidemic. It killed Pericles, the leader of Athens, 45,000 Athenian citizens, and 10,000 freedmen and slaves; it demoralized the Athenians, leading to their defeat.[69] Various scholars have attributed the outbreak to typhus, scarlet fever, bubonic plague, smallpox, measles, or anthrax, but the issue is still unresolved.[70]

If cities are a characteristic of civilizations, then Rome, a city of up to a million inhabitants at the center of a vast empire that traded with distant lands, was especially vulnerable. During the course of its history, the city and its empire were increasingly afflicted with devastating epidemics. An unidentified epidemic under the emperor Nero (r. 54–68 CE) is said to have killed 30,000 people. In 79, an epidemic, possibly of malaria, decimated the population of the Campagna, a rich agricultural district that served as the market-garden of the capital. Then came the first of the great pandemics, the Antonine Plague or Plague of Galen (after the doctor who described it). According to Roman sources, it started among soldiers campaigning in Mesopotamia in 164–65. It reached Rome in 166 and spread from there to Gaul and to the Germanic tribes beyond the Rhine. It lasted, off and on, for thirty years and is said to have killed up to 3,000 people a day, reducing the population it afflicted by a quarter to a third. It so reduced the size of the Roman army that Emperor Marcus Aurelius offered land to the Germans in exchange for military service. Among its victims was Marcus Aurelius. As with previous epidemics, it has not been identified, though smallpox, anthrax, and typhus have been blamed.[71]

After that, the Roman world was relatively free of epidemics until 250, when another mysterious disease, known as Cyprian's Plague, broke out in Egypt and spread throughout the empire and as far as Scotland. Again, its symptoms—diarrhea and vomiting, burning throat, and gangrene of the hands and feet—afflicted the people of the empire for sixteen years. At its height, 5,000 people a day died in the city of Rome alone.[72]

In addition to these pandemics, diseases also ravaged Roman Britain and North Africa. There is evidence that the population of the Roman Empire began to decline in the third century and with it the Roman economy, especially in the west. A faltering economy and a shrinking population weakened the armies that guarded the northern frontiers against the increasing pressures of Germanic tribes. Indeed, many historians have attributed the decline and fall of the western Roman Empire as much to the diseases and the demographic collapse they caused as to the barbarian invasions. Diseases, of course, did not discriminate between Romans and immigrants but also afflicted the Visigoths in 383 and may have forced the Huns to retreat before Constantinople in 447 and to abandon Rome in 451.[73]

China also experienced epidemics. William McNeill presents a list of the known epidemics in China—both local and regional—from 243 BCE to

1911 CE, but without identifying the diseases involved. Despite the paucity of evidence, there seem to have been two clusters of epidemics, in 161–162 and in 310–312, the latter possibly smallpox, for the Chinese knew smallpox and blamed it on the barbarians of the steppe who periodically invaded northern China.[74]

Sixth-century Disasters

The eruption of Mount Vesuvius in 513 and the earthquake that destroyed Antioch in Syria in the year 526 seemed ominous to many in the ancient world.[75] These events were soon followed by a much more widespread disaster that began in 535. Describing the year 536–537 in Italy, Byzantine historian Procopius wrote:

> For the sun gave forth its light without brightness, like the moon, during this whole year, and it seemed exceedingly like the sun in eclipse, for the beams it shed were not clear nor such as it is accustomed to shed. And from the time when this thing happened men were free neither from war nor pestilence nor any other thing leading to death.[76]

John of Ephesus, historian and churchman, wrote about dreadful events in Mesopotamia: "The sun was dark and its darkness lasted for eighteen months; each day it shone for about four hours; and still this light was only a feeble shadow. . . . [T]he fruits did not ripen and the wine tasted like sour grapes."[77] Another contemporary, bishop Zachariah of Mitylene, wrote that in Constantinople "the earth with all that is upon it quaked; and the sun began to be darkened by day and the moon by night, . . . and, as the winter [in Mesopotamia] was a severe one, so much so that from the large and un-wonted quantity of snow the birds perished . . . there was distress . . . among men . . . from the evil things."[78] And in Italy, Roman senator Cassiodorus wrote: "The Sun, first of stars, seems to have lost his wonted light, and appears of a bluish colour. . . . We have had a winter without storms, a spring without mildness, and a summer without heat. . . . The seasons seem to be all jumbled up together, and the fruits, which were wont to be formed by gentle showers, cannot be looked for from the parched earth."[79]

Nor were reports of strange weather limited to the Mediterranean world. In Ireland, the Annals of Ulster reported a failure of bread in 536, and the

Annals of Innisfallen also reported a harvest failure lasting from 536 to 539. China, too, suffered unusual weather. The chronicles of northern China reported a massive drought in 535. The *Nan Shi* or *History of the Southern Dynasties* noted that in late 535, "yellow dust rained down like snow." In July and August 536, frost and snow killed the seedlings in northern China, followed by hail in September. In the year 537 a drought hit northern China, along with snow, causing several years of famine.[80]

Scientific evidence corroborates these contemporary reports. Irish oak rings show abnormally little growth from 536 to 545. In Britain, tree-ring growth slowed down in 535–536 and did not recover until 555. Rings from bristlecone pines in northern California showed a colder and drier climate than usual in 535–536, then again from 539 to 550. Tree rings from Sweden show that the year 536 was the second coldest in the past 1,500 years. Tree rings from the Altai Mountains in Central Asia and from the Austrian Alps show that the 540s were the coldest or second coldest decade of the century. Rings from Siberian, Tasmanian, and Chilean trees also show abnormally slow growth.[81]

Volcanoes and Comets

What could have caused such a phenomenon? Two hypotheses have been advanced. One is a series of volcanic eruptions. According to geophysicist Robert Dull, the eruption of the volcano Ilopango in El Salvador in 536, as attested by sulfate deposits in Greenland and Antarctic ice cores, caused "the greatest atmospheric aerosol loading event of the past 2,000 years."[82] More recently, geophysicists writing in *Nature* in 2015 have argued that the climate anomaly of 536–550 resulted from not one but two, possibly three, volcanic eruptions. The first one, in 535 or early 536, caused a dimming in the Northern Hemisphere that lasted eighteen months. A second one, in 539 or 540, was tropical in origin for it affected both hemispheres and caused a cold snap that lasted until 540. A third smaller but still substantial eruption occurred in 547.[83]

Others have blamed a comet rather than volcanoes. Evidence for this are spherules, tiny balls of condensed rock vapor found in ice core layers dating from the year 536. These spherules would most likely have come from the impact of a comet in the Gulf of Carpentaria between Australia and Indonesia.[84] A meteor shower associated with the comet 1P/Halley is thought to have caused a modest dimming in 533, and the profound dimming of 536–537 to have resulted from a combination of cometary dust, volcanic sulfate, and a low-latitude explosion in the ocean.[85] Further research will hopefully resolve this mystery.

Justinian's Plague

No sooner had the climate shock of 535–537 tapered off than the Mediterranean region, the Middle East, and Europe were afflicted by another natural disaster. In the year 542, an epidemic of plague broke out in Constantinople.[86] This was not just another of those pestilences misnamed "plague" that had previously afflicted the ancient world. This was the first appearance of the bubonic plague, the disease that would later return under the name Black Death.[87]

The bubonic plague bacillus is transmitted to humans by the flea that lives in the fur of rodents, especially the black rat. But the flea can also survive for a while in any warm moist environment, such as clothing. When the temperature lies between 15° and 20° C (59–68° F), the bacillus multiplies in the gut of the flea and prevents it from eating. The flea, desperately hungry, then bites anything within reach. Since black rats prefer to eat grains, hence like to live close to humans, their fleas readily jump from rats to humans.[88]

Once it has infected a human being, the disease incubates for one to six days before breaking out in buboes. Before antibiotics, the recovery rate was 20 to 40 percent; the emperor Justinian was one of the lucky ones who recovered. The bacillus can also spread from person to person through the air, causing pneumonic or septicemic plague, a highly contagious form that is 100 percent fatal. In crowded cities like Constantinople, it is likely to have spread this way as well as through rats.[89]

Procopius described the symptoms:

The fever was of such a languid sort from its commencement and up till evening that neither to the sick themselves nor to a physician who touched them would it afford any suspicion of danger. It was natural therefore that not one of those who had contracted the disease expected to die from it. But on the same day in some cases, in others the following day, and in the rest not many days later, a bubonic swelling developed; and this took place not only in the particular part of the body which is called "boubon" [groin], that is, below the abdomen, but also inside the armpit, and in some cases also beside the ears, and at different points on the thigh.[90]

According to Procopius, the epidemic was first noted in 540 in Pelusium, a port on the Mediterranean coast of Egypt.[91] How it got there is still uncertain. John of Ephesus wrote: "It began first among the peoples of the interior of the countries of the southeast, of India, that is, of Kush, the Himyarites, and others," but clearly he confused India, Kush (in what is now Sudan), and the Himyarites (a kingdom in southern Yemen).[92] Evagrius Scholasticus, a church lawyer writing fifty years after the events, asserted: "It was said, and still is now, to have begun in Ethiopia."[93]

Historian William McNeill suggested that it originated in the foothills of the Himalayas, between northeastern India and southern China, from where it traveled across India and up the Red Sea.[94] Almost all other modern historians believe that the plague originated in Central or East Africa, perhaps because the disease is now found among rodents in these regions.[95] More recently, geneticists investigating this epidemic have found evidence that it did not exist in Africa at the time but instead came from somewhere in East Asia, either China or Tibet.[96]

Once the plague reached Pelusium, it spread to Alexandria, Palestine, and Syria, generally following shipping routes. When it reached Constantinople in early 542, it is said to have killed 5,000 to 10,000 people a day for four months.[97] In Western Europe, it reached Genoa and Marseille in 543–544 and Britain and Ireland in 549. In the 570s and again in the 580s, the plague made its way up the Rhône and Saône rivers in Gaul. The plague returned to the northern Mediterranean lands in 558, in 588–591, and on average every nine to twelve years thereafter. After about 650, it disappeared from Europe, in part because travel and trade with the Middle East and North Africa diminished sharply after the Arab conquests.[98]

By then, it had spread to Mesopotamia and Armenia and the eastern borders of Iran. In 627–628, the "plague of Shirawayh" broke out at Ctsesiphon in Mesopotamia, then the capital of the Sasanian Empire. That same year an outbreak struck a Muslim army at Emmaus in Palestine. In Syria, Mesopotamia, and Persia, other outbreaks occurred in 634–642, 688–689, 706, 716–717, and 767. It recurred every ten to twenty years until its last reappearance in Mesopotamia and Syria in 747–750. Then it vanished for almost 700 years.[99]

Consequences

The climate shock of 535–537 and the plague that followed had momentous consequences for the Mediterranean world, for Europe, and for the Middle East. In 533, just before these events, Justinian had launched an ambitious Roman military campaign in an attempt to reconquer the western Roman Empire from the Germanic peoples who had taken it over. It almost succeeded. His chief general, Belisarius, quickly defeated the Vandals and incorporated North Africa into the Eastern Roman Empire. Belisarius then invaded Italy and fought a series of devastating wars against the Ostrogoths that lasted from 535 to 554.

The city of Rome, though much diminished from its imperial heyday, was still a major administrative and commercial center. Other towns in Italy still functioned. Farmers still planted their fields and harvested their crops. The inhabitants, including their Germanic overlords, spoke a dialect of Latin that eventually became Italian. It was a combination of the climatic shock of 536–537, the plague, and the war between the Ostrogoths and the Byzantine army that damaged Italy. The city of Rome, which had survived many invasions almost intact, was reduced to ruins, while most of the peninsula was so devastated that it did not recover for several hundred years. The mutual destruction of the Ostrogoths and Byzantines created a power vacuum in the western Mediterranean and opened the gates to far less romanized invaders like the Lombards in Italy and nomadic Berbers in North Africa.[100]

In the eastern Mediterranean, the disastrous weather of 536–537, the resulting harvest failures and famines, and the pandemic of plague that followed sharply diminished the population of the Byzantine Empire. Constantinople lost half its inhabitants. Just as Justinian was warring against the Germanic peoples in the west and the Sasanian Empire to the east, these catastrophes reduced his empire to a shadow of the Roman Empire at its peak. And on the steppes of Eurasia, a drought that lasted from 535 to 545 made the horse- and sheep-herding Avars flee westward, pushing Slavic tribes to invade the Byzantine Empire in 536–537 and several times thereafter, expelling the Byzantines from the Balkans.[101] When Muslim warriors emerged from Arabia a century later, the Byzantine Empire was too weak to prevent them from conquering Egypt, the Levant, and parts of Anatolia.

Nor were the consequences of the climate shock limited to western Eurasia. In China, the mid-sixth century was also a time of troubles. Poor harvests meant the government collected less in taxes, causing a political crisis. The

Northern Wei dynasty split into regional states, while the south fell into chaos. The turmoil of the period encouraged the spread of Buddhism. The resulting disruption may have contributed to the conquest of China by the Xianbei, a nomadic people from the north who established the Sui dynasty in 589.[102]

Conclusion

The period from the early Iron Age to the sixth century CE witnessed the rise of powerful states with large populations and great cities in Eurasia and around the Mediterranean. In the process, they transformed the landscape around them, cutting down forests, plowing up rain-watered lands, and moving people, food, and other products over long distances. Several regions came under the rule of great empires, in particular the Roman and the Han.

What seemed like the triumph of civilization over nature had a dark side, however. Demographers estimate that the human population shrank by 18 percent in the first seven centuries CE. The decline was a result of the increased vulnerability that civilization wrought. Large populations, close contacts between humans and animals, great numbers of people living in cities, and increased communications between regions made them vulnerable to sudden shocks in the natural world, in particular epidemic diseases and abrupt climate changes. Not until after 1000 did the population of Eurasia—a large majority of the world's population—recover and begin to grow again.

5

Medieval Eurasia and Africa

The disasters of the sixth century remained localized, for communications across the breadth of Eurasia were still very difficult and sporadic. Once these passed, the long-suppressed human urge to "be fruitful and multiply and replenish the earth and subdue it" reappeared. Eurasia and North Africa saw a resurgence of population growth, urbanization, and economic development, as well as a remarkable cultural flowering.

Though the achievements were human, the context was environmental. Part of this included a warming of the climate, which translated into an expansion of the arable land in Eurasia, and a decline in epidemics, which allowed the population to grow. But the new era that we call the Middle Ages was more than a renewal of earlier civilizations; it was, in many ways, revolutionary.[1] Regions conquered by the Arab armies saw a remarkable efflorescence of agriculture, trade, and urbanization. Western and northern Europeans expanded their croplands at the expense of forests and wetlands. In China, mass migrations from the north opened up the Yangzi Valley and beyond to settlement, water control, and economic growth. And Africa south of the Sahara saw growing cities and prosperous long-distance trade.

The history of Eurasia and Africa in the Middle Ages was not one of unalloyed triumph for humans, however. In the fourteenth century, two changes in the natural environment reversed the trend of the preceding seven centuries. One was a cooling of the climate that brought famines all across Eurasia from western Europe to China. The other was the Black Death, an epidemic of bubonic plague that decimated the peoples of Europe and the Middle East. In short, relations between humans and the environment were still in flux.

The Arab Empire

The regions that recovered first and most spectacularly from the troubles of the sixth century were the Middle East and the southern Mediterranean. In the Middle East, as the Byzantine and Sasanian empires were weakened by the plague, the climate shock, and their incessant wars, a new force arose: warriors from Arabia fighting under the banner of Islam. Not only did the Arabians succeed in conquering, in an astonishingly short amount of time, an empire stretching from Morocco to the Indus River, but they also ushered in one of the most creative and prosperous eras in the history of the regions they conquered.

The Arab Empire gets its name from the language of its rulers, for the conquering Arabian soldiers and administrators brought their language to North Africa and the Middle East. It then spread among the new converts to Islam enjoined to read the Qu'ran in Arabic. With the conquest came a new Islamic jurisprudence, a favorable attitude toward commerce, and a common currency. Trade flourished as Muslim merchant-missionaries carried their business, their religion, and their language not only throughout the new empire but even far beyond it, to Indonesia, West Africa, and around the Indian Ocean. In the early centuries, Muslims showed a great interest in the philosophy and sciences of the Greeks and encouraged innovations in engineering, navigation, and agriculture. They rebuilt and expanded cities such as Damascus and founded new ones such as Baghdad, Basra, Cairo, Tunis, and Cordoba. In short, they created a cultural and economic efflorescence the likes of which had not been seen since the time of Alexander the Great and his successors.

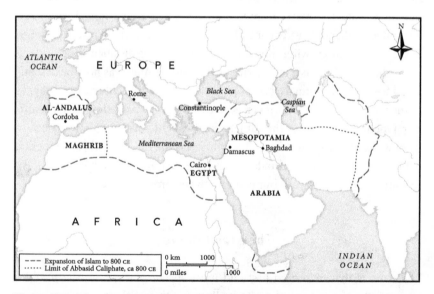

Figure 5.1. Map showing the expansion of Islam from Arabia to North Africa and the MiddleEast, and the limits of the Abbasid Caliphate to 800 CE.

The Islamic Agricultural Revolution

Much of the prosperity of the Arab world in those early centuries was based on a more intensive use of existing land. This intensification has been called an agricultural revolution, brought about by the opening up of free trade within the Middle East and North Africa and between these regions and South and Southeast Asia. This encouraged innovations in crops, some of them very important ones such as cotton, sugarcane, sorghum, hard (durum) wheat, and Asian rice (*Oryza sativa*). Other new crops were of culinary interest but lesser economic value, among them bananas and plantains, coconuts, watermelons, spinach, artichokes, eggplants, mangoes, sour oranges, lemons, and limes. Some of these were already known in a few places in the Middle East, but they were widely diffused by the Arabs. Others, new to the Middle East, were introduced from South and Southeast Asia. The latter were typically hot-weather crops that were planted in the summer in rotation with winter crops such as wheat and barley, hence doubling the productivity of the land and, in a few cases, even producing three crops a year.

To benefit from hot weather in dry lands, however, tropical plants needed irrigation. When the Arabs arrived, they found lower Mesopotamia reduced to a marshy quagmire and Egypt and North Africa in decline since the fall of the western Roman Empire. Employing engineers from Persia, they succeeded in repairing old irrigation systems and introducing new methods of capturing, storing, and distributing water, such as using dams, tanks, qanats (water tunnels), and water wheels. In historian Andrew Watson's view,

> In many, perhaps most, areas indisputable progress was made in both the quantity and quality of irrigation, so that by the ninth or tenth century virtually every part of the Islamic world had countless areas, great and small, which were heavily irrigated and into which the new crops could easily move—or had already moved. An environment fundamentally hostile to tropical and semi-tropical crops had been transformed into one in which, for a time at least, they were grown with astonishing success.[2]

Success came not just from crops but also from knowledge and institutions. During the zenith of the Arab Empire from the seventh to the eleventh century, agricultural manuals proliferated, as did works on irrigation. Experts wrote about the importance of fertilizing and crop rotation and of choosing the most appropriate crop for each type of soil, climate, and

Figure 5.2. Wooden waterwheels in Hama (Syria), which used the current of the river to lift water for irrigation. iStock.com/gertvansanten.

irrigation system. With a greater range of crops, they could use land more efficiently, for instance, by growing sorghum and hard wheat on dry land, and sugarcane, eggplants, coconut palms, and colocasia (or taro) on slightly saline soils. Rulers built botanical gardens, encouraged agricultural innovation, and passed laws favorable to irrigation and agriculture.

The new agriculture was very labor intensive. But as the land produced better and more valuable crops, prices remained stable and incomes rose, and so did the population. More people meant more villages, expanding cities, and a growing trade. Thus did the agricultural revolution create a cycle of prosperity and culture.

Agricultural prosperity was not continuous but underwent great fluctuations during this period. Having inherited the Middle East from the Sasanian Empire, the Umayyad Caliphate (661–750) restored some canals and encouraged agriculture in southern Mesopotamia. Whatever it achieved, however, was damaged by the plague epidemic of 749–750. The Umayyads' successors, the Abbasids (750–1258), moved their capital from Damascus in Syria to Iraq (the Arabic name for Mesopotamia), where they built a new city they called Baghdad. The prosperity of the Abbasid Caliphate was based on a revival of agriculture due to the use of waterwheels, cisterns, Archimedes screws, and windmills as well as to larger projects such as reservoirs, canals, and aqueducts. With many new crops available, prosperity came from intensive gardening, specialization, and trade rather than from high yields on large estates.[3] The Arabs who settled in Iraq repaired the Sassanian irrigation works and imposed an efficient administration on the agricultural sector.[4]

Conquests by the Arabs were not limited to Syria and Iraq but spread westward as far as Morocco and Spain and toward the east to Iran, which they took from the Sasanians. Upon settling in Iran, Arabs from Yemen invested in qanats to irrigate dry land and began producing cotton. As Muslims preferred cotton to silk and other textiles, a cotton boom developed during the ninth and early tenth centuries.[5]

Egypt experienced much political turmoil as leaders of various sects of Islam succeeded one another, nominally paying homage to the Abbasid caliphs in Baghdad. Despite political upheavals and military campaigns, Egypt grew wealthier as agriculture and trade flourished. After 1250, when the Mamluks, nominal slaves serving as soldiers, took over Egypt, they initiated enormous large-scale irrigation projects alongside the small-scale irrigation maintained by local landowners. Under Mamluk rule, the amount of arable land in Egypt increased by 50 percent, and Egypt became a major exporter of sugar.[6]

In al-Andalus, the Muslim part of Spain, the Arab agricultural revolution was a resounding success. Under the Romans and later the Visigoths, Spain had been a backwater valued for its mines, with few towns and an underdeveloped agriculture. When the Arabs and Berbers invaded Spain from Morocco in 711, they found a poor land rich in possibilities. Among those possibilities was the application of irrigation to the valleys of southern Spain with their rich soils and hot sunshine. The new rulers, themselves natives of hot dry lands, were keen to rebuild the Roman waterworks and to import new irrigation methods from Egypt, Mesopotamia, and Persia. Among the latter were reservoirs to hold water for the dry season, underground channels, and *norias* or wheels powered by the current of a river to lift water onto land. On a smaller scale, farmers built *saqiyas* or animal-powered water wheels and *shadufs*, buckets with counterweights, devices known in Egypt since the days of the pharaohs.

Taking advantage of the new irrigation methods, Spanish Muslims imported new crops, such as cotton and sugarcane long grown in the Middle East but never before planted in Europe.[7] The result was the richest

agriculture supporting the largest city in Europe—Cordoba—and a civilization as sophisticated as any in the Middle East.[8]

The Decline of Islamic Agriculture

The Arab Empire that stretched from Spain to Mesopotamia represented a
high point in medieval civilization. However, its foundation—agriculture
and especially water control—rendered it vulnerable to shocks, both political
and natural. In Watson's words: "When it was at its peak, then, Islamic agriculture was also at its most vulnerable. . . . [I]t could survive only in a complex environment protected by a government that was centralized, powerful
and sympathetic."[9]

Starting in the ninth century, political instability, corruption, and excessive taxation undermined the foundations of Middle Eastern prosperity. By
the late tenth century, rebellions wracked the heart of the caliphate, and the
Abbasid caliphs lost control over both local politics and distant provinces.
In Iraq, the major canals silted up and their associated irrigation works were
allowed to decay. Agricultural lands east of the Tigris were abandoned. The
population of the Diyala plain, an area of some 10,000 square kilometers,
declined from 600,000 in the seventh century to 50,000 in the twelfth
century.[10]

The social and institutional problems of the Middle East were exacerbated
by a deteriorating climate. In Egypt, the Nile, fed by the East African monsoon, was severely affected. From 950 to 1072, nine droughts lasted a total
of twenty-six years. In 967, the height of the Nile flood, normally 8.1 to 8.7
meters, dropped to 6.5 meters, leaving irrigation canals empty and fields
parched. From 1065 to 1072, a seven-year-long drought again ruined the
harvests and brought famine to the land.[11]

Drought also affected the Levant and Mesopotamia. Some winters, ice
formed on the Tigris and Euphrates rivers. Snow and hailstorms caused harvest failures that led to riots and sectarian wars. In normal times, surplus
crops from Egypt would have alleviated distress throughout the Middle East,
but this time, simultaneous droughts in both Egypt and the rest of the Middle
East compounded the disasters that afflicted both regions.[12]

The Iranian cotton boom ended in the mid-tenth century, as a century-
long cold period descended on Iran, northern Iraq, Central Asia, Anatolia,
and Russia. Then, starting in the eleventh century, came the invasions.

Turkish nomads from Central Asia, driven by the increasingly cold climate, migrated to Iran, ending the Arab dominance. After that, the Middle East was invaded time and again by Seljuk Turks, European Crusaders, Mongols, Timurids, and Ottoman Turks, invaders who damaged irrigation works or neglected to maintain them. Many were pastoral nomads with little understanding of complex agriculture or appreciation of the legal institutions and intellectual foundations that supported it.[13] By the twelfth century, the salinization of southern Iraq was very advanced. In 1258, the most brutal of the invaders, the Mongols, conquered Mesopotamia, sacked Baghdad, and overthrew the last Abbasid caliph. Their depredations have often been blamed for the demise of the Arab Empire, but it was already declining long before that.[14] As the Arab agricultural revolution perished, Iraq became a backwater on the global stage. By the sixteenth century its population had dropped to about a million; not until the twentieth century did it return to the level it had reached under the Sasanians.

Like Iraq, al-Andalus was vulnerable to political disruptions. In the early eleventh century, the Caliphate of Cordoba disintegrated into small rival kingdoms called *taifas*; in 1085, Saharan Berbers called Almoravids invaded Spain; other Berbers called Almohads invaded in 1147; and the Christian *Reconquista* finally ended Muslim rule in Spain in 1492. For a while, the agricultural base of al-Andalus persisted, because, unlike Iraq, it did not depend on massive government-operated irrigation systems but on small-scale irrigation works run by farmers. These works lasted as long as there were farmers who knew how to operate them. What finally damaged the flourishing agriculture of al-Andalus was the Christian rulers' preference for pastoralism and their decision to rid their kingdom of Muslims, ending with mass expulsions between 1609 and 1616.

Europe in the High Middle Ages

Climate historians call the period from the ninth through the twelfth century the Medieval Climate Anomaly, a time when western and northern Europe enjoyed unusually mild rainy winters and long hot summers. What caused the change in the climate of Europe was the coincidence of a solar maximum, unusually low volcanic activity, and exceptionally warm westerly winds.[15] Temperatures rose 0.5° to 1°C (0.9° to 1.8°F), while rains diminished by 10 percent, perfect weather for cereals and grapes. The growing season for cereals lasted up to three weeks longer than it had in Roman times, and few May frosts threatened the growing crops—though occasional hard winters interrupted the cycle. Overall, the climate of western and northern Europe between 700 and 1300 was more favorable to agriculture than it had been before.[16]

As the climate warmed, western and central Europe prospered. Good harvests meant fewer deaths, causing the population of Europe to increase from 30–35 million in 1000 to 70–80 million in 1347.[17] To be sure, this happened slowly, for Europe was long afflicted by the depredations of the Huns and the Vikings, people much more given to destruction than the semi-Romanized Germanic tribes that had preceded them. Not until the eleventh century was it really safe to farm and build. But from the beginning of the second millennium until the early fourteenth century, western and central Europe entered into a blossoming historians call the High Middle Ages. The Medieval Climate Anomaly did not cause this efflorescence, but it enabled it. In the words of medievalist Norman Cantor:

> The rise of political entities, legal systems, learning, urban living, and com-
> mercial productivity in medieval Europe from 800 to 1300 was made pos-
> sible by a warming climate and the absence of pandemics. . . . It was the
> benign climatic and biomedical environment that was most responsible for
> the rise of European power and wealth, the clearing of land, the revival of
> cities, and above all the expansion of the population base fourfold from 900
> to 1300.[18]

As with the decline of southern Europe, part of the reason was ecological. Western and central Europe were spared the plague epidemics that period-ically afflicted the Mediterranean and Middle East, and other diseases also seem to have been in abeyance. The unusually warm climate made it easier to grow crops north of the Alps and the Loire than it had been in Roman times. Farmers planted vineyards and pressed wine in England, Prussia, and southern Norway, regions where no grapes have grown since.

The warm climate extended far into the Atlantic Ocean. By the beginning of the eighth century, intrepid Viking mariners were sailing in small open cargo ships called *knarrs* to the Orkneys and the Shetlands, then to the Faroe Islands. By the end of that century, they had reached Iceland. In 986 Erik the Red left Iceland with twenty-five knarrs and 700 people. Fourteen of the ships and 450 settlers reached Greenland, along with horses, sheep, and goats. The

Vikings called the land green because they found trees growing there and meadows on which they could pasture their livestock and grow barley. They also found the Labrador Sea off the western coast to be free of ice during the summer, permitting ships from Norway to reach them every year.[19]

Medieval Technology

The Germanic peoples who drifted into the Roman Empire were farmers looking for new land to cultivate, as were those who stayed northeast of the old Rhine-Danube frontier. The newcomers operated small farms as tenants of their lords. They brought with them, or devised, new technologies to aid them in clearing the land and cultivating the thick moist soils of Atlantic Europe.

One of these technologies was iron. Already in Roman times, iron was much used for parts of plows, spades, sickles, and other agricultural tools. Abundant iron, in turn, permitted the development of a most important agricultural innovation: the moldboard plow. The scratch-plow that Roman farmers used had been unable to cope with the heavy wet clay of Atlantic Europe. In order to turn the soil and cut furrows deep enough to allow excess water to run off, a plow was needed that contained a vertical iron coulter or heavy knife to break the sod, an iron plowshare at right angles to the coulter to cut the sod at the grass roots, a moldboard to turn the slice of turf on its side, and wheels to hold up such a heavy implement.[20]

Such a plow, long known in northern China, was introduced to Europe in the sixth century by Slavic peoples from eastern Europe. The new plow spread gradually, because it demanded a radical change in both methods of farming and a source of energy. So heavy was this plow and so thick was the soil that on newly cleared land, eight oxen were required to pull it; once the sod had been broken and the soil loosened, only four oxen were needed. Since few farmers had that many animals, plowing demanded cooperation. Turning a team of oxen at the end of a field was difficult, so wherever the moldboard plow was introduced, the Roman-style square fields were slowly replaced by long narrow fields. Every time these fields were plowed, the soil grew higher in the middle than along the edges, allowing water to drain off. Such a change did not take hold overnight; for a long time, there were mixtures of old and new methods and tools. But gradually the moldboard plow and strip fields came to predominate in the heavier soils of northwestern and central Europe.[21]

As early as the reign of Charlemagne (768–814), European farmers began introducing a balanced system of animal and crop production known as the three-field rotation.[22] In the traditional two-field rotation, a field grew crops one year and was left fallow the next; in other words, only half the fields produced crops at any one time. Under the new three-field system, one-third of the fields were planted in winter wheat or rye, another third in spring crops like oats, barley, peas, lentils, and other legumes, and the last third were left fallow to recover their fertility. Under this system, the land produced one-third more than under the two-field system, labor was distributed more evenly over the year, and diversified crops reduced the chance of crop failure and gave people a more balanced diet.[23]

The shift to a three-field system produced a burst of yields that lasted fifty to a hundred years, then tapered off. As forests, with their high biomass (the amount of living matter per unit of area) and low productivity (the amount of growth per year), were replaced by the low biomass and high productivity of short-lived crops, the ecosystem became unstable, requiring energy inputs both mechanical (tilling and weeding) and chemical (fertilizing). Farmers replenished the fertility of their fields by draining nutrients from forests, heaths, and grasslands in the form of ashes from firewood and manure from animals left to browse in nearby woods or in fields after the harvest. As the arable land expanded and encroached on them, the hinterlands became less and less productive, causing a decline in the fertility of the fields. Medieval agriculture at its richest was still a very tenuous enterprise, with yields of cereals only four times the number of seeds sown.[24]

Another crucial technological advance was the introduction of horses into agriculture. Until the Middle Ages horses could be used only to ride or to pull light carts and chariots, because the neck-strap that attached the animal to the cart choked it if it had to pull too heavy a load. The horse-collar, introduced from China in the eighth or ninth century, shifted the load onto the shoulders of the animal, allowing it to pull four or five times as much as the older yoke-harness without choking, even heavy loads such as moldboard plows or four-wheeled wagons. At the end of the ninth or the beginning of the tenth century Europeans also began to use iron horseshoes, which saved the horses' hoofs from the excessive wear that occurred in wet climates.

Horses were as strong as oxen but moved faster and could work one or two hours longer a day, an important consideration in a temperamental climate when plowing, sowing, and harvesting had to be done quickly before the weather changed. In these circumstances, horses could do twice as much work as oxen. They could also walk farther than oxen, allowing farmers to live farther from their fields in villages of 200 to 300 inhabitants, instead of scattered in tiny hamlets. However, they were costly to keep. Whereas oxen could survive on hay and stubble, horses needed either large pastures or oats, a costly fuel, to maintain their strength. Thus, farm horses were best used with the three-field system, in which one crop fed the horses and the other fed the humans. The use of horses therefore spread gradually. Plow horses were a common sight in northern Europe by the end of the eleventh century, yet as late as the fourteenth century oxen still did most of the work.[25] Medieval historian Lynn White Jr. credited the new agriculture with bringing about "the startling expansion of population, the growth and multiplication of cities, the rise in industrial production, the outreach of commerce, and the new exuberance of spirits which enlivened that age."[26]

According to White, the new agriculture did more than stimulate the economic development of medieval Europe; it also changed people's attitude toward nature. Medieval Europeans came to see the natural world as a resource to be exploited and mechanical devices as a means of carrying out God's purpose: "Once man had been part of nature; now he became her exploiter."[27]

Creating the Netherlands

No place in Europe was so transformed by human action as the Netherlands. In the early Middle Ages, half of that country, a 50-kilometer-wide swath between the coastal dunes and the interior highlands, was covered with peat and drained by many rivers and streams. Humans had lived on the edges of this vast peat bog since Neolithic times, raising livestock, farming, and fishing.

Reclamation began in the ninth century. The counts of Holland, bishops, and local authorities granted lands to colonists. Farmers living in raised areas began to drain water from surrounding peat bogs to create new farmland. Digging ditches to drain the land required intense collective efforts. The Rijnland, a 900-square-kilometer area, was largely reclaimed by 1300 and there farmers raised cattle and grew barley, rye, and flax. By 1370, the settled area of the Netherlands had more than doubled. Cities like Haarlem, Amsterdam, Delft, Gouda, and the Hague were built on the new land.

The effort required was huge and costly, not just once, but in perpetuity. Because peat shrinks when it dries, peat bogs, once drained, subsided until they were below the water table, causing waterlogging and flooding. Removing peat as fuel made the problem worse. Disputes arose between communities as upstream villages dumped excess water onto their downstream neighbors. In the thirteenth century, the counts established the Drainage Authority of Rijnland to enforce cooperation among the villages of the region in maintaining ditches and canals. In the fifteenth century, waterlogging became so severe that communities built windmills to pump water out of their fields; by the end of that century, Holland was dotted with them.

To practice agriculture in a land lower than the water table required not just elaborate machinery but also sophisticated hydraulic engineering. As the land sank below sea level, dikes were built to hold back the high tides. But dikes were not sufficient protection against the occasional storm. Monster storms that blew in from the North Sea every few decades swept away dikes and villages and covered reclaimed farmland with clay and sand. Yet, after every storm, the Dutch returned to rebuild their farms and villages.[28] Historian Peter Hoppenbrouwers called it "a cat-and-mouse game between merciless nature and human inventiveness."[29]

Deforestation and Colonization

Between 800 and 1300, the natural world of western and central Europe with its forests, marshes, and heaths was transformed into the land of farms, meadows, villages, and towns. Powerful lords offered self-government and lower taxes to pioneers willing to settle new lands. New towns with names like Villeneuve, Neuville, or Neustadt sprang up. Benedictines, Cistercians, and other religious orders opened clearings deep inside forests and allotted forest lands to colonists. As the climate warmed, founders of monasteries moved onto lands previously too marginal to farm. Monks were the most active agents of deforestation, equating trees with paganism. As cultural geographer Clarence Glacken explained:

> St. Benedict himself had cut down the sacred grove at Monte Cassino, a survival of pagan worship in a Christian land. St. Sturm "seized every opportunity," says the monk Eigil, his biographer, "to impress on them [the pagans] in his preaching that they should forsake idols and images, accept the Christian faith, destroy the temples of the gods, cut down the groves and build sacred churches in their stead."[30]

The colonization movement was especially pronounced north of the Rhine and Danube and east of the Elbe River. In the twelfth century, 150,000 to 200,000 German, Flemish, and Dutch settlers moved east into lands sparsely populated by Slavic speakers, lands that were to remain predominantly German until 1945. Once there, they farmed. In Poland, the proportion of arable land almost doubled from 16 percent in 1000 to 30 percent in 1500. In Silesia (now southern Poland), the proportion multiplied ten to twenty times between 1150 and 1350.[31] In the words of agrarian historian Slicher van Bath, "It was an expansion comparable to that in North America many years later," an opinion echoed by forest historian H. D. Darby: "What the new west meant to young America in the nineteenth century, the new east meant to Germany in the Middle Ages."[32]

Farming was not the only activity that transformed the land, nor were forests the only kind of landscape affected. In the fourteenth century, England north from Cambridge to Liverpool became ranch country, raising sheep for the export of wool or herds of cattle to supply meat to the wealthy who had previously subsisted on venison.[33]

Cutting trees and clearing land was never a simple matter. Peasants valued forests as sources of firewood, berries, mushrooms, herbs, and honey as well as the places where their swine could forage for acorns.[34] Private property rights competed with communal rights. Kings and nobles maintained hunting parks where they protected deer and other game animals for their sporting pleasure.[35] Felling trees therefore created conflicts between different classes and between their values.[36]

Nonetheless, deforestation proceeded apace. Peasants and monks cleared trees on the edges of their fields, then uprooted the stumps, a difficult task requiring the strength of several oxen. Two tools were especially important. One was the axe; the other was the saw, introduced in the eleventh century, which accelerated deforestation and wood consumption.[37] By the thirteenth century, water-powered sawmills provided lumber for the booming cities and villages.[38]

Historical geographer Michael Williams estimated that around 500 CE, four-fifths of western and central Europe were covered with forests. Eight hundred years later, only half remained, with most of the clearing occurring after 1100.

To be sure, the amount of deforestation varied from region to region. In Italy, the forests recovered between the end of the Roman Empire and the Renaissance. In France, the forest cover declined from 30 million hectares in

800 to 13 million in 1300. The Domesday Book listed 15 percent of England as wooded in 1085; by the thirteenth century, 10 percent or less was still wooded. Spain was once so wooded it was said that a squirrel could travel from end to end without touching the ground. In Russia, peasants abandoned the grasslands and opened clearings in the forests to escape the steppe nomads, but their numbers were small and the forests of Russia so vast that the consequences were not felt until much later.[39] In western and central Europe, the consequences of deforestation were felt as early as the eleventh century. Bears, wolves, beavers, martens, and other wild animals became rare. Land eroded and silt filled lakes and estuaries. Near cities, shortages drove up the price of wood. To brew enough beer and bake enough bread for the 80,000 inhabitants of London in 1300 required between 29,000 and 32,000 tons of wood per year. As early as the thirteenth century, Londoners were burning coal, smothering their city in smoke on cold days.[40]

The Middle Ages also saw the beginnings of forest management and deliberate reforestation. Kings and nobles established game preserves and hired forest guards to protect one of their great privileges. In the mountains above Florence, the city government planted silver firs for construction. To supply timber for its shipyards, Venice took over oak, beech, and fir forests in its Italian possessions in the 1470s and established a forest administration to manage them. Where wood was running short, landowners started coppice plantations to supply poles for construction, firewood, and charcoal for iron smelting.[41]

All in all, the economic development, population growth, urbanization, and cultural flowering of western and central Europe in the period 800–1300 came at the expense of forests, wetlands, and wild animals. By the end of the thirteenth century, the benign climate that had sustained the efflorescence of the High Middle Ages in Europe was about to end. As in the Arab Empire, human societies were about to be tested by two natural disasters: climate change and disease.

CHINA

In an article entitled "The Coldward Course of Progress" published in 1920, distinguished American sociologist S. Colum Gilfillan proposed, as a law of history, that "progress" migrated from hot countries like Egypt and Mesopotamia to temperate ones like Greece and Rome and finally to cold ones like western and northern Europe. By progress he meant population growth, urbanization, and economic development as well as culture. Like almost all Europeans and white Americans at the time, he was thinking of Western civilization. Had he considered the rest of the world, he would have realized that in China, precisely the opposite had happened: throughout recorded history, the Han Chinese people, and with them their culture and way of life, were moving south.[42]

The Yellow River and North China

Though the Yellow River lay at the heartland of Chinese civilization, farmers had good reason to fear it and flee if they could. The North China Plain was a difficult land to farm because of the erratic monsoons. By one calculation, between 620 and 1619, one year out of five suffered a severe drought that caused famine among the people.[43] Even worse were the years when the river broke through the levees and flooded the countryside, washing away fields, villages, and inhabitants. Between 618 and 907, the river breached its levees on average once every ten years; from 907 to 960 the average rose to once every 3.6 years, and from 960 to 1126 to once every 3.3 years.[44]

Some floods were catastrophic. Between 1048 and 1128, the Yellow River crested its banks many times. In 1048 it broke through its northern bank 400 kilometers from the sea and shifted its bed hundreds of kilometers to the north. In the process, it captured several smaller rivers, filling their beds with silt, causing more flooding. For years thereafter, the river ranged across the North China plain, flooding every two years. During the flood of 1056, the government tried to contain the river with new dikes, but failed when the raging waters swept away all the material and carried 300,000 men to their deaths. Another great flood in 1068 killed a million people and destroyed whole cities and China's military defenses against the nomads.

Not all floods were natural. In times of warfare, as during the Five Dynasties (907–960) and the Northern Song (960–1126), opposing armies often breached the levees in order to thwart their enemies. Natural disasters were made worse by the state, which built levees on the right bank of the river to protect the province of Henan (Chinese for "south of the river"), site of the Northern Song capital Kaifeng, thereby allowing the river to flood Hebei ("north of the river").[45]

The worst man-made flood was that of 1128. That year, as nomadic warriors from Manchuria were ravaging northern China and threatening Kaifeng, the Chinese general Du Chong deliberately breached the levees to slow down their advance. Once out of its bed, the river divided in two, with one part flowing south of the Shandong Peninsula. In an effort to control the floods with new levees, farmers recruited by the government made bundles of sticks, reeds, bamboo, grass, and earth. In the process, they cut down every tree and bush still standing in the nearby provinces, completing the defor-estation begun long before; even mulberry bushes and trees on ancestral graves were felled. The resulting erosion caused more silting and still more flooding. Finally in 1288, the Yellow River moved entirely south of Shandong and settled in the bed of the Huai River, hundreds of kilometers south of its original outlet, leaving thousands of square kilometers of land devastated. There it remained until 1853–55, when it reverted to its former bed north of Shandong. In historic times, it has changed its bed ten times, each occur-rence with disastrous consequences.

The great floods of the eleventh and twelfth centuries ruined the soil of the North China Plain for centuries to come. Unlike the Nile, which deposited a fresh coat of fertile soil on its floodplain every year, the Yellow River silt was sandy, rocky, and sterile, and its floods turned arable land into sandy wastes scoured by dust storms.[46] For good reason, the Chinese called it "China's Sorrow," "the Ungovernable," and "the Scourge of the Sons of Han."[47]

Migration to South China (618–1127)

Farmers were not only fleeing from the north, they were also attracted by the potential wealth of southern lands that enjoyed a warm rainy climate suitable for growing rice. The first migrations of farmers to the south began during the Han dynasty (206 BCE–220 CE) and accelerated during the pe-riod of wars and turmoil known as the Six Dynasties (220–581) that followed the Han. These early migrations were sporadic and spontaneous, flights of refugees rather than frontier settlements organized by governments or pow-erful lords. Whereas under the Han only a quarter to a fifth of the Chinese people lived in the south, by 1080, two-thirds lived there and only one-third in the north.[48]

The Sui dynasty that reunited China in 581 began building the 1,900-kilometer-long Grand Canal to transport rice from the south to supplement the millet and wheat grown in the north. In the process, it interrupted the natural eastward drainage of the Huai River between the Yellow and the Yangzi rivers, depriving a 120- to 160-kilometer-wide band of farmland of fresh water and turning once fertile soil into saline lakes and marshes.[49]

After the Sui, the Tang dynasty (618–907) brought stability and prosperity to China for 200 years and under their rule the population rose.[50] During their early years, the Tang restored the Grand Canal and the state granaries. By the mid-eighth century, the canal system allowed grain to be shipped from the lower Yangzi valley to the capital at Chang'an, located for strategic reasons near the barbarian frontier.

What had delayed the migration of people from northern China to the south for 2,000 years after the first states arose along the Yellow River was the difficult terrain of the south, with its swamps and steep hillsides, and the long process of mastering the skill of growing rice in artificial ponds or paddies.[51] Diseases were an important deterrent as well, in particular the prevalence of malaria, dengue fever, filariasis, and schistosomiasis. Chinese authors, who confused various diseases and misunderstood their causes, attributed the unhealthy conditions in the south to "venom-laden air," "miasmas," or qi (demons).[52]

Unlike earlier spontaneous migrations, the Tang encouraged powerful landowning families to lead groups of farmers in massive drainage and land reclamation projects in southern China. They captured mountain streams with dams and dikes and retained their water in artificial ponds and tanks. As a result, according to Sinologist Mark Lewis, "the Tang had the fewest problems with flooding in the lower Yangzi of any dynasty in Chinese history" and there was a "massive increase in the amount of land being worked throughout southern China and its productivity."[53] The area witnessed an increase in the use of water buffalo and in new kinds of harnesses and plows with adjustable plowshares. Multi-cropping became common, as farmers planted three crops of rice every two years, along with beans and vegetables. To keep up the yields under continuous cultivation, farmers increased their use of manure and other fertilizers. In the process, the lower Yangzi Valley was transformed from a natural wetland into the richest farmland in China, and the population of the region multiplied several fold.[54]

While the advances under the Tang were undeniable, there was still much undeveloped land in the Jiangnan (the delta of the Yangzi). Farmers still left land fallow every other year and planted rice by broadcasting seeds rather than transplanting seedlings. In short, the south, and the Yangzi Valley in particular, were still far from their potential productivity.

Then, in the ninth century, a series of droughts afflicted northern China, causing hunger and undermining Tang rule. There followed a time of chaos called the Five Dynasties (907–960) followed by the Northern Song dynasty (960–1127), during which ecological and political difficulties in the north accelerated the migration of people to the south.

The Southern Song (1127–1279)

Fleeing the Jurchen who had conquered northern China by 1127, the Song rulers retreated beyond the Huai River and established their capital at Hangzhou, south of the Yangzi Delta, where they remained until 1279. To cope with a much larger population living in a much reduced territory and with the needs of a much larger army, the Southern Song government undertook an agricultural revolution so dramatic it has been compared to the Green Revolution of the mid-twentieth century.[55] It published instruction manuals, provided seeds, and offered low-interest loans and tax rebates to farmers. It established military colonies and settlements on state lands for refugees and landless peasants. Under the direction of landowners and businessmen, migrant farmers undertook the reclamation of the Jiangnan, transforming that vast region of salt-marshes into the most densely populated and productive farmland in China, comparable in size, complexity, and productivity to the Netherlands. As in the Netherlands, the marshes were transformed into polders, artificial fields surrounded by dikes.[56]

Transforming flat wetlands into productive rice paddies meant building dikes and sluice gates to control the level of the water, canals to bring in water, and ditches to remove the excess. Moving water in or out of fields required water-lifting devices such as swapes, waterwheels powered by streams, or treadle-pallet pumps moved by human muscles. Even more labor-intensive was the construction and maintenance of terraces built on the slopes of mountains to allow farming and to control erosion. Tunnels, aqueducts, channels, and bamboo pipes carried water to and from the terraces according to need. Farmers began using new devices such as harrows to break up clods, rollers to smooth the soil, and seed drills that mixed seeds with dung. Rice paddies were fertilized with pond silt, ashes, and the excrement of humans, pigs, chickens, and ducks. There even developed a commerce in human excrement from the cities; that of the rich, whose diet contained more protein, commanded a higher price than that of the poor. In addition to rice, peasants

planted tea and mulberry bushes. Silk began to replace hemp as a staple textile. As canals were improved and barge transportation increased, regions
specialized and trade in tea, sugar, and silk flourished.[57]

Among the contributions that the Southern Song government made to agricultural development, the most important was the transfer of Champa rice
from Vietnam to the Yangzi Valley in 1012. By order of the emperor, seeds of
this drought-resistant variety were distributed to farmers, while extension
agents called "master farmers" instructed them on cropping practices, tools,
fertilizers, and irrigation methods. Champa rice allowed farmers to plant
rice on hillsides and on marginal lands. It also ripened in sixty to a hundred
days, so that farmers could grow two crops of rice a year, or one of rice and
one of wheat, or even, in the tropical south, three crops. By the twelfth century, farmers had developed dozens of varieties, adjusted to different micro-
climates and economic conditions. In the same paddies where rice grew, they
also grew lotuses and water chestnuts and raised fish and turtles. Champa
rice more than doubled the area of rice culture in China. Sophisticated agricultural techniques such as using waterwheels, transplanting rice seedlings,
double and triple cropping, and fertilizing produced far greater yields than
growing rice in naturally flooded areas.

As a result, during the Southern Song the population of southern China
rose to over 20 million households, or around 100 million people, while the
population of the north stagnated or declined. So rich was agriculture under
the Southern Song that it supported luxury crafts and a large international
trade centered on Hangzhou, the largest and richest city in the world at the
time. Though in later centuries, China experienced vicissitudes of all sorts,
the agricultural system perfected under the Southern Song proved almost as
durable as that of the ancient Egyptians.[58]

Nonetheless, there were prices to pay. One was the deforestation of the
Yangzi Valley and neighboring regions similar to what had taken place in the
north many centuries before, with its resulting soil erosion and silting. Along
with the forests went the wild animals; by the end of the Song, tigers and elephants had vanished from all but the most inaccessible parts of southern
China.[59]

Another was what Sinologist Mark Elvin has called technological lock-
in: "the committing for an indefinite future of the use of a proportion of
income and resources simply for the maintenance of existing hydraulic systems, if the previous investment in construction and maintenance was not
to be lost."[60] In wet-rice agriculture, unlike dry-land farming, yields were

proportional to the amount of labor available. More land and bigger crops both demanded more labor and fed more people. Merely keeping up with the growth of population required constant investments of labor in the sophisticated hydraulic systems that underpinned intensive wet-rice farming, keeping living standards low.[61]

The history of China represents, in Elvin's memorable phrase, "3,000 years of unsustainable growth." By perfecting the technologies of water control and by strenuous and unremitting toil, the Chinese were able, for a time, to increase the carrying capacity of the land on a par with their growing population, but in doing so, they made their society vulnerable to external shocks.

The Mongol Conquest

This vulnerability became clear when the Mongols attacked North China in the early thirteenth century. The Mongols were nomadic herders of sheep, goats, horses, and cattle who regularly moved in search of grass for their herds. They were also avid hunters for whom hunting was not only a means of procuring food but also good training for warfare.[62] They were sensitive to the climate. The connection between climate change and the Mongol conquests is still being debated. Some have argued that a drought gave the Mongols an incentive to look south to better pastures in China.[63] Others have argued that, on the contrary, the climate of the steppe north of China was unusually warm and rainy in the early thirteenth century.[64] In either case, it was not climate change but the genius of Genghis Khan as an organizer and commander that catapulted the Mongols from disparate and unruly tribes into world conquerors.

The Mongols had little understanding of the farming people of China. In the Mongol language, as Sinologist Owen Lattimore explained,

> The term "hard" is used of Mongols and the term "soft" of Chinese. These terms do not stand only for physical robustness, but for the mental "hardness" of the man who lives in the saddle and makes his camp where he pleases, as against the moral "softness" of the man who is in bondage to the land he tills or the merchandise in which he deals, to his goods and his comfort, the safety of his roof and his walled town.[65]

The Mongols were contemptuous of the Chinese and even considered removing them from the North China Plain. Their goal was ecological: to transform North China into a grassland where they could graze their animals. According to René Grousset, "Chingis-khan's generals pointed out to him that his Chinese subjects were of no possible use to him, and that it would be better to slay them down to the last inhabitant, and so at least have the benefit of the soil which could be converted to grazing."[66] Though the Mongols never achieved this goal, they came close. In order to starve out cities during their initial invasion, they devastated the surrounding countryside. The area around Zhongdu, the capital of the Jin dynasty, was strewn with the corpses and bones of uncounted dead. These massacres eased slightly after 1217 when Genghis Khan recruited Chinese administrators to help him tax the surviving inhabitants. Conditions improved even more after the Mongols defeated the Jin in the 1230s and the Southern Song in 1279. By then, however, the population of North China had fallen from roughly 50 million in 1195 to around 8.5 million in 1235.[67]

Genghis Khan's grandson Kubilai Khan (r. 1260–1294) sought to become emperor of China and founded the Yuan dynasty. More benign than his predecessors, he preferred to conquer southern China and tax its inhabitants rather than destroy it. Despite his relative benevolence, many southern Chinese fled still further south to the Lingnan (China's southernmost provinces), a region then inhabited by hunter-gatherers and swidden farmers. At first, the Chinese migrants settled in the hills and on islands off the coast because the lowlands were swampy and malarial. Gradually, by building sea walls and embankments in the estuary of the Pearl River and letting soil that eroded from the hillsides fill the spaces, they turned that estuary into a delta with over 100,000 hectares of new farmland.[68]

Thus did China exemplify a pattern seen in Mesopotamia. Water control allowed a sophisticated civilization with a dense population to flourish. But it also made the region vulnerable to shocks, both natural—as in the North China Plain—and political—as in the Mongol conquest.

Africa South of the Sahara

During the period corresponding to the European Middle Ages, to the Abbasid Caliphate in the Middle East, and to the Tang and Song dynasties in China, cities and states arose in Africa south of the Sahara Desert. Much of their later prosperity rested upon their contacts and trade with the Islamic world to the north. The region that benefited most from these contacts was the Western Sudan, between the Sahara and the forests along the Guinea Coast.

In West Africa, the climate depends on two factors: the distance from the Gulf of Guinea and the North Atlantic Oscillation, the irregular shifting of rains north or south. Along the coast, ample rain and warmth produce a rainforest similar to that of equatorial Africa or of Amazonia. As one travels north, the climate becomes progressively drier. First the deep forest gives way to open forest, then to savanna, then to thin grassland called Sahel, then to semi-desert, and finally to the completely barren Sahara. This pattern was not permanent but shifted north or south. By 300 CE, the Sahara, which had once been green and lush, had become desert. South of the Sahara, besides the annual rainy and dry seasons, the climate oscillated between wetter and drier periods, depending on air currents in the North Atlantic. Some years, the rains failed entirely.

Each ecological zone favored a particular kind of human occupation. The rainforest remained thinly inhabited, but the open forest and savanna attracted farmers. The grasslands of the Sahel were occupied by cattle and goat herders. After the third century CE, the semi-desert and the desert oases were the domain of camel herders. Rivers, in particular the Niger, were home to fishing peoples. When the climate became wetter, herders moved north with the expanding grasslands, and farmers followed them into lands previously too dry to farm, though the soils were poor. When the climate turned dry again, peoples, their animals, and their way of life shifted south again. Sometimes there was symbiosis, for instance, when herders herded their cattle onto farmland after the harvest, and the cattle enriched the soil with their manure. But population movements also created conflicts. Peoples and their animals were also affected by the disease ecology of each zone. The wetter regions harbored malaria, yellow fever, and sleeping sickness. Nagana (animal trypanosomiasis) was prevalent along riverbanks and wetlands during the rainy season; horses were especially vulnerable, hence rare.

The real connections between peoples were the result of trade, for each zone had something to exchange for the products of another zone. Trade within West Africa long predated contacts across the Sahara. Towns near the Inland Niger Delta, such as Jenné-jeno and Gao, arose where crops from the more humid south and fish from the river were exchanged for copper and salt mined in the desert to the north.[69]

Two great changes transformed the region: the introduction of camels to the Sahara in the third century and of Islam to North Africa in the seventh and eighth. Before camels, the Sahara was an almost impenetrable barrier. Domesticated in the Arabian Peninsula, camels were introduced to North Africa in late Roman times. Able to survive for eight to ten days without water and carry up to 200 kilograms, these animals opened the Sahara to human occupation. Berber people, living on the fringes of the desert and in oases, took up camel herding and long-distance trade. For the first few centuries, such trade was sporadic and had no impact on the economies of either side of the desert.[70]

That changed after Arab armies invaded the Maghreb (western North Africa) in 643. By the eighth century, many of the Berbers of the region had intermarried with Arabs and converted to Islam. Eager to engage in trade with the Western Sudan and especially to acquire gold, they founded cities on the very northern edge of the desert: Qayrawan in what is now Tunisia, Tahert and Wargla in Algeria, Sijilmasa in southern Morocco. From there, caravans of several hundred camels crossed the desert to cities on the southern edge, such as Kumbi-Saleh, Awdaghust, Jenné-jeno, and Gao. The Sahara was no longer a barrier.

The most important item shipped north across the desert was gold, which was panned in the forests to the south of the Western Sudan and for which there was an insatiable demand in the Middle East, Europe, India, and China. Much of the growth of trade in late-medieval Eurasia was based on gold flowing from West Africa across the Sahara. Also very important was the export of slaves from the Western Sudan to North Africa and beyond. Unlike the Atlantic slave trade that followed, most of the slaves transported across the western Sahara were women purchased as domestic servants and concubines. Besides gold and slaves, the caravans also carried kola nuts, a stimulant in demand among Muslims forbidden to consume alcohol. Going in the other direction were salt and copper from the Saharan mines and hides and leather goods, glassware, textiles, ceramics, metal weapons, and other craft objects from North Africa.[71]

The End of the Medieval Climate Anomaly

From the ninth to the thirteenth century, centuries of good weather had contributed to the efflorescence of agriculture, urbanization, and culture throughout Eurasia. In Europe, it corresponded to the High Middle Ages. In China, those centuries correspond roughly to the Song dynasty, and in the

Middle East, to the Abbasid Caliphate, the peak of prosperity in both regions. Then, starting in the late thirteenth century, the climate of northern Eurasia began to worsen. Summers grew cooler, autumns came sooner, and winters grew longer and harsher. Thus began a period called the Little Ice Age.

Europe

In Europe, glaciers advanced down the valleys of the Alps. Off the coast of Iceland, sea ice, rare since 1000, became common. Fierce storms blew in from the North Sea in 1251 and 1287, killing thousands along the coasts of the Netherlands and Denmark. After the warm summers of 1284–1311 came years of bad weather. The winters of 1309–1312 were abnormally cold, with ice pack covering the North Atlantic between Iceland and Greenland.[72] The spring and summer of 1314 brought 155 days of rain to Germany and southern England. The winter was brutal and the following spring it again rained more than usual, causing floods and storm surges in coastal areas. The summer of 1315 was cool and overcast. The year 1316 was worse, with rains throughout the spring, summer, and fall, and a winter so cold that it froze the Baltic Sea. After yet another rainy year, the winter of 1317–1318 was the harshest of all and lasted from November to Easter. During the years 1318 to 1322 the weather was less severe, but periodic storms battered the continent. And in the winter of 1322–1323, the Baltic and parts of the North Sea froze over. To the anonymous author of the *Chronicle of Malmesbury*, the cause of the disastrous weather was clear: "Therefore is the anger of the Lord kindled against his people, and he hath stretched out his hand against them, and hath smitten them."[73]

The results were disastrous. Grain production in northern Europe dropped by one-third compared to thirteenth-century harvests. Molds, rusts, and mildew attacked the crops. Ergot blight caused convulsions and hallucinations among people and animals who ate infected rye. Because wood and peat were too wet to burn, salt production by boiling brine plummeted, hence herring could not be preserved by salting, depriving whole populations of their main source of protein.[74]

As hay could not dry properly, there was not enough fodder for animals during the winter, so their owners slaughtered them. At the same time, a cattle epizootic that people called "murrain" (probably rinderpest) broke out in central Europe in 1314, then spread to France, the Low Countries, and the

British Isles, reaching Ireland in 1321. In England, it killed two-thirds of the bovine herd within eighteen months. The mortality among livestock was catastrophic, reaching 90 percent; their loss caused a shortage of draft animals and manure, resulting in poorer harvests in the following years.[75]

Crop failures and the death of livestock caused the Great Famine, the worst in European history. Hungry people ate their seed corn, then roots, acorns and nuts, and even bark; some were reduced to cannibalism. In the cold winters, they succumbed to respiratory diseases. Children born during the famine suffered from stunted growth and weakened immune systems, leaving them vulnerable to diseases. Compounding the effects of the famine, wars broke out in Scandinavia, southern Germany, Austria, and France. All in all, in Europe north of the Alps and Pyrenees, over 30 million people were affected and the population declined by 10 to 15 percent.[76]

While humans suffered, the rest of nature recovered. Land reclamation from marshes and peat bogs stopped, as did the clearing of forests. When farmers in northern Europe abandoned their villages and farmlands, trees grew back. Marginal lands that had once been farmed became pastures for the few sheep and cattle that survived.[77]

Most dramatic of all was the collapse of the European settlements in Greenland where Norsemen had come to farm and raise cattle 300 years earlier. By the late twelfth century, the warmth that had sustained the small settlements on the southwestern coasts was ending. Farming ceased, pasture for livestock became rare, ice blocked the coast more frequently, and the survivors faced attacks from Inuit who were accustomed to Arctic conditions. Forced out by the cold and the Inuit, the Norse abandoned the Western Settlements in the mid-fourteenth century and the Southern Settlements soon thereafter.[78]

China

In the course of the fourteenth century, China experienced thirty-six years of exceptionally severe winters, more than in any other century in recorded history. Climate historian Hubert Lamb calls the year 1332 "one of the greatest weather disasters ever known, alleged to have taken seven million human lives in the great river valleys of China."[79] Between 1308 and 1325, the weather was cold and wet, then dry but progressively colder from 1352 to 1374. Though the Mongol-dominated Yuan dynasty (1271–1368), like all other Chinese dynasties, claimed the Mandate of Heaven, the forces of nature did not cooperate. Every year but one had colder than usual temperatures. There was a major famine on average every other year from 1268 to 1359, throughout most of the dynasty.[80] The official histories of the Yuan dynasty record famines in almost every year of the reign of the last Mongol emperor Toghun Temür, from 1333 to 1368.[81]

In the 1340s, revolts began breaking out in protest against the misery of life and the high taxes the Mongols imposed on their unruly subjects. The official biography of Zhu Yuanzhang, the leading rebel and founder of the Ming dynasty that replaced the Mongols in 1368, began with the words: "The year 1344 was a time of droughts, locusts, great famine, and epidemics." Even after the Mongols were ousted, the bad weather continued unabated. The winters of 1453 and 1454 were so harsh that the Yangzi estuary froze over and snow lay one meter thick in the Yangzi delta; and again in the winter of 1477 the canals froze and trade came to a stop.[82] Yet the Ming, less hated than the Mongols because they were Han Chinese, survived until 1644.

The shift to a colder climate was not the only natural disaster that afflicted the peoples of Eurasia. Even more disastrous was the deadliest epidemic to affect the continent before the twentieth century: the Black Death.

The Black Death

The Black Death swept through Europe and the Middle East in the mid-fourteenth century, killing a third to a half of their inhabitants, perhaps more. While the symptoms of this disease are well known, many questions remain about its causes and origins.

The Black Death in Europe

The standard story is that the Black Death began among the Mongols of the Golden Horde who were besieging the port of Kaffa in the Crimea, a city with a large Italian community. According to legend, before departing, the Mongols threw some corpses over the walls of the town. More likely, some rats made their way under, through, or over the walls into the town, where they infected others. As the Italians fled Kaffa in their ships, they carried with them infected rats, spreading the disease to Constantinople in May 1347.[83]

From Constantinople, the epidemic spread to southern Europe in 1348, to central and western Europe in 1349, to northern Europe in 1350, and to eastern Europe and Russia between 1351 and 1353. It spread more rapidly and more widely than Justinian's plague because northern and western Europe were much more developed and densely populated than they had been in the seventh century. Furthermore, ships regularly sailed from the Mediterranean to England and the Low Countries; starting in 1291, they did so even in winter.[84] The epidemic returned in 1361–1363, and at regular intervals thereafter, but less virulently than the first time.

Everywhere the epidemic appeared, the population plummeted. Most sources say the population of Europe dropped by a third, but historian Ole Benedictow has calculated that 50 to 65 percent of the European population of about 80 million died over the course of the fourteenth century.[85] In England, the population dropped from 6 million to 3 million or less, and it did not recover until the seventeenth century. As the population shrank, labor shortages developed. Peasants abandoned the estates of their landlords to seek better wages elsewhere; as a result, the incomes of surviving peasants rose while those of landowners fell. Not until the mid-fifteenth century did the population and food prices begin to rise again. Thereafter, outbreaks were local and mainly urban, as in London in 1665 and Marseille in 1720.[86]

The plague changed the environment of Europe as well as its population. One out of every four or five English villages was abandoned. Marginal lands became pastures for the sheep that provided wool for the export trade. As fields were abandoned and construction ceased, forests recovered. Tree-ring analysis shows a rapid growth of new oak trees in Ireland, Germany, and elsewhere.[87] What was terrible for humans was often good for the rest of nature.

The Black Death in the Middle East

As horrific as the epidemic was in Europe, its population and economy recovered more quickly than in the Middle East. The reasons are complex, involving the destruction of Baghdad and the collapse of the Abbasid Caliphate at the hands of the Mongols in 1258, later invasions from Central Asia, the decline of irrigation and agriculture, and the conservative trend of Islamic culture that followed. Though Middle Eastern wars were no worse than the ravages Europeans inflicted upon each other during this period, the environmental consequences of the Black Death were very different from those in Europe.

From Constantinople, the epidemic spread to Egypt and Iraq. The Moroccan traveler Ibn Battuta encountered the disease for the first time in Aleppo, Syria, in 1347 or 1348.[88] It also spread to Arabia, Morocco, Yemen, and Iran. Thereafter, it returned every few years until the end of the eighteenth century. Between 1347 and 1517, there were fifty-five outbreaks of the disease in Egypt and fifty-one in Syria. During the most virulent outbreak, 1429 to 1430, the population of the Middle East declined by over one-third and did not recover for centuries.[89]

Egypt was particularly hard hit. The Nile and the many irrigation canals made it easy to ship grain, hence also rats, throughout the country. The network of canals, dikes, and sluices that the Mamluk regime had constructed since 1260 required constant maintenance. When the epidemic hit, the population shrank, from about 4 million in 1300 to about 3 million in 1500.[90] This created a labor shortage, which in turn caused a decline in canal maintenance. Declining yields, hence a shortfall in taxes, weakened the government and made it unable to afford the repairs needed after the annual Nile floods. Without the usual control over the flood waters, some fields received too much water and others not enough. Harassed by their landlords, peasants revolted or fled to the cities, abandoning their fields. As Egyptian agriculture entered a downward spiral, some of the richest farmlands were taken over by Bedouin sheep herders who were less affected by the Black Death because they had no grain or permanent houses to attract rats.

Historian Stuart Borsch has contrasted the Egyptian case with that of England. Before the Black Death, both countries had about the same area and population. In England, the Black Death reduced the population by more than the arable land; as a result there was more land per person, farm wages rose, and the standard of living of the poor improved. Egypt's complex and delicate irrigation system, however, made the amount of arable land proportional to the labor invested in it. When the epidemic struck, the amount of productive land declined as much as, or more than, the population. Much the same happened in Iraq, as some of the key elements in the irrigation system built by the Sasanids centuries earlier, the transverse canals linking the Tigris and Euphrates rivers, were abandoned during and after the epidemic.[91] As the amount of irrigated land declined, so did the harvests and the amount of food available per person.

Black Death in Africa?

The spread of the epidemic to Africa is still shrouded in mystery. For centuries, West Africa had been in contact with the Mediterranean by trans-Saharan camel caravans. In the ninth century, the city of Jenné-jeno near the Niger River in today's Mali had grown to several thousand inhabitants. Later, other cities such as Timbuktu, Gao, and Mopti flourished in that great river valley, thanks to productive agriculture and extensive trade between the forests to the south and the Sahara Desert and the Mediterranean to the north. When Mansa Musa, the king of Mali, undertook a pilgrimage to Mecca in 1324, he awed the people he encountered along the way with his wealth.

In 1400, however, Jenné-jeno was abandoned; according to archaeologist Roderick McIntosh, "The prosperity of the first millennium was succeeded by massive migrations and depopulation on a breathless scale."[92] Further, he noted: "One cannot read these global tales of famine, war and pestilence without wondering how possibly the Middle Niger could have escaped the waves of Black Death.... All this is speculation."[93]

Similarly, archaeologists Gérard Chouin and Christopher Decorse noted that the settled landscape of southern Ghana was suddenly abandoned in the mid-fourteenth century, an event "traumatic enough to alter people's way of life within a generation, wiping out the structures of a centuries-old agrarian order.... Only one event can explain such a large-scale phenomenon: the occurrence of the Black Death or Great Plague.... It is plausible that it reached the forests of West Africa in the mid-fourteenth century."[94]

Causes of the Black Death

What caused the Black Death? Until the 1980s, the consensus among historians was that it was bubonic plague. This pathogen was transmitted to humans by the flea *Xenopsylla cheopsis* that infected commensal rats that caught it from wild rodents.[95]

Then in 1984 physician Graham Twigg argued that the Black Death was not bubonic plague but anthrax, a cattle disease that also infected humans. Several other scholars have also claimed that bubonic plague was not the cause of the Black Death. Demographer Susan Scott and zoologist Christopher Duncan concluded that the Black Death "was probably viral in nature," but they could not identify the virus in question.[96] Similarly, historian Samuel Cohn has argued that the Black Death was too virulent and spread too fast to have been the bubonic plague. In that case it could have mutated into pneumonic plague, an even more virulent variation that is transmitted directly from human to human by breath, which is always fatal.[97]

Of course, one disease does not prevent another from attacking the same victims, so there may well have been an outbreak of anthrax at the same time as the plague. However, both contemporary sources and all other authors who have written on the Black Death described the buboes that broke out on its victims, a symptom no other disease produces. Furthermore, if anthrax prevailed in northern Europe, that begs the question of what could have caused the Black Death in southern Europe, the Middle East, and elsewhere.

The answer is beginning to come from the work of geneticists who have analyzed the DNA found in the bones and teeth of victims of the Black Death. In 2000, Didier Raoult and his colleagues analyzed the DNA found in the teeth of a child and two adults who died in Montpellier, in southern France, during the epidemic. In it, they found evidence of *Yersinia pestis*, but not of anthrax or Rickettsia bacteria.[98] In 2007, Michel Drancourt and his colleagues confirmed that *Y. pestis* caused the Black Death. In 2010, Stephanie Haensch and her colleagues also found *Y. pestis* in human skeletons taken from mass graves in northern, central, and southern Europe. People who were buried in the plague years 1347–1348 definitely died of bubonic plague; those buried before 1347 did not.[99]

Where Did the Plague Originate?

More vexing is the question of the geographical origin of the Black Death. The first written evidence of the plague comes from the tombs and writings of ancient Nestorian Christians near Lake Issyk-Kul in Kyrgyzstan, where Russian archaeologist Daniel Abramovich Chwolson noted abnormally high death rates in the years 1338–1339.[100] Its next appearances were in 1345 or 1346 in Astrakhan and Saray on the lower Volga and among the Mongols of the Golden Horde who were besieging the port of Kaffa on the Black Sea, from which it spread to Constantinople and beyond.[101] But before it got to Lake Issyk-Kul, where did it come from?

In 1832, Justus Hecker, the father of medical history, suggested China as the birthplace of bubonic plague, a hypothesis later taken up by historian

William McNeill.[102] Recent estimates of the Chinese population show a decline from 110–120 million in 1200 to roughly 75 million in 1300.[103] While the Mongols who ruled China between the Song and the Ming were notoriously brutal, no amount of brutality could account for such a demographic disaster, nor could bad weather, floods, or famines. McNeill blames the demographic collapse between the Song and Ming dynasties on a combination of war and pestilence: "Disease assuredly played a big part in cutting Chinese numbers in half; and bubonic plague, recurring after its initial ravages at relatively frequent intervals, just as in Europe, is by all odds the most likely candidate for such a role."

McNeill traces the focus of the plague to the Himalayan borderlands between China, Burma, and India, where the disease had long been endemic among rodents and which the Mongols tried to conquer in the 1250s:

> Our hypothesis, therefore, is that soon after 1253, when Mongol armies returned from their raid into Yunnan and Burma, *Pasteurella pestis* [i.e., *Yersinia pestis*] invaded the wild rodent communities of Mongolia and became endemic there. In succeeding years the infection would then spread westward along the steppe, perhaps sporadically assisted by human movement, as infected rats, fleas, and men inadvertently transferred the bacillus to new rodent communities.

Though the Chinese records show nothing unusual before 1331, that year there was an outbreak in Hebei province that is said to have killed nine-tenths of the population, followed by an even more widespread disaster in 1353–1354. According to McNeill, "What seems most likely, therefore is that *Pasteurella pestis* invaded China in 1331, either spreading from the old natural focus in Yunnan-Burma, or perhaps welling up from a newly established focus of infection among the burrowing rodents of the Manchurian-Mongolian steppe." The greatly increased caravan traffic along the Silk Road during the Mongol period that permitted Marco Polo to visit China also allowed the plague to move across Asia from caravanserai to caravanserai between 1331 and 1346.[104]

There is no question that China suffered from frequent epidemics. In 762, an epidemic is said to have carried off more than half the inhabitants of the southeastern port cities. Another one that began in 832 and lasted ten years was even worse.[105] From the tenth to the twelfth century, epidemics struck every two to five years. According to historian Mark Elvin, the one that afflicted the city of Kaifeng in 1232 "is said to have carried off, in between fifty and ninety days, from 900,000 to 1,000,000 of the inhabitants."[106] The official history of the Yuan dynasty recorded serious epidemics in 1344–1345, 1356–1360, and 1362.[107]

But what diseases caused these epidemics? Chinese sources are frustratingly vague on the subject. Chinese writers described the epidemics they witnessed as *yi* (epidemic) or *da yi* (great epidemic), without specifying its symptoms. Not until the late nineteenth century did the Chinese language have a word for bubonic plague.[108] One of the reasons for the uncertainty is that Chinese doctors attributed diseases to bad weather. Faced with an epidemic, the response of the Song officials was to seek relevant books in the Imperial Family Archive, such as the 800-year-old *Treatise on Cold Damage Disorders*, and have them copied by the Bureau of Revising Medical Texts and handed out to physicians.[109]

Chinese medical texts described the symptoms of bubonic plague, such as lumps on the body, for the first time in 610. An epidemic in 636 may have been plague, for it followed the "plague of Shirawayh" that devastated Mesopotamia and Syria, regions with which China was in contact through the Silk Road. It may have recurred south of the Yangzi in 762 and may have been the cause of the epidemic of 832 that may also have spread to Korea and Japan.[110] The evidence, however, is too thin to identify these epidemics with any confidence. Medical texts written during the Song and Yuan dynasties describe a disorder whose symptoms (swollen lymphatic glands, coughing blood and phlegm) resemble those of the plague, but they describe it as new and not widespread. Other evidence comes from eighteenth-century encyclopedias, not the most reliable of sources.[111]

Most Sinologists are uncomfortable with the McNeill thesis. Noting the demographic catastrophe of the fourteenth century, Robert Marks wrote: "Historians of China just cannot be sure whether an epidemic caused the deaths, and if so, whether the disease was bubonic plague."[112] Other Sinologists agree that there is no evidence in the Chinese written records that bubonic plague affected China in the medieval period.[113] In a recent article, historian Paul Buell pointed out that during the early fourteenth century, the Mongols who ruled China had in fact very little contact with the Mongols of Russia or those of Iran, preferring instead to trade with the countries of the Indian Ocean. While there may have been cases of the plague in China, there is no evidence of a plague epidemic like that of western Eurasia.[114]

Geneticists and the Origin of the Plague

Between Lake Issyk-Kul and the Crimea stretches the great steppe of central Asia and southern Russia, across which the plague and its vectors made their way between 1339 and 1347. Several authors have therefore identified this steppe as the original source of the disease.[115] More recently, another source has been identified: the Qinghai-Tibetan Plateau between Mongolia and Tibet proper. This high and semi-arid region is home to gerbils and other wild rodents. During the early fourteenth century, the climate was unusually rainy, causing vegetation to flourish and wild rodents to multiply, along with their fleas and plague bacteria. When drought returned in the early 1340s, these rodents migrated in search of food and their fleas jumped to commensal rats and from them to members of the Golden Horde.[116]

Pinpointing the Qinghai-Tibet Plateau as the proximate source of the Black Death rules out the Chinese epidemic of 1331. That does not mean, however, that the bubonic plague avoided China. As geneticists have recently shown, the bacterium *Yersinia pestis* probably evolved somewhere in China over 2,600 years ago and perhaps as early as 6,500 years ago. From there it spread on several occasions to wild rodent populations in Tibet, Central Asia, and elsewhere.[117] Thus, the disease did originate among rodents in China after all. In the fourteenth century, however, the epidemic we call the Black Death started in Central Asia. If it reached China, no evidence known so far indicates that it triggered an epidemic in that country. For the time being, the epidemiological history of China awaits geneticists willing to extract DNA from centuries-old cadavers, as has been done in Europe.

Conclusion

The environments in which humans lived over a thousand-year period differed enormously in climate, topography and soils, native plants and animals, and much else. And human societies differed almost as much as the environments they inhabited. What conclusions, then, can be drawn from such disparate sources?

The period from the early seventh to the late fifteenth century began with an expansion of settled agriculture and urban civilization in the Middle East and North Africa, in China, in Europe, and in sub-Saharan Africa.

Among the important causes of the advance of humans in these regions were two natural phenomena: the Medieval Climate Anomaly and the retreat of epidemics. But the change was also the result of human actions: the transfers of crops, new agricultural methods and technologies, and policies and institutions that benefited agriculture, the mainstay of civilization throughout the Eastern Hemisphere.

In this period, as in earlier times, reliance on a benign climate and sophisticated institutions masked an increased vulnerability to natural and political shocks. Several such shocks undermined the prosperity of earlier times: droughts and cold weather in the Middle East, the Mongol invasions, a colder and more erratic climate in Europe, and the plague in Europe and the Middle East. Though horrific for the people who suffered through them, these setbacks did not push Eurasia and Africa into new "Dark Ages" but only caused a hiatus in the expansion of humans at the expense of the rest of nature.

6

The Invasion of America

In almost every textbook, 1492—the year Columbus first reached the New World—is celebrated as a major turning point in the history of humankind. In the history of the natural world, however, the important turning point is not 1492 but 1493, the year Columbus returned to the Antilles with settlers, plants, and animals. This inaugurated the encounter between the New and the Old World biota that environmental historian Alfred Crosby called the Columbian Exchange.

Biological transfers were nothing new in world history, yet two aspects of the Columbian Exchange stand out. One is the unprecedented volume of the transfers. More plants, animals, and pathogens have crossed the oceans in the five centuries since Columbus than in thousands of years before, thereby greatly accelerating the biological homogenization of the planet. The other aspect of the exchange is its lopsidedness. A comparatively small number of life forms migrated from the New World to the Old; and of these, only a few plants—maize, potatoes, tomatoes, cacao, and manioc—made a substantial difference to the land and peoples of the Eastern Hemisphere. In contrast, the plants, animals, and pathogens that came from the Old World radically transformed the Americas, overwhelming and displacing many native species. Not just individual species, but the entire New World biota was vulnerable to invaders from the Old World.

The American Holocaust

Europeans did not conquer the Americas on their own; the diseases they brought with them helped. The result of the encounter was the most horrific disaster that has ever befallen the human race.

Historically, small populations of foragers and isolated farm communities tended to be relatively free of diseases. When an epidemic did break out in a small group with limited contact with outsiders, the disease would usually burn itself out, leaving only immune survivors. Large populations in developed agricultural regions, by contrast, were incubators of diseases. In villages, sewage could contaminate drinking water and rodents and insects could spread diseases. Close contact with domesticated animals allowed pathogens to jump to humans; thus measles, tuberculosis, and smallpox may have mutated from cattle diseases, influenza from pigs or ducks, and malaria from apes.[1] In tropical regions, forest clearings provided new habitats for *Anopheles* mosquitoes that carried malaria. Cities increased the risks of diseases, for itinerant merchants, soldiers, and officials carried pathogens as well as goods and ideas. Before the nineteenth century, urban areas had higher death than birth rates, and their populations could only be sustained by a constant influx of migrants from rural areas. If a population was sufficiently large, a disease could survive long enough for a new generation of children to be born without immunity. In some cases, diseases that had once been epidemic could become relatively mild childhood diseases.

This was particularly true of the Eastern Hemisphere, the birthplace of many dangerous diseases, among them smallpox, measles, typhoid fever, bubonic plague, yellow fever, malaria, and influenza. After millennia of inter-regional contacts, the Old World formed one gigantic and relatively homogeneous disease pool. Within it, sub-Saharan Africa formed a special disease pool because it was the home of yellow fever and falciparum malaria, both of them transmitted by mosquitoes.[2]

When an infectious disease broke out among a people who had never encountered it before, the results were more calamitous than epidemics in experienced populations. Such "virgin soil" epidemics often killed adults, leaving fields untended and children and the elderly without food and care. Some infected but still healthy people tended to flee the scene of an epidemic, spreading the disease far and wide. As often happened, people sickened by one disease were vulnerable to other diseases, so a population could suffer from several simultaneous epidemics. In encounters between peoples, those who had experience with infectious diseases possessed a tremendous advantage over those who had not.[3] This is what happened when Europeans and Africans met Native Americans.

The Americas before 1492

The population of the Americas before 1492 is, at best, an estimate based on uncertain data. Furthermore, the numbers are fraught with political significance: the greater the number, the higher the mortality, hence, the worse the disaster that befell the Indians after the encounter. Early estimates ranged from 8 million to 112 million. In 1992, historical demographers William Denevan, John Verano, and Douglas Ueberlaker narrowed down the possible range to between 43 and 72 million, about the same as that of Europe.[4] More recently, demographer Massimo Levi-Bacci advanced the idea of a smaller population, "perhaps around 30 million, and so equal to about one-third of the population of Europe at the time."[5] Whatever the numbers, the American population was substantial, with concentrations in parts of Mexico, Central America, and Peru comparable to those of the more densely populated regions of China, the Middle East, and Europe.

Thus the idea entertained by Europeans that large parts of the Americas, such as North America and Amazonia, were "wildernesses" was an illusion caused by the collapse of the American Indian population after the Europeans arrived. As William Denevan explained:

> The Native American landscape of the early sixteenth century was a human-ized landscape almost everywhere. Populations were large. Forest compo-sition had been modified, grasslands had been created, wildlife disrupted, and erosion was severe in places. Earthworks, roads, fields, and settlements were ubiquitous.[6]

Not only were Native Americans numerous, but they were also healthier than Europeans, Asians, or Africans. One native of Yucatán remembered: "There was then no sickness; they had no aching bones; they had then no high fever; they had then no smallpox; they had then no burning chest; they had then no abdominal pain; they had then no consumption; they had then no headache. At that time the course of humanity was orderly. The foreigners made it oth-erwise when they arrived here."[7] That was certainly an exaggeration, but not far from the truth. The reason Native Americans were healthier than the peo-ples of the Eastern Hemisphere is because they descended from the pioneers who had crossed over from northeastern Siberia, a trek that must have weeded out the sick. The disease filter of Beringia also prevented disease-bearing insects (except lice) from accompanying them. Once established in the New World, the Native Americans had few domesticated animals and the few they had, such as the llamas of the Andes, never formed herds large enough to sustain diseases.[8]

Native Americans suffered from such afflictions as yaws, syphilis, tuberculosis, polio, hepatitis, and intestinal parasites.[9] They also experienced iron-deficiency anemia, degenerative joint diseases, and dental pathologies—debilitating but not life-threatening. Their immune systems were also remarkably uniform; as a result, viruses could spread more easily among them than among peoples with more diverse immune systems. In short, the first Native Americans left their descendants more vulnerable to Old World diseases when the encounter between the hemispheres finally happened.[10]

There were also differences among Native Americans based on their way of life. Agricultural peoples who lived in permanent and often crowded settlements harbored diseases more readily than hunter-gatherers. As in the Eastern Hemisphere, people who moved from foraging to growing crops increased in numbers but declined in health, stature, and longevity. As a result, farmers and city dwellers were especially vulnerable to imported diseases.[11]

Hispaniola and the Caribbean

The first place Spaniards settled after 1492 was Hispaniola. Like all the islands of the West Indies, it was covered with lush tropical rainforests on its windward side and scrub or grasses on its leeward side; though birds and animals were abundant, there were no large herbivores or predators; iguanas, introduced from South America, were the largest land animal. The inhabitants, called Tainos, had migrated from South America and practiced a simple mixed economy. They grew maize, beans, squash, manioc, sweet potatoes, peanuts, and tobacco on small plots called *conucos*, and fished and hunted green sea turtles, manatees, iguanas, and small rodents called *hutias*.[12]

Bartolomé de Las Casas, the first Spanish historian of the Americas, asserted that there were over 3 million inhabitants of Hispaniola; scholars have since debated this and put the figure at anywhere from 100,000 to half a million.[13] What is more important than the original population is what happened to these people after the Spanish arrived. According to official Spanish enumerations, in 1508 there were 60,000; by 1518–1519 only 18,000 were still alive; and by 1542 the population had been reduced to 2,000. The chronicler Gonzalo Fernández de Oviedo y Valdés, who estimated the original population at a million, wrote: "Of all those, and of all those born afterwards, there are not now believed to be at the present time in this year 1548 five hundred persons, children and adults, who are natives and are the progeny or lineage of those first." Even those few were soon to die, until there were none left.[14]

What caused this holocaust? The violence of the Spanish conquerors who enslaved the Tainos and forced the men to dig for gold and the women to provide food for the invaders certainly played a part. But Spanish greed and brutality, and the resulting social chaos among the Indians, cannot account for all the deaths, because the Spaniards depended on native labor. Furthermore, in later years they treated African slaves and Filipino peasants no more kindly than they had the Tainos, without an ensuing extermination. The difference can only have been diseases. Which diseases were responsible for the first epidemics is not known, for the Spanish descriptions are unclear and disease symptoms can change over time, but influenza, typhus, or measles are the most likely causes.[15]

Then came smallpox, among the most virulent and contagious of all diseases for non-immune populations, and one that produces unmistakable symptoms.[16] It reached Hispaniola in December 1518. It had been delayed because the whole cycle, from infection to death or immunity, took a month, less time than a transatlantic voyage. Hence, it could only spread if several people contracted the disease sequentially while crossing the ocean. Once smallpox arrived, it claimed a third to half of the surviving population of Hispaniola within a few months. From there it spread quickly to Puerto Rico, Cuba, and other Caribbean islands, as infected but still healthy persons fled, carrying the virus with them. Because the Indians had no inherited resistance, most of those it infected died. Other diseases, the disruption of their societies, and food shortages claimed many others. Altogether, the population of the islands declined by about 90 percent.[17]

Mexico

Smallpox had reached Mexico soon after the conquistador Hernán Cortés entered the Aztec capital Tenochtitlán in November 1519. Cortés had just been forced out of Tenochtitlán when the epidemic broke out among the Indians. When he returned in the summer of 1521, the disease had killed half of the Aztecs, including their leader Cuitláhuac and other warriors, leaving the rest weakened. The Spaniards entering the city on August 13 found it littered with bodies. A native Mexican told the Franciscan friar Bernardino de Sahagún,

> The illness was so dreadful that no one could walk or move. The sick were so utterly helpless that they could only lie on their beds like corpses, unable to move their limbs or even their heads. They could not lie face down or roll from one side to the other. If they did move their bodies, they screamed in pain. A great many died from this plague, and many others died of hunger. They would not get up to search for food, and everyone else was too sick to care for them, so they starved to death in their beds.[18]

The Spaniards saw their victory, and their own immunity, as a sign from God. A follower of Cortés, wrote: "When the Christians were exhausted from war, God saw fit to send the Indians smallpox, and there was a great pestilence in the city."[19]

Smallpox also afflicted the Indian tribes allied with Cortés, persuading them to let him choose their leaders and organize their armies. Even after the epidemic had petered out for lack of fresh victims, the miseries of the native Mexican peoples continued. A drought damaged the harvest. Other epidemics swept through the surviving population: measles in 1531–1532, typhus in 1545–1548, mumps in 1550, measles again in 1563–1564, typhus again in 1576–1580, measles yet again in 1595.

The epidemic of 1545–1548 was the most disastrous, with death rates of 60 to 90 percent among those infected. Another in 1576 reduced the surviving population by half, leaving only 1.2 to 2 million alive. Some scholars have identified the cause of these epidemics as *cocoliztli* (a Nahuatl word meaning "great sickness"), a hemorrhagic fever that originated in the valleys of central Mexico and was spread by rodents.[20] Evidently, the Spanish population was minimally affected. Making things worse, between the 1540s and the 1580s, Mexico suffered from the worst drought of the past 500 years.[21]

Several disease outbreaks occurred simultaneously, and at other times secondary diseases—pneumonia, influenza—attacked the sick and weakened. The exact identification of these diseases is uncertain, yet it is estimated that the Mexican people suffered fourteen epidemics between 1520 and 1600, and that the population dropped from roughly 14 million in 1519 to about 1 million by the end of that century, a decline of 93 percent.[22] Only the Tainos of Hispaniola and other inhabitants of the tropical lowlands suffered a worse fate.

South America

Having conquered central Mexico, the Spaniards turned south, toward Guatemala, Yucatán, and Panama, then toward South America. Wherever they went, they brought smallpox with them. In most places, it killed half the population or more, making it the Spaniards' most powerful ally. When Francisco Pizarro landed in Peru with his little band of 180 men and 37 horses, he found that smallpox was already rampant, causing the death of the Inca Huayna Capac and his son Ninan Cuyochi and precipitating a civil war between the followers of his sons Huáscar and Atahualpa. In the midst of death and confusion, the Spaniards seized the opportunity to capture and kill Atahualpa and establish their dominion over what had been, just weeks before, the largest and most powerful empire in the Americas. Like the people of Mexico, those of the Andes suffered multiple epidemics, some of them simultaneously.[23]

From Peru, epidemics spread throughout South America. On the Pampas of Argentina and Paraguay, the Spanish conquistadors found the native Indians numerous and friendly at first, then hostile. These were not members of highly organized states like the Aztec and Inca empires, but nomadic hunter-gatherers who lived in small groups at a distance from one another. Diseases spread more slowly among them than among dense agricultural populations. The first major epidemic reached them in 1558–1560, killing thousands of Indians but not a single Spaniard. Thereafter, epidemics followed one another throughout the seventeenth and eighteenth centuries. Thomas Falkner, an English traveler who visited the area in the eighteenth century, wrote:

> The two nations [Araucanos and Pehuenches] were formerly very numerous, and were engaged in long and bloody wars with the Spaniards . . . but they are now so much diminished as not to be able to muster four thousand men among them. . . . The small pox also, which was introduced into this country by the Europeans, causes a more terrible destruction among them than the plague, desolating whole towns by its malignant effects.[24]

In Brazil, the epidemiological situation was similar, though the geography differed. When the Portuguese first settled on the coast in the early sixteenth century, they found the area inhabited by about a million Tupi Indians, hunter-gatherers and swidden farmers who grew manioc, peanuts, beans, tobacco, and maize on land that they cleared and planted for two or three years, then left fallow for twenty to forty years. Smallpox broke out among them in 1562–1565 and again in 1613, 1621, and 1641–1642. As the native population shrank, the Portuguese began importing slaves from Africa to work in the sugar plantations, and the slaves in turn brought more smallpox. By the end of the sixteenth century, the Tupi had practically disappeared within 300 kilometers of the coastal towns.[25]

North America

Epidemics preceded the first Europeans in North America as they had in South America. Before Columbus, southeastern North America had been densely populated by agricultural peoples we know as the Mississippian culture or "Moundbuilders." When the first European explorer, Hernando de Soto, arrived in 1539–1542, diseases had already broken out. By the time French explorers reached the Mississippi Valley around 1700, there were few villages and fields left; a Frenchman who visited Natchez remarked: "Touching those savages, . . . it appears visibly that God wishes that they yield their place to new peoples."[26]

European whalers and cod fishers had visited New England many times before the first permanent settlers arrived on the *Mayflower* in 1620. In 1616–1619, the diseases they brought with them sparked an epidemic—perhaps typhus, pneumonic plague, or chickenpox—among the native peoples. Several coastal villages were left empty. Smallpox also appeared in northeastern North America in 1616. Starting in 1619, it reduced the Pequot population of Connecticut by three-quarters. In 1633–1634, it killed three-quarters of the Indians between Maine and southern Connecticut, and in 1639–1640 it killed half the Huron people.[27]

The Indians' first reaction was to tend to the sick, but they soon learned to flee; as one English settler said: "So terrible is their apprehension of an infectious disease that not only persons, but the Houses and the whole Towne takes flight."[28] This, of course, only helped spread the disease more quickly from village to village. In 1634, John Winthrop, the first governor of the Massachusetts Bay Colony, saw the disaster as a God-given real-estate opportunity: "For the natives, they are neere all dead of small Poxe, so as the Lord hathe cleared our title to what we possess."[29]

Smallpox recurred constantly into the eighteenth century and beyond. In 1763, when Ottawa Indians threatened Fort Pitt, a trader on the scene reported: "Out of our regard to them, we gave them two Blankets and a Handkerchief out of the Small Pox Hospital. I hope it will have the desired effect." This act was sanctioned by the British commander in chief in North American, Sir Jeffery Amherst, along with any "other method that can serve to Extirpate this Execrable Race."[30]

The smallpox epidemic had serious side effects. Secondary diseases like influenza and pneumonia, tuberculosis, measles, typhus, and dysentery attacked the weakened survivors. The population of the Huron and Iroquois confederations fell by half. It is estimated that the Indian population of New England declined from 70,000 to 12,000, and that Vermont and New Hampshire were virtually depopulated. Even those who remained healthy were often broken in spirit; as one Englishman said of the Indians near Plymouth, Massachusetts: "Those that are left, have their courage much abated, and their countenance is dejected, and they seem a people affrighted."[31]

Gradually, the imported diseases spread throughout North America. In Canada, French settlements along the St. Lawrence River in the early 1600s spread measles, influenza, and smallpox to the native populations. Later in the century, French fur traders carried the disease into the Great Lakes and beyond.[32] In the Southeast, soon after the first English settlers began enslaving the Indians, smallpox broke out in 1696–1700, reducing the native population by 1715 to such an extent that the English turned to African slaves to work on their plantations.[33]

Smallpox reached the nomadic peoples of the Great Plains from the 1770s on, killing most of those it infected and decimating the tribes that had dominated the region for centuries. Horseback-riding warriors traveled fast, carrying pathogens with them. Just as a worsening climate had reduced the population of bison—nomadic Indians' main source of food—an epidemic in the 1770s and 1780s killed 50 to 80 percent of the Cree, Arikara, Mandan, and Indians, and other tribes.[34] On the West Coast as elsewhere, disease preceded the Europeans; when George Vancouver explored Puget Sound in 1782–1783, he found piles of bones and survivors with pockmarked faces.[35] The epidemics continued, with increasing intensity, into the early nineteenth century. In 1837–1838, a river steamer brought smallpox up the Missouri River, igniting a pandemic that spread from New Mexico to northern Canada and from the Mississippi to the West Coast. Several once-powerful tribes were decimated, and some vanished entirely. As one observer wrote in 1838:

> The warlike spirit which but lately animated the several Indian tribes, and but a few months ago gave reason to apprehend the breaking-out of a sanguinary war, is broken. The mighty warriors are now the prey of the greedy wolves of the prairie, and the few survivors, in mute despair, throw themselves on the pity of the Whites, who, however, can do but little to help them. . . . Every thought of war is dispelled, and the few that are left are as humble as famished dogs.[36]

The human consequences of these epidemics were, in the words of historian Shawn William Miller, "an unprecedented human disaster that staggers the imagination."[37] As a result, the Americas were effectively depopulated, reaching a nadir of 4 to 5 million people in the mid-seventeenth century. Of this number, less than a million lived north of the Rio Grande, and 1 to 1.6 million in Mexico. Gradually, as immigrants arrived from Europe and slaves from Africa, and as the surviving Native American population became more resistant to the alien diseases, the population of the Americas began to recover. But it was not until the mid-nineteenth century that the population returned to what it had been before Columbus. As for Native Americans, there are still fewer today than there were in 1491.

What was an unmitigated disaster for humans was an opportunity for other forms of life in the New World. Not since the Black Death had germs and viruses found so many fresh victims. Animals, too, multiplied, especially newcomers like the pigs, sheep, cattle, and horses that had accompanied the conquistadors. And so did plants, especially imported weeds and grasses that proliferated in the fields that humans had once tilled. So did trees, as forests reclaimed the land.[38] Anyone arriving in Amazonia or North America or the Mayan lands around 1650 with no knowledge of the past would have thought—as did most settlers and romantic writers—that this was a world of nature, an untamed wilderness begging to be conquered, where the few humans scattered in it hardly made a difference. Only in Mexico, Central America, and western South America were there still dense populations of farmers and city dwellers—although much diminished from their pre-1492 numbers.

Plants and Pathogens

When Columbus reached Hispaniola in 1492, he and his companions were surprised to find that the natives did not grow any of the crops they were familiar with, but instead grew maize, squash, beans, and other plants unknown to Europeans. Hence, on their second voyage to the New World, the Spaniards brought with them wheat seeds, cuttings of grapevines, and olive trees—the trinity of Mediterranean staples—along with barley and rye and garden vegetables such as chickpeas, radishes, onions, carrots, cauliflower, garlic, and lettuce. They also brought Asian and Middle Eastern fruits known to Europeans: peaches, melons, figs, apricots, oranges, bananas, and others. With the exception of chickpeas and bananas, these plants did poorly in the tropical climate of the Caribbean.

The Europeans were surprised that in Mexico the Indians grew maize, beans, and squash, but also tomatoes, chili peppers, peanuts, pineapples, tobacco, and cacao trees. Wheat grew well in Mexico, but grapevines and olive trees did not. Not until the Spaniards reached Peru and Chile, where the Indians grew maize, cotton, and other staples in the lowlands and potatoes and quinoa in the highlands, did they find a climate suitable for grapevines and olive trees, as well as wheat.

The Portuguese experience in Brazil was similar to that of the Spaniards in the Caribbean. The native Indians cultivated cassava or manioc, peanuts, cashew trees, and other tropical crops. No European crops grew well there, but South Asian crops—yams, mangoes, coconuts, bananas, and citrus fruits—with which the Portuguese were familiar from their voyages to the east, did. From Africa came several other domesticated plants, among them sorghum, pearl millet, sesame, okra, and guinea pepper. In the tropics, African grasses—Angola grass, Bermuda grass, and Guinea grass—proved to be hardier and more productive fodder for cattle than native grasses. Then, in the early eighteenth century, the Portuguese introduced an Ethiopian plant for which Brazil was to become famous: coffee.

Two other crops played an important role in the history of the Americas: sugarcane and rice. Though Europeans imported Asian or white rice (*Oryza sativa*) for their own consumption, African or red rice (*Oryza glaberrima*) was brought to America by African slaves sometime in the seventeenth century. It had the advantage of tolerating poorer soils than its Asian counterpart and did not require paddies or the transplanting of seedlings. Ships that picked up slaves on the Guinea Coast purchased rice along with yams, sorghum, and millet to feed the slaves on the transatlantic voyage. Surplus rice that had not been milled and eaten on board ship was planted and harvested by the surviving slaves. By the late seventeenth century, African rice was grown in the Carolinas, Suriname, Brazil, and other tropical and semi-tropical areas.[39]

Not all the plants that came with the invaders were deliberately introduced. Weeds—opportunistic plants that quickly took hold—spread by themselves. In the West Indies, daisies, ferns, thistles, nettles, sedge, plantago, and many other weeds accompanied the invaders and found the soil and climate to their liking. In Peru, domesticated European garden plants like turnips, mustard, mint, endive, and spinach went feral and spread widely. In southeastern North America, peach trees proliferated far beyond the orchards in which they were planted.

The grasses of eastern North America and northern Mexico and the Pampas of South America had never been grazed by horses, cattle, or sheep. As they disappeared under the heavy grazing and hard hooves of the invading ungulates, they were replaced by hardier alien plants such as ragweed and bluegrass. On the Pampas, cardoon thistles (wild artichokes) grew up to three meters high, forming barriers impenetrable even to horses. Charles Darwin, visiting the region in 1833, wrote: "I doubt whether any case is on record of an invasion on so grand a scale of one plant over the aborigines." In short, alien plants behaved in the soil of the Americas like the alien pathogens in the bodies of the Native American peoples.[40]

Sugarcane

Of all the plants imported after 1492, none had as profound an impact on the economy, the peoples, and the land of tropical America as sugarcane (*Saccharum officinarum*).[41] Cultivating it is a time- and labor-intensive process. Sugarcane is a grass that can grow twice as tall as a man. Once the cane is cut, the juice has to be processed quickly before it loses its sweetness. Processing means extracting the juice by crushing the canes between rollers, then boiling it in a succession of copper kettles to evaporate the water, and finally pouring the resulting syrup into clay cones to allow the molasses to drip out, leaving sugar crystals behind.

What drove the sugar economy and its associated ecological changes was the growing European addiction to sugar. During the Middle Ages, sugar was so rare among European Christians that it was prescribed as a medicine. Spanish settlers discovered that it could grow well in the Canary Islands after their conquest in 1402, as did the Portuguese on the island of Madeira, settled in 1420. From then on, the demand for sugar kept growing, especially as Europeans became addicted to tea, coffee, and cocoa. From an average of less than 2 kilograms of sugar per person per year, consumption rose to 11 kilograms in the eighteenth century and to 18 kilograms in the mid-nineteenth century, most of it coming from the Americas.[42]

When Columbus returned to the Caribbean in 1493, he brought sugarcane cuttings to Hispaniola. The climate and soil of the island proved ideal for the plant. Emperor Charles V ordered sugar masters and mill technicians transferred from the Canary Islands to Hispaniola. By the 1530s there were thirty-four mills on the island. At the same time, the Portuguese began growing sugarcane in northeastern Brazil. By 1600, Brazil had 500 mills producing 10,000 tons of sugar a year. From 1630 to 1654, the region around Pernambuco was occupied by the Dutch, who tried, but failed, to make a profit from the sugar business. After that, Brazilian sugar production declined in the face of competition from the Lesser Antilles.[43]

The first successor to Brazil as the premier sugar producer in the Americas was Barbados. This small island, once forested, was stripped of its trees to provide fuel for the mills and to make room for sugarcane fields and pastures for the cattle that powered the mills and provided manure. By the late seventeenth century, Barbados was among the most densely populated agricultural regions of the world, with 186 inhabitants per square kilometer, two-thirds of them African slaves. Barbados sugar mills were more efficiently run than those in Brazil, and their methods were copied on other Caribbean islands held by the British and the French. In specializing in a single crop, however, the planters became totally dependent on the import of slaves, animals, food, construction materials, and even animal feed and fuel for the furnaces.[44] Deforestation caused soil erosion on hilly land. As bare soil slid downhill after every rain, the land of Barbados became so gullied that slaves were made to carry soil back up the slopes to restore the eroded fields. Nonetheless, by 1700, one-third of the once-arable land in Barbados was unrecoverable. For lack of fuelwood, planters began burning the bagasse, the cane stalks left after all the juice had been extracted, and even coal imported from England.

So lucrative was sugar, and so hard on the land, that competition from larger islands soon eclipsed Barbados. By the beginning of the eighteenth

Figure 6.1. The mill yard in a sugar plantation in Antigua (West Indies) in 1823, where sugarcane was crushed to extract the juice in making sugar. Wikimedia Commons.

century, Jamaica, a British island ten times the size of Barbados, briefly became the world's premier sugar producer. Then, after the French acquired the western third of Hispaniola from Spain (renaming it Saint-Domingue), it became the number one producer, with a population in 1791 of 32,650 whites, 28,000 black freedmen, and 480,000 slaves. [45]

The mills' voracious appetite for land and fuel had a significant impact on the islands' environments. The destructive process began on the mid-Atlantic island that Portuguese settlers called Madeira (meaning "wood"), because they found it completely covered with forests. The first thing they did was set fire to the forests in order to clear the land for farming. The fire burned for eight straight years. By the mid-sixteenth century, there was so little wood left that the inhabitants gave up sugarcane and turned instead to growing grapes for "Madeira" wine. By the late seventeenth century, on the island named for its wood, hardly a tree was left standing.[46]

Much the same happened in the land named after the brazilwood tree. Before the Portuguese arrived, the Atlantic coast of Brazil was covered by a forest stretching several hundred thousand square kilometers. Sugarcane planters cleared large areas by burning the trees. When the fertility of the soil was depleted, they moved on. Sugar mills were also voracious consumers of firewood. Boiling cane juice to make a ton of sugar required about a hundred tons of wood.[47] A typical mill consumed one huge tree every hour. Between 1550 and 1700, Brazilian sugar mills consumed 1,000 square kilometers of virgin forest plus 1,200 square kilometers of secondary and mangrove forests, and another 2,400 square kilometers between 1700 and 1850. After the trees were felled, pigs, cattle, and goats ate the saplings, preventing the regrowth of the trees. The sugar economy also consumed land and forests for secondary purposes. Sugar mills needed oxen and mules to pull carts and turn the rollers, horses for the masters and foremen, and cattle for beef to feed the workers and tallow for candles. In many places, cane fields occupied only one-third of the land, the rest being devoted to pastures, woodlots, and slave gardens. Enormous ranches occupied the interior along the São Francisco River.[48]

Sugarcane plantations had further environmental consequences. The replacement of native forests by sugarcane fields destroyed the habitats of native animals. In the West Indies, settlers hunted iguanas, agoutis, and opossums almost to extinction. On the beaches of Cuba and the Cayman Islands where hawksbill and green sea turtles came to lay their eggs, 13,000 or more of these animals were captured every year; by the eighteenth century, they were almost extinct. Likewise rats, stowaways from the Old World that had jumped ship on arrival in the Americas, destroyed the eggs of native birds and young monkeys. On small islands where native species were most vulnerable, untold numbers vanished. Deforestation also led to the desiccation of the local climate, as rainwater ran off instead of being absorbed in the soil and transpired by the leaves of trees. Planters and colonial administrators were well aware of the climate change they were causing and the need for costly irrigation systems.[49]

Deforestation, desiccation, and soil depletion explain why the center of sugar production shifted from Barbados and the smaller Antilles to Jamaica, then to Saint-Domingue, and finally, in the nineteenth century, to Cuba.

Mosquitoes and Malaria

Sugar had further consequences on the fate of the Caribbean. Two diseases introduced from the Old World—malaria and yellow fever—thrived in the sugar plantations and had deleterious effects on peoples and politics.

Malaria is caused by a parasitic micro-organism called a plasmodium. Two species in particular affected the Americas: *Plasmodium vivax*, which is debilitating but rarely fatal, and *Plasmodium falciparum*, which is far deadlier. Both are especially dangerous to newcomers and small children, but they confer a limited resistance to those who survive a first attack and are regularly re-infected. Since malaria suppresses the immune system, infected persons were vulnerable to other infectious diseases such as typhus, dysentery, and yellow fever.

Malaria is environment-specific. The mosquitoes that transmit it are common in the Americas as well as in the Eastern Hemisphere. The female mosquitoes drink the blood of cattle, mules, and other livestock as well as humans, and lay their eggs in any stagnant water bathed in sunlight, such as ponds, ditches, and canals. Thus, the environmental changes wrought by plantation agriculture—deforestation, irrigation works, and rice paddies—suited the malaria-carrying mosquitoes perfectly.

Malaria affected Europeans and Africans very differently. Tropical Africans, over thousands of years of evolution, had developed immunity to *vivax* and a resistance to *falciparum* called sickle-cell trait, which blocked the plasmodium from invading the red blood cells. While *vivax* was found in western Europe, *falciparum* was a tropical pathogen that required a temperature of 19°C (66°F) to survive and 22°C (72°F) to reproduce during the short life span of a mosquito. Hence the disease spread only in tropical and semitropical areas. In North America, *falciparum* malaria reached as far north as the Mason-Dixon line; beyond that, only *vivax* malaria, imported from Europe, became endemic.[50]

Yellow Fever

Another insect-borne disease, yellow fever, may have killed fewer people, but it struck greater terror in New World populations than malaria because it appeared at random moments and was even more selective in its victims. Among children, yellow fever was normally a mild disease that, once acquired, conferred lifelong immunity. A non-immune adult who contracted it, in contrast, suffered three or four days of high fever and intense pain, followed by either recovery or—in up to 85 percent of cases—jaundice, internal and external bleeding, black vomit, coma, and death.

Yellow fever originated among African forest monkeys and was spread by the mosquito *Aedes aegypti*. This insect cannot survive under 10° or over 40°C (50–104°F), prefers temperatures between 27° and 31°C (81–88°F), and seldom flies more than a few hundred meters. It was not native to the Americas but stowed away on ships bringing slaves from Africa. In the Caribbean, it found its ideal niche. Its eggs hatched in the pots, cisterns, barrels, and buckets in which people stored water during dry spells, and in puddles during the rainy season. Only females drank the blood of humans and monkeys, but both males and females ate sucrose, which they found in abundance in cane stalks and in the clay pots in which sugar mills stored sugar while the molasses dripped out.

Like malaria, yellow fever attacked Europeans much more frequently than Africans. Though contemporaries did not know it, this was because Africans had almost all been infected as children and had thereby become immune. Children born in the West Indies who were infected, whatever their race, also acquired immunity if they survived. Hence, the people most vulnerable to the disease were adult newcomers from Europe.[51]

Yellow fever did not break out in the Americas until the 1640s because it required bringing together an infected person, several non-immune passengers, and sufficient mosquitoes on a ship crossing the Atlantic, with a large enough population of non-immune humans upon arrival in a tropical setting. This happened on Barbados in 1647, where it killed 6,000 people, one-seventh of the population. Then it flared up in Guadeloupe, St. Kitts, Jamaica, and Cuba, and on the Yucatán Peninsula. In Havana, 536 whites died, but only 26 blacks. In general, the disease preferred the smaller islands where sugar mills provided an excellent incubator of *Aedes aegypti*. Then, after having killed or immunized most of the non-immunes, it disappeared for many years. It reappeared in Brazil in 1685 and in the West Indies in 1690, causing most surviving whites to flee and plantation owners to switch from indentured Europeans to African slaves.[52] Thereafter it recurred periodically, not only in the tropics but also in the port cities of North America and Europe such as Québec in 1711, Dublin in 1726, and Philadelphia in 1793, brought by ships from the Caribbean during hot summers.[53]

Political and Social Consequences

The diseases that came from the Eastern Hemisphere in the fifteenth and sixteenth centuries have been called the "swords of civilization," because they benefited the invaders at the expense of indigenous peoples. Malaria and yellow fever had the opposite effect. Local populations that had been infected as children and survived were much less vulnerable than newcomers. Hence these diseases shielded the local populations from invaders. As environmental historian John McNeill has shown, this had momentous strategic consequences. The two diseases, especially yellow fever, protected the Spanish empire from its rivals England and France long after Spain had lost its naval supremacy. In McNeill's words, "Without it [yellow fever], Spain might well have lost much of her American empire in the eighteenth century."[54] It avoided that fate by thwarting one invasion after another.

In 1655, England sent an invasion force of 9,000 men to conquer Jamaica, a poorly defended backwater of the Spanish empire. Within three weeks, some 3,000 men were sick; after six months, only 3,720 were still alive, of whom 2,000 were "sick and helpless."[55] It was the last successful major invasion in Caribbean history before the nineteenth century. After that came a string of failures: attacks on Guadeloupe, Martinique, Saint-Domingue, Cartagena, and Darien in the late seventeenth century; on Portobelo, Cartagena, Santiago de Cuba, Havana, and Nicaragua in the mid-eighteenth; and, most famously, Napoleon's attempt to reconquer Haiti in 1802–1803. All ended in massive loss of lives and ignominious withdrawals. Most of the victims were conscripted soldiers and press-ganged sailors.[56]

These fiascos speak less to the comparative strength of the European powers and more to the protection from disease carried by people born in the West Indies, regardless of race. The other consequence of malaria and yellow fever was ethnic. The first Europeans to invade the Antilles thought of using Indians as forced laborers. When the Indians died out, planters turned to European indentured servants. Only after malaria got a foothold in the Caribbean did importing Africans become the preferred solution to the labor shortage. Malaria and yellow fever together tipped the demographic balance in the Caribbean toward Africans and their descendants. To a lesser degree, the same was true of the semi-tropical colonies of North America. Wherever malaria and yellow fever were prevalent—in Virginia, Carolina, and Georgia, Spanish Florida, French Louisiana, and the colonies of Central and South America bordering the Caribbean Sea—Africans had higher odds of survival than Europeans, and both higher than Native Americans. As deaths among slaves outnumbered births, sustaining the slave population depended, until well into the nineteenth century, on constant imports of new slaves from Africa. In the long run, the result was the Africanization of most of the Caribbean islands and the large percentage of people of African descent in the surrounding regions.[57]

Animals, Native and Invasive

Animals abounded in the New World, but very few had ever been domesticated. Most of the large herbivores in the Americas had vanished soon after humans had arrived, and those that remained—moose, elk, and bison—were found only in North America and were never domesticated. Other parts of the Americas possessed substantial vacant niches. The largest animals to be domesticated were two South American camelids: the llama, which provided meat and carried goods, and its smaller cousin, the alpaca, which produced fine wool. Elsewhere, Native Americans had small hairless dogs, guinea pigs, turkeys, and muscovy ducks. These New World animals were of much less use to Indian societies than the many different domesticated animals found in the Eastern Hemisphere.

Pelts and Hides

When Europeans arrived, they found American animals of little economic value, except for fish and mammals whose furs or hides could be sold for a profit. Settlers and traders sought the skins of seals, moose, elks, and deer and the furs of foxes, martens, raccoons, bobcats, wolves, badgers, and other carnivores.

No wild animals were in as great a demand as beavers. Like humans, beavers transformed landscapes. They ate the bark, leaves, and twigs of willow, aspen, poplar, birch, and alder trees. They also used the trunks of trees they felled with their massive incisors in order to build lodges in streams and lakes, where they lived in large family groups safe from predators. The dams they built to block the flow of streams formed ponds that gradually filled with silt and vegetation, creating "beaver meadows," bogs, and wetlands that controlled floods and supported a large diversity of plants, animals, and birds.

Before the Europeans arrived, there were an estimated 60 to 100 million beavers in North America. The Indians hunted beavers, but not enough to deplete their numbers. On the western prairies, they had learned to avoid decimating the beaver populations because beaver ponds were their only source of water during dry spells.

This changed with the arrival of English and French traders and trappers. From the late 1500s to the 1840s, beaver hats were in great demand among fashionable Europeans. As European beavers were soon hunted to extinction, the trade turned toward the far larger population of North American beavers (*Castor canadensis*). Most highly prized were beaver pelts that had been worn for a year or two by Indians, as this removed the coarse guard hairs, leaving only the soft and dense wooly undercoat. By the mid-seventeenth century, beavers were being hunted to extinction in southern New England and along the St. Lawrence River, as were most other fur-bearing animals. Indians joined in the hunt as a means of obtaining European knives, hatchets, kettles, and wool blankets, but also guns, gunpowder and shot, and rum and brandy. By the mid-seventeenth century, Algonquian hunters had severely depleted northeastern North America of these animals. From there, the trade moved west into the lands of the Huron, the Mohawk, and other tribes. French *coureurs des bois* carried the trade west to the Great Lakes, south into the Mississippi Valley, and north to Hudson Bay. The number of beaver pelts harvested rose from 32,000 in 1627 to an average of 122,000 pelts per year in the period 1700–1763. Then the number plummeted to 4,087 a year between 1830 and 1849.[58]

The other animal in great demand was deer, common to the forests of eastern North America. Eastern woodland Indians had traditionally hunted deer for their meat and their hides, which made supple but durable clothing and footwear. With the European demand for deerskins and the sale of muskets to Indians, the number of deer they killed soared. In a good year, a Creek hunter could bring in seventy-five deerskins in a season. In the southeast, exports of deer hides reached 250,000 to 300,000 per year in the 1760s and '70s, dropped to 175,000 per year in the 1780s, and collapsed in the early 1800s. Likewise, moose were also severely depleted in the northeast in the mid-seventeenth century.[59] What saved these animals from total extermination was the decimation of the Indians by alien diseases and the rising imports of factory-made cloth starting in the late eighteenth century.

Old World Animals in the Caribbean

On Columbus's second voyage to Hispaniola in 1493, he brought with him a menagerie of European barnyard animals: horses, cattle, donkeys, pigs, sheep, goats, chickens, cats, and dogs. Rats came along uninvited. When these animals were let loose to forage for themselves, there ensued an unprecedented explosion of animal life.

Between 1493 and 1512, several hundred head of longhorn Spanish cattle survived the trip across the Atlantic. More tolerant of the tropical climate than other European livestock, they proliferated. In a land of rich pastures and no predators or diseases, cows calved after one year instead of three or four years as in Europe. Not only could they turn grass into meat, milk, and leather, but they could also be used as draft animals. Within a few years cattle ranchers were exporting hides to Spain.

Wherever they could find water and shade, swine did even better than cattle. They ate almost anything apart from grass: fruits, vegetables, insects, eggs, and small animals. Healthy sows gave birth to ten or more piglets at a time. Efficient foragers, pigs converted one-fifth of what they ate into flesh that humans found nourishing and flavorful. Left to run without supervision, they turned feral and were hunted like wild animals. Mariners dropped off pigs on islands they visited, knowing that they would provide meat for later visitors or for shipwrecked crews and early settlers. When swine were introduced to Virginia, it is said that they "swarm like Vermaine upon the Earth. . . . The Hogs run where they list and find their own Support in the Woods without any Care of the Owners."[60]

These animals had a rapid and dramatic impact on the environments of the West Indies. Cattle trampled and overgrazed the native grasses, replacing them with scrub palms, guavas, acacias, cardoon thistles, and other cattle-resistant weeds, or leaving bare ground that eroded, forming gullies. Pigs feasted on the gardens of the Tainos, contributing to their demise. Feral dogs and cats became predators of the local hutias, iguanas, and birds. Black rats, stowaways on ships, destroyed crops and food supplies. Within a short time, animals replaced the native inhabitants of the islands and altered the ecosystems they invaded.[61]

Insects

Tiny animals brought over by Columbus and his successors also found the New World to their liking. There were many others besides mosquitoes. Starting in 1518–1519, a plague of stinging ants invaded the houses of the small Spanish community of Hispaniola and destroyed their crops of oranges, pomegranates, and cassias. These were tropical fire ants (*Selenopsis geminata*), probably native to the island. Later, in the eighteenth century, two other species of ants, the native *Pheidole jelskii* and the imported African

Pheidole megacephala, infested Barbados, Martinique, and Guadeloupe. These ants lived off the excrement of mealy bugs and sapsucking coccids that ate the sugarcane and had probably been introduced from overseas. In the words of entomologist Edward O. Wilson: "These pests, at first unopposed by any parasites or predators natural to them, bloomed into dense populations. The ants, profiting from the increased food supply, similarly flourished."[62]

Unlike mealy bugs and ants, honeybees were deliberately imported to the Americas. In 1622, a ship arrived in Jamestown, Virginia, with a hive of bees brought over to pollinate imported plants such as peaches, apples, and watermelons. From there, they spread partly by humans and partly on their own, reaching the Mississippi Valley by the early nineteenth century.[63]

Finally the lowly earthworm was perhaps inadvertently brought from Europe in the eighteenth century in a ship's ballast or in the soil around a seedling's roots. From the East Coast of North America they spread across the continent, causing an increase in soil nutrients and large shifts in the composition of forest vegetation.[64]

Beasts of War

Dogs and horses helped the Spaniards conquer a continent. The dogs the Spaniards brought to the New World were not the small hairless Chihuahuas the Indians were used to, but mastiffs, wolfhounds, and greyhounds. When messengers from the Aztec emperor Moctezuma returned to Tenochtitlán after meeting with the Spaniards who had landed on the coast in Veracruz, they reported seeing "very big dogs, with floppy ears, long hanging tongues, eyes full of fire and flames, clear yellow eyes, hollow bellies shaped like spoons, as savage as devils, always panting, always with their tongue hanging down, spotted, speckled like jaguars."[65] The Spaniards used such dogs to inspire fear and to attack Indians in forests and rocky terrain where horses could not go, and the invaders did not hesitate to feed these dogs the flesh of Indians.[66]

Horses were more difficult to breed in the tropics than other domesticated Old World animals. Like humans, they sweated in hot weather and therefore needed to drink a lot of water. When ships carrying horses across the ocean were becalmed in the mid-Atlantic doldrums for weeks at a time, the crew had to sacrifice their horses for lack of fresh water; as a result, this part of the ocean became known as the "horse latitudes." In the tropical climate of the

Caribbean, horses reproduced slowly. In 1501 there were twenty to thirty on Hispaniola, and by 1503 there were still only sixty or seventy of them. Yet they were extraordinarily important to the European invaders, not only as means of transportation and weapons of war, but also as status symbols: on a horse, a Spaniard was a *caballero*, a gentleman. The Spanish crown established breeding farms in Hispaniola and later in Cuba to supply horses for the expeditions to the mainland.[67]

When Cortés first arrived in Mexico, he brought sixteen horses with him. Pizarro, in his attack on the Inca Empire, had twenty-seven. So valuable were they that one was worth sixty times as much as a sword and entitled its owner to a much larger share of the treasure the invaders found. The horses, the largest animals the Indians had ever seen, filled them with dread, not only because of their size, but also because they could outrun any Indian sentry or scout, allowing the Spanish to attack by surprise.[68] The Inca Garcilaso de la Vega recalled the Incas' first encounter with horses:

> And so nothing convinced them to view the Spaniards as gods and submit to them in the first conquest so much as seeing them fight upon such ferocious animals—as horses seemed to them—and seeing them shoot harquebuses and kill enemies two hundred or three hundred paces away.... They took them for sons of the Sun and surrendered with so little resistance as they did.[69]

Horses and Indians

The Spaniards did not long maintain their monopoly on the use of horses. The highlands of central and northern Mexico proved to be an ideal environment for these animals. By 1550, there were 10,000 of them. This was also where the Chichimec Indians, semi-nomadic hunters and gatherers, learned to capture and ride horses. Once mounted, they waged a fifty-year war against encroaching Spanish silver miners.

Before the Indians of the Great Plains of North America had horses, they lived very poorly, cultivating maize, beans, and squash in river valleys and supplementing their vegetarian diet with very difficult annual hunts for bison. The first horses were brought to New Mexico by Spanish missionaries in the early seventeenth century. Though the government of New Spain prohibited the sale of horses to Indians, some were lost, stolen, or exchanged for animal hides or slaves. Even then, many years elapsed before they reached the Plains and before enough Indians learned to ride them to become what were later called "horse Indians." There were horses in Texas by the 1680s, in the southern Plains by the early eighteenth century, and as far west as the Rocky Mountains and as far north as Saskatchewan by the late eighteenth century. Once the Shoshone, the Comanche, and other mountain peoples had obtained horses, usually by capturing mustangs (feral horses), and learned to break and ride them, they descended into the Plains and became full-time bison hunters. From then on, they dominated the Plains and kept the European-Americans at bay until well into the nineteenth century.[70]

The same pattern is found in the history of South America, especially in the Pampas of Argentina. The first Spaniards, led by Pedro de Mendoza, landed on the south bank of the Río de la Plata in 1536 with 2,000 men and 71 horses. The local Querandí Indians were friendly until Spanish demands provoked a rebellion that drove the Spaniards out. When the next expedition under Juan de Garay landed in 1580, they found the Pampas taken over by the descendants of horses either left behind by Mendoza or coming from earlier European settlements in Paraguay, Chile, or Brazil.[71]

The Pampas were paradise for horses, with few herbivorous competitors, no predators, and no diseases to restrain their reproduction. In this environment, they multiplied rapidly. According to a Spanish historian, in the early seventeenth century in northern Argentina there were wild horses "in such numbers that they cover the face of the earth and when they cross the road it is necessary for travellers to wait and let them pass, for a whole day or more, so as not to let them carry off tame stock with them" and that the plains around Buenos Aires were "covered with escaped mares and horses in such numbers that when they go anywhere they look like woods from a distance."[72] In the mid-eighteenth century, an English visitor wrote: "There is likewise a great plenty of tame horses and a prodigious number of wild ones. . . . The wild horses have no owners, but wander in great troops about these vast plains. . . . During a fortnight, they continually surrounded me. Sometimes they passed by me in thick troops, on full speed, for two or three hours together."[73]

Sheep and Cattle Ranching

Once they conquered the native populations, the Europeans settled down to rule and exploit their new empire. In New Spain (their name for Mexico), the Spanish settlers tried to reproduce the economy of Spain. When farming became more problematic as the Indian population shrank, they turned to raising cattle and sheep. Sheep had done poorly in the Caribbean but were well adapted to the drier climate of Mexico and were valued for their wool, sheepskins, and meat. The first viceroy of New Spain, Antonio de Mendoza, imported merino sheep to propagate the interior of the country; sheep were later introduced into Peru, Chile, and Río de la Plata (Argentina).

Thanks to the work of environmental historian Elinor Melville, a great deal is known about sheep raising and its environmental consequences in a region of Mexico called Valle del Mezquital, north of Mexico City. This valley was once a mosaic of fields, woods, and native grasslands densely populated by Otomí Indians. They irrigated their fields with an intricate system of terraces, dams, and canals fed by springs. The sheep the Spaniards introduced into the valley multiplied rapidly, eating the native grasses and the Otomís' crops. As the Indians died of diseases or fled the labor demands of the Spanish landowners, the area devoted to ranching rose from 2.6 percent in 1549 to 61.4 percent in 1599. By the end of the sixteenth century, the native population had shrunk by 90 percent, replaced by sheep, shepherds (most of them African slaves), and sheep dogs.

The ranchers encouraged the multiplication of sheep beyond the carrying capacity of the land. From the mid-1560s to the end of the 1570s, the average size of flocks rose from 3,900 to 10,000 and the total sheep population increased to several million. The excess numbers destroyed the native plants and the irrigation systems of the Otomí, leaving bare soil that was then invaded by cacti, thorny shrubs, cardoon thistles, and mesquite bushes that resisted browsing animals. Where the forests disappeared, the land eroded. Flock size dropped by 63 percent and animal density by 72 percent, and the remaining animals were smaller than before. To save their remaining sheep, ranchers began to move their flocks to Michoacán during the dry season. After that, the migration of sheep allowed pastoralism to settle at a lower but sustainable level until the eighteenth century.[74]

The other animals raised on Latin American ranches were cattle. The long-horn cattle brought over from Iberia were adaptable to various climates and provided meat, hides, traction for mills and carts, and tallow for candles. In central Mexico, ranchers moved cattle from the interior during the wet summers to the more humid Gulf Coast in the dry season. In the Valley of Oaxaca, cattle trampled the fields and ate the crops of the Indians to such an extent that Viceroy Mendoza wrote to the king of Spain: "May your

Lordship realize that if cattle are allowed, the Indians will be destroyed."[75] In northern Mexico, cattle are said to have doubled in number every fifteen years; as in many other places, they roamed free and survived on their own. As Europeans poured in to exploit the silver deposits of the sierras, ranchers raised up to 150,000 head of cattle to feed the miners. In the Pampas of South America, where there were no native herbivores to compete with them, cattle also multiplied rapidly. In 1619, the governor of Buenos Aires estimated that 80,000 could be killed every year for their hides without decreasing the herds. By the late eighteenth century, the Río de la Plata was exporting a million hides a year. After the Portuguese introduced livestock into the region of São Paulo and the São Francisco Valley of Brazil in the 1530s, cattle ranches replaced Indians and trees. The *llanos* or grasslands of Venezuela supported an estimated 140,000 head of cattle by the mid-seventeenth century.[76]

Mines and Forests

The Spanish conquistadors came looking for gold. Instead, they found silver in prodigious quantities. The silver they found in Mexico and the Andes, and the gold and diamonds the Portuguese later found in Brazil, kept these two nations in the ranks of the European powers for two centuries. The wealth of the Iberians came at the expense of the regions and peoples from which this bonanza came.

Silver in Spanish America

Between the sixteenth and the early nineteenth century, New Spain produced 50,000 metric tons of silver and 800 tons of gold, along with lead, copper, and other non-ferrous metals. In the 1520s and '30s, silver brought Spanish settlers into the Sierra Madres, mountain ranges running parallel north and west of Mexico City, where they founded the cities of Guanajuato, San Luís Potosí, and Zacatecas and hundreds of smaller mining centers. Likewise, it was silver that brought Spanish prospectors in the 1540s to the Cerro de Potosí, a mountain 4,000 meters up in the Andes. From that huge mine, which began operating in 1546, they extracted 41,000 metric tons of silver. In its heyday, the city of Potosí was the largest in the Americas, even larger than Madrid, Paris, or Rome.

Silver was extracted from the ore by amalgamation, a process toxic to humans and the environment. The ore was first ground into powder, then mixed with mercury, water, salt, and iron filings. To extract the silver from Potosí required tons of mercury. Fortuitously, the greatest mercury deposit on Earth was discovered at Huancavelica in central Peru in 1563 and produced 100,000 tons of the metal in the three centuries after 1572. Several weeks after the silver and mercury had amalgamated, heating the amalgam released mercury vapors, leaving pure silver behind. Though some mercury was recovered, much evaporated. From the mid-sixteenth to the mid-nineteenth century, the mines of Mexico consumed 150 metric tons of mercury a year. Miners died within a few years of mercury poisoning, silicosis, and other mining hazards. Around the silver mines, vegetation was also poisoned, as were fish downstream from the mines. The effects of mercury poisoning were also felt in the mines of Huancavelica, from which came most of the mercury used in the Americas.[77]

Though the worst impact of silver mining was local, the miners' demand for wood affected the environment over a large area. Most of the wood the miners consumed was used to speed up the process of amalgamation by heating the mixture of ore and mercury and to separate the mercury from the silver. It was also used to shore up mine shafts and to heat miners' dwellings and cook their meals. In a dry land like north-central Mexico, trees grew slowly, and reforestation was hampered by browsing animals. According to calculations by Daviken Studniki-Gizbert and David Schecter, to produce one ton of silver required 6.332 square kilometers of forest. Between 1558 and 1804, the officially registered mines consumed 315,642 square kilometers of forests, an area larger than Poland or Italy—and this number did not include wood use in the many small unregistered mines. It was the *carboneros* or charcoal-makers, more than the miners, who turned the forests of north-central Mexico into scrub lands and sheep pastures.[78]

Gold and Diamonds in Brazil

Mining also affected Brazil, but in a different way. In 1695, gold was discovered in gravel beds in the region of Minas Gerais (Portuguese for "mines all around") north of Rio de Janeiro. A gold rush soon attracted thousands of Portuguese immigrants and their African slaves to the area. By the 1720s, half the colonial population of Brazil was living in Minas Gerais, either mining or farming and ranching to provide food for the mining camps and trading posts. Miners diverted rivers and burned trees to get at the gravel deposits, while farmers and ranchers practiced a rough form of slash-and-burn agriculture.

Then came the discovery of diamonds in the 1720s. For several years, a diamond rush caused the price of diamonds in Europe to collapse. To protect its income, the Portuguese crown ordered the diamond districts evacuated and allowed in only a small number of miners, their African slaves, and armed guards. This slowed the growth of population as well as of land clearing. Despite these restrictions, by the time the boom ended in the early nineteenth century, much of Minas Gerais had been turned into a barren plateau pockmarked with the remains of the gold and diamond mines.[79]

Brazil's Atlantic Forest

The mining districts were not the only places in Brazil where Portuguese civilization brought deforestation. When the Portuguese first arrived, they found the Atlantic forest stretching from Recife in the north to Rio in the south, an area 2,300 kilometers long by 100 wide. Among the broadleaf evergreen trees that grew in the forest were the *Caesalpina echinata* or brazilwood tree that furnished a valuable red dye. Portuguese and other European ships came to buy brazilwood, jaguar skins, and slaves from the Tupi, in exchange for iron axes and other trade goods. In the sixteenth century, they bought 8,000 metric tons of brazilwood a year. Soon, however, diseases reduced the Tupi population by over nine-tenths, slowing down the attack on the forest.

In 1605, the Portuguese crown imposed a monopoly on the brazilwood trade, giving a few traders the right to sell 600 tons of the wood a year in order to keep prices high. Though some landowners burned their trees rather than let the officially designated contractors take them, the overall effect was to slow down the cutting of the trees. Likewise, the government encouraged latifundia, or huge estates, as a means of preventing the poor from becoming independent small farmers. In the words of environmental historian William Miller, "For four centuries, and sometimes longer, the powerful successfully locked nature away from the disinherited masses. . . . And in some cases it was the very greed of kings and the avarice of landholders that limited the production of colonial goods and, thereby, the despoliation of American nature." As in the case of diamond mining, a royal monopoly stood in the way of economic development, hence of environmental destruction.[80]

New England

The development of New England followed a different path from that of Mexico, Brazil, or other parts of the Americas. Here, the English settlers found no precious stones or metals, nor did they create a mining or plantation economy based on forced labor. Instead, they came to create an idealized vision of English village life free from government interference in their religious beliefs. Their first goal was to farm the land, which meant clearing trees. They preferred the soil under forests of hickories, ashes, maples, and beeches. Oak and chestnut forests had poorer soils, and farmers avoided land covered with conifers and thorny bushes. As elsewhere, the ashes of burnt trees and vegetation fertilized the soil for a year or two, after which it had to be fertilized with fish refuse and the manure of barnyard animals.

Agriculture was the main, but not the only, reason people felled trees. Houses were built of wood, often of logs, as were fences. Typical rural homeowners burned thirty to forty cords (109 to 145 cubic meters) of wood a year in their fireplaces, a consumption that struck visitors from Europe as unbelievably wasteful. Wood, turned into charcoal, fed the iron furnaces that dotted the landscape. Making 1,000 tons of pig iron required 2 to 5 square kilometers of forest; by the late seventeenth century, iron-making was consuming 6,000 square kilometers of forest a year.

New England farmers were not self-sufficient but needed to trade, and what they sold was the natural resources of the land. Once the fur trade petered out, the important item of trade was wood. In some areas, lumbering became the main economic activity, for New England proved rich in the kinds of trees needed by shipbuilders and carpenters. White oak was used for ships' timbers and planking and barrel staves for West Indian sugar and Madeira wine; black oak for underwater ships' timbers; cedar for wood shingles, clapboards, fence posts, and other outdoors uses; and pitch pine for pitch and turpentine. The ashes of broadleaf trees were refined into potash, used in making glass, bleach, gunpowder, and other industrial products. White pines were especially valuable as ships' masts. By the mid-seventeenth century, as England was becoming a major naval power, it began to deplete the supply of Scotch firs from the Baltic and shifted its attention to the white pines of Maine and New Hampshire. The Royal Navy reserved the best trees, those that grew to 2 meters in diameter and 36 to 60 meters high; as in Brazil, the monopoly was often violated by the colonists. By the early nineteenth century, the white pines were almost depleted.[81]

Though the goals and motivations of the English settlers differed from those of the Iberians, their impact on the environment was just as profound. The first consequence of their arrival, as in the rest of the Americas, was the collapse of the Indian population. This, in turn, ended the Indian practice of using fire to keep forest clearings free of underbrush in order to attract deer and other animals. As the Indians died, underbrush took over their fields and forest clearings and the wild animal population declined. Deforestation by English farmers changed the climate, which became hotter and drier in the summer and colder and windier in the winter. As snow melted earlier and more quickly in open fields than in forests, flooding increased. The flow of streams and springs became more irregular, and swamps developed where mosquitoes could breed.[82]

Amazonia

While the environmental changes wrought by Europeans in those regions of the New World they occupied have been studied, what happened out of sight of European observers was in fact more important. They saw Indians dying, but diseases spread far beyond the frontiers of European exploration and occupation. And as Indians disappeared, the rest of nature filled the voids they left. Terraces collapsed. Meadows and fields returned to forest. Monumental buildings became overgrown with vegetation, as in Guatemala and the Yucatán Peninsula where the Mayan civilization had once flourished. As the forest grew in place of fields and over buildings, it greatly increased the biomass of the region.

Among the most important changes is what happened in Amazonia. In the fifteenth century, the Spanish explorers who descended the Amazon River saw large towns and many people, but found no gold or silver. Unlike the peoples of the Andes or Mesoamerica, the inhabitants of the Amazon had no stone to build with, so they left no great monuments behind. Whatever they built was of earth and wood, materials that deteriorated quickly or returned to the soil as soon as they were abandoned; their remains and their burial sites have also vanished. When European-Americans began to explore the

Amazon rainforest again in the nineteenth century, they found few traces of ancient sites and only a tiny number of Indians living from hunting, fishing, and gardening. In short, what had once been the site of a well-developed civilization had returned to the forest. The result was an enormous increase in the biomass of Amazonia, more than enough to counteract the deforestation of Mexico, New England, and other areas occupied by Europeans. As trees took the place of people, they left the New World more forested in 1800 than it had been before Columbus arrived.[83]

Conclusion

Examining the Americas between 1493 and the early nineteenth century from a biocentric point of view reveals that a few Europeans benefited disproportionately from the invasion, as did a few European nations, while many others—indentured servants, press-ganged sailors, and conscripted soldiers—suffered or lost their lives in the process. For Africans, the invasion of the Americas was a tragedy. And for Native Americans it was a calamity unlike any other in the history of the world. Overall, for human beings in the Americas, it was a disaster, leaving fewer people in the early nineteenth century than before Columbus arrived.

For other living beings, the invasion brought dangers and opportunities. In some areas, native grasses were replaced by invasive plants and weeds. Many Old World animals, especially ungulates, flourished. Some areas—central Mexico, New England, Brazil's Atlantic coast, some Caribbean islands—were deforested. But elsewhere, nature rebounded and trees returned to places once cleared by humans. All in all, where humans retreated, nature advanced.

Yet the environmental history of the Americas cannot be told in isolation from the rest of the world, for the impact went both ways, as an examination of the Old World environments shows.

7

The Transformation of the Old World

While the New World was being altered by the introduction of new diseases, plants, and animals and by the collapse of the indigenous population, the Old World in the early modern period was undergoing two momentous transformations, one natural and the other man-made. The natural transformation was the Little Ice Age, a centuries-long period of cold climate that caused crop failures and hunger and contributed to political and religious upheavals throughout Eurasia. The man-made transformation was the transfer of new crops from the Americas—maize, manioc, potatoes, and many others—that provided more food and led to a growth in the Eurasian population, thereby compensating for some of the effects of the Little Ice Age. The growing population, in turn, demanded more land at the expense of wetlands and forests.

The Little Ice Age

The expression "Little Ice Age" originally referred to the growth of glaciers in the Northern Hemisphere during the sixteenth and seventeenth centuries. Since then it has been broadened temporally to include the entire period between the late thirteenth and the mid-nineteenth centuries. Within that long stretch of time, some historians emphasize a phase of more extreme weather from the late sixteenth to the early eighteenth century.[1] In either case, "ice age" refers to unusual fluctuations in the weather, with many abnormally cold winters, cool summers, droughts, and floods that caused havoc in Europe, China, the Middle East, and elsewhere.

Global Cooling

The Little Ice Age was a global phenomenon. After the 1590s, average temperatures in the Northern Hemisphere, especially in Europe, the Middle

East, and China, were cooler than before 1300 or during the twentieth century, and even more so in the higher latitudes of Scandinavia and Siberia.[2] Climate historian Hubert Lamb called the period 1550–1700 "the coldest regime . . . at any time since the last major ice age ended ten thousand years or so ago."[3]

Beginning in the 1560s, the Middle East suffered through a series of freezing winters and, from 1591 to 1596, the longest continuous drought in six centuries, followed by several cold wet winters. The winter of 1620–1621 was so cold that the Bosphorus froze over.[4] China experienced not only cooler temperatures but also severe droughts in several provinces.[5] In Europe, winters became longer and colder, often with exceptional snow accumulations. Glaciers advanced, while lakes, rivers, and canals froze over. Drift ice isolated Greenland and Iceland for months every year and even interrupted shipping in the North Sea.[6] In sub-Saharan Africa, the Sahel bordering the Sahara Desert enjoyed increased humidity until the early seventeenth century. Rain was more plentiful, the level of Lake Chad was 4 meters higher than its mid-twentieth-century mean, population grew, and kingdoms were more stable than at other times. Then began the great drought that has continued to this day.[7]

Though it lasted for centuries, the Little Ice Age reached its nadir in the seventeenth century. Europe suffered an unusually cold winter in 1620–1621, followed by cold wet summers in 1627 and 1628, followed by rain, then drought, from 1629 to 1632. In China, subtropical Fujian province was hit by a heavy snowfall in 1618. India suffered a drought in 1630–1631, then catastrophic floods in 1632. From the 1640s to the 1660s, conditions were worse. Exceptionally wet weather and floods affected Europe from southern Spain to the Netherlands. Droughts and famines hit West Africa and Angola. In 1641, the Nile reached its lowest level ever, while drought hit northern China.[8] Winters in the Northern Hemisphere were 1 to 2°C (1.8 to 3.6°F) colder than they had been. Russia, India, East Africa, and southern and western Europe experienced cold summers and storms, as did eastern North America. The winters of 1641 and 1642 were exceptionally cold in New England, Scandinavia, and East Asia, and the summers of 1641 to 1643 were also among the coldest. Droughts afflicted Indonesia, China, Mexico, Africa, and Europe. The weather remained cold, off and on, until the early eighteenth century.[9]

The Evidence

One source of information about the changing climate is documents left by people living at the time who observed these changes. In Europe, as the temperatures dropped, the growing season was reduced by five weeks, leading to more frequent crop failures. The maximum altitude of cultivation was lowered by almost 200 meters and the northernmost limit for vineyards retreated 500 kilometers from its medieval maximum. In the Alps, glaciers advanced. In northern Europe, lakes and rivers froze and even, in places, the sea; in early 1658, the army of Charles X Gustavus of Sweden, cavalry and all, marched across the ice from Jutland to Copenhagen. Around Iceland, where there had been little or no sea ice during the Middle Ages, ice blocked the fishing fleets most winters from the 1460s on. As the sea temperature dropped, schools of herring migrated south from the Norwegian Sea to the North Sea.[10] Pieter Bruegel the Elder documented the cold weather in

Figure 7.1. *Hunters in the Snow* by Flemish artist Pieter Bruegel the Elder (1565) during the Little Ice Age. Bridgeman/Kunsthistorisches Museum, Vienna.

his painting "Hunters in the Snow" (1565); other Dutch artists represented snowy landscapes and ice skaters on frozen canals.[11]

Backing up the evidence left behind by humans are new forms of evidence from natural archives. Ice cores taken from glaciers in Greenland, Antarctica, the Andes, and Tibet and tree-rings throughout the Northern Hemisphere show colder weather than during the Medieval Climate Anomaly or the twentieth century.

Causes of the Little Ice Age

Contemporaries saw in these natural events proof of divine displeasure at the sinful behavior of humans, such as adultery, sodomy, and theatrical performances. To atone for such sins, some engaged in witch-burning, anti-Semitic pogroms, and self-flagellation. Others, more secularly inclined, blamed unusual natural events such as eclipses, earthquakes, and comets.[12]

Today, experts on climate change offer four scientific explanations: orbital forcing, variations in solar radiation, volcanic eruptions, and changes in atmospheric greenhouse gases. Orbital forcing refers to changes in the Earth's orbit around the sun; when the Earth's elliptical orbit becomes more elongated and the Earth is farthest from the sun, the amount of sunshine reaching it diminishes. This is one of the long-term processes that caused the "big" ice ages. In the centuries of the Little Ice Age, the cooling it produced would have been 0.1–0.2°C (0.18–0.36°F) in the Arctic, and much less in the temperate latitudes.[13]

Also affecting the amount of sunshine the Earth receives are variations in the amount that the sun emits. These changes produce visible effects. Thus from 1645 to 1715, the aurora borealis, a spectacular display of light in the high latitudes, was rarely seen, while the sun's corona, normally brightly visible during a lunar eclipse, was replaced by a dull reddish ring around the moon. In 1716, the astronomer Edmund Halley saw an aurora borealis for the first time in sixty years of observation.[14]

Sunspots—dark stains on the surface of the sun—are another visible sign of solar activity. While sunspots are darker than the rest of the sun, their edges are considerably brighter; the more numerous and the larger the sunspots on the surface of the sun, the more energy it emits. Nowadays as many as fifty to a hundred sunspots may be observed in any given year. Between 1460 and 1550, a period astronomers call the Spörer Minimum, there were far fewer.

Likewise, during the Maunder Minimum from 1645 to 1715, there were years when few or no sunspots were seen. John Flamsteed, England's first Astronomer Royal, noted in 1684: "These appearances, however frequent in the days of Scheiner and Galileo, have been so rare of late that this is the only one I have seen in his [the sun's] face since December 1676."[15] The absence of sunspots therefore meant that the Earth received less sunshine, contributing to the 1° to 2°C (1.8° to 3.6°F) drop in the winter temperatures of the Northern Hemisphere.[16]

Erupting volcanoes also cooled the atmosphere. Between 1584 and 1610, there were twenty-six major volcanic eruptions; that of Huaynaputina in southern Peru in 1600 was one of the largest ever recorded. Between 1638 and 1644, twelve volcanoes had major eruptions, including Mounts Komagatake in Japan in 1640, Kuwae in Vanuatu, and Parker (or Melibengoy) in the Philippines in 1641–1642. Several others followed at the end of the century: Krakatoa in 1680, Serna in 1693, and Amboina in 1694, all in Indonesia, and Hekla in Iceland in 1695. A century later, the eruption of Laki in Iceland in 1783–1784 affected the climate as far as East Africa, and that of Tambora in Indonesia in 1815 famously caused a "year without summer" in eastern North America. These eruptions contributed to global cooling in several ways. By ejecting ashes and dust high into the stratosphere, they created a dust veil that darkened the sun and reddened the sky. Among the minerals they ejected, sulfur dioxide formed sulfate aerosols that reflected the sun's radiation back into space. This decline in sunshine also allowed sea ice to form in the Arctic Ocean and drift southward, disrupting ocean currents and increasing the sea's ability to reflect sunlight for several years.[17]

In recent years, climatologists have subjected these natural explanations for the Little Ice Age to critical analysis. As William Ruddiman points out, orbital forcing accounts for only half of the global cooling observed. The lack of sunspots during the Spörer and Maunder minima made a difference. But the change in the amount of sunshine reaching the Earth was too small to account for the global cooling. Volcanic eruptions blocked solar radiation very effectively, but only for short periods of time, and these cannot account for the long-term effects of the Little Ice Age.[18]

That leaves a fourth factor: carbon dioxide in the atmosphere, which blocks the escape of heat from the Earth into outer space. Ice cores from the Antarctic show a long-term rise in CO_2 since the beginning of agriculture, with three downturns: a long but shallow one between 200 and 600, a short one from 1300 to 1400, and a long and deep downturn from 1500 to 1750.

Periods of minimum CO_2 correspond to the coolest periods of the Little Ice Age, and periods of maximum CO_2 to warmer eras, such as the Medieval Climate Anomaly. The decline in the concentration of CO_2 in the atmosphere therefore had a stronger effect on global cooling than the other factors. This was true of earlier, "big" ice ages as well as the Little Ice Age.[19] However, the Little Ice Age differed in one significant way: the concentration of CO_2 decreased much faster than in previous ice ages. What could have caused the sharp drop? If natural causes alone do not explain the Little Ice Age, can there have been another, unnatural cause?

The Ruddiman Thesis

In his book *Plows, Plagues, and Petroleum*, Ruddiman argues that the same factors that caused the slow global warming that started 10,000 years ago—deforestation, herding, and agriculture—also contributed to the Little Ice Age. He notes the correlations between changes in populations and in CO_2: a decline in CO_2 from the first to the eighth century corresponds to the epidemics of the Roman Empire; then a rebound from the mid-eighth to the mid-fourteenth century—the Medieval Climate Anomaly; then a drop in CO_2 during and after the Black Death of the fourteenth century; then, after a weak rebound, a more precipitous decline during the American pandemics of the sixteenth and seventeenth centuries; then a rise in both population and CO_2 after the mid-eighteenth century that has continued to this day. The Little Ice Age, then, would correspond to a period following the Black Death and the American pandemics.

The causal links between pandemics and global cooling was the abandonment of farms and pastures that followed a drop in population and thereby allowed forests to recover. In most places, trees grew fast enough to achieve the biomass of full forests within fifty years, for young trees absorb carbon dioxide much faster than old ones. The removal of CO_2 during episodes of reforestation was much more rapid than the gradual increase in CO_2 that occurred during the spread of agriculture from 10,000 years ago until the eighteenth century.

The abandonment of farms and the recovery of forests in Europe during and after the Black Death have been well documented. The decline in the population of China between the Mongol invasion and the establishment of the Ming is also clear, but there is no evidence of reforestation; instead,

Ruddiman suggests that there was a decline in coal burning, hence in the emission of carbon dioxide. In the Americas, where the population dropped precipitously in the sixteenth century and did not recover until the late eighteenth century, forests grew back in North and Central America and in the Amazon basin. These reforestations and the decline in coal burning, he argues, account for the decline in CO_2 in the atmosphere, which in turn explains the global cooling that ensued.[20]

Since Ruddiman proposed this thesis in 2003, many scholars have debated it in the pages of scientific journals.[21] Recently, environmental historian John Brooke found that connecting the depopulation and reforestation of the Americas to the Little Ice Age "has considerable merit."[22] Yet other climatologists have argued that the decline in atmospheric CO_2 in the sixteenth century does not adequately account for the cooling of the period, which must have had other causes, still to be determined.[23] In short, Ruddiman's thesis connecting population and climate in the early-modern period is, at this point, a tantalizing hypothesis and a challenge to future generations of climatologists and historians.

The Seventeenth-Century Crisis

Bad weather had severe consequences for human beings, especially those living in complex societies with dense populations, giving rise to what historians call the "general crisis of the seventeenth century." Historian Geoffrey Parker has found "more cases of simultaneous state breakdown around the globe than in any previous or subsequent age" and that "more wars took place around the world than in any other era until the 1940s."[24]

Crisis in Europe

The change in the climate affected much of Europe severely. A decline in average summer temperatures of 1° or 2°C (1.8° or 3.6°F), as happened in the third quarter of the seventeenth century, shortened the growing season by three to four weeks. The northern limits of the cultivation of vineyards, wheat, and olive trees shifted south. The cold also lowered crop yields by up to 15 percent, especially in eastern and northern Europe where growing seasons were already shorter than in the south and west. A cold spell during germination, a drought or frost during the growing season, or storms during the harvest could lower yields substantially. This raised the price of bread on which most people spent over half their income, causing hunger among the rural poor and unemployment and misery among urban artisans. Two or more harvest failures in a row caused famines, as happened in France, Scotland, and Scandinavia in the 1690s and Iceland in the 1740s. Poverty and hunger, in turn, made people vulnerable to diseases: typhus, typhoid, dysentery, and plague. Hunger caused children to grow up stunted; soldiers born between 1650 and 1700 were 2.5 to 3.8 centimeters shorter than those born after 1700.[25]

This was the century of Spain's decline. Unrealistic imperial ambitions, religious fanaticism, and economic and financial mismanagement marked the decline, but nature contributed to it as well when a prolonged drought caused food shortages in the late sixteenth and early seventeenth centuries. In the words of economic historian Jaime Vicens Vives: "Aridity meant the failure of many harvests, hence undernourishment, triumph of plague, and depopulation."[26] Crop failures occurred throughout Europe, especially in marginal regions like Norway, which lost half its farms between 1300 and 1665. Farms were also abandoned and people starved in Scotland and other parts of Scandinavia.[27] Everywhere, environmental, demographic, economic, and political crises interacted in complex ways. In France during the Fronde rebellion (1648–1653), peasants went on a rampage, destroying hundreds of thousands of hectares of forests. In Germany, the Thirty Years' War (1618–1648) reduced the population so much that farmlands were abandoned and forests grew back.

Crisis in the Ottoman Empire

During the sixteenth century, the Ottomans expanded their empire to include Egypt, Syria, Mesopotamia, North Africa, and parts of Arabia and the Balkans. A generally stable climate and the security that the empire provided favored the expansion of trade and the growth of population.

Starting in 1591, a six-year drought in Anatolia—the heartland of the Ottoman Empire—was followed by a major epizootic of cattle and sheep that ruined farmers, causing famine and a flight to the cities. War between the Ottoman and Austrian empires and oppressive taxation exacerbated the crisis, provoking a rebellion and outbreaks of banditry. Another drought in

1606–1608 brought starvation and even cannibalism to Anatolia and nearby provinces. The crisis resumed in the cold decades of the 1620s and '30s, reducing the population of those regions by half. It returned in the 1680s, thwarting the Ottoman hopes of defeating their long-standing Hapsburg enemy. As agricultural lands were abandoned, nomads with their herds of sheep replaced farmers; as in Mexico, it represented a victory of sheep over humans. Meanwhile the cities, especially Istanbul, suffered from outbreaks of plague, typhus, and dysentery. In the words of Ottoman historian Sam White, "Climate was a critical factor—perhaps *the* critical factor—in understanding the Ottoman crises of the seventeenth century."[28] The population of that region did not recover until the nineteenth century.[29]

The Plague in the Middle East

One of the most serious environmental problems facing the Middle East was disease. The peoples of the region suffered from numerous respiratory and gastrointestinal ailments and, along the Nile, from schistosomiasis, a snail-borne parasitic disease transmitted through water that can cause liver damage and kidney failure. But the disease that attracted the most attention was the plague.

In the sixteenth and seventeenth centuries, major plague outbreaks occurred in the Ottoman Empire on average every nine to eleven years, but irregularly and unpredictably. In Egypt, the plague struck mostly young women and children, contributing to demographic stagnation.[30] Some outbreaks were particularly widespread and devastating. One epidemic affected the Mediterranean coast from 1647 to 1656. Another spread from Egypt to Istanbul in the early summer of 1778, killing one-third of the inhabitants, then spread to the European provinces of the Ottoman Empire between 1779 and 1783 and finally to Syria and North Africa between 1784 and 1787. Yet another afflicted Egypt in 1791.[31]

The plague spread easily because the Ottoman Empire was a crossroads of communications, trade, and pilgrimages.[32] Often the disease's appearance coincided with changes in environmental conditions, such as an especially low or high Nile. Thus in 1790 the Nile rose so high that it flooded the streets of Cairo, damaged farmland, and destroyed stores of grain, bringing rats closer to humans. The following year, the Nile was exceptionally low, exacerbating the competition between rats and humans for the remaining supplies of food. Fleas, jumping from dying rats to starving humans, spread yet another outbreak of plague, killing between 1,000 and 2,000 humans (and untold numbers of rats) a day.[33]

Islamic plague treatises written during and after the Black Death of the fourteenth century maintained that the disease was a blessing of God and a means to achieve martyrdom. They rejected the idea of contagion and admonished the faithful to care for the sick, ensure that the dead had proper burials, and submit to God's will. Though they prohibited the faithful from fleeing plague-stricken areas, many did flee. Treatises written from the sixteenth century onward accepted the idea of contagion through putrid air and permitted the faithful to leave plague-stricken areas in search of clean air. Historian Alan Mikhail argues that Egyptians maintained a secular attitude toward the plague. They considered it to be a normal part of the natural world, like floods, droughts, and famines, and did little to prevent or combat it.[34]

In contrast, non-Muslims living in the Ottoman Empire viewed the plague as divine punishment. To save themselves, Greek Orthodox turned to the Virgin Mary and saints and offered masses. Those who could fled to the mountains, as did Armenians, Jews, and resident Western Europeans. Consuls representing the European powers were especially concerned, perhaps because memories of the Black Death were so deeply ingrained in the European consciousness; much of what we know of plague outbreaks comes from their reports. Some fled, while others shut themselves in their houses.[35]

Yet there was little they could do. In 1798, when Napoleon Bonaparte invaded Egypt, he imposed the latest European methods of dealing with the plague. Plague victims were forcibly removed and placed in quarantine. Their houses and the people who lived in them were also quarantined. Their clothes and sometimes their houses were burned. Separating the sick from the healthy took priority over caring for the afflicted. These measure offended Egyptians and alienated them from European rule but did little or nothing to prevent the disease. Not until 1838, when an Egyptian government imposed similar restrictions, did the incidence of plague begin to diminish.[36]

Crisis in China

The worst crisis of all was the one that engulfed China. Several cold decades— the 1610s, 1630s, 1650s, and 1680s—caused a decline in Chinese agricultural production, especially of rice. Droughts so severe that the Yellow River or its main tributaries dried up completely afflicted the North China Plain in 1639, 1640, 1675, and 1690—four of the seven recorded instances in the past 2,500 years.[37] Even in the sub-tropical south, as the growing seasons shortened, farmers shifted from three crops a year to two or from two to one; the years 1614–1615 were especially severe. The 1630s and '40s witnessed an accumulation of disasters: first torrential rains, then drought and locusts, then floods. Heavy frosts in Jiangxi province in southern China between 1646 and 1676 ruined the centuries-old orange groves.[38]

In addition to the harsh weather, China suffered from numerous epidemics. One of the most devastating, known as the "Big Head Fever," occurred in 1586–1588; but, as historian of diseases Helen Dunstan noted, "The typical gazetteer entry offers not the slightest clue as to the diagnosis of the disease it records."[39] Another serious epidemic in the 1630s and '40s called "Peep Fever" ("you had only to peep and you would catch the disease and die") may well have included cases of bubonic plague; as a gazetteer reported: "In the autumn there was a great epidemic. The victims first developed a hard lump below the armpits or between the thighs or else coughed thin blood and died before they had time to take medicine. Even friends and relations did not dare to ask after the sick or come with their condolences. There were whole families wiped out with no one to bury them."[40] This and other famine-related diseases—typhus and dysentery—ravaged northern China until 1644, then disappeared for over a century.[41] As a result of diseases and warfare, the population of China fell from around 150 million in 1600 to 120 million in 1650. By the late eighteenth century, bubonic plague was known and clearly described in Yunnan and later in eastern and southern China.[42]

Adding to the misery caused by the floods, droughts, and epidemics were uprisings, invasions, and civil wars. In most histories, the year 1644 marks the end of the Ming dynasty and the beginning of the Qing. In fact, the Ming-Qing transition began in 1629 with a major peasant rebellion, followed by the invasion of northern China by the Manchus and the proclamation of the new Qing Dynasty in 1644. Yet for years thereafter, as the Ming court fled south, loyalists fought on until 1683 when the last holdouts were defeated by the Manchu armies. For sixty years, as armies marched back and forth across the land, civilians who were not killed outright fled for their lives, abandoning their farms and towns only to die somewhere else.[43] As Sinologist Robert Marks recounts: "Briefly the 40-year period beginning in 1644 was one wrought by crisis: civil war, banditry and piracy, peasant uprisings, trade dislocations, and declining harvest yields caused by colder temperatures all combined to make life in those years uncertain at best, unlike any that had preceded or followed."[44]

Recovery from the General Crisis

It would be easy, but misleading, to attribute the political troubles of the time to the weather. Geoffrey Parker asks: "Could sudden climate change, and in particular intense cold and prolonged drought, have been the common denominator that caused the unprecedented wave of both wars and state breakdowns, with peaks that coincided with the El Niño clusters of 1640–42 and 1647–50?" His answer is that climate change did not cause but only exacerbated the political and religious tensions of the age that resulted in the crisis of the 1640s.[45]

Proof that the climate was only one factor among many is not hard to find. The decline of Spain was also the Golden Age of Holland; in spite of severe winters, Dutch people flourished, thanks to the trade they diverted from southern and central Europe, an influx of talented refugees, and wise government policies.[46] In China, the opening of new lands to cultivation and the spread of double cropping alleviated the decline in yields. In Japan, the same climatic conditions as in China caused famines in the 1630s and 1640s, but its government avoided the costly public works projects and imperial extravagances that ruined the Ming exchequer. Japan thereby suffered less and recovered more quickly than China from the crisis.[47] Difficult as the climate was, many societies proved resilient enough to rebound quickly once the crisis passed.

The recovery from the General Crisis is reflected in the number of people. While historical demographers are not in full agreement on population figures before the twentieth century, there is a general consensus about the trends. The population of the world was rising throughout the sixteenth century, then stagnated during the seventeenth, then began to rise again in the eighteenth. Even during the seventeenth century, the world's population did not drop, for the population of the Old World grew enough to compensate for the demographic collapse of the New World population.[48]

Thus, for all its severity, in the Eastern Hemisphere the seventeenth-century crisis was only a crisis, not a catastrophe as in the Americas or

Table 7.1. World Population (estimates in millions)

1400	374–390
1500	425–500
1600	498–579
1650	500–545
1700	600–679
1750	700–813
1800	900–990

a collapse as in the Mediterranean world of the sixth century. By the mid-eighteenth century, population and economic activities had recovered and were on a path to growth that has continued to this day. One reason for the remarkable recovery of Eurasian societies is the development of institutions—quarantines during epidemics, stockpiling of food for emergencies, better transportation—that alleviated the impact of natural disasters. But far more important was the rise in food supplies brought about by the introduction of new food crops from the Americas. It is largely thanks to these new crops that the peoples of the Old World avoided the worst effects of the Little Ace Age.

The Columbian Exchange, from the New World to the Old

The ships that carried animals, plants, and diseases from the Old World to the New also carried living beings from the Americas to the Eastern Hemisphere. In the Americas, the human population and some native plants and animals crashed, while Old World animals, plants, and pathogens thrived. In the Eastern Hemisphere, in contrast, it was humans and their domesticated plants and animals that benefited most from the Columbian Exchange, at the expense of the rest of nature.

Diseases are one example of the lopsided nature of the Exchange. While the Americas were invaded by a whole host of pathogens that attacked and decimated the native population, only one disease—syphilis—traveled in the other direction. To be sure, syphilis was a serious disease that caused agony, dementia, and death in its victims, but its mode of sexual transmission narrowed down the population of victims to soldiers, sailors, prostitutes, and other sexually promiscuous groups (and their spouses and offspring).

It also probably evolved to become less virulent over time. Hence its demographic impact was much less than that of the crowd diseases unleashed in the Americas.

The animal exchange was equally lopsided. Of the few animals that Native Americans had domesticated, none became important in the rest of the world, and certainly none went feral and took over great swaths of territory the way swine, cattle, horses, and other Old World domesticates did in the Americas.

New World Crops in the Old World

Likewise, almost no American plant spread out of control in the Eastern Hemisphere the way Old World plants did in the Americas. But plants that humans favored—that is, crops—proved to be very well adapted to the climates and soils of the Old World and hugely important to the nourishment of its peoples. Most succeeded by enhancing and diversifying the cuisines of Old World societies. Tomatoes, beans, peanuts, chili peppers, pineapples, pumpkins, squashes, guavas, papayas, avocados, cacao, and other plants too numerous to list made daily meals more varied and appetizing for peoples around the world. And tobacco, an American plant, contributed to the social life and personal pleasure, and to the illness and early death, of millions of people.

Among all the New World plants that spread widely around the world in the centuries since 1492, four not only enhanced the cuisine but also vastly increased the food supply of peoples on other continents, and hence allowed their populations to grow. These four plants—maize, potatoes, sweet potatoes, and manioc—succeeded not by competing with or replacing existing staples like rice, wheat, barley, and millet, but by growing successfully where the soil and climate were not suited to grains. They could also be grown on land that would otherwise have lain fallow and they employed farmers at times when traditional staples did not require their attention. Thus they increased the amount of food that farmers could grow.[49]

Of the four crops, maize or American corn (*Zea mays*) was the most widespread and had the largest impact. It is amazingly tolerant of environmental conditions, growing on hillsides and in sandy soils and in places where rainfall is too great for wheat and insufficient for rice. It has a shorter growing season than other grains and requires little weeding or other labor, but still requires at least six weeks of warm sunny weather. Per unit of land, it yields twice as much nutrition (including carbohydrates, fructose, and fat) as wheat, though it is deficient in niacin and proteins. It depletes the land of nitrogen unless it is grown along with beans (as American Indians grew it) or in rotation with nitrogen-fixing plants such as beans or clover.[50]

The sweet potato (*Ipomoea batatas*) is, despite its name, completely unrelated to the white or Peruvian potato. It originated in Mesoamerica or northern South America. Like maize, it grows well in poor soils. It ripens fast; in the tropics, it can yield two or even three crops a year. It produces three to four times more calories per hectare than rice and requires much less labor. It is immune to locusts, a major consideration in much of Africa and Asia. However, it does not tolerate frosts or droughts and thus is restricted to wet tropical or semi-tropical climates.[51]

Manioc (*Manihot esculenta*), also known as cassava or yuca, probably originated in the Amazon rainforest. It grows in many soils, even those that have been depleted by the repeated cultivation of other crops. It can withstand prolonged droughts that would kill other plants and is invulnerable to locusts. Its tuber remains underground until it is dug up and thus can be harvested at the convenience of the farmer months or even years after it is planted. However, it cannot tolerate frosts, hence it is limited to the tropics. In its raw state the tuber contains Prussic acid (hydrogen cyanide), a poison, and must be soaked in running water before it is cooked. It provides the most calories per hectare of any staple, but it consists of almost pure starch with some calcium and Vitamin C, but no protein, fat, or other vitamins or minerals. It is therefore often grown in conjunction with other crops like maize, plantains, and bananas.[52]

Potatoes are the fourth great American crop. This tuber (*Solanum tuberosum*) originated on the Altiplano of Peru and Bolivia. Unlike all other staples, the potato provides a balanced diet, with carbohydrates, protein, and essential vitamins and minerals. Furthermore, it is exceptionally productive, as three-quarters of its biomass is edible, compared to one-quarter the biomass of grains; thus it produces two to four times more calories per hectare than grains. It grows well in sandy loam and in climates too wet for grains. While it will keep for months in a cool dry place, warmth and moisture will cause the eyes to sprout and the tuber to soften and rot. Unlike grains or dried maize, it cannot be stored from one year to the next.[53]

In the Eastern Hemisphere, each of these four American crops (and others as well) found niches where traditional crops would not grow or provided low yields. They thus increased the total nutrition of its peoples by a third or more. Yet, despite their virtues, the new crops caught on slowly, because most people are set in their dietary ways and will experiment with new foods only when hunger forces them to.

Europe

European population figures are uncertain before the nineteenth century. In broad brushstrokes, we know the continent's population was growing slowly until the early seventeenth century, then stagnated until the early eighteenth century, then began to rise rapidly.[54] Among the factors that affected population size, disease was one of the most important. Though the plague never caused another pandemic like the Black Death of the fourteenth century, it returned periodically to different parts of Europe. Then it ended, partly because states imposed quarantines on plague-stricken cities and ships, and partly for unknown reasons. Smallpox became the most feared of diseases in the eighteenth century; though often lethal to those who contracted it, its effect on Europe was much more limited than on the New World. The rise in the European population after the mid-eighteenth century was due to a drop in the death rate rather than a change in the birth rate.[55]

Wetlands and Forests

Important as were diseases, the main cause of the stagnation or growth of the European population was food. Harvest failures were common until the mid-seventeenth century, then diminished in the eighteenth, although they recurred from time to time, as in the lean years of 1788–1789 leading up to the French Revolution. After the seventeenth century, food supplies began to rise faster than the population. Investments in agriculture by capitalist entrepreneurs, managerial innovations, and growing efficiencies in England and the Netherlands produced higher yields. A growing trade improved the allocation of resources as regions specialized in crops for which they had a natural advantage.[56]

Some of the changes affected the land. The Dutch reclaimed wetlands by installing windmills to drain water from polders and by improving the social organization of farming communities and the technical expertise of farmers. Peat beds, drained by mechanical means, provided fuel in a wood-deficient country and contributed to the prosperity of Holland during its Golden Age. In eastern England, an act of Parliament gave developers the right to drain the Fens, a marshy region inhabited by herders, fishers, and peat cutters. An area of 2,500 square kilometers was drained by windmills and turned into cultivated, albeit mosquito-infested, land.[57]

In many parts of Europe, deforestation continued, creating shortages of wood for fuel and construction. By the sixteenth century, England had few forests left other than parks and commercial woodlots and began importing wood. In Scotland, the proportion of land covered by forests fell from 10–15 percent in 1500 to 3 percent in 1815. In Ireland, forests declined from over 12 percent in 1600 to 2 percent in 1700. As in previous eras, most deforestation resulted from farmers expanding their fields. Much of the wood was used as fuel, especially during the cold winters of the Little Ice Age, not only to heat houses but also to make glass, bricks, ceramics, salt, lime, sugar, soap, beer, and other manufactured products; it was said that building a brick house consumed more wood than constructing a wooden house. Though iron makers were often blamed for the wood shortages, most of them relied on woodlots to provide a sustainable supply of charcoal for their furnaces. Shipbuilding was another drain on the forests, especially elms and oaks for hulls, and firs, pines, and spruces for masts and other parts. Great ships of the line, with which the naval powers fought their frequent wars, were especially costly; one such ship required 4,200 to 5,600 cubic meters of wood, or several thousand mature trees. In northern Europe, wood consumption varied between 1.6 and 2.3 tons per person per year. From 1650 to 1749, Europe lost between 18.4 and 24.6 million hectares of forests. As shortages intensified, near cities, the price of wood rose much faster than any other commodity prices. Yet the cries of a "wood famine" were often exaggerated by those who stood to benefit from government regulation.[58]

Venice was an especially voracious consumer of wood. In addition to the usual uses of wood, the city also built one of the Mediterranean's most powerful fleets. Just as important were the foundations of the buildings, for Venice was built in a lagoon, and all the buildings rested on wooden pilings. How many such pilings—logs driven into the mud to support the city—is not known. It has been estimated, however, that to support just one of Venice's famous buildings, the Basilica of Santa Maria della Salute, may have required some 100,000 pilings—in other words, a sizable forest.[59]

There were attempts at conservation, especially by governments concerned with the depletion of oak forests needed for shipbuilding. Venice reserved forests near the Adriatic but could not enforce its prohibitions effectively.[60]

French finance minister Jean-Baptiste Colbert introduced a law in 1669 that established royal forests patrolled by guards; as a result, French forests remained protected.[61] In tropical colonies like Mauritius, Saint Helena, the Cape of Good Hope, and Java where deforestation had an immediate impact on the soil and water, colonial governors attempted to prevent excessive cutting, with mixed results.[62]

Maize and Potatoes

The opening of new lands, improvements in efficiency, and the discovery of substitutes were not, in themselves, sufficient to keep up with the growth of population. As environmental historian Alfred Crosby pointed out, if the seventeenth-century crisis ended much more quickly than the crisis of the fourteenth century it was because "the Amerindian crops . . . provided . . . a delaying of disaster, a fending off of the Malthusian checks, until those tardy developments that historians have called the agricultural and industrial revolutions were under way."[63]

The two Amerindian crops that mattered most were maize in southern Europe and potatoes west and north of the Alps. Maize spread slowly among people whose culinary habits were conservative and had traditionally been attached to wheat. In Spain and Italy, the plagues and poor wheat harvests of the seventeenth century overcame people's dietary compunctions. In the Balkans, the colder climate delayed the spread of maize until the eighteenth century, when the poor began to eat maize in the form of polenta, despite the high incidence of pellagra, a niacin-deficiency disease that it caused. By the eighteenth century, peasants in northern Italy produced more maize than traditional grains like wheat and rye.[64]

In northern and western Europe, where it was generally too cool and rainy for maize, it was the potato that transformed people's diets. Potatoes reached the Canary Islands in the 1560s and Spain in the 1570s. Basque fishermen may have carried potatoes on their fishing expeditions to Newfoundland and, in the process, transferred them to the British Isles. In any event, potatoes were known but not widely grown in western Europe by the early seventeenth century.

In northern Europe, where open fields predominated, agriculture was governed by the seasonal rhythms of grain agriculture, and potatoes were grown in garden plots or in fields left fallow after the grain harvest. At first, as summer days in Europe were much longer than in the Andes, the plants produced more flowers than tubers. Not until farmers, after many decades of trial and error, obtained plants that produced large tubers did potatoes become adapted to the European seasonal cycles. Farmers also learned to grow nitrogen-fixing clover in rotation with potatoes to improve the yields of the nitrogen-hungry tubers.

Besides producing more food per hectare, potatoes had another advantage in an age when roving armies "lived off the land," that is, requisitioned farmers' stores of grain, leaving hunger in their wake, as happened in the disastrous Thirty Years' War in Germany in the early seventeenth century. Like manioc in Africa, potatoes could be left in the ground after they ripened, making them harder for soldiers to steal.

Nonetheless, for a century most Europeans denigrated potatoes as hog feed or food suited only for the very poor. It took over a century for potatoes to gain the approval of the powerful. Frederick the Great of Prussia (r. 1740–1786) is said to have ordered his peasants to grow and eat potatoes. In France, the chemist Antoine-Augustin Parmentier, who had survived on potatoes while a prisoner of the Prussians, touted the benefits of the tuber in his book *Examen chymique des pommes de terre* (1774). Only after the poor harvest of 1785, when potatoes staved off famine, did they become acceptable to most French people. Similarly, it was bad wheat harvests that convinced the Austrian and Russian governments to encourage farmers to grow potatoes.

By the early nineteenth century, potatoes had become the staple food of peasants throughout western and central Europe. The new food encouraged a rapid increase in the population. In Ireland, peasants discovered that an adult could survive and be healthy on a diet of five kilograms of potatoes a day and a bit of milk, and that less than one hectare of land planted in potatoes and one cow were enough to feed a family year-round. As a result, the Irish population rose from about 1.5 million at the end of the seventeenth century to over 3 million in the mid-eighteenth century and to over 8 million in the early 1840s. Nowhere else was the increase that dramatic, but the rest of Europe also saw its population rise, thanks in large part to Amerindian crops.[65]

Sub-Saharan Africa

Many diseases endemic to sub-Saharan Africa, such as yellow fever and falciparum malaria, were much deadlier to outsiders than to its indigenous inhabitants. The continent was therefore protected from the fate of the New World and its peoples. To Africans, however, avoiding the fate of Native Americans was only a partial blessing, since so many were enslaved and shipped to America.

Nonetheless the population of Africa grew thanks to new crops from the Americas. This mattered even more than in Asia or Europe because there were only a few domesticable plants indigenous to Africa: African yams (*Dioscorea opposita*), African rice (*Oryza glaberrima*), sorghum (*Sorghum bicolor*), teff (*Eragrostis tef*), and two condiments, Guinea pepper (*Piper guineense*) and kola nuts (*Cola vera*). In the centuries before Europeans reached the Americas, several Asian plants had made their way to East Africa, brought by Arab traders and Indian Ocean mariners, including Asian yams (*Dioscorea alata*), Asian rice (*Oryza sativa*), bananas (*Musa acuminata*), and taro (*Colocasia esculenta*). After 1500, many of these crops were transferred from East to West Africa by the Portuguese. More important were the American plants brought to tropical Africa by Europeans. They include all the domesticated plants found in the American tropics. Most were condiments, additions to the diet, or trade items.[66] But two—maize and manioc—significantly increased the nutrition of Africans, allowing their numbers to increase.

No one knows when maize reached Africa, but it must have been in the early sixteenth century, for the first documented mention of the plant growing in the Cape Verde Islands dates from 1540. By the mid-1500s it was found on the island of São Tomé in the Gulf of Guinea, and by the early seventeenth century it was grown alongside rice in the wetter parts of Guinea Coast and sorghum and millet in the drier areas of Senegambia. Flint maize from the Caribbean may have been carried from Spain to North Africa and from there across the Sahara to West Africa. A second transfer across the Atlantic was of floury maize, the kind grown in Mexico and the Andes that produced a softer starch and more abundant yields.[67] Though almost always grown with other crops, maize became a significant source of food in Africa by the eighteenth century.

The other important American crop introduced to Africa was manioc. The Portuguese found the plant growing in Brazil and imported it to the Congo estuary and the islands of São Tomé, Principe, and Fernando Po in the late sixteenth century. It yielded ten times more calories per hectare than millet, yams, or sorghum and grew even in rainforests that supported few domesticated plants. It was attractive during the years of the slave trade because people could leave it in the ground when they fled the slavers and harvest it when they returned to their fields. At first, it was eaten only by the Portuguese, but during the eighteenth century, it spread to Angola and Gabon and also to the lands bordering the Indian Ocean. In the nineteenth century it became a significant source of nutrition for people in the equatorial regions of Africa.[68]

China: Land, People, and Crops

The Little Ice Age affected China more deeply than any other part of Eurasia. Cold weather and political upheavals decimated the human population. The recovery, based on the cultivation of American crops, was more radical and in the process caused more deforestation than anywhere else.

Chinese population figures before the nineteenth century are notoriously unreliable. The consensus is that the Chinese population grew to about 150 million during the sixteenth century, then declined during the seventeenth century to around 120 million. As the population declined, so did the amount of land under cultivation.[69] During the Ming-Qing transition, lands were abandoned.

In the eighteenth century, when peace returned under the Qing, the population rebounded, tripling to about 400 million by 1800.[70] Lands that had been abandoned were repopulated, often under government orders. The government loaned seeds and animals and granted tax exemptions to entice migrants to relocate from the overpopulated lower Yangzi Valley to Hunan and Hubei provinces in central China. Until 1668, farmers were encouraged to settle in southern Manchuria. After that date, the Kangxi emperor (r. 1661–1722) forbade such migrations in order to preserve the environment and the Manchu way of life with its horses and martial virtues.[71] This exclusionary policy worked for a time, but in the eighteenth century, a rapidly growing population overcame the desires of the Manchu government to leave some regions free of Chinese peoples.[72] From the mid-eighteenth century on, the population of China began to increase. Sichuan, which had been almost emptied out during the time of troubles, increased its population fivefold between 1787 and 1850. Yunnan in the far south saw a doubling of population in that same period because of the opening of silver and copper mines and the growth of trade.[73] Other peripheral regions—Hainan and Taiwan islands, Tibet, and Xinjiang—were added to China by the Qing whose empire grew to double the size of the Ming's, opening up new lands to Chinese farmers.[74]

Traditional Agriculture

Not surprisingly in a land where four out of five people were farmers, changes in the numbers and distribution of people were reflected in changes in the amount of cultivated land. According to Geoffrey Parker, the amount of cultivated land in China dropped from 77 million hectares in 1602 to 27 million in 1645, then went back up to 40 million in 1685 and continued to increase thereafter.[75]

The increase in arable land took place in three ecological zones. One was northern China and (for a time) southern Manchuria, land with fertile soils but short growing seasons. There, both the Ming and Qing governments encouraged peasants to return to once-cultivated lands and put "wastelands" under the plow to increase the supply of food for the inhabitants of Beijing, the capital. Nonetheless, Beijing had to import half its food from the south via the Grand Canal. Since the canal crossed the Yellow River from which it received part of its water, maintaining it and protecting the North China Plain from the river's devastating floods required complex hydraulic engineering projects that cost 10 to 20 percent of the government's revenues.[76]

The second zone was the lowlands of central and southern China, potentially the most productive lands in China. In Hunan province south of the Yangzi, especially around Lake Dongting, farmers carved out rice paddies by enormous efforts. Dikes, dams, reservoirs, and human- or animal-powered pumps produced, for a time, rich harvests. Yet the farmers' encroaching on natural reservoirs such as the marshes, floodplains, and lakes that buffered the normal river fluctuations made the land more vulnerable to irregular but massive floods.[77] In the Yangzi Delta, ever-increasing intensification by peasants eager to make use of every scrap of land produced extraordinary yields per hectare and the highest farm-population density in the world at 500 inhabitants per square kilometer in 1750 (compared to 62 in the Netherlands). After that, the development of the region slowed down and the population ceased growing until the mid-nineteenth century; evidently, farmers had reached the maximum carrying capacity of the region with the technology at their disposal.[78] In the south, migrants from the north turned the Pearl River Delta into a rich agricultural region by reclaiming marshes and erecting embankments, dikes, canals, and other means of water control. Once they had created paddies, farmers planted fast-ripening varieties of rice twice or even three times a year.[79]

The third ecological zone transformed by agriculture under the Ming and Qing was the hillsides and mountain slopes of central and southern China that were not amenable to growing rice, wheat, and other traditional crops. What opened these lands to farming was the introduction of American plants.

American Crops in China

After having imposed order on the country by the late 1680s, the Qing government was eager to have farmers clear and cultivate new land. At the same time, the demand for land induced peasants to move into the hills above the Yangzi Valley. There they grew sweet potatoes and maize for subsistence, and tobacco, tea, sugarcane, and other crops for the market. Farther north, landless peasants, whom those with land called "shack people," practiced slash-and-burn agriculture, living in temporary sheds and growing maize that depleted the fertility of the soil in a couple of years.[80] Having filled the Yangzi hills, these migrants moved into the hills of southern China. As a result, the amount of cultivated land in China tripled between 1400 and 1850, from 25 million to 81 million hectares. Historian Sucheta Mazumdar called it China's second agricultural revolution, second only to the introduction of Champa rice.[81]

The most important new crop was the sweet potato. Introduced in the late sixteenth century, probably by Chinese merchants who obtained it from the Spanish in Manila, it was quickly adopted by the poor, for it yielded more food than other crops, required little labor, and grew well in marginal soils as long as it received enough sunshine and rain; in the tropical south, it could yield two or even three crops a year. By the end of the seventeenth century, it had become a staple in southern China and spread as far north as Shandong, southeast of Beijing.[82]

Maize was introduced to China sometime in the sixteenth century. It grew fast in soils unsuited for rice and produced more food with far less labor than wheat, millet, or barley. As it competed with sweet potatoes in the warmer regions of China, it was not extensively grown until the eighteenth century, when an expanding population using traditional technologies began to strain the carrying capacity of the land.[83]

Peanuts (*Arachis hypogaea*), though not a staple food crop, were attractive to farmers for they grew well in sandy soils along riverbanks and near the coast. They were often grown in rotation with sugarcane and other soil-depleting crops because they renewed the fertility of the soil by fixing nitrogen in their roots. Peanuts were also pressed for oil and the leftover meal was used as fertilizer or as feed for hogs.[84]

Of the four American crops, potatoes were the least important in China. They thrived best where the weather was cool but moist, a combination that was rare in China. Farmers grew them on mountain slopes beyond where sweet potatoes and maize would grow.[85] As a result, potatoes never became a major part of the Chinese diet.

Environmental Consequences of Expanded Agriculture

The opening of the hillsides to agriculture had serious ecological consequences. Tenant farmers, needing to maximize the yields they could obtain quickly, planted maize in straight rows in the topsoil of newly cleared land. When their crops had depleted the nutrients in the soil, they moved on. Heavy rains washed away the exposed humus, leaving behind soils of little agricultural value. The topsoil that flowed into the valleys silted up rivers, lakes, and rice paddies. After the topsoil had been washed way, the floods carried sand and stones into the valleys, ruining the land. The frequency of floods, on average one every two or three years during the Song Dynasty (960–1279), increased to twice a year on average under the Ming and six times a year under the Qing.[86]

Farming the hillsides also meant clearing them of trees. Besides farmers who practiced slash and burn agriculture, woodcutters, charcoal burners, papermakers, and miners also took trees, causing severe shortages of wood. Under the Ming, heavily populated areas, such as between the Yangzi Delta and the Shandong Peninsula, were entirely deforested. Timber merchants sought wood from ever more distant regions; even Yunnan, in the far south, was becoming deforested by the early sixteenth century.[87] In 1642, one of the last ancient trees in the Yangzi Delta, a three-hundred-year-old tree on the grounds of the tomb of the founder of the Ming Dynasty, was cut down for firewood. According to historian Timothy Brook: "When the Ming fell two years later, many believed that this desecration had brought the dynasty down. Perhaps they were right."[88]

The Manchus came from a heavily forested region that teemed with wildlife—tigers, bears, leopards, fish, birds, and pine nuts—that supplemented the food their farmers produced. When they conquered China, vast quantities of these wild products were sent to Beijing to satisfy the cravings of the new elite for the products of their homeland. From 1681 to 1820, the Manchus kept a 10,400-square-kilometer area north and west of Beijing as an imperial preserve for the autumn hunt. But after 1820 even that succumbed to the woodcutters' axes. By the early nineteenth century, these wild products were almost depleted.[89]

In much of China, wood became so scarce that people burned straw, pine needles, and dung as fuel. Writes forest historian Michael Williams: "One must conclude that throughout the seventeenth and eighteenth centuries the Chinese moved inexorably toward the almost total deforestation of their portion of the earth—leaving forests only in the remote wild mountainous parts, which were not suited for agriculture."[90]

Soil and trees were not the only natural elements affected by encroaching humans: tigers, once common throughout China, also lost their habitats and wild prey. Though some tried to survive by attacking villagers and their livestock, hunting decimated the survivors. Only when humans retreated did they recover. Thus, when the populations of the coastal provinces were forcibly removed between 1661 and 1669, the forest grew back: "Because of the relocation of the border, grass and trees have grown in profusion, and tigers have become bold." By the end of the eighteenth century, however, tigers had almost disappeared.[91]

Japan: People, Crops, and Forests

In the early modern period, Japan faced many of the same problems as China: civil war, a growing population, and severe ecological constraints. Yet the Japanese reaction to them was very different from that of the Chinese. The population rose rapidly from 12 million in 1600 to 26 million in 1720, before leveling off for the next century. Feeding the growing population required increasing both the amount of cultivated land and the yields on existing land. This was not easy because Japan is one of the most mountainous nations in the world, with only 12 to 15 percent of its area available for cultivation. Yet farmers and local officials were able to double the cultivated land from 1.5 to 2.97 million hectares between 1600 and 1720 by building levees, dikes, and embankments to control the rivers and by draining marshes. Their task was easier than in China because none of Japan's rivers was as large or erratic as the Yellow and Yangzi rivers.

On the available land, farmers grew mainly rice, but also wheat, barley, and millet. Of the American crops, the sweet potato was most enthusiastically adopted in the warmer parts of the country. The Japanese were slow to adopt white potatoes, but began growing them more extensively in the eighteenth century. Still, most of the food came from traditional crops grown ever more intensively by fertilizing the soil with human wastes, grass, and the ashes of burned leaves, bark, and twigs culled from the forests. An extensive literature on improved farming techniques contributed to the development of agriculture.[92]

The Japanese and Their Forests

As everywhere else, population grew and agriculture spread at the expense of forests. Yet the Japanese story diverges from that of China and Europe in important ways. Because of its rainy climate and steep mountains, Japan has always been among the most thickly wooded countries in the world. Yet forests were already under attack in early medieval times. From 600 on, the Kinai region, the first to be developed, experienced an extraordinary construction boom, almost all of it in wood. Monasteries, shrines, palaces, and mansions consumed the old-growth forests on the slopes in the area, especially high-quality Japanese cypresses. Building the largest monastery, the Todaiji, consumed 890 hectares of first-quality forest, as much 3,000 1950s-style houses. When it burned in 1180 it was rebuilt of timber, some brought from far away. Wooden buildings required replacement every twenty years or so because the pillars set into the soil rotted or attracted termites. Rebuilding the Ise shrine, as has been done every twenty-five years since 685, required 100,000 cubic meters of processed lumber each time. In addition to furnishing lumber, trees were also felled for firewood and for smelting iron, firing pottery, and boiling salt. Among tribal rulers, the succession to headship was marked by tearing down the old palace and building a new one. Ordinary houses, built of wood with sliding paper doors and tatami mats, were prone to fires; when the new city of Kamakura burned in 1219, it too was rebuilt out of wood.

Although the forests of the Kinai region were soon depleted, the forests in other parts of Japan continued to meet human needs.[93]

Until the late seventeenth century, Japan's growing population and increasing standard of living put pressure on the woodlands. In their incessant wars, the *daimyos*, or great lords, built over 200 fortresses between 1467 and 1579, mainly out of wood. The large fleets built by Hideyoshi Toyotomi in his ill-fated attempts to conquer Korea in 1592 and 1597 also consumed vast quantities of lumber. The end of the civil wars and the establishment of the *bakufu*, or military dictatorship, after 1600 led to further demands for wood. Tokugawa Ieyasu (1542–1616), the first of the Tokugawa dynasty of shoguns, built three massive castles at Edo (now Tokyo), Nagoya, and Sunpu that required 280,000 cubic meters of wood, or 2,750 hectares of prime forest.[94] He and later shoguns also demanded massive amounts of timber as tribute from the daimyos.

After 1600, peace brought prosperity, especially to the growing cities. Edo, a small town of 30,000 inhabitants in 1600, grew to over a million by 1720, potentially making it the largest city in the world at the time. Osaka and Kyoto grew to over 300,000 in that same period.[95] Periodically, fires swept through these wooden cities. In 1657, a fire destroyed two-thirds of Edo, killing 100,000 of its inhabitants. Rebuilding the city required over 10,000 hectares of prime forests. Nor was this the only fire; substantial fires burned parts of Edo on average every 2.75 years.[96]

By 1660, the nation began to suffer from an increasing scarcity of wood. Japan's great virgin forests on the three main islands were gone, and loggers began to exploit the forests of Hokkaido, the distant northern island, then thinly inhabited. At this point, the government took action to reverse the trend. An official wrote: "The treasure of the realm is the treasure of the mountains. Before all is lost, proper care must be taken. Destitution of the mountains will result in the destitution of the realm."[97] Forests were closed to cutting. Wood and other forest products were rationed. Forest rangers were appointed and households were given incentives to care for trees. Tree plantations were set up to regenerate the forests. Thus was the supply of wood carefully controlled.[98]

Such a policy would not have worked if the demand for wood had not also declined. An unprecedented phenomenon—a culture of austerity and frugality—transformed Japanese life. In construction, sawn boards replaced whole logs. People began to build houses with light wooden frames and sliding walls of paper and flooring of *tatami* or straw mats. Furniture was either built-in or light and portable, so that a room could be used as a parlor, a dining room, or a bedroom, depending on the time of day. In cold weather, instead of heating an entire room, people used a *kotatsu*, a charcoal heater placed under a low table covered with a quilt, to keep their legs and feet warm. To warm up their beds in cold rooms, they used an *anka* or charcoal bed heater. And they cooked over a *hibachi* rather than a fireplace or stove.[99] Regulations and culture combined to reverse the deforestation of Japan. According to forest historian Michael Williams: "In stark contrast to China, Japan is one of the most densely forested countries in the world today, and the seeds of that plenitude were sown during the sixteenth century."[100]

Conclusion

Between the late fifteenth and the early nineteenth centuries, empires, trade, and peoples became interconnected across the world, but so too did the rest of nature: diseases, crops, wildlife, forests, and even the climate. In the process, the peoples of the Americas suffered a calamitous decline, while many plants and animals, both native and invasive, flourished. In the Eastern Hemisphere, the consequences were diametrically opposite: the populations of humans and their chosen crops rose at the expense of forests and wildlife. Yet, as so often in the past, expansion brought humans close to the carrying capacities of their environments and increased their vulnerability. Some societies escaped these constraints, while others did not.

8

The Transition to an Industrial World

By the end of the seventeenth century, humans had made great advances at the expense of nature. In the Eastern Hemisphere, the human population was larger, and the areas occupied by forests and natural grasslands were smaller than ever before. Yet, in their struggle with the rest of nature, the victory of humans was not assured. Epidemics had reduced the populations of the Americas by nine-tenths; the Little Ice Age had provoked subsistence crises throughout Eurasia; and the plague still hobbled the Middle East. From the eighteenth century to this day, humans have prevailed. They have multiplied as never before, have subdued the Earth, and have finally achieved dominion over (almost) every living thing. The causes of this ascendency are varied and complex, but one stands out: the Industrial Revolution.

The Industrial Revolution is one of the most thoroughly studied phenomena in history. Historians of technology have analyzed the new machines and processes. Economic historians have investigated the emergence of manufacturing, transportation, and finance as key sectors of the economy. And social historians have analyzed the impact of the new technologies and organizations on social structures and on industrial workers, women, and children, groups that bore the brunt of the transformation. The environmental consequences of industrialization, however, have received far less attention.

This chapter focuses on two sites of early industrialization and their environmental consequences. One—Great Britain—was the first nation to be industrialized, starting in the late eighteenth century. At the time, it was already a densely inhabited country, with much of its population living in towns and cities and engaged in commerce and traditional manufacturing. Most of the land had long since been deforested and transformed into fields or pastures. In short, it was a highly developed country.

The other case is the United States. At the time of its independence from Britain, it consisted of a band of states along the Atlantic seaboard occupied and thinly populated by Europeans and by African slaves, most of whom were engaged in farming. Beyond the Appalachian Mountains lay a vast continent even more thinly populated by Native Americans, most of them hunter-gatherers, with pockets of agriculture. In other words, the continent was comparatively underdeveloped. Industrialization spread quickly to the United States, in part because so many of its people shared a common language and culture with Britain, and in part because the land possessed seemingly infinite resources.

Industrialization also spread to other parts of the world: to western Europe in the mid-nineteenth century, to Russia and Japan in the late nineteenth century and subsequently to China, India, and Latin America. Great Britain and the United States illuminate the environmental impacts of industrialization on two contrasting environments—a highly developed society and a land of abundant resources. Other countries' industrializations are examined in later chapters.

Industrial Technologies

At the heart of the British Industrial Revolution lay a series of technological innovations—iron and cotton manufacture, coal, steam engines, steamboats, and railways—that revolutionized manufacturing and transportation. They also consumed resources and produced environmental effects at a far higher rate than ever before.

Coal

Already in the sixteenth and seventeenth centuries, as Britain's population grew, it suffered from increasing shortages of wood. While France and Japan sought a solution to deforestation in conservation, Britain turned to coal as a substitute fuel. In Britain, abundant coal was found near the surface and close to coasts and navigable rivers, making it cheaper than firewood. From the late sixteenth century, while the well-to-do continued to use wood, the poor were cooking and heating their houses with inexpensive "sea-coal" transported from Newcastle-upon-Tyne. By the mid-seventeenth century, Londoners depended on coal to survive the winters; when the Thames froze over or coal shipments were delayed due to storms or high tides, many froze to death. In response to this demand, shipments of coal to London increased thirty times between 1550 and 1700.[1]

Thanks to coal, Britain devoted its land to farming rather than to forests and woodlots. Besides heating homes, coal was used in making bricks, salt, gunpowder, soap, beer, glass, lime, and other products. By the 1660s, reverberatory furnaces allowed the smelting of copper and lead with coal. Salt boilers were especially voracious consumers of coal, for they required six to eight tons of coal to make one ton of salt; by 1700, salt manufacturing consumed 300,000 tons of coal, 10 percent of the total output. Coal production rose from 227,000 tons per year in the 1560s to 2.64 million tons per year in 1700, to 10 million tons in 1800, and 189 million tons in 1900.[2]

Cotton

The first technology that was revolutionized in the eighteenth century was the manufacture of cotton cloth. This fabric was in great demand because it was softer than linen and cooler than wool; it could be dyed or printed in fade-resistant colors but was much less expensive than silk. Until the eighteenth century, much of it came from India. As demand in Britain grew, cotton imports threatened the wool industry. Under pressure from that industry, Parliament banned the wearing of printed calicoes in 1721 but allowed the import of raw cotton, providing an incentive to inventors to create devices that would hasten the production of yarn.[3] Yet, for most of the eighteenth century, cotton manufacturing evolved slowly. The reason was biological.

Of the many varieties of wild cotton, only four have been domesticated. In the Old World, *Gossypium herbaceum*, originally from Africa and Arabia, spread to India and the Middle East, while Indian *Gossypium arboreum* spread to China. Both species produced short-staple cotton that could be spun by hand but not by machine. In the Americas, *Gossypium barbadense*, a long-staple variety known as sea-island or pima cotton, originated in Peru and spread throughout South America and the Caribbean, while *Gossypium hirsutum*, a medium-staple variety, was found in Mexico. What caused the difference between Old World and New World cotton was that the New World varieties had twice as many chromosomes, hence twice the chances of mutations for longer fibers. The industrial manufacture of cotton yarn required not only new kinds of machines but also supplies of medium- or long-staple fibers.[4]

During the eighteenth century, Britain imported increasing amounts of sea-island cotton from its colonies in the West Indies and the southeastern coast of North America. Yet supplies were limited by environmental conditions, for sea-island cotton requires an exceptionally long growing season—at least 250 frost-free days—and ocean breezes. Until the end of the eighteenth century, cotton manufacturers in Britain paid premium prices for this variety of fiber, because it produced a finer yet stronger thread than short-staple cotton and was much better suited for mechanization.[5]

The story of the cotton machines that spawned the Industrial Revolution in Britain has been told many times.[6] John Kay's flying shuttle, invented in 1733, halved the work involved in weaving but still required skilled labor. In 1764, James Hargreaves invented the spinning jenny, a machine that could spin many threads at once, albeit uneven ones suitable only for the weft (the thread that is drawn back and forth across the loom). Five years later, Richard Arkwright's water frame was the first machine to produce thread strong enough for the warp (the longitudinal threads on the loom) that could replace the linen warp used until then. Samuel Crompton's self-acting mule, invented in 1779, produced thread that was both strong and fine. These machines led to a hundredfold increase in thread production. Weaving was mechanized after 1815 by Edmund Cartwright's power loom.

The new machines required more power than human beings could generate. That power came from waterwheels and, from the 1830s on, increasingly from steam engines. Waterwheels and steam engines, along with the manufacturers' desire to control their labor force, led to the construction of huge cotton mills. One observer described the mill as a "great oblong ugly factory, in five or six tiers, all windows, alive with lights on a dark winter's morning, and again with the same lights in the evening; and all day within, the thump and scream of the machinery, and the thick smell of hot oil and cotton fluff and outside the sad smoke-laden sky, and rows of dingy streets and tall chimneys belching dirt, and the same, same outlook for miles."[7] In those mills, men, women, and children toiled under appalling conditions.

Yet these mills produced astonishing results. Raw cotton imports to Great Britain rose almost two-hundredfold, from 2.3 million kilograms in 1780 to 425 million kilograms in 1856. Spinning 45 kilograms of raw cotton that took Indian artisans 50,000 hours required only 300 hours on Arkwright's machine, or 135 hours on an early nineteenth-century self-acting mule. Between 1790 and 1812, the price of cotton yarn fell by 90 percent.

Iron

Before the eighteenth century, coal could not be used to smelt iron ore because it contained sulfur and phosphorus that contaminated the metal. The first industrially useful method was devised by Abraham Darby in 1709, using coke (coal heated to drive off impurities) in place of charcoal; even then, it was not until the 1780s that coke-iron became competitive with charcoal-iron. The results were enormous: British production of pig iron (the raw material produced by a furnace) rose from 17,250 tons in 1740 to 2.7 million tons in 1852. As smelting a ton of iron, if made with charcoal, required 50 cubic meters of wood or the sustainable yield of 10 hectares of forest, smelting that much iron with charcoal would have required almost twice as much land as existed in all of Great Britain.[8] Even as Britain's coal production rose, the share used to smelt iron increased even faster, from almost none before 1750 to almost a fifth in 1830.[9]

The Steam Engine

Coal was necessary but not sufficient to bring about an industrial revolution in Britain. For that to occur required a byproduct of coal mining: the steam engine. Already at the end of the seventeenth century, many coal mines had reached the water table, preventing further digging. Using horses to pump water was very costly. Only a few mines could be drained by carving a tunnel to a lower level. Inventors attempted to create a device that would use heat to pump water. The first to build a commercially successful engine powered by burning coal was Thomas Newcomen, whose Dudley Castle Machine, a device as large as a house, began pumping water from a coal mine in 1712. Since broken pieces of coal were practically free at the mine-head, it did not matter that in this engine, 0.7 percent of the energy contained in the coal moved the piston, while 99.3 percent went up the chimney. Newcomen's engines were soon copied throughout Europe. For the first time, a fossil fuel was used as a source of mechanical energy.[10]

The next major leap in technology occurred between 1765 and 1788 when James Watt, an instrument maker at the University of Glasgow, devised several inventions that raised the efficiency of his engine to 4.5 percent, allowing it to power machinery that required a smooth delivery of rotary motion. By the early nineteenth century, steam engines were becoming common, both in mines and in a host of new industries.

The environmental impact of mining extended far beyond the mines and mills to the cities and countryside of Britain. Surface mining left great pits in the ground. Even after surface deposits were exhausted, underground mining left piles of stones and other residues on what had once been farmland. Mines drained water from wells and springs and polluted streams with minerals and heavy metals. Coal dust covered nearby fields and meadows. Abandoned mines caused subsidence, undermining the land above that sometimes caved in. Transporting coal by ox or horse-drawn cart raised the cost so much that after 1750, mine owners built canals between their mines and the nearest port or city, changing the landscape.

Railways

The most spectacular application of steam engines was the railway. Even before steam power, mine owners in Britain had built tramways of wooden planks or iron plates to move coal-laden carts more easily than on rutted roads. The first commercial application of steam to land transportation was the railway from Stockton to Darlington, in northeastern England, built in 1825 to carry coal as well as passengers. William Huskisson, president of the Board of Trade, expressed the excitement of the age when he wrote: "If the steam engine be the most powerful instrument in the hand of man to alter the face of the physical world, it operates at the same time as a powerful lever in forwarding the great cause of civilization." Ironically, he became the very first person to die in a railway accident.[11]

As Huskisson noted, railways altered the face of the physical world. They cut across the land, not following the terrain but taking the straightest route possible. To smooth out the landscape, engineers designed tunnels, bridges, and embankments. Railways also consumed natural resources at a prodigious rate. Each kilometer of track required 180 tons of iron just for the rails, plus 1,640 ties, consuming over 3 hectares of forest. Even after being treated with creosote, ties had to be replaced every seven years on average. Historian Michael Williams estimated that in the 1840s, railways used 20 million ties or 41,000 hectares of forest per year, a number that rose to 857 million ties or 3.4 million hectares of forest per year by 1900.[12]

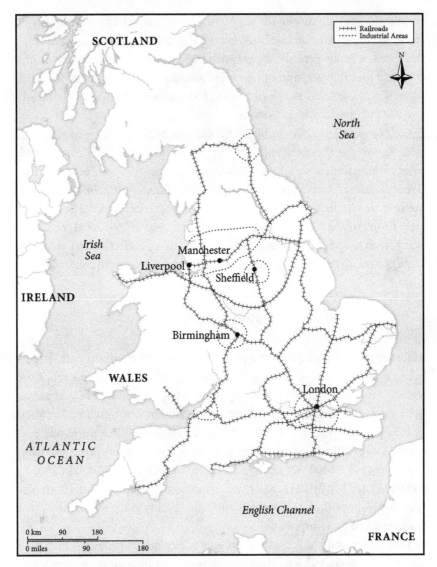

Figure 8.1. Map of England during the early Industrial Revolution (ca. 1850) showing industrial cities and major railroad lines.

Urban Environments in Great Britain

Long before the Industrial Revolution, European cities were so unhealthy that only a constant influx of healthy rural people prevented their populations from shrinking. Streets were narrow, houses were poorly built, and the poorer neighborhoods were vastly overcrowded. Horses, cows, pigs, chickens, and other animals befouled the streets. Chamber pots were emptied into cesspools periodically emptied by scavengers. Tanning, bleaching, dyeing, paper making, and other trades produced revolting effluents. Rain, when it came, washed away some of the filth but turned the unpaved streets into a blend of mud and excrement.

London was especially polluted because its buildings were heated with coal. As early as the thirteenth century, when the inhabitants complained, royal commissions were appointed to investigate the thick pall of smoke that enveloped the city on cold days. Complaints declined during the Black Death but returned in the sixteenth and seventeenth centuries.[13] As the population grew from around 60,000 in 1534 to around 530,000 in 1696, coal consumption rose in proportion, from 20,000 metric tons a year in the mid-sixteenth century to close to 450,000 metric tons in the late seventeenth century.[14]

Industrial Cities

Industrialization multiplied the problems of urban environments, largely because manufacturers, seeking the proximity of customers and suppliers, attracted workers, causing urban populations to grow faster than ever before in history. London grew from 1.1 million inhabitants in 1801 to 2.7 million in 1851 and to 6.6 million in 1901, by far the largest city the world had ever seen.[15] Other British industrial cities grew as fast or faster: Liverpool from 77,000 in 1801 to 400,000 in 1851 and Birmingham from 73,000 in 1801 to 250,000 in 1851.[16] By 1851, half the population of Britain lived in cities, fifty years before any other European country and seventy years before the United States reached that proportion.

With the coming of industries, urban environmental problems increased faster than the population. Much of the growth in population took place in new cities or in the surroundings of old towns. Until well into the nineteenth century, the lack of housing standards or zoning ordinances meant that builders built as cheaply as possible, which was all that industrial workers and their families could afford. What shocked visitors most was the new housing. In 1844, Nassau Senior, the foremost English economist of his day, wrote:

> These towns . . . have been erected by small speculators with an utter disregard to everything except immediate profit. . . . In one place we saw a whole

street following the course of a ditch, in order to have deeper cellars (cellars for people, not for lumber) without the expense of excavation. Not a house in this street escaped cholera. . . . The streets are unpaved, with a dunghill or a pond in the middle; the houses built back to back, without ventilation or drainage, and whole families occupy each a corner of a cellar or of a garret.[17]

Sewage and Garbage

As in older cities, human waste polluted the nearest streams and rivers, but animals also contributed to the sewage problem. Before refrigeration was introduced in the late nineteenth century, dairies were located in cities, which meant that cow urine and manure flowed into the nearest river. Horses were essential to urban life, pulling wagons, the carriages of the rich, and the horse-drawn streetcars that increasing numbers of people used to travel the ever-longer distances in growing cities. London employed tens of thousands of horses, each of which produced 8 to 15 kilograms of manure, a total of a thousand tons or more every day. Much of this was collected and sold to nearby farmers for fertilizer, but much also ran into the Thames. To these traditional effluents industrialization added new ones from engine coolant water, textile mills and dye works, and coal washing. Government action lagged behind the growing needs of cities. Not until 1878 was the first legislation passed to curb river pollution, but it was not enforced until the twentieth century.

By the 1840s, as rivers became toxic, anglers complained that fish had disappeared. Tidal rivers, like the Thames at London, sometimes left sewage-laden water stagnant for days; when the level of the Thames dropped during the heat wave of 1858, leaving sewage to ferment in the hot sun, Parliament had to adjourn for a week. This event persuaded the municipal governments of London and other English cities to install sewers, thereby transferring the pollutants from cesspools to rivers. Though London had completed its trunk sewers by 1864, for decades thereafter many houses still used cesspools and outdoor privies.

Garbage also accumulated. Food scraps and other organic matter were eaten by pigs and dogs that roamed the streets. In coal-heated cities, ashes piled up until they were carted off to be dumped outside the city limits; since this service cost money, in poorer neighborhoods the ashes simply accumulated. Horses were so overworked and mistreated that many died in harness or were killed when they broke a leg, and their carcasses were often left in the streets for days. Not until the last third of the nineteenth century did municipalities come to grips with these problems of too-rapid urbanization.[18]

Air Pollution

In the industrializing cities, air pollution became progressively worse in the nineteenth century. In the 1880s, during the winter months London received only one-sixth as much sunshine as small country towns. Some of the pollution was in the form of soot and ashes that fell to the ground. While in the pleasant spa town of Malvern, 2 tons of solids fell per square kilometer every year (most of it windblown leaves and soil), 15 tons of solid matter fell on London and 21 tons on Attercliffe, a suburb of industrial Sheffield.[19]

Much of that pollution came from domestic grates and fireplaces, where coal smoldered for lack of sufficient oxygen. But increasing amounts of coal smoke issued from smelters, coke ovens, pottery kilns, blast furnaces, steam engines, and railroad yards, especially when unskilled workers did not ensure that all the coal was thoroughly burned. Some industrial establishments built tall smokestacks to disperse smoke and particulates over a large area. Locomotives produced more smoke at ground level than stationary engines because their smaller fireboxes and higher drafts burned coal less efficiently. In rail yards, many locomotives and switching engines worked closely together, emitting a pall of smoke near the centers of cities.[20]

Soot and smoke so dirtied the air that people began to wear dark clothes and carry black umbrellas. Coal consisted not only of carbon—which turns to carbon dioxide when it burns—but also of various minerals, especially sulfur. Whereas anthracite, the purest form of coal, contains 0.1 percent sulfur, the much more common bituminous coal contains up to 2.9 percent sulfur. In a fire, the sulfur turns to sulfur dioxide, which escapes into the atmosphere and reacts with rainwater, producing sulfuric acid. Rain containing diluted sulfuric acid turned limestone (calcium carbonate) into gypsum (calcium sulfate), which crumbled when wet; as a result, many limestone buildings began to erode. Coal smoke and soot also attacked vegetation and damaged interiors.[21] Hydrochloric acid, another dangerous pollutant, was a byproduct of the manufacture of sodium carbonate used in many industrial processes. Released as a gas, this acid produced a stench like that of rotten eggs and damaged trees and grass.[22] Legislation to curb air pollution had a long history, as did appeals to manufacturers to exercise voluntary restraint. But they could not be enforced because much of the pollution came from domestic hearths, where burning coal was the only source of heat until well after the Second World War.[23]

Urban Diseases

Pollution and other urban problems caused disease and death rates to soar. In the mid-nineteenth century, England's life expectancy at birth was 39.5 years, 3.2 years lower than it had been in 1581.[24] Half the children born in Manchester died before the age of five and three out of five before the age of ten; as a result, life expectancy at birth was only seventeen years, reflecting the horrendous mortality caused by smoke, polluted water, and unsanitary food.[25]

Until the mid-nineteenth century, medical authorities attributed diseases to miasmas, or foul smells from putrefying organic matter. Physicians often confused various diseases or failed to specify the causes of diseases, or remained unconcerned about pollutants they could not smell. They did not consider smoke unhealthy; some even thought of it as a disinfectant.

Then came the cholera pandemics of the mid-century. To contemporaries, cholera was the most shocking of all diseases. If untreated, it killed half its victims in a particularly gruesome way. An infected person could seem perfectly healthy one minute, then suddenly collapse with terrible stomach cramps, copious vomiting, and watery diarrhea, until the body lost half its weight and shriveled, with death by dehydration following within a couple of days, or even a few hours.

Cholera had been endemic in India for centuries. An outbreak in Bengal in 1817 spread to Southeast Asia in 1819–1820, to Russia in 1823, to western Europe in 1831, and to the British Isles and North America in 1832. What allowed it to spread was the introduction of steamships, which shortened the time for long-distances travel from months to weeks, less than the incubation period of the disease. During that first pandemic, the general reaction to the outbreak of cholera in Europe and North America was, as in centuries past, to see cholera as a divine punishment for sin. By the time the disease returned to Britain between 1849 and 1854, public health was becoming a political issue and a topic of interest to the scientific community.

The physician Dr. John Snow, while investigating an outbreak of cholera in a mine near Newcastle, had observed that the miners, who had no means of staying clean underground, ate food contaminated with feces. He publicized his findings in 1849 in a pamphlet *On the Mode of Communication of Cholera*. In 1854, while investigating the cause of the epidemic in London, he found a correlation between the incidence of cholera in a neighborhood served by a water company that obtained water from sewage-contaminated sections of the Thames, while neighborhoods served by another company, which obtained is water from uncontaminated sections of the river, were less affected.[26] This was the first major breakthrough in epidemiology. Yet the implications were not accepted right away. Only after yet another pandemic, between 1863 and 1875, did wealthy industrialized nations begin to invest in the costly infrastructure needed to supply clean water to their cities.[27]

Since then, this once-feared disease has practically vanished from Western nations, although it is still endemic in parts of India and Africa and periodically breaks out in South America. Whereas over 150,000 Americans are said to have died of cholera between 1832 and 1849, only eleven Americans died in the sixth pandemic, which reached the United States in 1911. In the words of medical historian Charles Rosenberg: "The cholera pandemics were transitory phenomena, destined to occupy the world stage for only a short time—the period during which public health and medical science were catching up with urbanization and the transportation revolution."[28]

Cholera and other water-borne diseases like typhoid can be blamed on fecal contamination of drinking water rather than specifically industrial conditions, but respiratory diseases were a direct result of air pollution. After the last cholera epidemic in Britain in 1866–1867 had passed, bronchitis became the number one cause of death in factory towns, followed by tuberculosis.[29] Rickets, a deformation of the bones caused by a lack of vitamin D, was endemic in the industrial towns of northern Europe where smoke often obscured the sunshine; parents could not follow the traditional practice of taking babies out of doors in the winter, as there was seldom enough bright sunlight for their skin to make the amounts of vitamin D they needed to be healthy.[30]

Manchester

No city epitomizes the Industrial Revolution as much as Manchester. Dubbed "Cottonopolis," it was a city of dramatic successes and shocking horrors.

A small market town of some 25,000 inhabitants in the mid-eighteenth century, Manchester grew to 75,281 in 1801 and to 303,382 in 1851. Early cotton manufacturers like Richard Arkwright were attracted to it because they encountered few restrictions on business. The city was close to the coal mines of Lancashire and to Liverpool, where ships bringing raw cotton from the Americas unloaded and where poor Irish immigrants disembarked. With twice the rainfall of London, Manchester's moist climate was well suited to manufacturing yarn that tended to break in drier places. In this favored location, cotton mills proliferated, from 26 in 1802 to 66 in 1821 and to 182 in 1838.[31]

Three rivers, the Irwell, the Irk, and the Medlock, provided both water power to turn the machinery and a handy place to dispose of wastes. Cotton mills dumped coal ash and cinders into the waterways, along with arsenic and other chemicals used in textile finishing. Every few years, heavy rains caused the rivers to burst their banks, flooding low-lying areas where workers lived.[32] Hugh Miller, an English geologist of the mid-century, described the Irwell:

> The hapless river—a pretty enough stream a few miles up, with trees overhanging its banks and fringes of green sedge set thick along its edges—loses caste as it gets among the mills and print works. There are myriads of dirty things given it to wash, and whole wagon-loads of poisons from dye houses and bleachyards thrown into it to carry away; steam boilers discharge into it their seething contents, and drains and sewers their fetid impurities; till at length it rolls on—here between tall dingy walls, there under precipices of red sandstone—considerably less a river than a flood of liquid manure.[33]

The extremely rapid growth of a largely impoverished population created a housing and sanitation crisis. In working-class neighborhoods, people got their drinking water from wells or from the rivers, while their wastes accumulated in privies and cesspools that were emptied twice a year. Those who could afford to do so moved to suburbs south of the city, where the air and water were cleaner. Not until 1878 did the municipal government begin the necessary public works to supply clean water to the whole city.[34]

So unprecedented were the conditions in Manchester that they attracted international attention. Perhaps most famous is the report of Friedrich Engels in his *Condition of the Working Class in England in 1844*:

> One day I walked with one of these middle-class gentlemen into Manchester. I spoke with him about the disgraceful unhealthy slums and

drew his attention to the disgusting condition of that part of town in which the factory workers lived. I declared that I had never seen so badly built a town in my life. He listened patiently and at the corner of the street at which we parted company, he remarked: "And yet there is a great deal of money made here. Good morning, Sir!"[35]

American Cotton

American industrialization followed a different path from that of Great Britain. While the Northeast of the United States emulated British industrialization, the South, the Midwest, and the West were, to European Americans, vast storehouses of commodities there for the taking. Unconstrained by the traditions and vested interests that prevailed in Europe, Americans devoured the natural resources they encountered. In a land of seemingly inexhaustible abundance, there was little need to save for the future. In the nineteenth century, the federal government encouraged white settlers by giving them plots of land west of the Appalachians. Though nineteenth-century European Americans often spoke of "progress" or "the advance of civilization," theirs was an economy of plunder, the very opposite of sustainable development. Not until the end of the nineteenth century, when governments caught up with the pioneers, were restraints gradually put in place.

The New England Cotton Industry

At the time of their independence, Americans were familiar with Britain and its cultural and technological accomplishments. Some had visited the British Isles and others had immigrated from Britain, bringing with them British skills and ideas. It is not surprising that the new technologies would find fertile soil in America. Thus the cotton industry soon made its way across the Atlantic. In 1790, Samuel Slater, an apprentice to Richard Arkwright who had memorized the design of Arkwright's spinning machines, emigrated to Rhode Island, where he opened a textile mill. Soon cotton mills proliferated. From the beginning, the new cotton industry was dependent on water power, which was cheap and abundant in the region.

In 1813, several Boston businessmen built an integrated cotton mill—including carding, spinning, weaving, and other machines—on the banks of the Charles River in Waltham, Massachusetts. Once the Charles River was dammed and put to use, their Boston Manufacturing Company moved to take over the Merrimack, a river with eighteen times the volume of the Charles. On its banks, the company built mills at Lowell and Lawrence, Massachusetts, and at Manchester and Nashua, New Hampshire. By mid-century, forty mills used the Merrimack's waters. Water also powered grain mills, forges, sawmills, and other industries throughout New England. In the process, these industries dumped organic wastes, dyes, resins, soaps, and various chemicals into the rivers from which they got their power. Their dams prevented fish from swimming upriver to spawn. New England rivers had once teemed with salmon, sturgeon, smelt, shad, alewives, and other fish, but by the mid-century, the fish were gone.[36]

Cotton in the South

Had it not been for a breakthrough in the biology of the cotton plant, the industrialization of cotton would soon have reached a dead end. Long-staple sea-island cotton, the raw material best suited for mechanization, grew poorly away from the ocean, while short-staple cotton was unsuited to the new machinery. American planters resolved this dilemma by adopting the Mexican species *Gossypium hirsutum*, which they named "upland cotton" because it grew well even far from the sea. In fact, it flourished throughout the southeastern United States as far west as Texas, anywhere with 200 frost-free days and 500 millimeters of rainfall in the spring and summer. Its fibers were of medium length, appropriate for machinery. It produced larger bolls, allowing slaves to pick five times more than cotton with small bolls.[37]

The cotton industry transformed the landscape of the American South. The huge expansion of cotton cloth production in Britain and New England depended, until 1860, on imports from the American South, where cotton was grown on slave plantations. In response to the booming demand for raw cotton and the high prices it commanded, white planters moved with their slaves from Georgia and the Carolinas into Alabama, Mississippi, Louisiana, and Tennessee from 1800 to the 1840s. Slave plantations were nothing new in North America, but by the late eighteenth century, the traditional crops—tobacco, indigo, and rice—were no longer profitable, and many people expected slavery to die out.[38] Instead, the cotton boom revived slavery.

The United States had something that Great Britain did not, namely, a vast continent brimming with boundless resources. Traditional societies had always placed restraints on activities that depleted certain resources: royal or aristocratic hunting preserves, peasant rights to pasture their flocks and herds on common lands, taboos against killing certain animals. Even in a newly capitalistic nation like Great Britain, access to resources involved negotiations between sectors of society claiming vested interests. In the new United States, in contrast, traditions and vested interests were weak; the only real obstacles were the geography of the land itself and the resistance of the Native Americans.

As cotton plantations moved west, slaves cleared the land, turning long-leaf pine forests into fields. Cotton exhausted the nitrogen in the soil, one of the essential nutrients, in two years. Maize, the South's other main crop, was grown to feed humans and their pigs but exhausted the fertility of the soil even more thoroughly than cotton.[39] Plantation owners raised little livestock and did not find it worth the cost of collecting and spreading manure; as Thomas Jefferson once said: "We can buy an acre of new land cheaper than we can manure an old one."[40]

Figure 8.2. Black workers picking cotton under the eye of a white overseer (Louisiana, 1890). iStock.com/duncan1890.

By the 1840s, most of the virgin land suited for cotton had been used up, just as the price paid for raw cotton was collapsing. When the long boom ended, farmers began taking measures to restore or maintain the land. They experimented with crop rotation, alternating cotton, maize, and cowpeas (or black-eyed peas), a legume that restored the nitrogen in the soil and could be fed to hogs. To make cotton grow year after year in the same fields, they applied guano imported from South America and phosphate mined in South Carolina.[41] After a surge in yields, cotton production began a long decline because these fertilizers, unlike crop rotation, left the soil poor in nitrogen. As once-fertile land in Georgia and Alabama was abandoned, cotton growing moved ever farther west. The result was the worst soil erosion and environmental destruction in the history of the South. The Cotton Kingdom eventually stripped the South of an estimated 25 cubic kilometers of topsoil.[42] Left behind were vast areas of denuded land, soon eroded into gullies and fit, in many places, only for yellow pines that could tolerate poor soils.[43]

The Midwest

This vast region, stretching between the Appalachian Mountains and the 98th meridian, includes the Great Lakes and the valleys of the Ohio, the upper Mississippi, and the lower Missouri rivers. Forests once covered the north and its eastern edge. West of Indiana, much of the land was covered with a tall-grass prairie. Both forests and grasslands were blessed with rich soil and sufficient rainfall for agriculture. The European Americans who moved there in the early to mid-nineteenth century hoped to establish family farms, idealized versions of European yeoman peasantry. To do so, they had to transform the ecosystem from woods and natural prairie to arable land. In the process, they removed forests and wildlife, often permanently.

For a long time, white Americans living on the East Coast who wanted to move into the lands west of the Appalachians were restrained, first by British policy and then by the War of Independence. Starting in the 1790s, pioneers made their way across the mountains. Many came from New England where the soil was thin and filled with rocks, the growing seasons were short, and transportation away from rivers or the coast so difficult that many farm families had to be practically self-sufficient.[44]

The incentive to move west was especially acute in and after 1816, the "year without a summer," which followed the eruption of the Indonesian volcano

Tambora. By the following spring, the Earth was covered with a dust veil, which the poet Lord Byron immortalized with these words:

> The bright sun was extinguish'd, and the stars
> Did wander darkling in the eternal space,
> Rayless, and pathless, and the icy earth
> Swung blind and blackening in the moonless air;
> Morn came and went—and came, and brought no day.[45]

A dry fog settled over northeastern North America. In June, July, and August 1816 there was frost or snow in New England, New York State, New Jersey, and Quebec. Crops were either stunted or entirely ruined and much livestock was lost. Some farmers migrated to the new mill towns of Connecticut and Massachusetts, while others relocated to what promised to be better land in Ohio and beyond.[46]

Steamboats and Railways

In early nineteenth-century America, roads were few and in poor condition. In a land of rugged terrain and abundant rainfall, rivers and the sea provided the easiest routes for travel and trade. It was an American, Robert Fulton, who built the first commercially successful steamboat. In 1807, his *North River* (better known as the *Clermont*) steamed up the Hudson River from New York City to Albany in thirty-two hours and back in thirty—a trip that would have taken several days each way by stagecoach. Soon steamboats were plying all the major rivers in the northeastern states.[47] Within a few years, they appeared in the Midwest. Fulton's *Orleans* steamed from Pittsburgh to New Orleans in late 1811. In 1816, Henry Shreve launched the *Washington*, a new kind of steamboat with a flat bottom and a shallow draft, propelled by a high-pressure engine on deck and a stern wheel—the first of the classic Mississippi riverboats. Boats of this design were able to navigate year-round, even when the Ohio and Mississippi rivers were very low.[48]

Thanks to steamboats, the Midwest gained better communication with the Gulf of Mexico than with the geographically closer eastern states. Travel across the Appalachian Mountains was difficult and lengthy, and carrying goods was prohibitively expensive. The harvest failures of 1816 and the migration of settlers across the mountains provided an incentive to dig a canal linking Albany on the Hudson River with Buffalo on Lake Erie, above Niagara Falls. The Erie Canal, 584 kilometers long through what was then called "trackless wilderness," took nine years to build. When completed in 1825, it opened the Midwest to European settlement. Though too slow for passengers, it lowered the cost of carrying freight across the Appalachian divide by 97 percent.[49]

As the white population of the Midwest grew to 3.5 million in 1830 (25 percent of the white American population) and to 8.3 million (36 percent) in 1850, so did the need for transportation. From 69 steamboats on the Ohio and Mississippi rivers in 1820, the number rose to 557 in 1840 and to 750 in 1850. At their peak, these boats, steaming through what was then a forested land, consumed 1,942 square kilometers of forest a day, depleting the wood supply for miles along the major rivers.[50]

Soon after the Erie Canal was opened to traffic, the railway broke through the Appalachian Mountains into the Midwest. Construction proceeded rapidly. The first railroad reached Chicago in 1848, the same year as the opening of the Illinois and Michigan Canal that linked the Great Lakes with the Mississippi watershed. Chicago became the hub of midwestern transportation and, for a time, the fastest growing city in the world. Only twenty-one years later, a railroad connected the Midwest with California.[51]

By then, the United States surpassed all other nations in its enthusiasm to build railroads. In 1840, the country's 4,500 kilometers of track exceeded all of Europe's. By 1860, the United States had 49,200 kilometers of track, and by 1900 it had 311,200 kilometers of track, more than Europe including Russia. All of these rail lines consumed gargantuan amounts of iron ore, coal, and limestone for rails and equipment and wood for ties and fuel.[52]

Farms

Midwestern farming was not a simple matter of transferring European or New England methods to a new place. Land that was not forested was covered with prairie grasses that had grown there for thousands of years and sent a thick mat of roots several feet into the soil. To cut the sod, farmers hired professional prairie breakers with large teams of horses pulling steel plows. These plows, manufactured by John Deere from 1837 on, proved ideal for the thick soils of the Midwest; in 1857 his company sold over 13,000 of them. By then the wheat harvests were so abundant that farmers began purchasing another machine, McCormick's mechanical reaper, designed to harvest a large field quickly in a region prone to fickle weather. Within a generation, the Midwest became the breadbasket of the United States, producing most of America's wheat and maize and a few other crops such as rye, barley, oats for horses, and hay for cattle. Complementing the production machinery were new technologies of storage and transportation: grain elevators, dedicated railroad cars, trains of barges on the big rivers, and freighters to carry the bounty to Europe.[53] By the end of the nineteenth century, the simple ways of the pioneers were yielding to a more complex agriculture, with multiple crops, several kinds of livestock, and a variety of machines drawn or powered by horses. After the mid-twentieth century, these integrated farms were increasingly replaced by industrial farms with tractors, silos, steel plows, combines, and chemical fertilizers. Each farm, each region in fact, specialized in one or two crops to be exported to distant markets—Kansas produced wheat, Iowa maize and pigs, Texas beef—and imported most other things from far away. What had been a very complex ecosystem was simplified to the extreme, the outdoor equivalent of an industrial manufacturing plant.

Forests

The upper Midwest—Michigan, Wisconsin, and Minnesota—was once covered with a great forest. When pioneers first arrived in the region, they saw it as an obstacle to progress. Alexis de Tocqueville, after visiting Michigan in 1831, wrote that Americans were "insensible to the wonders of inanimate nature and they may be said not to perceive the mighty forests that surround them till they fall beneath the hatchet. Their eyes are fixed upon another sight . . . peopling solitudes and subduing nature."[54] Or, as nineteenth-century historian and horticulturist Francis Parkman put it more bluntly, forests were "an enemy to be overcome by any means, fair or foul."[55]

Forests were more than an obstacle, however; they were also a treasure chest of that most essential raw material, wood. The forests of the northern Midwest comprised many species of trees: elms, basswoods, and sugar maples along the prairie; hemlocks, birches, and maples farther north; then, even farther north, the great conifer forest that extended into Canada. Especially sought after were white pines, tall straight trees with smooth, even wood, excellent for making beams and planks.

Wood had many uses, none as important as fuel. In the mid-nineteenth century, steamboats burned 14.5 million cubic meters of wood, and railroad locomotives another 11 million cubic meters; in addition to these figures are fuel for stationary steam engines and for heating houses and for industrial processes. Well into the late nineteenth century, railways and riverboats in the American West burned wood rather than coal.[56]

Wood was also the essential raw material for construction. In heavily forested areas, settlers built log cabins or houses with heavy posts and beams with mortise-and-tenon joints. But on the prairie where trees were scarce, such extravagant consumption was out of the question. Solving the problem was the balloon frame, a method of construction that used planks instead of logs. With a saw, a hammer, nails, and modest skills, one man could build a house out of lumber precut to standard dimensions. By the mid-century, it was the most common form of construction in America. Using balloon frames, entire towns sprang up on the prairie almost overnight. By the late nineteenth century, mail-order catalogs offered kits for houses, churches, stores, barns, and other buildings shipped out by railroads and ready to assemble. Wood was also used to build boats and ships, buckets, barrels, water pipes, and furniture, even to pave sidewalks and "plank roads" in muddy areas. The railroads of America were not only fueled but also built of wood, for they consumed enormous amounts of wood for rolling stock, bridges, trestle viaducts, and cross-ties. By the 1890s, they needed 73 million cross-ties a year, both for new tracks and for existing ones that had to be replaced every five to eight years.[57]

With wood so abundant and cheap, it is not surprising that Americans used it to smelt iron long after the British had switched to coal. As late as the mid-century, most iron was still made with charcoal. Making 1 ton of iron required 1.5 cubic meters of charcoal which, in turn, used 21.7 cubic meters of wood. Forest historian Michael Williams calculated that as late as 1880, wood provided 60 percent of US energy and that Americans consumed over twice as much wood per capita as Britons.[58]

Like everything else in nineteenth-century America, harvesting and processing wood was transformed from a craft into a heavy industry. In the early nineteenth century, lumber mills were using circular saws to cut 12 to 19 cubic meters of lumber a day; steam-powered saws raised their productivity to 95 cubic meters but turned much of the wood into sawdust. The introduction of balloon frame buildings created a demand for standardized sizes of lumber for doors, windows, stairs, and other parts as well as planks and joists. With mass production came consolidation, as small lumber mills gave way to giant companies like Weyerhaeuser that owned vast tracts of timberland dotted with lumber camps. Much logging took place in the winter when roads covered with ice made transporting the logs easier. In the last quarter of the century, lumberjacks laid small-gauge rails to haul logs to the nearest river or lake, moving the rails every year as the forests were depleted. On rivers and lakes, logs from a variety of sources and landowners were collected into gigantic booms or rafts covering up to 2 hectares. Booms from Michigan and Wisconsin were pulled by steamboats to mills in Chicago to be processed into lumber. So inefficient were the methods and extravagant the use of wood that much of it was wasted, to the shock of visiting Europeans.[59]

As long as old-growth forests remained, the lumber companies showed no interest in scientific management or sustainable development. By the end of the century, almost all the commercially valuable pines were gone.[60] In the three Great Lakes states, over 20 million hectares of forest were laid bare by clear-cutting, leaving behind a wasteland of stumps, dead branches, and other debris.

In this wasteland, passing locomotives, lightning, and misguided attempts to burn stumps ignited huge fires. In 1871, a fire devastated a million hectares in Michigan. According to Williams, "fire probably consumed about as much timber every year as reached the mill."[61] Fires also destroyed the thin layer of humus and whatever saplings had managed to grow in the cutover areas. Unlike in New England and upstate New York, where the clear-cutting of forests had left land suitable for dairying and agriculture, in the Great Lakes states the land under the pine forests was sandy and infertile and the growing season of 100 to 130 frost-free days too short for farming. When lumbering declined, the mill towns and the land around them were abandoned, leaving only barren ground and stunted bush. It would be a hundred years before the northern Midwest was once again forested, this time with secondary growth trees, a pale shadow of the virgin forest that once covered the land.

The Great Plains

West of the 98th meridian, rain and snowfall diminishes and becomes more erratic. Here the tall-grass prairie gave way to shorter blue gamma and buffalo grasses that could withstand floods and droughts. Going west, agriculture became progressively riskier and finally impossible. This was once the land of the bison.

Bison and Indians

Of the many large animals that had once roamed North America, few survived the great Pleistocene extinction. One of them was the bison. On the prairie, bison became the keystone species of North America upon which the grasses, other animals, and Indians depended. Bison ranged from Georgia to Montana and into Canada and Mexico. Their population fluctuated considerably, with 24 to 30 million being a conservative estimate.[62] The very behavior that had saved the bison thousands of years before—moving in large herds and reproducing prolifically—protected the species from wolves and human hunters. This behavior proved to be a weakness, however, when the Indians acquired horses, and a fatal one when whites came with rifles.

With the horses they acquired during the eighteenth century, the Plains Indians no longer had to hunt on foot or gather roots and berries in the river valleys. Instead, the men became full-time hunters. From bison, they obtained not only meat but also hides from which the women made bedding, clothing, moccasins, and tepee covers; bones for fuel; sinew for bowstrings, thread, and snowshoes; and other parts that satisfied almost all their needs.[63]

Before the nineteenth century, hunting probably did not affect the bison population. That changed in the early nineteenth century with the arrival of white merchants offering to trade whiskey, blankets, clothing, horse gear, knives, firearms, and ammunition for bison robes and tongues, then considered a delicacy. The rest of the animal was often wasted. Until 1848, the trade was brisk, for unusually high rainfall caused the grass to grow well and bison to proliferate. One trading company, the American Fur Company, bought 45,000 bison robes in 1839 and 110,000 in 1847. The new market economy also affected Indian society. As the Indian population doubled between 1820 and 1840, intertribal warfare spread throughout the Plains. The most warlike Indians—the Comanches in the southern Plains, the Cheyenne, Kiowa, Sioux, and others farther north—raided one another's camps to steal horses and to kidnap the women and girls they needed to process the skins. A peace treaty in 1840 between the Comanches and Kiowas, on the one hand, and the Cheyennes and Arapahoes, on the other, led to more intensive hunting, putting more pressure on the bison herds.[64]

Figure 8.3. Native Americans hunting bison using horses, spears, and arrows, 1868. iStock.com/Grafissimo.

Indian hunters were not the worst threat to the bison, however; their horses were. In the winter, the bison retreated to river bottoms, especially the "Big Timbers" along the Platte, Arkansas, Republican, and Smoky Hill rivers, where they found shelter from the wind and snow and enough grass and twigs to carry most of them through the cold months. The Indians, who owned between five and thirteen horses apiece, needed the same habitats. Their horses, numbering between 100,000 and 150,000, trampled the ground and ate the saplings and the twigs and bark of the cottonwood trees that lined the riverbanks. So did the estimated 2 million feral mustangs that roamed the Plains, competing with the bison for forage. The bison, expelled from their winter grounds, retreated to poorer areas exposed to life-threatening blizzards, where many perished.[65]

Then came a dry spell, with droughts in 1849, 1855, and the 1860s, the worst in the century. Less grass meant fewer bison, but other pressures affected their population as well. The Indians preferred to kill young bison cows because their meat was more tender and their hides produced softer, more luxurious robes. Removing them however, reduced the herds' ability to reproduce. Nor could the herds move east into the tall-grass prairie, increasingly occupied by white farmers. The bison were also affected by anthrax, brucellosis, tuberculosis, and parasites they acquired from infected cattle. As the bison population declined, so did the harvest of robes; in 1859 the

Figure 8.4. Pile of bison skulls waiting to be ground up for fertilizer, mid-1870s. Wikimedia Commons.

American Fur Company bought only 50,000, fewer than half as many as a decade earlier.[66]

The White Invasion

Meanwhile, European Americans were pouring into the Great Plains. After having been confined to the East Coast and the Appalachian Mountains for 200 years, they occupied the Ohio River valley in the 1820s, the Mississippi River valley in the 1850s, and the western half of the North American continent by 1880. In part, what explains this sudden conquest is the growing European and European American demand for farmland and for the resources of the West: grain, meat, hides, timber, and minerals, especially the gold that was discovered in California in 1848, causing a rush of pioneers across the West. This demand had always existed, but what changed was the new balance of power between whites and Indians, the result of industrial technologies and imported diseases.

One such technology was the steamboat. In 1819, the *Western Engineer* steamed up the Missouri River to Council Bluffs (now in Iowa). In the early 1830s, a steamer belonging to the American Fur Company reached the mouth of the Yellowstone River. Steamers brought not only fur traders and prospectors but also US Army troops sent to fight the Indians who resisted the invasion. Before 1866, only half a dozen steamers reached the Yellowstone each season. The following year, however, thirty-nine steamboats arrived, carrying 10,000 passengers.[67] Steamboats changed the ecology of the land as they consumed the trees that grew along the riverbanks in an otherwise tree-less region, while their passengers and crew hunted bison for food.

The other industrial technology that opened up the western United States was the railroad. After the Civil War, the victorious Union government encouraged the building of railroad lines in the West with land grants and subsidies. The railroad companies, in turn, enticed settlers with cheap land and extravagant promises. Almost all the new lines ran east-west, connecting the populous East with the booming West Coast. Railroads were, in the words of William Cronon, "a knife in the heart of buffalo country."[68]

The white invaders overcame the resistance of the Indians with yet another new technology: breech-loading rifles. Before the 1840s, Indians and whites had possessed similar firearms—single-shot muskets and rifles that took a long time to load and could only be loaded standing on the ground—while Indians could fire off many arrows accurately while riding a galloping horse. Beginning in the 1840s, soldiers, prospectors, and settlers acquired revolvers and breech-loading rifles and carbines that could be fired and reloaded on horseback. After the American Civil War, the West was flooded with new and even more powerful firearms, especially Remington and Winchester re-peating rifles that could fire many rounds in a few seconds. Indians acquired such rifles too, but had difficulty obtaining enough of the factory-made am-munition these guns consumed. At one of the last violent encounters be-tween Indians and white soldiers at Wounded Knee, South Dakota, in 1890, the army used machine guns against the Indians.[69]

But the Indians were dying anyway from diseases brought by whites. Like the Aztecs and Incas centuries earlier, the Plains Indians were vulnerable to smallpox, which broke out in epidemics in 1816, 1818–1819, and 1837–1838. Then came cholera, brought by ship from India to Europe and Europe to North America, by steamboat up the midwestern rivers, and by migrants on the Oregon Trail in 1849–1850. Some Indian tribes lost three-quarters of their members; others disappeared entirely.[70]

The Bison Holocaust

Wars and diseases were not the only disasters that befell the Plains Indians in the last half of the century; so was starvation caused by the decline in the number of bison. Whites moving west followed the Platte and Arkansas rivers. What had been a trickle until 1840 became a flood after 1848; about 185,000 came between 1849 and 1852, and nearly 300,000 by 1859, just along the Platte, the first section of the Oregon Trail. Not only people traveled the route but also their horses, cattle, mules, and sheep—eleven animals per person by 1853—all of which grazed and browsed their way across the land, destroying the vegetation that had once sustained the bison herds in winter. In 1858–1859, after gold was discovered near Denver, the rush intensified and, with it, skirmishes between the US Army and the Plains Indians. The Union Pacific Railway, laying tracks along the Platte River in 1867, used up the last trees in what had once been a forested valley.[71]

According to historian James Shaw, "The bison population west of the Mississippi at the close of the Civil War numbered in the millions, probably in the tens of millions."[72] That is when the bison, already under pressure of hunting by Indians, became the favorite prey of white hunters as well. One reason was their commercial value. Bison leather was especially sought after to make belts for industrial machines. Railroads shipped fresh bison skins by the thousands to tanneries in the east. In 1870–1871 tanners in Philadelphia perfected a way to tan bison hides into strong supple leather using hemlock bark. In turn, the tanneries depleted entire forests of eastern hemlocks for their bark rich in tannins. In historian Andrew Isenberg's words, "A spasm of industrial expansion was the primary cause of the bison's near-extinction in the 1870s and early 1880s."[73]

The government was of two minds on the subject. In June 1874, Congress voted to save the bison. But President Ulysses S. Grant, pressured by his generals who were campaigning against the Indians in the West, refused to sign the legislation.[74] The army even gave ammunition to bison hunters; as bison hunter Frank H. Mayer wrote: "The army officers in charge of plains operations encouraged the slaughter of buffalo in every possible way. Part of this encouragement was of a practical nature that we runners appreciated.

It consisted of ammunition, free ammunition, all you could use, all you wanted, more than you needed."[75]

But government policy was of little importance compared to the enthusiasm with which hunters killed bison for "sport." Bison hunting attracted not only professional hunters but also amateurs from the East Coast, Mexico, and Canada, even English aristocrats. Some hunters killed 50 to 100 bison in one morning. The most famous of them, William "Buffalo Bill" Cody, boasted of having killed 4,280. Some "sportsmen" fired their rifles from the windows of trains as they passed herds of bison, leaving the corpses to rot.[76]

The slaughter continued throughout the 1870s and '80s. At its peak, one bison merchant in Kansas shipped 400,000 hides in one season. In 1873, there were no more bison in Kansas; by 1878, they had disappeared from Texas; in Montana, they vanished in 1883. That year, 40,000 hides were shipped; the next year, only 300 arrived.[77] Left behind were hundreds of thousands of dead bison. Colonel Richard Dodge witnessed the carnage: "Where there were myriads of buffalo the year before, there are now myriads of carcasses. The air was foul with a sickening stench, and the vast plain, which only a short twelvemonth before teemed with animal life, was a dead, solitary, putrid desert."[78]

The flesh on the carcasses rotted or was eaten by scavengers, but the bones remained. Homesteaders collected and sold them to fertilizer merchants. By the late 1880s, even that trade was over. In 1886, Spencer F. Baird, secretary of the Smithsonian Institution, sent taxidermist William Temple Hornaday out west to kill a hundred bison and bring back their skins, skulls, and skeletons for the museum's collection. A survey in 1889 counted 200 bison in Yellowstone Park, but poachers reduced that number to 23 by 1902. All in all, some 300 to 600 survived in Canada and the United States. Only a few voices were raised expressing concern.[79]

The Cattle Kingdom

What replaced bison and Indians were cattle and cowboys, the stuff of Western movies. In his classic history of the Great Plains, Walter Prescott Webb describes the origin of the cattle kingdom. It began in southern Texas before the Civil War, when cattle ran wild and reproduced prolifically on the prairie vacated by the bison. By 1860 there were between 3 and 5 million head of cattle in Texas. During and shortly after the Civil War, there was no market for them. Then came the railroad, reaching Sedalia, Missouri, in 1861 and later Abilene, Kansas, and other railroad towns. In huge drives led by teams of cowboys, thousands of animals were marched to the railheads to be shipped to slaughterhouses in Kansas City, St. Louis, and Chicago to feed the insatiable demand for beef in the East and the Midwest, even in Great Britain after the introduction of refrigerated ships in 1879. By 1871, Abilene alone was shipping 750,000 head of cattle a year.

Cattle, however, were not adapted to surviving on the Plains north of Texas. In the winters, when bison could clear the snow with their snouts and horses with their hooves to uncover grass, cattle stood and starved to death. Half the cattle in Kansas and Nebraska died in the winter of 1871–1872. Allowed to graze at will on the open range, they overgrazed and many starved. If they encountered farmland, they trampled or ate the crops. Furthermore, these were longhorn cattle of Mexican origin, producing tough low-grade beef, a breed that could not be improved since they mated freely on the open range.[80]

Barbed wire, invented in 1873–1874, alleviated some of these problems. Before barbed wire, building a wooden fence on the prairie like those in the East was prohibitively expensive. Farmers quickly adopted the new fencing material to keep cattle off their fields. Cattle owners resisted at first but later saw that it would allow them to breed their animals selectively instead of letting nature take its course. In response to the demands of the market, they could substitute imported breeds for the Texas longhorns.[81]

Barbed wire turned the open range into ranches. In the 1880s, cattle numbers grew fast. Wyoming, which had 90,000 head of cattle in 1874, boasted over half a million in 1880, as did Montana. Metal windmills, another invention of the period, could pump water for the animals, though feed remained a problem. Cattle grazed on the tastiest grasses, like bluestem, until these were replaced by inedible ones like ironweed, goldenrod, and Canadian thistle. The soil, once stripped of grass, became prone to wind erosion. Not only bison but also pronghorns, quail, and other native animal populations were massively depleted. And the problem of snow remained; as many as nine out of ten cattle died in the blizzard of 1884–1885. Ranchers, eager to profit from the huge demand for beef, stocked their ranches with more cattle than the land could feed, thus creating an unsustainable kind of animal husbandry that damaged the land, in some places irreparably. Even a century later, after decades of scientific investigations and government regulations, ranching has had, in environmental historian Donald Worster's words, "a degrading effect on the environment of the American West."[82]

The Far West

West of the Great Plains lay an immense region of mountains and deserts. This region—two-fifths of the contiguous states and territories—was, with few exceptions, too dry to farm. Only a narrow strip of the Pacific Northwest received enough precipitation for reliable rain-watered agriculture. Elsewhere, farming demanded irrigation. The Indians had known this, as had the Spaniards and Mexicans who settled among them, bringing experience from their homelands. Anglo-Americans, however, had no experience with irrigation.

Irrigation

The first Anglo-Americans to create farming communities in the Far West were the Mormons who migrated to Utah in 1847 to escape harassment for their religious beliefs and customs. By 1850 they were growing potatoes, wheat, rye, oats, maize, and hay on 6,500 hectares of irrigated land. Forty years later, their farms covered 107,000 hectares of land, almost all of it irrigated, and supported 200,000 people. By then, thanks to the railroad that reached Utah in 1867, they had become commercial farmers participating in the greater American economy.[83]

Though they accepted some forms of private property, the Mormons operated their irrigation system communally under a theocratic elite. This arrangement conflicted with the American ideas of separation of church and state and of free enterprise. Other Anglo-Americans of the time rejected bureaucratic control over a resource as indispensable as water and sought a method of water control compatible with either grassroots democracy or with free-market capitalism. In their search, they advanced two alternative doctrines: riparian rights and prior appropriation. Riparian rights asserted that any landowner whose property abutted a body of water had the right to use that water as long as his actions did not affect others' use of the same water. The doctrine of prior appropriation, in contrast, asserted that whoever was first to seize control over a body of water could claim it as personal property. In a land of scarce water, only prior appropriation could induce capitalists to invest in large irrigation systems.

California had both land and water, albeit not in the same place. The valleys of the Sacramento and San Joaquin rivers that together form the

Central Valley contained the richest farmland in the nation. But from March to November, the Pacific High Pressure Zone diverts the clouds toward the Pacific Northwest and British Columbia, leaving the heart of California bathed in sunshine during the growing season. With so little rain, only wheat could grow; in the 1860s, farmers took advantage of the surging demand in the East and in Europe to plant up to 154,000 hectares of wheat. This bonanza faded when Canada, Argentina, Russia, and India began exporting wheat. Faced with this competition, Central Valley farmers turned to crops that demanded irrigation.[84]

Irrigation in California proved more complex and contentious than elsewhere—for legal rather than ecological reasons. Riparian rights were implicit in the 1850 state constitution, but the courts, influenced by powerful vested interests, favored prior appropriation. The "California Doctrine" that emerged from these trials was a hodgepodge. The San Joaquin and King's River Canal and Irrigation Company and the Kern County Land Company purchased 133,000 and 167,000 hectares of land, respectively, built canals, sold land to farmers, and engaged in endless lawsuits. State and federal engineers provided surveys but otherwise had no authority over free enterprise in land and water. Finally in 1887, the state passed the Wright Act that allowed farmers to form cooperative irrigation districts. By 1890, California had 406,000 hectares of land under irrigation and was shipping fruits and vegetables by the trainload to the rest of the nation.

Gold

California's transformation from a distant frontier of scattered Indians and Spanish missions into the cornucopia of America was bedeviled by a far more pernicious kind of exploitation. Unlike agriculture, potentially a sustainable form of development, mining was by its very nature a plunder of the Earth's capital. A mining law of 1872 opened federal lands in the West to mining with almost no restrictions. In a mining rush, prospectors excavated what they could and moved on, leaving behind ghost towns and empty mine shafts.

As soon as news reached the eastern United States that a prospector had found a gold nugget in the American River in 1848, the trickle of migrants making their way to California turned into a flood. Unlike the Spanish conquistadors who were motivated by both Christ and gold, the

Forty-Niners (as the gold seekers were called) craved only gold. Pioneer farmers knew that years of toil lay ahead of them before they became comfortably well off, but the Forty-Niners hoped to quickly find the gold nuggets that would make them rich, or at least earn enough to live well. Eighty thousand came in 1849 alone. All told, 300,000 Anglo-Americans crossed the continent on the Oregon Trail, along with 1.5 million horses and cattle. The non-Indian population of California surged from 14,000 in 1848 to 380,000 in 1860.[85]

The first Forty-Niners scoured the riverbeds of northern California for gold nuggets and flakes washed down from the Sierra Nevada, using picks and shovels, pans, and sluice boxes. But the easy surface gold was soon found, and the flow of the rivers varied from floods in the spring when the mountain snow melted to a trickle in the summer. A few companies used imported Chinese workers to divert rivers and dig up the dry riverbeds, but panning and sluicing were too time-consuming and the results too meager to interest wealthy investors.

What changed all that was hydraulic mining, a new technology introduced in 1852. Water, shot through gigantic nozzles called water cannon or monitors at up to 160 kilometers per hour, demolished entire hillsides in a few hours, flushing boulders, rocks, gravel, sand, and dirt—along with gold flakes—into long wooden sluices. The gold, being heaviest, settled at the bottom, while everything else was washed away. Every week or two, the flow was stopped to allow workers to pick the gold out of the sluice boxes. Mining a given volume of gravel with a stream of water cost one-hundredth as much as mining with a pan. To supply the mines with water, these mining companies built dams high in the mountains. By 1883, their reservoirs held 215 million cubic meters of water. Ten thousand kilometers of ditches, tunnels, and aqueducts carried the water to the mining camps. The profits were sufficiently lucrative to attract investors from the East Coast and Great Britain to hire thousands of workers and buy all the equipment needed.[86] In less than a decade, gold mining was transformed from an artisanal craft into an industrial-scale undertaking.

Environmental Impacts

The impact of hydraulic mining on the landscape around the mines astonished contemporaries. The *Sacramento Daily Union* reported in 1854 that under a powerful jet of water, a hillside "melts before them, and is carried away through the sluices with almost as much rapidity as if it were a bank of snow."[87] A visitor reported later: "The effect of this continuous stream of water coming with such force must be seen to be appreciated; wherever it struck it tore away earth, gravel, and boulders.... It is impossible to conceive of anything more desolate, more utterly forbidding, than a region which had been subjected to this hydraulic mining treatment."[88]

Much worse were the downstream consequences. Spring floods washed boulders and gravel down the Yuba, Bear, American, Feather, and Sacramento rivers. Sand and soil were carried farther downstream, clogging riverbeds with muddy effluents and spreading out over the adjacent farmland. Unlike the rich topsoil they covered, these deposits were so deficient in phosphorus and nitrogen that nothing would grow on them. Towns and farmers built levees to contain the flood, but the levees often broke or leaked. By 1874, the Yuba River had risen 5 meters. The next year, a flood finished off the remaining farms in the Yuba Valley. In 1879 alone, hydraulic mines dumped 10 million cubic meters of debris into the Feather River. Altogether, a study in 1891 reported that over 16,000 hectares of farmland in the valleys of the Yuba, Bear, and Sacramento rivers had been destroyed by hydraulic mining.[89]

Farms were not the only environments affected. Before the gold rush, the gravel bottoms and cold clear waters of the rivers coming down from the Sierra Nevada had attracted salmon that spawned in them. It is estimated that Indians harvested 650,000 salmon a year without diminishing their numbers. By 1872 salmon had disappeared from these rivers. Mining also consumed forests, partly submerged behind the dams, partly felled for lumber to build sluices, flumes, and mining camps. The state agricultural society reported that one-third of California's accessible timber of value was already gone by 1870.

Mining brought other heavy industries to California. To produce the monitors, pipes, and other metal products that the mining industry required, as well as the railroads and steamboats that served a growing population, coal and iron ore were mined and foundries built, especially around Sacramento. Lung diseases caused by pollution from the foundries became one of the leading causes of death in the state.[90]

Mercury and its effects were also byproducts of gold mining in California. As in the silver mines of Mexico and Peru, mercury was used as an amalgam to separate gold from the surrounding rocks and dirt. When the amalgam was heated, the mercury escaped as vapor, leaving gold behind. Mercury was mined in the form of mercuric sulfide, or cinnabar, a red mineral that, when heated, released vapors that then condensed as pure mercury. From 1845 on, much of it came from the New Almaden Quicksilver Mine near San Jose, California. At its peak between 1850 and 1885, this mine produced 771,000 kilograms of mercury each year and was, after the gold mines, the second most important industry in the state.

Mercury is a poison, especially in the form of vapors that escaped from cinnabar processing or was lost as amalgam in the gold fields. In water, microorganisms converted the metal into methyl mercury, an organic compound that traveled up the food chain and concentrated in human bodies, where it caused neurological symptoms popularly known as "mad hatter's disease." From the mid-1870s on, mercury and its byproducts were found in the waters of northern California.[91]

Miners and Farmers

Inevitably, industrial gold mining, as one of California's major consumers of water, was deeply involved in the litigation surrounding water and irrigation. In the 1850s, when the gold frenzy was at its height, the California Supreme Court gave hydraulic miners the same rights as lone prospectors; they could stake their claims even on land already settled by farmers and ranchers and appropriate rivers for their use, with no liability for damage to the property of others.

The pro-mining bias of the courts notwithstanding, by the 1870s, the damage to the farms downstream from the mines led farmers to sue the mining companies. After debris almost buried the cities of Marysville and Yuba City in 1875, their citizens formed the Anti-Debris Association to fight the Hydraulic Miners' Association in court. The suit dragged on for years, as the courts debated the merits of prior appropriation versus riparian rights, while the state assembly was torn between the supporters of mining and those of agriculture.[92] Finally, in 1884, a federal court ruled against the miners, forbidding further dumping of debris.[93] This victory for the farmers marked the rise of big agriculture and the decline of mining as the leading sector of California's economy. Water, that precious resource, was henceforth to benefit agriculture.[94] The worst aspects of hydraulic mining came to an end, sparing the valleys from further destruction.

Conclusion

Industrialization transformed Great Britain and the United States in two distinct ways. In an already settled and densely populated nation like Britain, it intensified the use of natural resources, expanded the cities, and caused substantial, and often harmful, changes to specific environments such as cities and mining regions. In the United States, the transformation was far more drastic. Industrial technologies encouraged people of European descent to enslave Africans, plow up the prairie, exterminate Indians and native wildlife, fell forests, and ravage the land in their search for valuable minerals. Some of these activities were sustainable, replacing one kind of exploitation with another, more intense kind. Others, however, were sheer plunder, seizures of non-renewable resources that left the environment impoverished for humans as well as the rest of nature.

Such was the demand for the products of industry and for lumber, wheat, meat, and other goods produced and transported using industrial technologies that the world entered a new era of mass production and consumption. The share of the Earth's resources appropriated by humans began a form of growth that statisticians call the "hockey-stick effect": after millennia of slow and fluctuating increases came a sudden rise to exponential growth that has continued to this day. Thus the area of the Earth devoted to cropland at the expense of forests and grasslands increased from 265 million hectares in 1700 to 537 million in 1850 and 913 million in 1920.[95] The amount of fresh water diverted for irrigation rose from 95 cubic kilometers per year in 1650 to 226 cubic kilometers in 1800 and to 550 cubic kilometers in 1900.[96] The amount of energy used by humans—most of it fossil fuels—also accelerated sharply in the mid-nineteenth century; coal production alone increased a hundred-fold from 10 million tons in 1810 to 1 billion tons in 1910. So did the resulting amount of methane and of carbon dioxide in the atmosphere.[97] In short, industrialization unleashed a rapid transformation of the natural world for the benefit of humans, especially those fortunate enough to command the world's industrial economies.

9

The New Imperialism and
Non-Western Environments

During the nineteenth century, the major European powers conquered Africa, South and Southeast Asia, and Oceania and put heavy pressure on China, while the United States dominated Central America and the Caribbean. The Latin American republics, nominally independent, became economic dependencies of the West.[1] Imperialism had a strong environmental component as well. In 1905, Sir Charles Eliot, the British commissioner of the East Africa Protectorate, expressed the attitude of a Western imperialist toward the natural world in these words:

> Nations and races derive their characteristics largely from their surroundings, but on the other hand, man reclaims, disciplines and trains nature. The surface of Europe, Asia, and North America has been submitted to this influence and discipline, but it has still to be applied to large parts of South America and Africa. Marshes must be drained, forests skilfully thinned, rivers be taught to run in ordered course and not to afflict the land with droughts and floods at their caprice; a way must be made to cross deserts and jungles, war must be waged against fevers and other diseases whose physical causes are now mostly known.[2]

As the twentieth century dawned, the nations of Europe and North America were busy "reclaiming, disciplining, and training nature" not only in their own lands but also in the non-Western world, especially the tropics. Yet it would be wrong to think that the environmental changes to the non-West in this period all derived from the interference of the West. Many parts of the non-Western world continued to be subject to autonomous environmental shocks, such as epidemics, droughts, and floods, and non-Western peoples initiated changes of their own.

Western Demand and Tropical Products

Among the motives for nineteenth-century imperialism, one of the most important was the Western demand for the products of the tropical and semi-
tropical regions of the world. Some were industrial raw materials. Most of the
raw cotton consumed by the cotton industry came from the American South
until 1860, when the outbreak of the Civil War forced manufacturers to seek
supplies in India and Egypt. To color textiles, they imported indigo and other
dyes from Central America and India. Palm oil from West Africa was used
to make soap and candles and to lubricate machinery. Twine and bags were
made of Indian jute or Mexican sisal. Natural rubber from Amazonia was
used to make waterproof clothing and, later, bicycle and automobile tires.
Guano from Chile and Peru was in great demand as a fertilizer. Cinchona
bark from the Andes was the raw material for the manufacture of quinine,
an anti-malarial medicine. And gutta-percha, the sap of trees growing in
Southeast Asia and the East Indies, was used to insulate submarine telegraph
cables.

Just as important were the tropical stimulants consumed in ever-
increasing quantities by the inhabitants of the industrial world. As industrialization encouraged the growth of the population and increased the
purchasing power of the Western nations, luxuries became necessities and
desires turned into addictions. In the nineteenth century, tea, once imbibed
only by the upper classes of Great Britain, became the beverage of choice
among the working poor. Coffee, a more powerful stimulant, prevailed on
the continent of Europe and in the United States. And drinking tea and coffee
meant consuming ever more sugar.

The growing demand for these and other products brought a tremendous increase in trade between the tropics and the industrializing West,
especially after the introduction of steamships and the opening of the Suez
Canal in 1869 lowered freight costs dramatically. The value of India's exports
increased fivefold between 1864–1868 and 1914. Overall, the volume of tropical exports increased threefold from 1883 to 1913, rising at 3.5 percent per
year, the same rate as industrial production in the West.[3]

Satisfying the Demand

How could the industrial nations obtain the tropical products their industries and consumers demanded? At first, the growing demand caused a rise

in prices, while supply lagged behind. Agricultural production could not change rapidly. Chinese farmers could not expand their production of tea fast enough to keep up with the Western demand, for tea bushes took several years to grow to maturity and suitable land was not readily available in a crowded country. West African middlemen had little incentive to demand more palm oil from farmers as long as they benefited from the high prices that came with restricted supplies.

What distinguished the New Imperialism of the nineteenth century from earlier episodes of empire-building was the transfer of selected Western scientific and industrial technologies to the non-Western world in order to increase the production and lower the costs of desired commodities and to improve transportation and communication between the West and the non-West.[4] These technologies had an impact not only on the politics and economies of the non-Western world but also on their natural environments.

To satisfy their demand for the products of tropical agriculture, the industrial powers brought two major changes to the tropics. One was the transfer of commercially valuable plants from their native habitats to new parts of the world: tea from China to India and Ceylon, coffee from Ethiopia to Brazil, cinchona from Peru to Java and India, rubber from Brazil to Southeast Asia, gutta-percha from Malaya and Sumatra to Java. The other was the replacement of natural forests with vast plantations to produce the desired products on a massive scale, aided by the application of Western science.

Western writers justified these changes on the grounds that the peoples of the tropics were derelict in their duty to the global economy. Thus John Christopher Willis, director of the Peradeniya Botanic Gardens in Ceylon (now Sri Lanka), wrote in 1909: "The northern powers will not permit that the rich and yet comparatively undeveloped countries of the tropics should be entirely wasted by being devoted merely to the supply of the food and clothing wants of their own people, when they can also supply the wants of the colder zones in so many indispensable products."[5]

Expeditions

Travelers had been transferring plants from one part of the world to another for millennia. But these early transfers were anonymous and not always intentional. European governments knew how much of their wealth came from sugar, tobacco, indigo, and other crops that grew in their tropical colonies. In the eighteenth century, motivated by a rising popular interest in science and hoping to find lucrative new crops, they began funding expeditions to distant parts of the world. Louis Antoine de Bougainville, sent to circumnavigate the world between 1766 and 1769, brought with him the botanist Philibert Commerçon. Joseph Banks, a gifted amateur botanist, accompanied James Cook on his first voyage to the Pacific in 1768–1771 and returned with a collection of seeds and descriptions of myriad plants hitherto unknown to Europeans. At the urging of Banks, the British government sent Captain William Bligh to the Pacific in 1787; though this expedition was cut short by a mutiny, Bligh returned in 1791–1793 and succeeded in transferring bread-fruit trees from Tahiti to the West Indies.[6] From 1799 to 1804, Alexander von Humboldt traveled throughout Spanish America under the patronage of the king of Spain; interested in all aspects of the natural world, he returned with descriptions of the American flora that inspired later explorers.

Thanks to the work of these and other explorers and to the classification systems devised by Carl Linnaeus and Georges Louis Leclerc de Buffon, botany attracted the attention of kings, aristocrats, and country parsons as well as the scientific elite. It also turned from an avocation into a profession based on the institution of botanic gardens.

Botanic Gardens

Botanic gardens originated in the pleasure gardens of royal courts and in the apothecary gardens in which physicians grew medicinal plants. Such were the Jardin des Plantes in Paris and Kew Gardens near London. Kew began its transformation into a research institution under Joseph Banks when he added his collection of seeds and plants to the herbarium as well as the manuscripts of Gerhard Koenig, a Danish physician who had collected specimens of plants in India from 1768 to 1785. By the 1870s Kew had become the world's foremost botanical research institution and the publisher of distinguished botanical journals and books, with a museum of economic botany and a network of fifty-four other botanic gardens, thirty-three of which were in the British Empire. By the end of the century, it boasted over a million species of plants in its gardens and herbaria.[7] In contrast, the Jardin des Plantes was much neglected during the French Revolution and the reign of Napoleon, to the dismay of French botanists.[8]

The urge to create botanic gardens followed European botanists into the tropics. Among the first was the Jardin de Pamplemousses, established in the Ile de France (now Mauritius) by Pierre Poivre in 1767 to grow plants from the East Indies and break the Dutch monopoly on spices. Other botanic gardens followed in Calcutta in 1768, Jamaica in 1793, Peradeniya in 1822, and in almost every other European tropical colony. Though often founded as recreational gardens around the governor's mansion or as sources of European vegetables for the resident whites, they eventually joined the global network of gardens that exchanged valuable plants and information throughout the tropics.

By far the most important of the colonial botanic gardens was the one founded by the Dutch at Buitenzorg (now Bogor) in Java in 1817. In the late nineteenth century, it was second only to Kew among the world's botanic gardens and boasted 15 European botanists and other professionals and 300 Javanese gardeners, along with a school to train Javanese agricultural extension agents and a laboratory for visiting foreign scientists.[9]

Plant Transfers and Experiment Stations

The original purpose of colonial botanic gardens was to import new species or varieties of plants from other parts of the tropics that might prove economically valuable for the colonial power and to exchange plants and seeds with other botanic gardens. They also provided seeds or seedlings to European planters and indigenous farmers and offered advice on growing and handling new plants. They were aided in this task by the invention in the 1830s of the "Wardian case" or terrarium by Nathaniel Ward, a London physician and amateur botanist. In a sealed glass case, he discovered, delicate plants that would die if exposed to salt spray or dry air on board a ship could survive long ocean voyages. Many important plants were transferred this way, including tea bushes from China to India and rubber trees from Brazil to Kew and thence to Southeast Asia.[10]

In the late nineteenth century, botanic gardens that collected a few examples of hundreds or thousands of plants were supplemented by a new kind of institution: agricultural experiment stations that specialized in a few species of plants or even a single species. Part of the impetus was to find the most productive variety or the most suitable soils, climate, and other conditions for each species. They were also motivated by the appearance of devastating plant diseases readily spread by improvements in steamship communication around the world and by exchanges of plant material by botanic gardens and planters.[11] To deal with these problems, colonial governments and planters' associations founded experimental farms that employed not only botanists but also agronomists, soil scientists, chemists, entomologists, mycologists, and plant pathologists. Among the most successful were the Tjikeumeuh Agricultural Experiment Station in Java (1876) and the Proefstation Oost-Java (1907), the Imperial Department of Agriculture in Barbados (1897), and the Agricultural Research Institute of Pusa (1903) and the Imperial Sugarcane Breeding Institute of Coimbatore (1912), both in India.[12]

Plantation Agriculture

What transformed the tropics most radically was the replacement of natural forests with commercial agriculture. In some places, this involved an intensification of peasant agriculture or its expansion into new land. Elsewhere, trade in agricultural commodities spawned plantations on the model of West Indies sugar estates, but on a larger scale. From the mid-nineteenth century on, the most profitable tropical crops were rice and perennials such as tea, sugar, cacao, rubber, and palm oil that demanded water year-round. To grow these crops, governments encouraged a massive migration of peoples from dry regions, where grains grew best, to regions where the natural vegetation was forests. World trade caused a more rapid deforestation than peasant agriculture ever had. Sugar required large-scale processing immediately after harvesting, for the cane juice lost its sweetness as soon as the cane was cut. Likewise, tea leaves needed to be dried as soon as they were picked, before they wilted and oxidized. However, there were no technical reasons for other crops to be grown on plantations, for indigenous farmers could have obtained the same yields as large planters. Rather, it was for political and economic reasons that plantation agriculture flourished in the tropics.[13]

Tea

In the nineteenth century, tea became the national drink of the British people. Annual imports of tea into Great Britain grew from about 45 tons in 1700 to about 14,000 tons in the 1830s and to 87,000 tons by the end of the century.[14] Until the 1830s, all the tea imported into Britain came from China. The East

India Company, eager to reduce its export of silver—the form of payment the Chinese demanded—hoped to find ways of growing tea in India. The superintendent of the botanic garden at Saharanpur reported that the foothills of the Himalayas were ideal for tea-growing. In 1848, after China's defeat in the Opium War had opened the interior of China to Western explorers, the company sent the botanist Robert Fortune there to obtain tea bushes. He returned to India with 2,000 plants in Wardian cases, several thousand seeds, and several experienced Chinese gardeners. The plants that survived the trip launched the Indian tea industry.[15]

Once it was clear that the hill country of Assam and northern Bengal could produce tea, Europeans bought land in a frenzy of speculation starting in 1859. By the end of the century, 764 large estates covered 253,000 hectares in the area around Darjeeling. In southern India, planters found the Wyanad Plateau and the Nilgiri Hills of Madras (now Tamil Nadu) ideal for growing coffee; by 1866 they brought untouchables from Madras to clear forests and drain wetlands covering 5,900 hectares in order to establish over 200 plantations. After a blight devastated the coffee bushes in the 1870s, they switched to tea. The same was true of Ceylon. There too, planters started with coffee, but when the coffee blight devastated their trees, they switched to other crops, especially tea. The first tea plantation on Ceylon was started in 1867. By the turn of the century, tea covered some 160,000 hectares and Ceylon exported ca. 40,000 tons of tea a year to Great Britain.[16]

This triumph of commerce was a defeat for nature, for the tea and coffee plantations had been carved out of the forests that had once covered Assam, the Himalayan foothills, and the hills of southern India and Ceylon. The result, as in other instances of deforestation in hilly country, was erosion and the silting of the nearby lowlands.[17]

Coffee

The coffee trees *Coffea arabica* and *Coffea canephora* (better known as *robusta*), which originated in Ethiopia, were introduced to Europe in the seventeenth century as rare plants, then carried to the Americas in the early eighteenth century. In Brazil, serious production began in the early nineteenth century.[18] Planters found that the trees grew best in southern Brazil's *terra roxa*, volcanic soil mixed with decayed vegetation found at an elevation of 1,000 to 2,000 meters in a region where the temperature never dropped below freezing or rose much above 27°C (80°F). In such soil, coffee trees matured after four or five years, then produced beans (really seeds) for fifteen to twenty-five years. After that, it was much simpler to slash and burn another piece of the forest than to fertilize once-used soil.

The first area to be exploited for coffee plantations was the valley of the Paraíba do Sul River north and west of Rio de Janeiro. Beginning in the 1850s, planters brought slaves to clear the forest. The slaves cut partway through the trunks of trees on the lower slopes of hills, then cut through the largest trees at the tops of hills. As vines covered the trees and linked them to one another, when a large tree fell, it brought down with it trees standing to the sides and downhill of it as well. As the fallen trees were aligned with the slope of the hill, coffee seedlings were planted in rows running up and down the hillsides between the fallen trunks and stumps. This way, the soil was easily accessible to the slaves who planted the seedlings and collected the beans but was also easily washed away by torrential rains.

By the 1890s, the virgin forest had been clear-cut, the soil was depleted, and the aging coffee trees no longer produced as they once had. After slavery was abolished in 1888, coffee production dropped by half. In short, the coffee boom was an unsustainable mining of the natural bounty of the soil, as in the Cotton Kingdom of the American South. As historian Stanley Stein wrote about Vassouras, one of the most productive counties in the region: "In one century, the município of Vassouras and the major portion of the extensive Parahyba [sic] Valley were the scene of a complete economic cycle which started with tropical forest and terminated with denuded, eroded slopes. Once exploited, the lands of an interior frontier were abandoned to grass, weeds, and cattle."[19] Or, as historian Shawn Miller put it, "Civilization devoured wilderness and spit it out. In a matter of decades the frontier went from a state of nature to the status of ruins. . . . The result was a spreading cancer, ravaging everything at its perimeter and leaving a black, dead core characterized by deforestation, erosion, and ghost towns."[20]

Nonetheless, Brazil benefited from the collapse of the coffee plantations in India and Ceylon in the 1870s and '80s.[21] As in the American South, where cotton growing moved west as exhausted soils were abandoned, so in Brazil did coffee growing move south to new lands west of São Paulo to be farmed by indentured European workers. Brazil thus retained its position as the world's largest producer of coffee, and coffee remained Brazil's main export. By the end of the century, some 3 million hectares had been cleared for coffee.[22]

Rubber

Natural rubber comes from the sap or latex that prevents insects from boring into the bark of tropical plants. Until the 1840s, Europeans and Americans considered this product nothing more than a curiosity, for it became brittle in the cold and sticky in hot weather. In 1839, Charles Goodyear discovered that heating raw latex in the presence of sulfur vulcanized it, making it elastic, waterproof, and impervious to temperature variations. Manufacturers used it to make waterproof boots and garments, rubber balls, condoms, gaskets for steam engines, and belting for machinery. Demand rose dramatically when John Dunlop invented the pneumatic tire in 1887, allowing the proliferation of bicycles and later of automobiles.

The most generous and consistent supply of rubber latex comes from *Hevea brasiliensis* trees native to the Amazon rainforest. These trees can be tapped on alternate days, producing a few grams of latex a day. So intense was the demand and so high the prices that entrepreneurs invaded the Amazon and used violent and brutal methods to force native Indians and immigrant workers to collect the latex. Each tapper was responsible for up to 200 hevea trees. Every morning, he slashed a cut in the trunks and attached a bowl to the bottom of the cut, then returned in the afternoon to collect the latex and dry it over a smoky fire in order to coagulate it. An expert tapper could produce between 200 and 800 tons of rubber a year this way.

The Amazonian rubber boom of the late nineteenth century is famous not only for the cruel treatment of the tappers but also for the immense wealth it brought to a few entrepreneurs and to the city of Manaus on the Amazon River. The production of rubber in Amazonia rose from 31 tons in 1827 to 2,673 tons in 1860 and to 26,750 tons in 1900. The impact on the environment is more ambiguous. If moderately tapped, trees could last fifty years or more, but if bled too hard, as happened during the drought of 1877–1879, they died.[23]

Hevea was not the only source of natural rubber exploited during this turn-of-the-century boom. *Castilla elastica*, a vine that grew in Central America and in the Putumayo region of the upper Amazon, was as ruthlessly exploited as were hevea trees in the lowlands. The same was true of the *Landolphia* vine tapped by Africans in the Congo Free State during the most violent period of European colonial rule. As in Amazonia, native tappers were forced to collect the latex under threat of death to their families or of having their hands chopped off if they did not bring back their quota. Under such pressure, the tappers destroyed the vines in order to extract the maximum amount of latex possible.[24]

Hevea rubber was Brazil's most important crop after coffee. The government, hoping to prevent competition, discouraged the export of seeds and seedlings. Great Britain and the Netherlands, meanwhile, were eager to obtain seeds and transfer this lucrative crop to their Asian colonies. The theft of hevea seeds from Brazil and their transfer to Asia is the most famous of all cases of botanical piracy. In 1873, Sir Clements Markham, head of the India Office's geographical department, and the Marquess of Salisbury, secretary of state for India, persuaded Kew Gardens to send missions to Brazil to obtain hevea seeds. In 1876, the British adventurer Henry Wickham succeeded in smuggling 70,000 hevea seeds out of Brazil. Of those that reached Kew, 2,700 germinated and 2,000 seedlings reached Ceylon in September 1876. Of these, twenty-one were shipped from Ceylon to the Singapore Botanic Garden. These twenty-one seedlings were the ancestors of all the hevea trees in Asia.[25]

For many years, these surviving hevea trees were ignored by the planters of Malaya, for it took six to eight years before rubber trees could be successfully tapped, and speculative investors were more attracted to tea. Then, in the early twentieth century, demand from the growing automobile industry caused a rise in the price of natural rubber. While rubber tappers ransacked Amazonia for rubber-bearing plants, planters in Southeast Asia suddenly found hevea trees attractive. Encouraged by Henry Ridley, the superintendent of the Singapore Botanic Garden, Tan Chay Yan, a Chinese planter in Malaya, put 17 hectares under rubber trees in 1896. He and other investors preferred to grow hevea trees on freshly cut forest land to take advantage of the natural fertility of the humus. To provide labor for the plantations, the colonial governments encouraged the immigration of indentured workers from the poorest regions of China and India, especially Tamils from southern India. The area occupied by hevea plantations in Malaya grew from 800 hectares in 1898 to over 200,000 in 1910 and to over 400,000 in 1914. By then, heveas occupied over 62 percent of the cultivated land in Malaya.[26] Other European colonies in Southeast Asia and the East Indies followed suit. The surge in exports of rubber from Asia put an end to the Brazilian wild rubber business.[27]

Like tea in India and Ceylon and coffee in Brazil, hevea plantations caused the most rapid deforestation in the history of Asia. Where the great diversity of rainforest trees (and their associated plants, animals, and insects) once flourished, there were now thousands of heveas all growing in straight rows on carefully weeded land.

Cinchona and Gutta-Percha

Two crops—cinchona and gutta-percha—occupied much less land than rubber, tea, or coffee, yet they contributed to the success of Western imperialism in the tropics. Cinchona, the bark of trees that grew wild in the Andes, was the source of drugs that protected Europeans from malaria. In 1820, French chemists extracted the alkaloids that made the bark effective. One of them, quinine, was soon manufactured in commercial quantities. As increasing numbers of Europeans traveled to malarial parts of the tropics, demand for the product soared. To meet it, Andean Indians went into the forests, felled the cinchona trees, and peeled off the bark. This practice shocked Europeans, who were convinced that the trees would soon be depleted. Though Europeans blamed South American Indians for threatening the supply of this valuable product, it was their own demand that led to the danger of depletion.[28] The Andean republics, recognizing the value of cinchonas, controlled the export of bark and forbade the export of seeds.

In the 1850s, the European governments with colonies in Asia and outposts on the coasts of Africa became increasingly aware of the importance of quinine to maintain the health of their citizens living in the tropics and the high cost and unreliable supplies of cinchona bark from South America. They therefore resolved to transfer cinchona plants to their colonies.

To get around the Andean prohibitions, the British government sent expeditions to the Andes to collect cinchona seeds. Richard Spruce, a botanist living in Ecuador, and Robert Cross, a gardener at the Royal Botanic Gardens at Kew, loaded almost 100,000 seeds and 637 seedlings of *Cinchona succirubra* (the "red bark" tree) onto a ship; 463 seedlings survived the three-month trip to India. The surviving trees were received by William McIvor, the superintendent of the Ootacamund Botanic Garden in the Nilgiri Hills of southern India. Under his care, these trees quickly multiplied. In 1866, cinchonas covered 20 hectares; by 1880 they occupied 343 hectares in government plantations and 1,619 hectares on private land. In Ceylon, cinchona growing was taken up by planters who had lost their coffee bushes to disease. At one time, over 25,900 hectares were covered with cinchona, but competition from the Dutch in Java caused these planters to switch once again, this time to tea.[29]

Meanwhile, Dutch botanists had started a cinchona plantation at Tjinieroean, a mountain valley in Java with a climate similar to that of the Andes, where they experimented with different methods of planting, cultivation, bark peeling, and seed germination. In 1865, Charles Ledger, an English trader living in Bolivia, smuggled out 20,000 seeds of yet another species, *Cinchona calisaya ledgeriana*, which he sold to the Dutch. This tree thrived in Java, where it produced the highest proportion of quinine of any cinchona. By 1916, 114 cinchona plantations in Java covered 15,500 hectares, and the Dutch captured 80 percent of the world market for quinine.[30]

The other crop, gutta-percha, is now virtually forgotten, but once it had a major impact on international trade and on some tropical forests.[31] The sap of *Isonandra, Palaquium,* or *Dichopsis* trees that grow only in Southeast Asia, it is a natural plastic that is impervious to many acids and alkalis and, most important, to saltwater. From the 1850s on, it was used to insulate submarine telegraph cables. The first transatlantic cable of 1857 contained 250 tons of gutta-percha. This required an enormous number of trees, for big trees 20 meters high produced no more than 312 grams of latex; even the largest trees produced less than 1,360 grams. By the early 1890s, the cable industry was consuming almost 2,000 tons of gutta-percha annually. In 1896, Eugen Obach, the chief chemist for a cable manufacturer, estimated that making the world's 304,169 kilometers of cables had required 32,000 tons of gutta-percha, and further construction would demand 3,000 tons a year, or the output of several million trees annually.

Because the sap of gutta-percha flowed very slowly, when native collectors in Sarawak, Borneo, or Malaya found a suitable tree in the forest, they chopped it down and cut rings in the bark to let the latex ooze out into holes in the forest floor. As the *India Rubber and Gutta Percha and Electrical Trades Journal* wrote in 1892, "the *modus operandi* might be compared to that of a butcher slaughtering a cow for the sake of her milk, instead of judiciously titillating her udder periodically."[32] Once felled, the trees were left to decay with most of the latex still in them.

As early as 1860 there were no gutta-percha trees left on Singapore Island; twenty years later few remained in Malaya, and collectors were combing the forests of Borneo and Sumatra. By 1891, prices had increased fourfold, and the *India Rubber Journal* warned that an impending shortage would endanger the cable industry and with it, the communications network upon which global trade and the security of the British Empire rested.[33]

Producing gutta-percha on plantations was even more difficult than growing cinchona trees, for the trees that produced it did not mature until they were twenty years old, too long a time to interest private planters. Yet governments that laid submarine cables wanted this product for strategic reasons. In 1882, a French chemist discovered a means of extracting gutta-percha by grinding up twigs and leaves of *Palaquium* trees and soaking the powder in toluene; trees could thus continue producing sap for several years, instead of only once. Finally, in 1885 the government of the Dutch East Indies opened a plantation in Java that began producing gutta-percha on a small scale in 1908. By then, there were few wild trees left in Southeast Asia.[34]

Forests and Game

Besides clearing land for crops, colonial rule also affected native plants and animals, especially in India. A growing population and the expansion of agriculture combined with the demands of the British to reduce India's once dense forest cover at an accelerated rate. Along the west coast, deforestation was already well under way before the mid-nineteenth century, as farmers cleared the low-lying coastal wetlands of mangroves in order to plant rice. Most of the Deccan Plateau was treeless by 1840. The British saw forests as an impediment to agriculture and as refuges for rebels and bandits. During the Napoleonic Wars, the East India Company expropriated the teak forests of Malabar to ensure supplies of wood for the Royal Navy. Supplying timber, however, was left to private entrepreneurs who operated for short-term profits and depleted the forests at an unsustainable rate.[35]

The coming of the railroad to India after 1854 intensified the felling of desired trees for cross-ties, especially teak (*Tectona grandis*), sal (*Shorea robusta*), and deodar (*Cedrus deodara* or Indian cedar). Railroad tracks in India required up to 1,250 ties per kilometer; those made of teak lasted fourteen years, those of sal and deodar thirteen, others six or seven years. Less durable woods ended up in the fireboxes of locomotives and river steamers. In the 1860s and '70s, the Indian railroads consumed a million ties or 28,600 hectares of forest per year; by the 1890s and early 1900s, 50,000 to 53,000 hectares of forest were felled for railroad use alone.[36] By providing access to remote regions, railroads also encouraged the felling of previously inaccessible timber.[37]

Scientific Forestry

The rapid depletion of valuable hardwoods raised alarms within the British colonial administration.[38] After the Rebellion of 1857, the new government of the British Raj began thinking of forests as tree farms rather than as timber mines. In this, they followed in the footsteps of France and the German states, countries that could not, like Britain, rely on imports of timber when local supplies ran short.

The French "Ordonnance sur les eaux-et-forêts," issued by King Louis XIV's minister of finance Jean-Baptiste Colbert in 1669, inaugurated the era of sustainable forestry. After the devastation of the Seven Years' War, the German states took measures to ensure future supplies of commercially valuable timber. To meet their needs, the governments of France and the German states founded schools to train professional foresters to survey their woodlands and determine the value of each species and its sustainable yield. They also created forestry departments staffed by state foresters with powers to enforce regulations. Their goal was to replace natural forests exploited by local people with state forests of "standard" trees of the same species and age that would be useful for industrial and construction needs.[39]

Forestry in India and Burma

In 1856, the government of India appointed Dietrich Brandis, a German biologist turned forester, to manage the teak forests of Pegu in Burma by overseeing the logging of older trees and the planting of new ones on a sustainable basis. Impressed by Brandis's success in Burma, Governor-General Lord Dalhousie established the Indian Forest Service in 1865 and appointed him Inspector-General of the Forests of India. To assist him, Brandis brought two German foresters, Wilhelm Schlich and Berthold Ribbentrop, and other staff trained in Germany or France. Brandis was a conservationist concerned with maintaining forests to benefit Indian farmers and the "household of nature" as well as the railroad companies. Like many people at the time, he believed that forests prevented desiccation as well as erosion and silting.[40]

By the mid-1870s, the law of 1865 that had established the Indian Forest Service was challenged by those who wanted to increase the production of desirable timber. After fierce debate and over Brandis's objections, the government passed the Indian Forest Act of 1878, a much tougher law that essentially requisitioned India's forests for state use; it was followed by the similarly stringent Burma Forest Act of 1881. These marked a shift from multi-use conservation to commercial forestry.[41]

These acts divided all forests into three kinds: forest reserves; protected forests that allowed local villagers a few rights; and, in rare cases, village woods. By the end of the century, of 234,000 square kilometers of state-owned forests (covering 20 percent of British India, excluding the Princely States), 90 percent were "reserved," that is off-limits to everyone but government foresters. On them, the foresters carried out inventories of trees and supervised the planting and maintenance of valuable species like teak, sal, deodar, and the fast-growing chir pine (*Pinus roxburghii*).

From a commercial point of view, the results of scientific forestry were impressive. The percentage of the government's revenues generated by the Forest Service rose from less than 1 percent in the 1880s to almost 3 percent after 1910, and the amount of timber sold rose from 17,000 cubic meters in 1886–1887 to 116,100 cubic meters in 1913–1914.[42] Socially and environmentally, however, the results were decidedly negative. To protect the reserved forests, the foresters prohibited the gathering of firewood and other forest products, hunting and fishing, and pasturing cattle, sheep, and goats. Impoverished villagers, barred from gathering firewood, burned manure instead of using it to fertilize their fields. Scientific forestry therefore represented an attack on their traditional way of life and undermined the village economy. Not surprisingly, this provoked widespread resistance from peasants and tribal foragers, who retaliated with arson and banditry.[43]

The Forest Service encountered resistance from other sources as well. Timber merchants, eager for quick profits, wanted access to the trees. Planters demanded land. Villagers cut trees for fuel and to expand their fields. Revenue officers wanted more land for agriculture in order to increase the tax base. Other branches of the government, needing wood for construction and railroad ties, pressured the Forest Service to harvest more trees. Until the outbreak of war in 1914 the Forest Service managed to protect the forests of India from these pressures.[44] Had the forests of India been exploited by commercial enterprises as in the United States or by local woodsmen as in China, they probably would have declined faster than they did under government control.

Although the government foresters may have slowed down the rate of depletion of the forests, they changed their composition. The Forest Service favored pine, cedar, and teak at the expense of other species used by villagers. In the foothills of the Himalayas, mixed deciduous-conifer forests gave way to pure coniferous stands, and in the mountains overlooking the west coast of India, mixed forests were replaced by teak plantations.[45]

The Hunt

By the nineteenth century, hunting served several purposes. The rural poor hunted to obtain meat and other animal products. Hunters also protected villagers and their livestock from dangerous carnivores and tried to stop wild herbivores from eating their crops. Then there was the elite hunting of European aristocrats, Manchu warriors, Indian maharajas, and other noble hunters eager to prove their manhood by besting wild animals. Such forms of hunting had been practiced for millennia and were probably sustainable, if only because it was in the interest of the elites to preserve their way of life.[46]

Unlike these was the hunting that arose in response to the voracious Western demand for products of the hunt and colonial Europeans' craving for trophies. One prized object of the hunt was ivory. Europeans and Americans wanted billiard balls, cutlery handles, combs, and ornaments of various types made of ivory. The fashion for chamber music in well-to-do homes led to the proliferation of pianos with half their keys made of ivory.

In the mid-nineteenth century, Egyptian merchants imported over 100 tons of ivory a year, much of it from East Africa, where 4,000 elephants were killed annually. As the colonial powers extended their sway over Africa in the late nineteenth century, professional big game hunters and wealthy tourists came to kill elephants for sport, using newly invented high-powered breech-loading rifles. As a result, by the first decade of the twentieth century, elephants had almost disappeared from West Africa, and the South African herds had been drastically thinned. In East Africa, 12,000 elephants were killed every year in the early 1880s, yet they survived in greater numbers because the African population was smaller and the European penetration came later than in West and South Africa; yet here too they disappeared from the coastal areas and retreated to the least accessible parts of the continent.[47]

In India, hunting was a sport, not a business. Wild elephants, if caught young, could be tamed, and so many survived. Before the Indian Rebellion of 1857–1858, British efforts were aimed at ridding their territories of threats to the tax-paying villagers, such as tigers, wolves, and bandits. In the Princely States, the maharajas pursued their traditional and very elaborate tiger hunts from the backs of elephants, with beaters to drive the prey into the open. British officials, eager to kill tigers that had attacked peasants, often joined the hunts.

After 1858, Indians in British India were disarmed and restricted to muzzle-loading smoothbore muskets that could not kill an animal more than 80 to 100 meters away. Only Europeans could possess modern high-powered rifles that could kill at a great distance. The goal of the British administration was threefold: to provide sporting opportunities to British officials and their friends; to exterminate tigers, lions, cheetahs, and wild dogs; and to reduce the numbers of antelopes, deer, and wild water buffalo that ate farmers' crops. After 1878, most of India's forests were also off limits to Indian villagers and tribal peoples who had traditionally supplemented their diet with game.[48]

In the last third of the century, the British adopted the extravagant tiger hunting expeditions of the Mughals and the maharajas. Viceroys were famous hunters; indeed, enthusiasm for hunting may have been a prerequisite for high office. With its elaborate code of conduct, "pig-sticking," hunting wild boars with spears, became a favorite pastime among military officers and lower-ranking government officials, especially foresters. Hunting in India had become a ritual of dominance; as historian John MacKenzie explains: "Like most of the invented traditions of the late nineteenth century, hunting represented an increasing concern with the external appearance of authority, the fascination with the outward symbols serving to conceal inner weakness."[49]

To be sure, the British rulers did not intend to eliminate big game entirely. Like elite hunters throughout history, they wished to conserve enough game for the viceroys, senior officials, army officers, royal visitors, wealthy tourists, and Indian princes to enjoy hunting and collecting trophies into the indefinite future. By the end of the century, as many species of animals were becoming rare, officials blamed their disappearance on "wild, unruly" indigenous forest dwellers. To protect the shrinking number of wild animals, the Indian government issued licenses and imposed other restrictions. Despite these measures, the hunting spree depleted the fauna of India. In the Central Provinces, the blackbuck (*Antelope cervicapra*) and its main predator, the cheetah, were gone, as were lions and rhinoceroses. Bears, wolves, and leopards were reduced in numbers. Tigers survived in remote or inaccessible ares, such as the Sundarban wetlands of lower Bengal. Gaurs or Indian bison (*Bos gaurus*) and wild water buffaloes (*Bubalus arnee*), seriously depleted in a rinderpest epidemic of 1896–1897, became rare. In 1800, India had been a land of great forests and abundant wildlife. By 1914 it had become largely populated by humans and controlled by a repressive government.[50]

Irrigation

For agriculture, the great advantage of the tropics—abundant sunshine—is offset by serious problems with the supply of water: either excessive or insufficient, and often erratic. Hence the importance of water control in so many civilizations, from ancient Mesopotamia to the contemporary American West. In the nineteenth century, the Western demand for tropical agricultural products made colonial rulers increasingly aware of the need to manage water.

Irrigation in India

India is famously subject to monsoons. Most years, winds from the Indian Ocean bring torrential rains to the sub-continent between June and September, but when they fail, drought and famine ensue. To compensate for their unpredictability, the people of the sub-continent have long built irrigation systems. In the eighteenth century, as the Mughal Empire deteriorated, so did the many canals the Mughals had built to bring water to Delhi and the surrounding lands along the Ganges and Yamuna rivers. When the East India Company took over that region in the early nineteenth century, they found only one, the Hasli Canal built in the seventeenth century, still carrying water, though badly in need of repair. Elsewhere, land lacking water had reverted to bush and scrub.

The company assigned the task of repairing canals to officers from the Bengal Artillery. They surveyed the Mughals' Delhi Canal, defunct since 1753, and rebuilt it between 1815 and 1821 under the name Western Jumna Canal. The Eastern Jumna Canal, begun in 1830, irrigated 366,000 hectares by 1837.

Artillery engineers had no experience in hydraulic engineering, nor were there any textbooks they could follow, so they learned by trial and error. In some places, they made the slope of a canal too steep, causing the water to scour its banks and undermine bridges and embankments. In other places, it was too shallow and the canal silted up. Without proper drainage, excess water caused waterlogging and salinization of irrigated land. The work proceeded by fits and starts whenever the company had funds after a good harvest had brought in sufficient taxes. Then came the drought of 1837–1838, when the crops failed except on the 366,000 hectares of newly irrigated land. Company officials, for whom a drought meant a drastic shortfall of tax revenues and a sudden rise in the cost of relief, realized the need to build canals before disaster struck again.

In 1841 the Court of Directors of the East India Company tentatively approved the construction of the Ganges Canal. It was to be 290 kilometers long, then split into two 274-kilometer-long parallel branches with many distributaries, making it the largest irrigation project in the world. The engineer in charge, Captain Proby Cautley, who had had only one year of training as an artillery cadet before being sent to India, returned to Europe to study the Caledonian Canal in Scotland and the canals of the Po Valley in Italy. The canal Cautley built, and that was inaugurated in 1854, flowed too fast, scouring its bed. This problem was not resolved until the 1870s. Yet in the drought of 1865–1866, the crops that grew on the land it irrigated were said to have saved 2.5 million lives.[51]

During the 1850s and '60s, irrigation projects proliferated, buoyed by the exaggerated optimism of canal builders who dangled the prospect of dazzling profits before starry-eyed British investors. In the end, however, many of these projects turned out to be very unprofitable, as farmers had no need for irrigation in years of adequate rainfall. Yet the government was attracted to the idea of irrigation on lands that could not support agriculture without it, such as parts of the Punjab that received only 240 to 400 millimeters of rain in good years, and sometimes as little as 150 millimeters. In spite of its aridity, the Punjab, a region of easy gradients and fertile soils, was especially promising. In order to undertake projects there without popular interference, the government of India passed the Northern India Canal and Drainage Act of 1873, giving itself the right to claim and control all irrigation, navigation, and drainage in the region.

The British engineers suffered occasional setbacks. One challenge was the extreme swings in weather. At one place, the Lower Ganges Canal had to be carried over the Kali Nadi River on an aqueduct. After interviewing older farmers in the area, the engineers built the aqueduct to withstand a 4-meter rise in the river, which would carry 500 cubic meters of water per second. In 1884 the river rose 7 meters, carrying 1,100 cubic meters per second, and tore out part of the canal. The next year it carried 4,000 cubic meters per second, eight times the anticipated maximum, and swept away not only the aqueduct but also bridges over 240 kilometers. Repairing this canal and related works required the work of thousands of men for many years.

Until the 1870s, irrigation projects were expected to bring a profit to the government in the form of water fees and higher taxes, or at least to break even. During the famine of 1876–1878, over 5 million Indians starved, as did the oxen that pulled the carts that would have brought food from other provinces. The government realized the need for "protective" works designed to irrigate land in dry years, along with "famine railways" that would transport food to areas in need. By the late 1880s, therefore, the government was ready to undertake new and bigger projects, especially in the Punjab. By 1895–1896, government projects in the Punjab alone irrigated 5.43 million hectares, more than in all of Egypt. What had been a very thinly populated region was filled with carefully structured and regulated agricultural colonies, turning Punjab into the richest grain-producing region of India and India into a major food exporter.[52]

Rice and the Wetlands of Southeast Asia

Not all of the new commercial crops came from plantations; many of the agricultural commodities that entered world trade were produced by indigenous smallholders, from indigo in Bengal to hevea rubber in Malaya to palm oil and cacao beans in West Africa.[53] Of all the export crops produced by smallholders in the tropics, none was as important as rice.

Rice had long been the staple crop throughout South and Southeast Asia and nearby islands. Most of it was produced in paddies that required carefully controlled irrigation. What changed after the mid-nineteenth century was explosive population growth. In Java, the population rose from around 7 million in 1830 to 28.4 million in 1900, yet the Netherlands East Indies government required farmers to alternate sugarcane with rice.[54] Meanwhile, millions of Chinese, Indians, and Javanese migrated to the new plantation frontiers in Malaya, Sumatra, and Ceylon. In these new areas, labor was devoted to the production of export commodities. Western demand for tropical commodities thus created a secondary demand for food for the workers in plantations and mines. After the Second Anglo-Burmese War of 1852, the British encouraged Burmese farmers to settle in the Irrawaddy Delta, a rich alluvial land covered with forests of mangroves (*Heritiera fomes*) and kanazo trees (*Baccalaurea ramiflora*). By the end of the nineteenth century, the population of lower Burma had risen from 1.5 to 4 million, most of them farmers who turned 12,000 square kilometers of rainforests and wetlands into rice paddies. By the early twentieth century, Burma had become the leading rice exporting country in the world, annually shipping 2.5 million tons of rice, mainly to Malaya, Indonesia, and Ceylon.[55] On a lesser scale, the same happened in Cambodia and in the Mekong Delta of southern Indochina.[56]

Another major rice exporter was Siam (now Thailand). Under a treaty signed with Great Britain in 1855, Siam was allowed to remain independent, but in exchange, foreigners were permitted to trade directly with the Siamese. Enticed by the profits of international trade, the government and wealthy landowners created a network of canals in the delta of the Chao Phraya River and brought in Thai and Chinese farmers to grow rice.[57] With them, the transformation of the major river deltas of Asia from wetlands and mangrove forests into irrigated rice paddies was almost completed. In the process, not only were natural areas transformed, but myriad species of birds, fish, and other wild animals also lost their habitats.

Irrigation in Egypt

Until the nineteenth century, Egyptian farmland had been irrigated by the basin system in which the annual flood of the Nile was impounded by temporary barrages to allow water to soak into the soil and deposit silt, then allowed to flow down to the next basin, and so on, until it reached the sea. This system, which had proved more or less sustainable for thousands of years, was seen as retrograde by Muhammad Ali, the viceroy of Egypt from 1805 to 1848, who had been very influenced by the French during their short occupation of Egypt.

Muhammad Ali's goal of modernizing Egypt along European lines required funds, which the country could obtain only by selling cotton. Between

1817 and 1821, Louis Alexis Jumel, a French adviser to Mohammad Ali, introduced the long-staple *barbadense* cotton for which Egypt became famous.[58] Cotton, however, required water in the summer months when the Nile was at its lowest. The first attempt to bring water to the cotton fields involved dredging canals up to 6 meters deep to reach the level of the Nile at its lowest. This project, begun in 1816, proved a failure because each new flood filled the canals with silt, and clearing the silt was not just costly and difficult but had to be done at the very time farmers needed to tend their crops. The project was abandoned in 1825.

In 1831, Muhammad Ali named the Frenchman Louis Linant de Bellefonds chief engineer of public works in upper Egypt. Linant suggested the construction of two barrages on the two branches of the Nile, the Rosetta and Damietta, at the head of the delta just below Cairo. The goal was to impound water and distribute it to the feeder canals in the delta when it was needed by the growing cotton crop. The work began in 1833, then, after an interruption of several years, was renewed in 1843. The barrages, built in haste under orders from the viceroy, developed cracks when they were tested in 1861 and could not be used to irrigate the delta as Linant had hoped.

After Muhammad Ali's death in 1849, Egypt continued to rely on exports of cotton to support the extravagant expenditures of the government. Modernization schemes and the construction of the Suez Canal, completed in 1869, sank the country ever deeper into debt. Finally in 1882, a political crisis gave Britain an excuse to invade and occupy Egypt. A year later, Sir Evelyn Baring, the British consul general, appointed Colin Scott-Moncrieff undersecretary for public works.

Before coming to Egypt, Scott-Moncrieff had worked on the Western Jumna and Ganges canals and taught at the Thomason Civil Engineering College, India's foremost engineering school. He and several associates he brought with him from India set out first to repair the barrages at the head of the delta, a task they accomplished in 1890. Thereafter, the lands of the delta received five times more water year-round than they had previously received in the low season. Yet most of the Nile's water still ended up in the Mediterranean Sea. With a flow that varied from 225 to 14,000 cubic meters per second, the Nile irrigated only one crop a year. The engineers knew that, in theory, the Nile could provide a constant 900 cubic meters per second year-round. In Egypt's climate, that would permit two crops a year or even five crops every two years, a promise that would be fulfilled in the twentieth century.

DISEASES

Westerners, and especially colonial officials, prided themselves on their accomplishments in transforming the societies, economies, and environments of the regions they affected. Thus between 1860 and 1936, the government of India called its annual reports *Statement Exhibiting the Moral and Material Progress and Condition of India*. Yet India's material—let alone moral—progress represented more wishful thinking than reality, for the subcontinent, like the rest of the non-Western world, remained vulnerable to floods, droughts, epidemics, and other natural shocks. In the tropics and in the monsoon world, the tug-of-war between humans and nature was as undecided as ever.

Among the most dramatic examples of the power of nature, then as in the past, were epidemics. Yet the spread of diseases around the world in the nineteenth century was exacerbated by increased trade and migrations, new modes of transportation, and colonial wars. They led, belatedly, to scientific advances in sanitation that would lead, in the twentieth century, to cures and preventions.

Cholera

Known in India since ancient times, cholera outbreaks generally occurred at the beginning of the rainy season, when fecal matter that had accumulated during the dry season was carried by rainwater into village wells and tanks. What had been a regional epidemic became a pandemic in 1817. That spring, as Bengal experienced extraordinarily heavy monsoon rains, an unusually virulent strain of the pathogen *Vibrio cholerae* infected hundreds of thousands of Bengalis. From India cholera spread to Ceylon in 1818–1819 and to Southeast Asia and the East Indies in 1819–1820; in Java, over 100,000 died. It reached southern China in 1820 and northern China in 1822. It also spread westward to Arabia in 1821 and to Syria, Iraq, and southern Russia in 1822-1823. In most places, it lasted for a few months, then disappeared.[59] Cholera infected between 1 and 2 percent of the population in the West and 10 to 15 percent in the poorer parts of the tropics, killing half of those it infected. Though it was fatal to fewer people than smallpox or the plague, it was the most terrifying of all epidemics because of the horrible way in which its victims perished.

A second pandemic broke out in 1829. This time, it headed west, to Persia and the Caspian Sea in 1829, to Russia and Europe in 1830–1831, and to the British Isles and North America in 1832. It also reached South America in 1835, and North Africa and again Europe in 1837. In Egypt it was particularly deadly, killing 36,000 in Cairo out of a population of 250,000, and 150,000 nationwide out of a population of 3.5 million. In some places, it lingered on or recurred periodically until 1854.

A third pandemic began in India in 1852 and quickly spread to Russia and the East Indies, then to China and Japan, Europe, the United States, North Africa, and Latin America. In Russia, over 1 million died. In Japan it was especially severe because the country had just opened to foreign trade in 1854; Tokyo alone lost between 100,000 and 200,000 people to cholera.[60]

Cholera did not disappear with the discovery of the bacillus and its mode of transmission but has reappeared many times since. The scientific understanding of this disease, as well as typhoid, led to expensive public health measures such as water chlorination and sewage systems and treatment plants that only wealthy countries could afford. In the poorer parts of the world, especially in the tropics, cholera continued to present a threat. In India, it remained a major scourge, causing an estimated 15 million deaths between 1815 and 1865 and 23 million between 1865 and 1947, especially among the rural poor and during droughts, when people drank water from contaminated wells.[61] As late as 2010, cholera still caused thousands of deaths every year, almost all of them in poor, mostly tropical, countries such as Haiti.[62]

The Plague

The other disease that connected nature with humans was the plague, which aroused great fear in the West because of memories of the Black Death. The third plague pandemic originated in southern China. Western Yunnan, with its rugged mountains and humid tropical climate, was the natural habitat of both the yellow-chested rat *Rattus flavipectus* and the rat flea *Xenopsylla cheopsis*. Yunnan was on the periphery of China, with few inhabitants until a copper mining boom in the eighteenth century brought in Han migrants from other regions. It was also an ideal environment in which to grow poppies beyond the reach of the Qing bureaucracy. In the late eighteenth and early nineteenth centuries, as Chinese increasingly became addicted to opium, their demand was met not only by foreigners smuggling it into Guangzhou from India but also by farmers in Yunnan. Hence an active trade network developed between Yunnan and the coastal cities of southeastern China. Along with the traders and miners came rats, fleas, and the plague bacillus. The spread of plague was greatly accelerated by a Muslim rebellion in southern China between 1856 and 1873 and by the resulting movements of troops and refugees.

The disease reached Guangzhou in 1894. As with cholera, the elite blamed the poor; the American consul in Hong Kong attributed the plague to "the unspeakable filth in which thousands of the natives have lived in utter indifference to sanitary laws."[63] From Hong Kong, the plague spread quickly, thanks to modern steamships. It reached Bombay in 1896, where the British authorities imposed draconian measures to prevent contagion. They did not yet know about the links between the bacillus *Yersinia pestis*, rats, and fleas. Their reactions to the disease, therefore, resembled those of European municipal governments during the Black Death: searching and disinfecting houses (and sometimes setting them on fire), compulsory isolation of the sick and separation from their families, and hasty burials in mass graves without any of the traditional funerary customs. Fearing the authorities more than the disease, many Indians fled to Karachi, Calcutta, and other cities, which only accelerated the diffusion of the epidemic. By 1914, the plague in India had claimed over 8.5 million lives.[64] It also spread to Madagascar, Cape Town, Tangier, Dakar, Honolulu, and other ports around the world. Everywhere it was met with quarantines, racial segregation, and other measures designed to protect if not the entire population, at least the resident Europeans.[65]

Plant Diseases

Human diseases like cholera and the plague were not the only ones to spread around the world thanks to steamships and railroads. Diseases of plants and animals had an impact on humans as well. It was a plant disease, *Hemileia vastatrix*, that destroyed the coffee bushes of Ceylon in the 1870s. Another famous case was the grape blight caused by the aphid *Phylloxera vitifoliae* that devastated almost half of the vineyards in France in the 1860s and 1870s, forcing farmers to re-graft their vines onto American root stock.[66]

Problems with the supply of coffee and wine may have annoyed consumers and harmed regional economies, but the blight that destroyed the potato crop in Ireland in the 1840s devastated a nation dependent on the crop.[67] The climate of Ireland, with its mild rainy weather and long growing season, was ideal for potatoes, One variety, called Lumper, was a prolific plant that could produce almost 15 tons of potatoes per hectare, more than any other crop. As a result of its abundance, the Irish population had doubled during the early nineteenth century from about 4 million in 1800 to over 8 million in 1841.

In June 1845, farmers in Belgium noticed a new disease of potatoes. This was *Phytophthora infestans*, a fungus-like growth that caused the tuber to turn into a putrid, stinking mass. It probably came with a shipment of seed potatoes imported from Peru, the original homeland of potatoes, to replace those affected by dry rot, a minor disease. By July the blight had spread to France and the Netherlands, by September to England, Wales, and Scotland, and by mid-October to Ireland. It was the fastest-spreading of all diseases.

A year later, in the autumn and winter of 1846, the potato crop failed entirely and the weather turned exceptionally cold. By mid-1847, a million Irish had died from hunger. The blight returned in 1849, along with an outbreak of cholera. All the while, Ireland exported grain to Britain, a trade that the British government encouraged under the banner of free trade, while discouraging imports of food and charitable donations to alleviate the famine.[68]

Rinderpest

Animals also suffered from diseases spread by trade, steam transportation, and empire-building. The most prominent case is rinderpest in Africa. Rinderpest had periodically broken out among cattle in Europe throughout the eighteenth and nineteenth centuries until the 1880s. It also afflicted the cattle in India, where it caused serious losses to herders and dairy farmers. It reached Egypt in 1841 and later became endemic there. In 1887–1888, Italian colonists in Eritrea imported infected zebu cattle from Aden or Bombay, as did the British in their campaign in Sudan in 1889–1890. German colonists in German East Africa (now Tanzania) later acquired infected animals from the same source. Over the next ten years, the epizootic spread from East Africa to western and southern Africa, killing 90 percent or more of the cattle it infected. This brought starvation to Ethiopia and to pastoralists like the Maasai who depended on their animals for their entire subsistence. It also killed 80 to 90 percent of elands, giraffes, kudus, warthogs, and several species of antelopes. Elephants, hippopotamuses, rhinoceroses, wildebeests, zebras, and water bucks seemed immune, however. In some places, grasslands, emptied of animals, returned to bush. Though game soon recovered, by the early twentieth century sub-Saharan Africa had fewer large animals than it had had in centuries.[69]

El Niño aClimate Shocks

Much the non-Western world, and especially those parts that were most heavily populated, is found in the monsoon belt of Asia stretching from the Indian sub-continent east to Indonesia and north to eastern China. The monsoons create extremely seasonal weather. From June through September, as the sun heats the northeastern part of the Indian sub-continent, rising air sucks moisture-laden winds from the Indian Ocean into South Asia, bringing heavy rains. Then, when the land mass of Asia cools off, the winds reverse, blowing cool dry winds over South Asia. In Southeast and East Asia, the rains come a little later and move northward, reaching the North China Plain by late July. Warmth and heavy summer rains contribute to a very productive agriculture, supporting a high population density.

However, the monsoon rains are unpredictable, sometimes failing for an entire year, and on rare occasions, for two years or more. These failures are related to a periodic climate shock called El Niño, best known for bringing heavy rains and floods to the west coast of South America, a region that is otherwise among the driest on Earth. It is frequently accompanied by drought in East Asia, the Indian sub-continent, and northeastern South America, with dire consequences for humans. Such events happened seven times in the nineteenth century.[70]

Crops, animals, and people are highly vulnerable to the unpredictable nature of the summer monsoons. In order to survive, the people in affected regions have devised elaborate methods of controlling water, bringing it to where it is needed, storing it for dry years, channeling rivers to prevent floods, and draining wetlands. Some of the methods used to manage water date back to prehistoric times. Most are local, the work of farmers and townspeople. Others are the result of gigantic government projects involving enormous expenditures and massive amounts of labor. The advent of Western imperialism in monsoon Asia in the late eighteenth and nineteenth centuries brought Western scientific engineering to bear on the problems of water

control. In the process, it also created new disturbances in the relations between the inhabitants and the environments in which they lived. One was the
erosion of traditional authority, especially in China, leaving the population
vulnerable to climatic disturbances. Another was the imposition of Western
rule in India and Southeast Asia, accompanied by the Westerners' belief in
their ability to "reclaim, discipline, and train nature."

The Chinese Crisis

Beginning in the early nineteenth century, China entered a severe crisis that
lasted a century and a half, the result of an erratic climate, a deteriorating
environment, foreign intervention, and the weakening of the Qing state.[71]
Between 1750 and 1850, the population more than doubled, from 200 million
to 430 million—growth as fast as that of Europe but without the safety valve
of emigration or the benefits of industrialization. In the words of Sinologist
Robert Marks: "By the nineteenth century, the Chinese agro-ecosystem had
probably reached the limits of its ability to capture and funnel energy and
nutrients to the human population, and hence had placed a limit on the size
of the population."[72] The crises of the mid-century caused the population to
level off; by 1900 it had only reached 436 million.[73]

Part of the problem was the deforestation of much of China. Forests that
had long been protected—as imperial hunting preserves, temple and monastery woods, or village commons—succumbed during the nineteenth century to the desperate need for land and wood. Even Manchuria, long kept off
limits to Han Chinese by the ruling Manchus, were finally opened to Han
migrants in 1860; as a result, its population soared from 2.5 million in 1820
to 17 million in 1910.

Yellow River and Grand Canal

The region hardest hit by environmental degradation was the North China
Plain. Not only was it vulnerable to the periodic floods of the Yellow River, but
it was also where the Grand Canal crossed the Yellow River. This demanded
tremendously complex engineering and presented the state with an insoluble
dilemma. To prevent floods on the North China Plain, it built and reinforced
levees along the Yellow River. And to supply the capital and the armies that

protected China from its northern neighbors, it maintained the Grand Canal between the Yangzi Valley and Beijing. But the canal took its water from the Yellow River. Sinologist Kenneth Pomeranz explains:

> Without effective control of the Yellow River, the canal was useless. Too strong a current would block or flood the canal; dike breaks upstream would lead to too little water being fed into the canal, and perhaps water trying to enter the canal bed at the wrong places. And because the canal was the "throat of Beijing" . . . allowing it to be blocked was unthinkable.[74]

By the late eighteenth century, shoring up the levees along the Yellow River was becoming more and more difficult. The cost of maintaining them strained the budget of the Qing state, leaving it vulnerable to political or environmental shocks. After a particularly horrendous flood in 1801, the state stopped seeking a permanent fix and settled for limited maintenance, allowing some flooding but saving money for future disasters. In 1824–1826, a flood tore breaches in the dikes, flooding eastern Jiangsu province and damaging the Grand Canal.[75]

The Qing state might have recovered from these natural calamities but for a series of political disasters that diverted its attention and resources. First came the Opium War of 1839–1842, during which the British sent steam-powered gunboats up the Yangzi River to its junction with the Grand Canal, thereby cutting the capital off from its sources of grain and forcing the government to cede Hong Kong and several treaty ports, pay a large indemnity, and allow the unlimited import of opium.

As the government grew increasingly unable to afford the cost of maintaining the Yellow River dikes, floodwaters broke through in 1841, 1842, and 1843, and again in 1851, 1852, and 1853. By the mid-nineteenth century, the buildup of silt had lifted the river as much as 12 meters above the North China floodplain, so that it could no longer be induced to return to its old channel. In 1851, some of its waters began to flow north of Shandong. By 1855, the entire river had switched back to where it had once been. It was the worst environmental disaster since the fourteenth century. Thirty counties were flooded to a depth of 7 to 10 meters. What had once been farmland was turned into shallow wetlands and remained so for thirty years.[76]

From 1851 to 1864, the government was fighting the Taiping Rebellion, in which 20 to 30 million people are said to have perished and vast areas of land, especially in northern China, were devastated. At the same time, the Nien

Rebellion in northern China and Muslim uprisings in southwest and north-east China helped exhaust the government's resources.

The depredations of marauding armies and the collapse of civil authority between 1854 and 1878 contributed to the destruction of China's few remaining forests, as rival armies set them on fire to deny them to their enemies. Timbers to rebuild the Summer Palace in Beijing after British troops had burned it down in 1860 had to be imported from Oregon, for no trees large enough could be found in China.[77]

Along with the domestic calamities came another foreign attack, the Second Opium War of 1856–1860. Deprived of revenues from China's richest regions, the Qing state could neither maintain the Yellow River dikes nor keep the Grand Canal functioning. It abandoned the Grand Canal and did little to protect what was left of the Yellow River floodplain. A part of China that had once supported prosperous farms went into a steep decline, both demographically and environmentally.[78]

1876–1878 and After

The calamities that befell China did not end with the reestablishment of Qing rule in the 1860s. The rebels were vanquished and the foreign demands temporarily met, but China, like India, remained exposed to natural disasters.

One of the worst was the El Niño of 1876–1878 that caused a drought affecting the entire belt of lands from the East Indies to northeastern Brazil.[79] In China, the drought spread throughout the north, causing a harvest failure. Between 9.5 and 13 million out of about 80 million inhabitants died of starvation and another million fled to Manchuria. The crisis was especially severe in China because the government, weakened by rebellions and wars, was too poor to help the hungry, and were there not any forests, marshes, or other natural reserves left to which they could turn for emergency foods when their crops failed.[80]

In Egypt, the year 1877 was marked by an exceptionally low Nile flood, as were subsequent El Niño years, although perennial irrigation mitigated the damage.[81] In India, about half the sub-continent was afflicted with a drought that lasted two years. The Madras region received about a quarter of its usual rainfall. Though food was available in some parts of the sub-continent, it could not be transported to regions of food deficit because the bullocks that pulled the carts—India's traditional means of transporting freight—were also starving. Bengal was hit by a cyclone that drowned 100,000 people and killed another 100,000 from disease or famine. The government of India was powerful, but the British rulers were largely indifferent to the famine, as they had been in Ireland, on the grounds that providing relief would violate the principles of free trade.[82]

Nor was this the last of the natural disasters of the late nineteenth century. In China, the Yellow River flood of 1887 is said to have drowned 900,000 people and left 2 million homeless. The drought and famine of 1896–1897 in India killed 5 million and that of 1899–1900 another 1.25 million or more in Bengal alone.[83] Despite the myriad changes brought about by Western attempts to control natural forces through industrial methods, the non-Western world, especially the monsoon belt of Asia, was still vulnerable to the forces of nature.

Conclusion

The Industrial Revolution had a powerful effect on large parts of the non-Western world, especially in the tropics and monsoon Asia. The cause of environmental change was the Western demand for the products of the non-Western world, not only traditional imports like sugar, tea, and coffee, but also guano, cinchona, rubber, gutta-percha, and other products previously almost unknown in the West. As the non-Western parts of the world were drawn ever more tightly into the Western-dominated networks of power and commerce, their environments were transformed to satisfy the Western demands. Those parts of the world that were under direct colonial rule were most affected, but even those that were independent, like China, Siam, and Brazil, felt the influence of Western pressure.

It is instructive to compare the impact of Western industrialization on the West and on the non-West. In western Europe and North America, the forces of industrialization sharply increased the human impact on nature through clear-cutting forests, plowing up the prairie, building cities, strip mining, decimating wildlife, and polluting the air and water. Yet at the same time, Western science and technology developed measures such as urban sanitation, flood prevention, and hurricane relief to mitigate the power of nature to harm humans, measures that were out of reach of poorer countries.[84] Environments throughout the rest of the world were also transformed by irrigation, plantations, railroads, and cities, their forests cut down, and their wildlife decimated. Their peoples increased in numbers, but their living standards did not improve, and they remained as vulnerable as ever to droughts, floods, and epidemics. In the tug-of-war between humans and the rest of nature, some people living in the industrializing nations were clearly winning; but for those living elsewhere, the contest was still undecided.

10

War and Developmentalism in the Twentieth Century

In 1900, large parts of the world were still untouched or only lightly changed by humans: the Arctic, the great forests of the far northern and equatorial zones, even grasslands and deserts. By 2000, these regions were being invaded and transformed, and only Antarctica remained a true wilderness. Human actions were even changing the oceans and the atmosphere.

Three forces conspired to cause these transformations during the twentieth century. One was the quadrupling of the human population, from about 1.6 billion in 1900 to roughly 6 billion in 2000.[1] The second was the wide dissemination of powerful new forms of energy (electricity, oil, and nuclear power), new materials (concrete, steel, chemicals), and new machines (automobiles, aircraft, and many others). And the third was the extraordinary growth in the world economy—interrupted but not reversed by the Great Depression—that outpaced even the growth in population, multiplying the production and consumption of goods and services eighteen times over, or 4.8 times per person per year over the course of the century.[2] Together, these three forces have changed the planet and are continuing to do so.

To analyze the forces that propelled these changes, this chapter stresses the actions of governments driven by wars and revolutions and ideologies of development to undertake vast, sometimes pharaonic projects. The subsequent chapter will look at how consumers and private enterprises have contributed to the transformation in the environment. These actions are not unidirectional, for nature is neither a passive victim nor an innocent bystander to the actions of humans but an active agent causing devastating epidemics and disastrous droughts and dust storms.

Warfare

Long before the twentieth century, the urgency of war made belligerents cast aside all other considerations: the sanctity of human life, of course, but also respect for nature and the need to conserve resources for the future. Navies were voracious consumers of timber; much of the deforestation around the Mediterranean, for example, can be traced to the construction of warships from ancient times to the nineteenth century. Armies on the march consumed food and requisitioned animals. Armies also practiced scorched-earth tactics in enemy territory, deliberately destroying croplands, animals, forests, and cities and poisoning water supplies with animal carcasses. To deprive their enemies of cover, the Romans felled a wide swath of forests on either side of their roads in Gaul, as did the Russians in the Caucasus and the British in India in the early nineteenth century. During the Taiping Rebellion of 1850–1864 in China, both sides devastated the lower Yangzi region. In 1864, during the American Civil War, General Philip Sheridan's army destroyed farms and woods in the Shenandoah Valley. Later, during the Indian Wars of the late nineteenth century, the US Army encouraged the killing of bison to deprive the Plains Indians of their main food supply. In short, environmental destruction as a military tactic has been practiced for millennia. What changed in the twentieth century was the vastly increased means of destruction at the disposal of armed forces.[3]

The First World War

In the fall of 1914, when the armies of Germany, France, and Great Britain came to a standstill in the trenches of the Western Front, the traditional goals of warfare—to seize enemy territory and to destroy the enemy's ability or will to resist—were upended. When capturing territory and taking prisoners became impossible, the goal became to kill enemy soldiers by obliterating the very landscape on which they stood.

To escape the hail of bullets and artillery shells, soldiers on both sides dug ditches in the ground, then connected them into parallel networks of permanent trenches stretching 500 kilometers from the Swiss border to the North Sea. Periodically, soldiers were ordered "over the top" into the no-man's land between the lines, where they got tangled in barbed wire, felled by machine gun fire, or torn apart by exploding shells. What had been a landscape of woods, fields, and villages became a chaos of mud, metal, and corpses, reeking of excrement, explosives, and decaying flesh. The living cowered underground while the dead occupied the surface. For those who attempted to survive in this ghastly terrain, cold weather and frequent rain made life even more miserable. Yet some forms of life not only survived but flourished: rats found abundant food eating the dead and lice fed on the blood of the living.[4]

Worse was to come: gases that did not strike soldiers directly but poisoned the air they breathed. During the second battle of Ypres in April 1915, the Germans released chlorine gas, killing 6,000 French and colonial troops and blinding thousands of others, including many Germans injured while releasing the gas. Chlorine, heavier than air, made its way into the trenches, forcing soldiers attempting to escape the gas to expose themselves to enemy gunfire. After that experience, both sides began to experiment with poison gases and equipped their troops with gas masks. Chloropicrin, though much less toxic than chlorine, penetrated gas masks and caused nausea; when a soldier removed his mask to vomit, he would inhale other, more poisonous gases. Phosgene, even deadlier than chlorine, was mixed with the latter. Mustard gas, the most widely used of the war's toxic gases, did not kill immediately but blinded soldiers and caused a slow, agonizing death from internal

Figure 10.1. Corpses of soldiers killed while crossing No Man's Land between the trenches on the Western Front during World War I. The Bridgeman Art Library.

and external bleeding. It also remained in the soil for weeks or months, making entire areas dangerous long after the battle was over. Poison gases also killed all animals that came into contact with them, even the rats and lice that tortured soldiers on the front. Though 190,000 tons of chemical gases were used during the war, poison gas killed only 88,498 out of 8 million soldiers and injured 1.2 million out of 15 million. Gas caused great suffering but did not change the outcome of the war, for soldiers on both sides suffered equally.[5]

At the end of the war, a strip of land a few kilometers wide across northern France and western Belgium lay devastated, the earth chewed up, trees and other vegetation killed, and buildings destroyed, leaving a landscape of shell holes, barbed wire entanglements, concrete bunkers, uprooted tree trunks, and human corpses and body parts. For years thereafter, unexploded ordnance lay buried, threatening to blow up whoever disturbed it. A hundred thousand hectares of arable land were so devastated that they could not be restored to farming but instead were planted with trees. In the German-occupied areas of France and Belgium, 200,000 hectares of forest were so badly damaged that they had to be cut down and reforested after the war.

The Influenza Epidemic

While humans were becoming more adept at massacring one another, nature could still outperform humans at this grisly task. Military deaths in World War I totaled 6.8 million as a result of combat, plus another 3 million from accidents, mistreatment in prisoner of war camps, and diseases other than influenza. The war also caused approximately 6 million civilian deaths from malnutrition, disease (again, other than the flu), and the Armenian genocide.[6] In contrast, the influenza pandemic of 1918–1920 (misnamed the "Spanish" flu because its effects were first publicized in Spain) infected a quarter of the world's people, killing an estimated 24 to 40 million people and making it one of the deadliest natural disasters in human history.[7]

A normal, mild form of influenza began spreading in the spring of 1918, followed by a far more virulent kind in the late summer. The latter variety was unusual in two respects: unlike ordinary flu, which affected mainly the very young and the very old, it was especially likely to infect young adults. It killed 8 to 20 percent of infected soldiers. Its symptoms were more extreme than doctors had ever seen: high fever; delirium; bleeding from the nose, ears, and mouth; and cyanosis, as the skin turned blue from lack of oxygen. Its victims often died very suddenly, in a matter of hours or days.

Where this flu strain began is still being debated, but its military origins are unambiguous. In early September, the virulent flu was first noted in Brest, France, where American soldiers were disembarking. Camp Devens near Boston, where recruits were being trained for combat before being shipped to France, suffered 14,000 cases, or 28 percent of its population, of whom 757 died. From there it reached Étaples, a training center in France, and Fort Riley in Kansas. It spread rapidly among the troops on both sides of the Western Front. In the appalling conditions in camps and hospitals, one out of every six American servicemen fell ill. The disease spread particularly fast in the trenches of the Western Front, where soldiers lived crowded together in mud and waste. Of the soldiers in the American Expeditionary Force, 227,000 were hospitalized with combat wounds and 340,000 with influenza. Behind the lines, doctors and nurses were overwhelmed by the number of sick and wounded and were themselves prone to succumbing to the disease.[8]

The Central Powers were especially hard hit. The German High Command had counted on the spring offensive of 1918 to bring them victory. But when 1.75 million German soldiers fell ill and twice as many as were wounded, the offensive stalled. The third wave of influenza, in October 1918, led to the collapse of morale and discipline in the German army and eventually altered the state's ability to govern effectively. Similarly, the outbreak of influenza occurring that fall weakened the Austro-Hungarian Empire.[9]

Once it began, the flu spread to civilians around the world. In May 1918 it reached India, where it killed 17 to 20 million people. In Russia, 7 to 10 percent of the population are thought to have died. In the United States, it killed 540,000. The epidemic reached China by the Trans-Siberian railroad and by ship through the treaty ports, and it may have killed as many as 4 million people. In Japan, over 200,000 people died. It reached Cape Town by ship from Europe, then moved inland by railroad to British Central Africa, then down the Congo River by steamboat. In equatorial Africa it killed an estimated 82,000 out of a population of about 2.9 million. Indigenous peoples were hardest hit; some Pacific islands lost one-third to one-half of their population. And, quite forgotten amid the human carnage, untold numbers of horses, pigs, moose, baboons, and other animals also died of influenza. In the words of historian Alfred Crosby, "Nothing else—no infection, no war, no famine—has ever killed so many in so short a time."[10]

Colonial Wars

The end of fighting between the great powers did not bring an end to wars in colonial areas. Here, the industrialized nations brought to bear two innovations, airplanes and gas, that proved to be very valuable against less advanced, poorly armed peoples. Poison gas proved tempting to use against peoples who could not retaliate. With aircraft and toxic chemicals, modern industrial nations had found quicker and cheaper means of bringing destruction to the very environments that supported their enemies' civilian populations.

After the Great War, Great Britain seized Mesopotamia from the Ottoman Empire. When the people of Mesopotamia rebelled in June 1920, Great Britain had to send in 100,000 troops and eight squadrons of airplanes to repress the uprising. This campaign cost more than Britain, exhausted after four years of warfare against Germany, could afford. To lower the cost of controlling the region, Colonial Secretary Winston Churchill suggested that the Royal Air Force develop gas bombs, "especially mustard gas, which would inflict punishment upon recalcitrant natives without inflicting grave injury upon them."[11] The Air Staff responded that gas bombs were "non-lethal, but were not innocuous. They may have an injurious effect on the eyes, and possibly cause death."[12] After protests by the Colonial Office cancelled plans to use gas, the British relied upon local militias and ordinary bombs against recalcitrant villages.[13]

The Spanish army, in its campaign in the Rif Mountains of northern Morocco, was not deterred by the bad publicity that surrounded the use of poison gas. After suffering a decisive defeat at the hands of Riffi tribesmen in July 1921, the Spanish government turned to Hugo Stoltzenberg, a German chemical manufacturer, to produce mustard gas for use in the Rif. It also purchased phosgene and chloropicrine gas shells from the French arms manufacturer Schneider.[14] By 1923, Spanish warplanes began dropping gas bombs on Riffi towns on market days and burned crops with incendiary bombs during the harvest season. While the bombing caused extreme suffering among the civilian population, it did not deter the rebels, who learned to avoid towns and hide in caves during the day.[15]

Italy, after a poor showing in World War I, sought to recover its self-esteem in Africa. In 1936, Benito Mussolini sent a huge army into Ethiopia, a nation that had inflicted a humiliating defeat on Italy in 1896. When the Ethiopians resisted, Mussolini ordered his air force to use poison gas. Appearing before the League of Nations, Ethiopian emperor Haile Selassie described the new method of warfare:

> Vaporizers for mustard gas were attached to their planes, so that they could disperse a fine, deadly poisonous gas over a wide area. From the end of January 1936, soldiers, women, children, cattle, rivers, lakes, and fields were drenched with this never ending rain of death. With the intention of destroying all living things, with the intention of thereby insuring the destruction of waterways and pastures, the Italian commanders had their airplanes circle ceaselessly back and forth. . . . This horrifying tactic was successful. Humans and animals were destroyed. All those touched by the rain of death fell, screaming in pain. All those who drank the poisoned water and ate the contaminated food succumbed to unbearable torture.[16]

The Second World War

By the late 1930s, many feared the world was headed toward another conflict marked by trenches and deadlock, aerial bombardments, and poison gases. What happened took the world by surprise.

The Second World War engulfed far more of the planet than had the First World War. Because the fighting spread out over huge swaths of land and sea, only bombed cities suffered as much damage as the Western Front had. Furthermore, none of the belligerents used poison gases (except in the German annihilation camps), perhaps because they had proven ineffective and counterproductive in the previous war. Yet here too, armed forces did not limit themselves to attacking enemy forces but tried to weaken them by damaging the environment on which they depended. The results were catastrophic for civilians and for the natural world.

One particularly deadly form of environmental warfare was flooding. In June 1938, as the Japanese forces were fighting their way into China, Generalissimo Chiang Kai-shek, the leader of the Chinese Nationalist government, ordered the breaching of the Yellow River dikes to slow the Japanese advance. That summer was one of the wettest on record and the river, enclosed by massive dikes, had already risen several meters above the land. So thick were the dikes that explosives were useless, and soldiers were ordered to cut them with hoes and spades. Once the water began to flow, it widened the breach and advanced down the North China Plain at a rate of 16 kilometers a day, eventually flooding an area of 70,000 square kilometers. Between 500,000 and 1 million peasants, about to harvest their crops, drowned or died of hunger or illness. Another 2 to 4 million fled, but the neighboring Nationalist-held provinces were just as poor. After the Rape of Nanjing in late 1937 and early 1938, no Chinese dared flee to the Japanese-occupied areas. Those who returned after the fighting ended found their villages and farms washed away, their irrigation channels eroded or silted up, and their land turned into hard, baked mud. It was, in the words of one historian, "perhaps the most environmentally damaging act of warfare in world history."[17]

The Pacific islands caught in the war were badly damaged. By their nature, small islands have fragile ecologies, with much lower chances of recovery than large bodies of land.[18] On islands that were used as bases, the Japanese and Americans mined coral reefs to build airfields and harbors; some have never recovered. Ships that leaked oil or tankers that were sunk contaminated lagoons, reefs, and beaches. Birds that nest on the ground were killed off by newly introduced pigs, dogs, and rats. On islands such as Tarawa, Saipan, and Iwo Jima that suffered intensive bombardments before troops landed on their beaches, forests were destroyed, animals exterminated, and the land cratered. Once the fighting moved on, wrecks of planes, ships, vehicles, and other military materiel littered the landscape and nearby waters and were left to rust away.[19]

Forests at War

Forests played a part in both world wars as sources of timber and as obstacles to fighting men. With their very survival at stake, nations set aside established notions of sustainable harvesting. In France, over 500,000 hectares of forest were felled, damaged, or burned in each world war. While the forests of Germany remained largely intact, the Germans severely depleted those of the areas they occupied. In World War II, 20 million hectares of forests were destroyed in Nazi-occupied sectors of the Soviet Union, in part to deprive partisans of hiding areas.[20] Great Britain consumed half of its few remaining forests. To help the war effort, loggers exploited the forests of the Americas. With steel and aluminum in short supply, gliders, boats, barrels, buildings, and other constructions were made of wood wherever possible. British Mosquito fighter planes, for example, were built of Sitka spruce from the Pacific Northwest and balsa from Ecuador.

The war also affected the forests of Asia. The Japanese exploited the forests of Burma, Java, and the Philippines. The British logged the forests of India, especially those of Assam, for construction and railroad building. The most heavily cut were Japan's once magnificent forests, especially as its empire shrank. Pines were destroyed in a futile attempt to extract motor fuel from their roots. From 1941 to 1945, 3.6 million hectares of Japan's forests were logged, of which 2.6 million were clear-cut.

The impact on forests continued long after the end of the war. The Japanese cities that were firebombed in the war were made of wood, and their reconstruction required enormous quantities of lumber. In Europe, the bitterly cold winter of 1945–1946 and the damage to coal mines and railroads meant that every available scrap of wood, even trees in the parks, was taken for fuel. As after World War I, damaged old-growth forests were mostly replaced by single-species tree plantations.[21]

The Vietnam War

In Vietnam, the United States military faced a particularly challenging environment. Much of the inhabited part of the country consisted of rice paddies, and tanks and trucks could scarcely maneuver in these. Another large area was mountainous and forested, again posing great difficulties to motorized forces and making it difficult for aircraft to spot enemy soldiers. And the Mekong Delta contained large areas of mangrove forests. In these terrains, the Vietcong, or communist guerrilla forces, moved stealthily on foot or in small boats.

To fight against the environment as well as against the communists, the United States used an unprecedented arsenal of sophisticated weapons. To clear woods in strategic areas, the army employed huge tractors called "Rome plows" that could knock down and crush large trees; altogether, 325,000 hectares were cleared this way. Elsewhere, the American forces relied on airpower: bombers, gunships, transports, and helicopters. Between 1964 and 1975, the United States dropped 14 million tons of bombs and shells on Vietnam, more than in both world wars combined. Unlike in World War II, in which most explosives were dropped on cities, in Vietnam, they were targeted at rural areas, where they were much less effective.[22]

Most controversial was the use of defoliants. In the course of the war, the United States sprayed 72 million liters of Agents Orange, Blue, and White

and other defoliants with colorful names. Spraying began in 1962, even before the United States was officially involved in combat. The amounts used
peaked between 1966 and 1968, then declined, ending in mid-1970, when
defoliants were banned from use in the United States. Their manufacturers,
the Dow, Monsanto, Hooker, Alkali, and Hercules chemical companies,
described them as harmless to humans and animals, although internal
memos later released showed that they knew that 65 percent of the defoliants
contained dioxin, a chemical that causes miscarriages, birth defects, and
cancer. Agent Blue, designed to destroy rice crops, contained arsenic. It was
sprayed on rice paddies, gardens, and orchards in communist-held territories in which between 2.1 and 4.8 million people were living, forcing them to
flee to government territory.[23]

According to a National Academy of Sciences report in 1974, over 1 million hectares or 10.3 percent of the inland forests of South Vietnam were
sprayed. In the forests, 10 percent of the trees died after one spraying, and up
to half after multiple sprayings. As in all tropical forests, the nutrients were in
the vegetation above the ground while the soil itself was sterile. Once an area
was deforested, the ground turned to laterite, a hard, impermeable red soil
on which it was difficult to plant new trees or crops. Defoliants also damaged
over 15,000 hectares of South Vietnam's and Cambodia's hevea rubber trees.
Over 100,000 hectares, or 36 percent of South Vietnam's mangrove forests,
were sprayed. Mangroves were far more vulnerable than other trees; often
one spraying was enough to kill them, along with the fish, shellfish, birds, and
other animals that lived among them, leaving mudflats open to erosion by
storm surges.

The Nuclear Arms Race

The connections between warfare and the environment involved more than
just combat. Before the Second World War, they entailed testing weapons,
building fortifications and military camps, and conducting maneuvers.
Since then, they have included making and testing nuclear weapons, with far
graver consequences.[24]

The bombs that leveled Hiroshima and Nagasaki caused 200,000 human
deaths, most from the blast and fire, but about 30,000 from the acute radiation that penetrated bodies kilometers from the epicenter.[25] No doubt, animals living in or near those cities suffered equally. The radiation from the two

A-bombs quickly dissipated, and just weeks after the end of the war, people began returning to rebuild.

Nuclear testing caused more extensive environmental damage than the two atom bombs used in warfare. The United States tested nuclear weapons in New Mexico, Nevada, and the Marshall Islands. The United Kingdom tested its weapons in the Australian desert, France in the Sahara and in French Polynesia, the Soviet Union in Kazakhstan and in the Arctic, China in Xinjiang, India in the Thar Desert, and Pakistan in Baluchistan. All of these sites remained off limits to people for decades after the testing and were toxic to animals as well.

Even more dangerous were the nuclear processing and storage areas where radioactive wastes accumulated for decades. At the Hanford site in eastern Washington state, nine nuclear reactors and five plutonium processing plants left behind 200,000 cubic meters of high-level liquid wastes and 710,000 cubic meters of solid wastes, contaminating 520 square kilometers of groundwater. Water used to cool the reactors made the fish that lived in the Columbia River radioactive. Winds carrying highly radioactive iodine-131 into Idaho, Oregon, Montana, and British Columbia contaminated pastures, the cows that grazed on them, and the milk they produced.[26]

The Soviet Union built plutonium-producing reactors and bomb manufacturing plants near Kyshtym in the southern Ural Mountains and at Tomsk in Siberia. In September 1957, high-level nuclear wastes improperly stored in uncooled tanks overheated and exploded, releasing 70 to 80 metric tons of highly radioactive particles into the air. The resulting radioactive cloud fell on an area of over 1,500 square kilometers between the cities of Sverdlovsk and Chelyabinsk, forcing the evacuation of at least 100,000 people and sickening scores of them with radioactive poisoning. Though the Soviet regime imposed a total blackout on news of the disaster, Western intelligence agencies knew of it but kept it secret to avoid stirring up public hostility to nuclear power. Other radioactive wastes dumped into Lake Karachay in the southern Urals made it the most polluted spot on Earth, a place so radioactive that anyone there would get a lethal dose of radioactivity within an hour. In the 1960s, the lake dried out and winds carried radioactive dust that irradiated half a million people.[27] Each of the plutonium-producing complexes in the United States and the Soviet Union released 200 million curies of radioactivity into the surrounding environments, twice as much as the Chernobyl nuclear plant meltdown in 1986.[28]

Wildlife at War

In war, there are winners and victims among animals as well as among humans. In countries short of food, edible animals, even migratory songbirds and zoo animals, were sacrificed for the welfare of humans. In some places, wild carnivores, such as the bears, wolves, and wolverines of Norway, increased in numbers while humans were occupied killing one another. According to wildlife biologist Ronald Nowak, "Large predatory animals traditionally increased in numbers during times of war when men were more concerned with killing each other than with hunting wildlife. Wolves, for example, are said to have multiplied in Europe during the Thirty Years and Napoleonic wars and to have made remarkable comebacks during World Wars I and II."[29]

Indochina was once known among big-game hunters for its elephants, rhinos, crocodiles, and tigers, as well as deer and pheasants. Starting in 1940, one war after another decimated its wildlife population. Defoliants did not kill animals directly but damaged their habitats and caused genetic abnormalities.[30] War was good for tigers, however. In 1970, zoologist E. W. Pfeiffer and ecologist A. K. Orians wrote:

> They have learned to associate the sounds of gunfire with the presence of dead and wounded human beings in the vicinity. As a result, tigers rapidly move toward gunfire and apparently consume large numbers of battle casualties. Although there are no accurate statistics on the tiger populations past or present, it is likely that the tiger population has increased much as the wolf population in Poland increased during World War II.[31]

Wildlife Refuges

Wars and preparations for war have left damaged and contaminated areas scattered around the world, though some have subsequently been restored. In the United States, the notorious Rocky Flats Nuclear Arsenal in Colorado, the scene of many antiwar demonstrations and lawsuits over radioactive contamination, has been largely decontaminated and, since 2005, turned into a National Wildlife Refuge, open to animals but closed to humans because of the remaining radioactive materials. Similarly, the US Navy's firing range on Vieques Island off Puerto Rico was turned over to the Fish and Wildlife Service as a National Wildlife Refuge in 2005.[32]

These are small areas, however, compared to the borderlands between East and West Germany and between North and South Korea. During the Cold War, a no-man's-land 1,393 kilometers long and 5 kilometers wide separated the German Federal Republic and the German Democratic Republic. A 500- to 1,000-meter-wide strip on the East German side was filled with land mines, electrified barbed wire fences, and watchtowers. When the two Germanies were reunited in 1990, the mines, fences, watchtowers, and other remnants of the Cold War were taken down. The strip, named the German Green Belt, became the site of over 300 nature preserves and three UNESCO biosphere reservations. There wildlife survives, including black storks, red kites, otters, and other endangered species.[33]

The Demilitarized Zone (DMZ), a strip of land 4 kilometers wide and 250 kilometers long, still separates the two Koreas. Once an area that was settled and farmed, this strip is now off limits to humans and studded with land mines, barbed wire, and other obstacles. It has become an inadvertent nature preserve, with environments ranging from wetlands near the coast to grasslands, forests, and mountainous highlands. Here, native wildlife flourishes. Migratory birds stop there on their way between Siberia and China or Southeast Asia, including endangered ones such as the red-crowned Manchurian cranes that would probably be extinct if they could not find a safe place to land.[34]

DEVELOPMENTALISM

The imperative to produce more and maximize a nation's economic growth did not cease with the ends of wars. Instead, it has become ingrained in the ideologies of all political parties. Politicians of all stripes justify economic development on nationalist grounds—to make their nation stronger or more prosperous. International organizations like the World Bank, the International Monetary Fund, and the United Nations Environment Programme also see development as an unalloyed good. Even economists, with very few exceptions, believe that growth is a good thing and always will be.[35] As John McNeill noted: "The overarching priority of economic growth was easily the most important idea of the twentieth century."[36]

Developmentalism, the ideology of economic growth, reflects both the goals of economic development and the means to achieve them. Among the goals, three stand out: improving the living standards of a nation's citizens; strengthening its military power; and enhancing its economic standing in the world. The means to achieve these goals range from a command economy entirely managed by the government to various combinations of private enterprise and government intervention.

Starting in the late nineteenth century, the Japanese deliberately built up their economy under the slogan "Rich Nation Strong Army."[37] In the West, the link between a nation's economy and its military strength became apparent during the First World War. After the war, the Soviet Union pioneered the idea of a government-managed economy controlled by five-year plans. The success of the Soviet Union in World War II and the communist victory in China led Mao Zedong to adopt an extreme form of command economy.

In contrast to the communist planned economies, the United States championed the ideal of a free-market economy, with government intervention when needed to build infrastructures or invest in research and development. The United States, a much richer country to begin with, developed its economy mainly for civilian consumers, but also, after 1940, for its armed forces. As for the other nations of the world, they adopted a combination of communist-style developmentalism and American-style consumerism, according to their circumstances and the leanings of their rulers.[38]

The environment, meanwhile, was easily ignored. Abundant natural resources were taken for granted; substitutes were found for those in short supply. Nations desperate to develop their military might or to alleviate poverty felt they could ill afford to jeopardize their economic growth for the sake of the natural environment. In democracies, popular opinion gave the highest priority to maintaining living standards; only in the last decades of the twentieth century was protecting the environment added to the list of priorities. For economists, the oceans, the air, the wilderness, and other non-human parts of the biosphere were "externalities," hard to value in monetary terms, hence easily ignored.

Soviet Attitudes toward Nature

After both world wars, most nations reverted to their prewar pursuits. But in others, revolutions demanded total mobilization, even in peacetime.

When development became a means of enhancing a nation's status in the world militarily or economically, other considerations, such as the natural environment, the health of a country's population, even the sustainability of that development, were easily ignored. The Soviet Union and the People's Republic of China represent two extreme cases of forced development.

The Soviet Union was born of warfare with Germany from 1914 to 1917, followed by a revolution and a civil war from 1918 to 1921. Even after the fighting stopped, the Bolsheviks believed they could accelerate their nation's economic development by harnessing the resources of nature as never before.[39] Leon Trotsky, Stalin's rival, expressed the more extreme Bolshevik attitude toward nature: "The present distribution of mountains and rivers, of meadows, of steppes, of forests and seashores, cannot be considered final. Through the machine, man in socialist society will command nature in its entirety. . . . He will point out places for mountains and passes. He will change the course of rivers, and he will lay down rules for the oceans."[40] Or, as a Communist Party slogan put more bluntly: "We cannot expect charity from nature. We must tear it down."[41]

Once Stalin had achieved control of the party and the state, he set out to transform the Soviet Union from a poor and backward nation into one of the world's great powers. To achieve military security in a world filled with hostile nations required armaments on a massive scale and therefore the industries to produce them. This meant creating, at breakneck speed, factories, steel mills, power plants, railroads, canals, mines, and other industries. The Five-Year Plans, starting in 1928, were the tools the party employed to catch up, in a few years, with the industrial nations of the West, regardless of the social or environmental costs. Rapid economic growth under state command continued throughout the Stalin era and long after his death in 1953.

Soviet Dams

In 1920, the Eighth Congress of Soviets proclaimed electrification as a leading Soviet goal, under the slogan: "Communism equals Soviet power plus electrification of the entire country." When Stalin took power, he demanded immediate results, regardless of the economic, human, or environmental costs. The first major effort was the Dneprostroi Dam, built between 1927 and 1932, that made the Dniepr River navigable from central Ukraine to the

Black Sea. Destroyed by retreating Soviet troops during World War II to keep it from being used by the Germans, it was rebuilt in 1947.

In October 1948, Stalin announced his Plan for the Transformation of Nature. Among its most ambitious projects was turning the 3,700-kilometer-long Volga River into a series of reservoirs. This gigantic effort involved building twelve major hydroelectric stations and many minor ones. The largest was the Zhiguli Dam that created the 6,450-square-kilometer Kuybyshev Reservoir, the largest in Europe and the third largest in the world. On the Don River, the Soviets completed the Tsimlyansk Reservoir and the Volga-Don Canal in 1952. These dams, reservoirs, and canals completely transformed central Russia, propelling the industrialization of the Soviet Union at a cost of 3.1 million hectares of farmland and another 3.1 million hectares of forests, in addition to tens of thousands of forced laborers.[42]

Environmental Costs

The Soviet regime's hyper-industrialization project resulted in millions of Soviet citizens being thrown into forced-labor camps or executed while millions more lived in poverty and terror. The communist creation of the second-largest and, for a time, fastest-growing economy in the world also came at steep environmental costs.[43]

As of 1989, of the 600 million hectares of cultivated land in the USSR, nearly half were seriously imperiled: 157 million hectares were saline, 113 million badly eroded, 25 million waterlogged, and 5 million covered by mine tailings, slag and ash piles, and city dumps. Because of excessive mechanization, chemical use, and poor farming practices, the amount of humus in the soil had dropped 20 percent in the Russian Federation and 9 percent in Ukraine, the "breadbasket" of the Soviet Union. Erosion was taking between 100,000 and 150,000 hectares of farmland out of service annually. Much of the irrigated land, which had increased by 445,000 to 525,000 hectares annually after 1970, was waterlogged or salinized.[44]

Most of the newly irrigated land was in the Central Asian republics of Tajikistan, Turkmenistan, and Uzbekistan, where the Amu Darya and the Syr Darya rivers were diverted to irrigate cotton fields. Between 1960 and 1989, as water from these rivers ceased to flow into the Aral Sea, the sea's level dropped by one-quarter, its volume by two-thirds, and its area by almost half.

By 2012, it had shrunk to 10 percent of its original size. The water that still flowed into it was laden with phosphates, ammonia, nitrites, nitrates, and chlorinated hydrocarbons, the runoff from cotton fields. Fish populations that had once sustained a rich fishing industry almost vanished. Boats were abandoned kilometers from the shores of the shrunken sea. Winds blowing over the newly exposed land carried salt and sand as far as Belarus and Afghanistan, sickening people, especially children.[45]

Lake Baikal, "the pearl of Siberia," also suffered from industrial pollution. An extraordinary natural phenomenon, it covered 31,722 square kilometers and was seven times deeper than the Grand Canyon. It contained 20 percent of the fresh water in the world; as the purest, cleanest natural water anywhere on Earth, it hosted a unique set of plants and animals. Taking advantage of the purity of the water and the abundance of timber in the vicinity, the Soviet government built a giant pulp and paper mill on its shores in 1966 and later other factories to produce paper and cords for the tires of bombers. In the process, these factories dumped 191,000 tons of inadequately treated wastewater into the lake. These toxic effluents killed off the zooplankton that were

Figure 10.2. Boats left stranded in the desert that used to be the bottom of the Aral Sea before it dried up, late twentieth century. iStock.com/DanielPrudeck.

the base of the aquatic food chain, causing algae to bloom and aquatic animals to die off.[46]

These were hardly the only cases of water pollution. Along the coast of the Black Sea between Odessa and Yalta, so filled with resort hotels, summer camps, and villas that it was called the "Riviera of Russia," the sea became too polluted to bathe in because of effluents from industrial plants and runoff from agriculture. In the Caspian Sea, the world's only source of fine caviar, the sturgeon population was decimated by toxins in the water flowing into it from the Volga River. In the Baltic, the Gulf of Riga and the Gulf of Finland also became polluted with industrial waste.[47]

The same was true of the air in Soviet cities. Older cities were ringed with factories. New ones were created near navigable rivers or to escape the Nazi invasion in 1941–1942. From the 1930s on, heavy industry and armaments manufacturing were the most important activities. In the drive to industrialize, environmental protection was not even taken into consideration. By the 1990s, only 15 percent of the urban population breathed air that was not harmful to health.[48]

All cities became polluted, but none more so than Norilsk, the northernmost city in the world. Located 500 kilometers above the Arctic Circle, it was founded in the 1920s to exploit the abundant nickel and copper deposits in the area. Its first inhabitants, confined to Gulag labor-camps, were political prisoners and *kulaks* (rich peasants) who were the objects of Stalin's ire during the collectivization of agriculture in the 1930s. The air around the metallurgical complex contained seventy-two times the maximum allowable concentration of sulfur dioxide. Workers had to wear masks over their faces. Men are said to have suffered the highest rate of lung cancer in the world, and children were the sickest in the Soviet Union.[49]

These and other incidents of ecological disaster all had a common theme: the militarization of Soviet society. During the Second World War, the nation's energies were fully committed to war, with half or more of its Gross Domestic Product devoted to military needs. But even in the 1980s, the military still absorbed 20 to 30 percent of the nation's production by its control of the nation's resources and its dominant position in the government. In their exclusive focus on production, the armed forces undermined not only the health of the Soviet economy but also the health of its people. Concerns about the impact of frenetic development on people's health were dismissed or suppressed. The health and survival of other living beings were hardly noticed.[50]

Mao and the Great Leap Forward

As in the Soviet Union, developmentalism in China arose from the crucible of wars: the communist rebellion in the mountains of Jiangxi (1928–1934), the Long March (1934–1935), the Sino-Japanese War (1937–1945), the civil war between Nationalists and Communists (1945–1949), and the Korean War (1950–1953). Long after the fighting ended, Mao Zedong continued to wage war against landowners and other class enemies and against the forces of nature that the Chinese people had struggled with for so many centuries. The methods used were very different from those of the Soviets because China was much poorer and more densely populated than the Soviet Union and because of the personality and ideas of Mao himself.

When Mao and the communists took over mainland China in 1949, their first policy was to emulate the Soviet Five-Year Plans. However, China had an overwhelmingly rural population with too small an urban or industrial base on which to build a Soviet-style proletarian revolution. Furthermore, Mao was too impatient for the deliberately planned and bureaucratically guided industrialization that the Soviets had pioneered. In 1958, therefore, he announced that China would surpass the United Kingdom in fifteen years by mobilizing the enthusiasm of the masses in a campaign called the Great Leap Forward.[51]

Among the most startling features of the Great Leap Forward were the production of iron and steel in backyard furnaces and the communization of agriculture, two subjects about which Mao relied on his intuition while rejecting technical expertise. To inspire the masses, Mao's followers used military terms such as mobilization, regimentation, attack, and redeployment. Headlines proclaimed "Chairman Mao's Thoughts Are Our Guide to Scoring Victories in the Struggle against Nature" and "The Desert Surrenders."[52] In 1960 Mao declared: "In another two years, by 1962, it is possible [for us to produce] eighty to a hundred million tons [of steel]. Approaching the level of the United States."[53] Producing steel in small village furnaces meant not only sacrificing all the iron implements in the village—pots and pans, bicycles, doorknobs, railings, hoes and other tools—but also cutting down every tree within reach to make charcoal. The Great Leap Forward consumed 10 percent of China's few remaining forests, mainly in northern Manchuria and in the extreme south.[54]

Mao's agricultural policies were equally destructive. To improve public health, peasants were mobilized to "eliminate the four pests," namely flies, mosquitoes, rats, and sparrows. A few years later, when the elimination of sparrows led to infestations of insects, Mao ordered: "Do not eliminate sparrows any more, and replace them with bedbugs, and now the slogan is 'eliminate rats, bedbugs, flies, and mosquitoes.'"[55] Based on the belief that plants of the same species would not compete for nutrients, farmers were ordered to plant seedlings up to ten times more densely than normal. Grasslands in China's northern and western provinces were plowed up in the belief that human efforts could compensate for the deficiencies of soil and climate. Trees were uprooted to make room for crops. Farmers were distracted from their work by endless political meetings.

The results were not slow in coming. Crops rotted in the fields. Of the meager harvests, much was confiscated by party officials while statistics were conjured out of thin air to satisfy the government's utopian demands. The famine that ensued between 1958 and 1961 was the most devastating in history, killing between 30 and 50 million Chinese.[56] Peasants, in their desperation, ate every bird, rat, and mouse they could catch, as well as seeds, roots, and the bark of trees.[57]

The Cultural Revolution

The results of the Great Leap Forward were so devastating that even Mao had to pause to let China recover. But in 1966 he announced a new plan to upend Chinese society: the Great Proletarian Cultural Revolution. This time he aimed his wrath at the educated. Bureaucrats, party members, and teachers were either executed or sent to perform manual labor in distant regions. The Cultural Revolution also affected the land. Fearing attack by the USSR after the Sino-Soviet split as well as by the United States or India, Mao ordered a "Third Front." This meant building industrial plants deep in the interior of China and ordering each province to become self-sufficient in food. The slogan "Take grain as the key link" encouraged commune leaders to cut down trees and plant grain in their place, even where grain did not grow well. Once again, it was the few remaining woods that paid the price. "Encircle lakes, create farmland" was another policy that encouraged farmers to chip away at the margins of lakes. Hebei province lost 740 of its 1,066 lakes, or 72 percent of its lake surface. Lake Poyang, China's largest, was reduced by one-fifth. Lake Dongting, the second largest, lost half its surface. These lakes, which had absorbed floods and released water gradually, could no longer function as overflow reservoirs, making floods even more devastating.[58]

In spite of the Great Famine, China's population grew from 583 million in 1953 to over 800 million in 1976. Without chemical fertilizers, the only way to feed that many people was to increase the amount of cultivated land. Yet between 1957 and 1977, China suffered a net loss of 29 million hectares of farmland, despite the reclamation of 17 million hectares of "waste land," either grasslands that were farmed for a few years, then ended up as deserts, or hillsides that quickly eroded. By the 1970s, China faced the possibility of another devastating famine.[59]

Amazonia

Besides the communist nations, others bought into the developmentalist agenda in order to achieve power, prestige, and prosperity at the expense of nature. In this, they were encouraged during the Cold War by the US government, which equated development with capitalism, and capitalism with freedom. The fate of Amazonia is a case in point.

Once the rubber boom of the late nineteenth and early twentieth centuries ended, Amazonia was forgotten. As the largest (almost) untouched tropical rainforest on Earth, it attracted few people other than adventurers, explorers, and scientists. In the 1960s, outsiders began to penetrate and transform the rainforest in a significant way. Some of the impetus came from the Andean republics. To prevent Brazil from taking over the entire rainforest, especially after oil and natural gas were discovered in the 1960s, Colombia, Peru, and Bolivia built a 3,500-kilometer-long highway between the lowlands and the eastern slopes of the Andes. Peasants from the highlands migrated into the area, followed by cattle ranchers. By 1987, they had cleared 6 million hectares of the Peruvian Amazon and 2 million hectares in Colombia.[60]

The Andean invasion was dwarfed by that of Brazil, which owned far more of Amazonia and had a much larger population than the Andean republics. Three groups participated: private entrepreneurs, the Brazilian government, and the poor. The generals who ruled Brazil from 1964 to 1988 tried to turn Brazil into a modern developed country, partly to reduce the crushing poverty of so many of its citizens and partly to turn Brazil into the dominant power of South America. Their developmentalist motivations were similar to those of their Soviet and Chinese Communist counterparts. The results were far different, however, for the generals did not control the Brazilian people but only unleashed them on the natural world of Amazonia.

Their means to open up the Amazon was by carving roads through the wilderness. The first one, the "Road of the Jaguar" from Brasilia to Belém, was begun in 1960 and paved in 1973. As peasants and ranchers fought over newly opened lands, the human population rose from 100,000 in 1960 to 2 million in 1970, while the number of cattle jumped from zero to 5 million. Counting this as a success, the generals undertook a bigger challenge, the 4,000-kilometer-long Transamazonian Highway paralleling the Amazon River through the rainforest. Their goal was to attract impoverished peasants from the drought-stricken northeast and dissuade them from moving to the overcrowded cities of the south by providing, in the words of General Emilio Medici, "a land without people for a people without land." In the end, only 8,000 families came instead of the millions the generals expected.

Those who ventured into the forest to farm were quickly disappointed. In tropical rain forests, almost all the nutrients in dead organic matter quickly decompose and are re-absorbed by living matter. Once exposed to heavy rainfall, the few minerals left in the soil are soon washed away. As a result, what little fertility the soil possesses disappears after the first harvest.[61] Newcomers, after felling and burning the trees on the plots they had acquired, found the soil soon barren of nutrients and the environment teeming with pests. When they failed at farming, they were evicted by wealthy speculators who turned the newly cleared land into cattle ranches.[62]

More successful, from the generals' point of view, was a new road linking south-central Brazil to the western states of Acre and Rondônia, regions that had previously been among the most remote in the world. By 1980, half a million migrants flooded the newly opened regions in a frenzy of lawlessness, violence, speculation, and corruption. On either side of the highway, feeder roads opened up the forest to loggers and farmers. By 1987, 51,000 square kilometers of Rondônia (22 percent of the state) had been deforested. Most of it was taken over by cattle ranches averaging 24,000 hectares in size, with the largest covering 560,000 hectares. But much newly deforested land soon degenerated into brush too tough even for cattle to graze.

Even after the attempts to colonize Amazonia with poor farmers had faded, the new roads opened the region to loggers using powerful machines to cut the trees and trucks to transport the logs. Mahogany and other tropical hardwoods are in great demand in the developed world. Because the valuable trees are scattered in the forest, logging them caused enormous collateral damage. To extract one tree per hectare of forest involved killing or damaging over half of the others.

More recently a new project has threatened Amazonia. In the 1960s geologists discovered the world's largest deposit of iron ore. The Grande Carajás Project included seven huge smelters to produce raw iron. With no coal deposits in South America, the smelters used charcoal, consuming 1,000 to 1,500 square kilometers of forest a year. These and other industries have led to the deforestation of large parts of the state of Pará.

Estimating the amount of Amazon deforestation has proved difficult. Forest historian Michael Williams estimated that 15,200 square kilometers of forest were lost each year from 1978 to 1988; by 2002–2004, the rate of deforestation had risen to 24,000 square kilometers (an area the size of Vermont) every year. Since 1960, 640,000 square kilometers of forest (an area the size of Texas) have vanished in Brazil alone. Out of a total Amazon forest of about 4.2 million square kilometers, it is a sizable proportion, not the apocalyptic destruction that some have predicted, but a very worrisome trend nonetheless. The Amazon forest is still the largest tropical forest in the world, with the greatest diversity of life. It also captures and recycles rain in enormous quantities. If it were turned into savanna, the change would have a profound, but incalculable, impact on the global climate.[63]

Figure 10.3. Part of the Amazon forest clear-cut to make room for cattle or agriculture. iStock.com/luoman.

Big Dams

Developmentalism—government efforts to boost economic growth—is not restricted to totalitarian regimes or to countries at war. Democracies like the United States and autocracies like Brazil under the generals in peacetime have felt the same urge, either to overcome economic downturns or to alleviate poverty and keep up with growing populations. These efforts were especially manifested in projects too vast for private enterprises to undertake, such as big dams and irrigation schemes.

Big dams have three main purposes: to irrigate dry land, to control floods, and to produce electricity. Irrigation was the original goal of water control. Even before the rise of civilization, farmers worked to bring water to their crops, especially in dry lands or in areas where the rains did not fall during the growing season. Sometimes they were able to get water from local wells, springs, or brooks, but more often water came from a distance and at the cost of massive amounts of communal labor. At the beginning of the twentieth century, 40 to 48 million hectares were irrigated worldwide, half in China using traditional methods.

In the twentieth century, irrigation projects exploded around the world, thanks to concrete and steel and powerful earth-moving machines, but also due to massive government efforts. In addition to agriculture, dams were also built to produce electricity. By 1985, the total amount of irrigated land had risen to 220 million hectares and by 2000 to 280 million hectares, half of it created after 1960. By the early twenty-first century, 35,000 major dams (and countless minor ones) interrupted the flow of 40 percent of all river water on the planet.

Big dams and irrigation projects also create massive environmental problems. Salinization affects about half the irrigated land and causes the loss of a million hectares of irrigated land a year. On some rivers, dams have a short life span for they soon fill up with silt. Electricity used to pump water up from aquifers faces diminishing returns as the aquifers become depleted.[64]

Dams in India

The British rulers of India in the nineteenth century were proud of having turned large parts of the Punjab from dry scrubland into productive farmland. Yet droughts in 1896–1898 and in 1899–1900 showed that their efforts were insufficient to avert widespread famines. In 1901, Viceroy Lord Curzon appointed a commission "to Report on the Irrigation of India as a protection against Famine." Its report, published in 1903, argued that further progress in irrigation would require storage dams and barrages on major rivers, at a far greater cost than any previously undertaken.

Beginning in 1905, the government undertook the world's most massive irrigation project up to that point: the Triple Canals Project. Connecting four of the five great rivers of the Punjab and irrigating the land between them required eight major canals and 12,000 kilometers of distributaries. The Triple Canals Project involved, for the first time, an entire basin with its many rivers. By the time they were completed in 1919–1920, the canals commanded 1.6 million hectares of land, of which 692,000 were irrigated each year.

After the end of World War I, the government of India once again took up irrigation projects with enthusiasm. It encouraged farmers to pump water from underground aquifers with electric or diesel-powered pumps. It also began building dams of reinforced concrete. One, the Sukkur Barrage over the Indus River built between 1923 and 1932, was capable of distributing up to 1,288 cubic meters of water per second over 2.2 million hectares of land. Land that had formerly been scrub had become productive farmland yielding two crops a year, usually rice or cotton in the summer and wheat or pulses in the winter. These irrigation works made India self-reliant in grain and fed a population that had tripled from about 100 million in 1700 to about 300 million in 1920.[65]

Since independence in 1947, the Republic of India has pursued irrigation projects with as much enthusiasm as the British Raj. The most pressing incentive has been the growth of population from 300 million in 1920 to 1.2 billion in 2011, not counting Pakistan (187 million) and Bangla Desh (142 million). Most striking has been the Narmada River Development Project in western India. This river, one of the largest in India, has an average annual flow of 41 billion cubic meters of water, 90 percent of which occurs during the monsoon months of June to September. The project, which is still under way, would involve 30 major dams and 135 medium dams and irrigate 4 to 5 million hectares of drought-prone land.[66]

The great irrigation projects have averted famines but have also created serious environmental problems. Large tracts of land became waterlogged or poisoned by salts rising to the surface. Canals, ditches, and waterlogged soil provided breeding sites for anopheles mosquitoes, leading to an increase

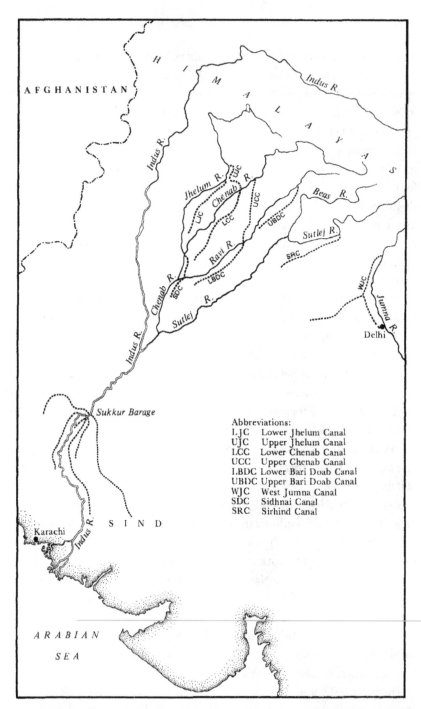

Figure 10.4. Map of the Indus watershed in northwestern India in the early twentieth century showing the major rivers and irrigation canals. From Daniel R. Headrick, *The Tentacles of Progress* (New York: Oxford University Press, 1988), p. 191.

in malaria and a deterioration in public health.[67] In Bengal and Bihar, where most natural drainage channels run north-south, the construction of roads and railroad tracks interrupted that drainage, causing waterlogging and aggravating the spread of malaria.[68]

Dams in Egypt

The building of irrigation projects in India made the country a laboratory of hydraulic engineering and a school from which this knowledge spread around the world. The first place British engineers applied their new knowledge was Egypt. There, they calculated, perennial irrigation would allow not one but two crops a year, or even five crops every two years. To achieve that goal, William Willcocks, a civil engineer who had trained in India, proposed in 1894 to build a great dam at Aswan to regulate the flow of the Nile. The dam he projected was to be 25 meters high and store 3.6 billion cubic meters of water. However, as it would drown parts of the ancient Temple of Philae, Lord Cromer, the British governor of Egypt, reduced the plan to a 20-meter-high dam that would store only 1 billion cubic meters. This first Aswan Dam, built between 1898 and 1901, created a 160-kilometer-long reservoir. As this smaller dam proved incapable of providing perennial irrigation to Upper Egypt south of the delta, it was raised in 1908–1912 to hold 2.5 billion cubic meters of water, drowning Philae. Meanwhile, other barrages were being built at Asyut, halfway between Aswan and Cairo in 1910, and at Esna near Aswan in 1912. Though they could not supply enough water for perennial irrigation when the Nile was very low, they did even out its flow over the course of the year.

These dams transformed Egyptian agriculture. More land could be irrigated than ever before, allowing it to support several crops a year. The area producing crops quadrupled, from 822,000 hectares in 1821 to 3.5 million hectares in 1917. Per capita, however, the result was not nearly as impressive, for the population of Egypt had grown fivefold, from 2.5 million people in 1821 to 12.7 million in 1917, leaving the majority of the Egyptian people worse off than before.

The hydraulic projects built during the British occupation of Egypt were but a prelude to the much more massive Aswan High Dam. Planning began almost immediately after the revolution of 1952 that brought Gamal Abdel Nasser to power. Nasser was determined to propel Egypt into the modern

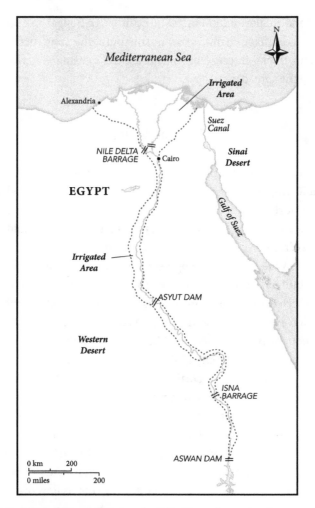

Figure 10.5. Map of Egypt showing the Nile River, the major dams, and the irrigated areas in the early twentieth century.

world by building a great dam that would provide abundant electricity and water for year-round irrigation, making it "a source of everlasting prosperity."[69] After much Cold War politicking, the dam was built with Soviet aid between 1960 and 1970. With a length of 3.8 kilometers and a height of 111 meters, it was one of the world's largest. It also created a reservoir, called Lake Nasser in Egypt and Lake Nubia in Sudan, that was 550 kilometers long and covered 5,250 square kilometers, with a capacity of 162 cubic kilometers.

The Aswan High Dam dramatically changed the Egyptian economy. It allowed the irrigation of 3.36 million hectares of farmland, increasing Egypt's irrigated land by a third. It prevented the damage that would have been caused by floods and droughts, such as the drought that devastated the Sahel in the 1980s. Yet improvements in agriculture have barely kept up with the growth of population, which reached 79.6 million in 2011, an eightfold increase in one century. Egypt, once the breadbasket of the Roman Empire, now imports food.

The Aswan High Dam has also had a mixed effect on the environment. The Nile used to carry sediment that enriched the farmland of Egypt, but now 134 million tons of sediment accumulate in Lake Nasser/Nubia every year, especially in its upper reaches. Though the lake will not be completely filled with sediment for several centuries, the volume of water that it holds will gradually diminish. Meanwhile, deprived of the Nile's nourishing silt, farmers must purchase chemical fertilizers. The fields watered year-round by the Nile suffer from salinization and waterlogging. A drainage system covering 2 million hectares, built at great expense, has reduced, but not eliminated, these problems. Lack of sediment also affects the Nile Delta, which is losing 1.8 square kilometers a year. As nutrients no longer flow into the Mediterranean, fisheries have declined. Perennial irrigation has also increased the incidence of schistosomiasis (or bilharzia), a parasitic disease transmitted by freshwater snails that live in canals.[70]

Egypt once had a sustainable economy. For 5,000 years, its farmers depended on the annual flood of the Nile to bring water and fertile silt to their fields. To be sure, basin irrigation could support at most 5 million people, and often less. Since the mid-nineteenth century, Egyptians have turned to progressively more massive, complex, and costly technologies to keep up with their growing population. Their dams and canals may seem permanent, but the silting of Lake Nasser, the erosion of the Delta, and the reliance on chemicals are, in the long run, ephemeral and unsustainable.

Dams in the American West

The American West is rich in land but poor in water. West of the 103rd meridian, the land receives a scant 300 millimeters of rain a year on average. Half of that evaporates, and only one-third is taken up by vegetation. Only in western Washington and Oregon and in northern California can farmers rely on rain for their crops. In the rest of California, fertile land is found in one part of the state and water in another. Despite the efforts of farmers and well-financed corporations and their lawyers, neither free-market capitalism nor grassroots democracy could handle the complex problems of water control over so vast an area. By the beginning of the twentieth century, 90 percent of the irrigation companies were in or near bankruptcy, and the states, landowners, and corporations had no choice but to appeal to the federal government, the only entity capable of commanding the waters of the entire West.

By the National Reclamation Act of 1902, the federal government took over the task of providing, at taxpayers' expense, water to private interests in the West. Until 1928, the Bureau of Reclamation carried out surveys and made plans.[71] Then, from 1928 to 1956, it carried out what environmental historian Donald Worster described as "the most elaborate hydraulic system in world history, overshadowing even the grandiose works of the Sassanians and the Pharaohs."[72]

The largest river in the Southwest that could be tapped for irrigation was the Colorado. To capture it and channel its waters to satisfy human needs, the Bureau of Reclamation built a series of nineteen large dams on the Colorado. Hoover Dam, built in the middle of a desert between 1931 and 1935, was, at the time, the largest in the world. The Colorado River Storage Project, signed into law in 1956, had an estimated cost of $1.6 billion; the cost of irrigating land in the upper basin would cost taxpayers up to $2 million per farm, perhaps five times as much as the farms were worth.[73]

A second major western hydraulic project of the twentieth century was the Central Valley Project in California. At the time it was begun in the 1930s, farmers in the San Joaquin Valley were irrigating their vegetable, fruit, and nut crops with water pumped up from the underlying aquifer, causing the water table to drop and salts to accumulate in the soil. The Central Valley Project's goal was to divert the Sacramento River from northern California to irrigate the water-deficient southern two-thirds of the Central Valley. It eventually included twenty-one dams yielding 8.6 million cubic meters per year to irrigate 1.5 to 3 million hectares of land.[74]

The third major hydraulic project in the West embraced the Columbia River and its tributaries, especially the Snake River. This watershed had the largest hydroelectric potential of any in the United States, with a flow twice that of the Missouri River and ten times that of the Colorado, but in a region with few inhabitants and very little demand. This did not deter the engineers who, by 1920, had proposed almost a hundred dams in the area. During the New Deal, President Franklin D. Roosevelt won congressional approval to establish the Bonneville Power Authority. In 1935, work began on the Grand Coulee Dam. After it was completed in 1942, it backed up a reservoir 82 kilometers long and 332 square kilometers in area. During World War II, 92 percent of the electricity produced by the Bonneville and Grand Coulee dams went into war production, mainly to make aluminum for warplanes. By 1957, hydroelectric plants in the Columbia River basin produced 82 percent of the total electricity in the Pacific Northwest.[75]

Following these major projects, the Bureau continued to build dams throughout the 1950s and 1960s. By 1976 it operated 320 reservoir dams, three diversion dams, 23,175 kilometers of major canals, and 55,715 kilometers of lateral canals, as well as power plants, pumping plants, pipelines, tunnels, and much else.[76]

As in all irrigation projects, there were undesired consequences. One of them was silting. Since so many dams were built within three decades of each other, they are all filling up with silt at the same time. The Colorado, one of the most silt-laden rivers in the world, carries almost as much silt as the much larger Mississippi, which it dumps below the Grand Canyon into Lake Mead, the reservoir formed by Hoover Dam. When it was built, Lake Mead's maximum capacity was almost 40 cubic kilometers, but by 2013 this had declined to 35.4 cubic kilometers, while its actual volume had shrunk to 16.6 cubic kilometers, due to a prolonged drought and the insatiable demand for water in the Southwest.[77]

Another problem is pollution. In the Central Valley in particular, the runoff of industrial and agricultural chemicals made the water dangerous to drink. Irrigation water was also slightly saline; as it flowed through fields, it evaporated, leaving salts behind. Drainage and desalinization projects only dumped the unwanted chemicals farther downstream. By the time the Colorado River reached Mexico, with which the United States had treaty obligations to provide a certain amount of freshwater, it required very costly purification. Even in wet years, barely a trickle of polluted water reached the Gulf of California.[78]

Yet another consequence of the West's development was the depletion of wildlife. Some was part of the general decimation of wildlife, especially the wolves, coyotes, and bobcats that threatened cattle and sheep. But hydraulic works also led to the end of the salmon run in the Columbia River and to a 97 percent decline in salmon in the rivers of the three West Coast states. Likewise, the irrigation of Central Valley land resulted in the decline of wetlands that had once supported large numbers of migratory birds.[79]

Dams in China

Despite its long history of irrigation and flood control, China in 1949 had fewer than ten large dams and a dozen medium-sized ones, some ancient. Mao was determined to build dams and irrigate as much of the nation as possible. By 1980 the Chinese had built 1,800 large and medium-sized dams, as well as tens of thousands of smaller ones. Many of these were hastily constructed by farmers untrained in hydraulic engineering. Others were large projects inspired by Soviet engineering.[80]

One such was the Sanmenxia Dam on the Yellow River. Between 1913 and 1938 the Yellow River had broken through its dikes seventeen times, so Mao decided to tame it by building a gigantic dam. Despite the warnings of hydraulic engineers that it would silt up quickly, the dam was built between 1957 and 1962. Because the land it flows through before it reaches the North China Plain is so badly eroded, the Yellow River carries three times more sediment than the Yangzi, with just a small fraction of its flow; in other words, it consists of liquid mud. The result was what the experts predicted. Within four years the dam was almost half blocked, its generators no longer functioned, and holes had to be pierced in it to allow the silt to escape. In short, it had become useless.[81]

The Huai, a river that flows between the Yellow River and the Yangzi, has a violent reputation. After their victory in 1949, the communists set out to "Harness the Huai." Among the many dams they built on that river and its tributaries were two massive ones, the Banqiao and the Shimantan. In August 1975, a typhoon swept through Henan province, dropping 1,000 millimeters of rain in three days, well beyond the worst-case scenarios imagined by the designing engineers. When the Banqiao Dam collapsed, a wall of water 6 meters high and 12 kilometers wide rushed down the valley. The collapse of the Shimantan Dam was almost as destructive. The rushing waters flooded 255 million hectares of land, killing 85,600 people officially (estimated as high as 230,000), and harming 11 million others. Nor were these the only dam failures. In Henan province alone, of the 110 dams built under Mao, half had collapsed by 1966. In all of China, 3,200 dams had collapsed by 1981, and many others needed repairs.[82]

Many parts of China, when they are not flooded, suffer from insufficient water for irrigation. Tapped by farmers, cities, and industries, the Yellow River often runs dry 300 kilometers before it reaches the sea. In recent years, farmers have turned to another source of water: the aquifer that lies under the surface. As electricity became available, they installed pumps to irrigate their fields. By 1990, half of China's 220 million hectares of irrigated land was watered by pumps. To get the water they need, cities and industries have also been mining the aquifer below the North China Plain. Beijing has been drawing water from up to 1 kilometer below the surface. As a result, the water table has been dropping by up to 1.3 meters a year; in some places it has dropped by some 50 meters since 1960. As the water table beneath them declines, Tianjin, Shanghai, and other cities have been sinking.[83]

The Three Gorges Dam

After Mao, China's experiments in extreme communism were replaced by capitalism, or "free-market socialism," as the Chinese government prefers to call it. But dam building is so costly and its returns so unpredictable that only governments can afford to build them. And big dams, in turn, boost the prestige of governments and political leaders. In China, where the excesses of Mao's regime left the population disillusioned with politics, the Communist Party needed a grandiose project to burnish its image, especially after the massacre in Tiananmen Square in 1989. In the words of party leader and president of China Jiang Zemin, "Man Must Conquer Nature."[84] That project was the Three Gorges Dam.

Building a dam across the Yangzi River as it makes its way through the Three Gorges was first suggested by Sun Yat-sen in 1912, then taken up by Chiang Kai-shek between the wars, then revived by Mao Zedong in the 1950s; in short, it has been a goal of every twentieth-century Chinese leader. As planned in the 1990s by Premier Li Peng and President Jiang Zemin, it was to be one of the largest dams in the world: 2 kilometers long and 185 meters high; it would be able to store 39.3 cubic kilometers of water. Started in 1992, it was completed in 2006 and filled to capacity in 2009. It formed a 600-kilometer-long reservoir that drowned 30,000 hectares of arable land, nineteen cities, and 326 towns, and displaced 1.3 million people.[85]

The primary goal of the Three Gorges Dam was to smooth out the uneven flow of the Yangzi and prevent floods. Indeed, when floods returned in 2009 and 2010, their water was captured by the dam and discharged during the following dry season. It was also designed to produce as much electricity as burning 31 million tons of coal a year, thereby reducing the production of greenhouse gases and other air pollutants. And it would enable 10,000-ton ships to reach Chongqing, hundreds of miles from the sea.

As in all mega-projects, there have been unanticipated drawbacks. One was silting, which was much more serious than expected because of deforestation around the dam itself and in the upper watershed of the Yangzi in Tibet. Pollution from towns and industries and agricultural runoff from farms that used to be washed out to sea now accumulate behind the dam, turning its water stagnant and murky. The dam also threatened several endangered species by destroying their habitats: the Chinese river dolphin, the Yangzi sturgeon, and the Siberian crane.[86]

Northern China has 42 percent of the population and 45 percent of the agricultural land but receives only 11.3 percent of the nation's freshwater. To remedy this imbalance, the government is constructing the "South-North Water Transfer Project," which is intended to divert some of central China's abundant water to the parched and densely populated north. It involves three parallel canals, one from Tibet to the upper reaches of the Yellow River; another from the Han River (a tributary of the Yangzi) through a tunnel under the Yellow River and on to Beijing; and a third one following the route of the old Grand Canal. The project is expected to cost twice as much as the Three Gorges Dam, displace some 350,000 villagers, and cause many unanticipated environmental problems.[87]

The problems of the Yangzi, Huai, and Yellow rivers are symptomatic of the way China's breakneck industrialization is taxing its water resources to the utmost. China receives 2,156 cubic meters of water from rain or snow per person per year, but it is so unevenly distributed that 260 million people do not have enough for their daily needs, and 700 million have access to water that is contaminated with human and animal wastes. As factories dump toxic effluents into rivers and streams, almost all the country's urban water supplies are too polluted to drink and, in the case of the Huai, too polluted for use even to irrigate crops.[88] China's hasty development has come at the expense of its people's health and to the detriment of nature.

Conclusion

Developmentalism—the urge to increase a nation's economy regardless of the cost—began as a consequence of total war. Even in peacetime, governments that had acquired totalitarian power over their citizens continued to press for maximum economic development in anticipation of future wars. The Soviet Union and China under Mao represent the two most extreme examples of maximizing production to the detriment of the environment and their citizens' health. As a regime, the Soviet Union's future ended in 1989, but its successor states still bear the scars of its single-minded pursuit of military power and unfettered economic growth. After Mao, the Chinese Communist leadership has avoided that fate by transforming its goal from permanent revolution to economic growth, but the means are similar: the maximum exploitation of natural resources, with minimal concern for the environment, human health, or long-term sustainability.

Totalitarian regimes are not the only ones to develop their economies with massive government-sponsored projects. So have democratic ones like the United States, authoritarian ones like Brazil under the generals, and colonial and post-colonial regimes like those of India and Egypt. For them, war and military might were not the main motivating factors. Instead, they pursued economic growth to raise the standard of living of their peoples or to satisfy nationalistic hubris. Regardless of their goals, they too jeopardized their natural environments.

11

Peace and Consumerism in the Twentieth Century

The twentieth century was indelibly marked by the explosion of mass consumption. In earlier civilizations, elites engaged in conspicuous, even extravagant, consumption. What was new in the last century was its spread to vast numbers of people. This phenomenon first emerged in the United States in the 1920s, then in western Europe in the 1950s, later in Japan and eastern Europe, then in Latin America, now in China and India.[1]

Mass consumption requires mass production, distribution, and marketing. These, in turn, require correspondingly massive amounts of natural resources and produce equally massive amounts of waste and effluents. The growth of the world economy in peacetime—in the 1920s and even more after World War II—did more to transform the global environment than all the wars in history combined.

This chapter analyzes four examples of mass consumption. In the first three of these, the United States led the way. Automobiles transformed the land, the economy, and the way of life of millions. Petroleum powered the American—and increasingly the global—way of life. Mechanized, chemical-based agriculture has kept up with a growing population and its increasingly carnivorous tastes. The fourth case study comes from China, which had once repressed mass consumption but since 1976 has become the world's fastest-growing industrial economy based on the mass production of consumer goods for both export and domestic markets.

The Automobile

Of all the new technologies of the twentieth century, none has impacted the environment as much as the automobile. In 1990, when Germany was reunified, among the first things that East Germans did with their newfound freedom was to buy an automobile—whatever they could afford, as long as it was made in the West. Twenty-some years later, the Chinese are buying cars by the tens of millions every year.

The reason cars are popular is not hard to find. Automobiles are the great liberators. They allow people to travel comfortably in any weather, at any time of the day or night, oblivious of public transit schedules. They give rural people and suburbanites access to cities, city dwellers access to small towns, and everyone access to the countryside. They are also a status symbol, a sign of their owners' wealth and/or good taste.

Even the most vociferous critics of the automobile and the most ardent proponents of public transit cannot envision banning cars entirely. The most that reformers have proposed has been to make cars safer, less polluting, and more fuel efficient and to relieve inner cities of congestion. Cars are increasingly taking over the planet, with severe consequences.

The American Automobile to 1940

The first automobiles were European handcrafted luxuries for the very rich. Henry Ford envisioned a different kind of customer: "I will build a car for the great multitudes. . . . But it will be so low in price than no man making a good salary will be unable to own one—and enjoy with his family the blessing of hours of pleasure in God's great open spaces."[2] It was Ford who opened up the era of mass automobility.

Between 1916 and 1939, the total number of cars in the United States rose from 1.5 million to 31 million, excluding trucks and buses.[3] Of the world's 32 million cars in the late 1920s, 80 percent were in the United States.[4] Ford's cars were sturdy and inexpensive; in 1930 one cost $300 at a time when workers in his factories made $5 a day. Soon, multitudes were buying cars, whether they could afford them or not; by 1926, three-quarters of all cars were bought on credit.[5] Americans were mortgaging their homes to buy cars. As one working-class woman told sociologists Robert and Helen Lynd: "We'd rather do without clothes than give up the car," and another one said, "I'll go without food before I see us give up the car."[6]

The Depression halted the purchase of new cars. In Muncie, Indiana, the town that the Lynds studied, the number of new car sales plummeted from 2,410 in 1929 to 556 in 1932. Yet car registrations and gasoline sales declined only slightly, then rose again. When the Depression hit, drivers continued to take to the roads; despite the expense of gas, the distance traveled by car actually rose, from 319 billion kilometers nationwide in 1929 to 348 billion in 1931.[7]

The Postwar Car Boom

During World War II, Americans regained their purchasing power but found their desires blocked by the exigencies of a war economy. Gasoline and tires were severely rationed. The number of private cars and the distance they traveled dropped while the number of trucks and buses increased sharply.

Starting in 1945, the pent-up demand—along with corporate marketing and government policies—produced the longest economic boom in history. Much of it revolved around cars. From 1945 to the mid-1970s, the United States produced two-thirds of the world's motor vehicles. As the number of new vehicles grew much faster than old ones were junked, motor vehicle registrations climbed in quantity from 27.5 million in 1940 to 49.2 million in 1950 to 108.4 million in 1970. The number of vehicles on the road increased much faster than the US population. In 1950, there was a motor vehicle for every 3.1 Americans. In 2010, there were 242 million motor vehicles in the United States (one for every 1.3 Americans), 32 million more than the number of licensed drivers, for many drivers owned several vehicles.[8]

The numbers of vehicles were not the only relevant factors; so were the kinds of vehicles being driven and their gasoline consumption. In the 1950s and '60s, car designs became ever more extravagant, with preposterous tail-fins, wraparound windshields, chrome decorations, and myriad color schemes. They were not only gaudy but huge. To propel these behemoths along the highway and to power their brakes, steering, automatic transmission, air conditioning, and other equipment required V-8 engines producing 200 or more horsepower. As a result, mileage plummeted; in 1949, a large Cadillac used 11.8 liters per 100 kilometers (20 miles per gallon); by 1973 the average American car consumed 17.4 liters per 100 kilometers (13.5 miles per gallon).[9]

Planned obsolescence, clever marketing, and factory-built defects persuaded American consumers to buy new cars every few years. The oil crises of the 1970s led many consumers to switch to more fuel-efficient foreign cars. But once the price of gasoline dropped again in the 1990s, sport utility vehicles and pickup trucks came into fashion. From 1982 to 2002, the number of new cars produced remained steady at around 8 million a year, but the number of new pickup trucks and SUVs soared from 2.6 million to 9 million, causing average fuel consumption per vehicle-mile, which had declined during the 1970s, to rise again.[10] The ever-growing number of motor vehicles in America meant that total annual highway fuel use continued to climb, from 418.3 billion liters in 1973 to 666.6 billion liters in 2007.[11]

Automobiles in Europe and Japan

Automobiles also transformed western Europe. During World War II, most civilian vehicles were destroyed or worn out, and scarce fuel was reserved for military and official vehicles. For a decade after the war, only the wealthy and well-connected could afford cars while most of the population traveled by bicycle, streetcar, train, or bus.

Then came the boom. By the mid-1950s, the western European economies had recovered. As wages rose faster than the costs of food and housing, a majority of the population had disposable income to spend on cars. Equally important, the cost of gasoline and diesel fuel (much used in Europe) declined by 20 percent (in constant dollars) between 1957 and 1973. The result was what environmental historian Christian Pfister called "the democratization of consumption." Even working-class families bought cars.[12]

Many observers noted—some disapprovingly—that western Europe was becoming Americanized. Yet it did so with restraint. Because motor fuel cost twice as much as in the United States, European cars were, on average, much smaller and more economical than the American cars of the time. Yet they multiplied. In West Germany, where there had been one car for every ninety-seven people in 1950, by 1970 there was one to every four inhabitants, as in France and the United Kingdom.[13] In 2008, the inhabitants of the European Union owned 256 million motor vehicles, more than in the United States, of which 223 million were private automobiles.[14]

Japan followed the western European example with a few years' delay. Japan, which had produced no passenger cars since the 1930s, started its auto industry in 1960 with 36,000 *kei* cars, tiny vehicles with motorcycle engines that could carry only two persons; meanwhile, 1.7 million motorcycles and scooters were built that year. By 1975, the Japanese auto industry was producing compact and economical cars that found a ready market overseas. That year it exported 1,827,000 cars, most of them to the United States where consumers, faced with a sudden jump in the price of gasoline, sought more fuel-efficient vehicles. In the 1990s, Japan dominated the world market, producing up to 13 million cars a year.[15]

In 2011 the world produced about 80 million motor vehicles, about 65 million of which were cars, for an accumulated total of 1 billion cars and about 200 million trucks and buses on the world's roads.[16] The automobile, its way of life, and the transformation of the land that it wrought reflected the will of the majority of people in the United States, Europe, and Japan and the wishes of most people in the rest of the world.

American Roads and Highways

During the first age of automobiles, roads lagged far behind the vehicles that used them. Henry Ford recognized this when he built his Model T to drive on rutted dirt paths as well as on paved roads. The state of Oregon found a means of financing road construction when it imposed a tax on gasoline, making users pay for the roads they used. From 1921 to 1930 the total length of all roads and streets increased by only 3 percent, but the length of paved and surfaced roads doubled. The federal government, besides imposing a numbering system for "US" highways, left road construction to the states and cities.

During the Great Depression, the federal government found that road building was a useful way to put the unemployed to work. From 1933 to 1942 it spent $4 billion this way, mostly on country roads and on scenic drives like the Blue Ridge Parkway. The opening of national parks to automobiles and expanded roadways enshrined the American ideal of enjoying the wilderness from the comfort of one's car.

Meanwhile, pressure was building to create limited access highways reserved for motor vehicles (even, in some cases, just for cars). Parkways near New York City were built to serve commuters from the wealthier suburbs. The Pennsylvania Turnpike, begun in the late 1930s, was designed to expedite traffic through the Appalachian Mountains.[17] After World War II, public demand for more and better roads, the powerful auto manufacturers and oil industry lobbies, and the growing funds for state and federal highways made it both imperative and possible to grow a road network proportional to the growing number of vehicles.

In the 1950s, many states in the Northeast and Midwest built toll roads for long-distance traffic. California, the nation's most car-loving state, chose to construct freeways instead. In and around New York City, the Triborough Bridge and Tunnel Authority built a comprehensive network of expressways for cars and trucks and parkways for cars alone. Finally, President Dwight Eisenhower and the United States Congress passed the Federal Aid Highway Act of 1956, funding a 66,000-kilometer network of limited-access expressways linking cities across the entire nation.[18]

These new roads, four and sometimes six or more lanes wide, cut a swath across the landscape. With lanes 3.6 meters wide on average, plus shoulders and medians, a six-lane highway could be 30 meters wide, not counting entrance and exit ramps.[19] They sliced through cities to provide—it was thought—high-speed access between inner cities and surrounding communities. Entire communities were razed, while others were severed from one another by the new arteries. In most cities, urban mass transit, already weakened by competition from automobiles, went into sharp decline as increasing numbers of commuters preferred to drive their cars, even in rush-hour traffic, rather than commute by bus or train. Los Angeles, the quintessential automobile city, devoted two-thirds of its land area to expressways, streets, parking lots, and other automobile-oriented uses, leaving one-third to houses, people, and nature. Once the interstate highways were built, according to automobile critic Jane Holtz Kay, the United States devoted 155,400 square kilometers (about the area of Georgia) to roads and parking lots.[20]

European Highways

Europeans built limited-access highways for automobiles before the United States did. The first one, a 36.5-kilometer-long *autostrada* or toll road built in 1924, allowed wealthy Milanese car owners rapid access to their homes or cottages in the lake district of northern Italy. By 1935, Italy had 378 kilometers of private toll highways. When Adolf Hitler came to power in Germany in 1933, his government undertook a massive highway expansion program, both for propaganda purposes and to facilitate military movements in preparation for war. After the war, Germany and Italy lengthened and improved their existing roads. The German Autobahn network, which was 2,128 kilometers long in 1950, had been lengthened to 8,000 kilometers by 1984 and to 12,845 kilometers in 2012.[21] Other western European countries also began building expressways; they reached 40,000 kilometers by the end of the 1980s and have continued to grow since then.

Suburbanization, Auto-Ecology, and the Land

In Europe and Japan, land near cities was much more expensive than in the United States. Furthermore, the governments of these countries deliberately encouraged the construction of high-density housing and discouraged suburban development with heavy taxes and restrictive zoning rules. In the United States, where land was much cheaper, cars and roads enabled suburban living. In newer cities like Los Angeles, single-family homes had already made up a large part of the housing in the 1920s. Elsewhere, suburbanization accelerated in the 1950s, when developers covered vast areas of exurban farmlands with thousands of similar (sometimes identical) single-family homes. These builders responded not only to a cultural predilection for more spacious housing but also to deliberate federally subsidized mortgages and tax deductions that privileged suburban homeownership over urban rental housing. Exacerbating the move to single-family homes was the exodus of whites from city neighborhoods into which African Americans migrated from the rural South. As a result, whereas 36 million Americans had been suburban homeowners in 1950, that number soared to 74 million in 1970.[22]

The low population density of suburbs made it financially impossible for public transit systems to survive on their own, and governments were loath to subsidize them. Hence, suburban living required an automobile, often one for each adult family member. The combination of automobiles, highways, and suburbs transformed America profoundly.

The new environment, which we might call auto-ecology, involved much more than highways and houses. A new kind of vegetation—lawns—begin to replace farmland in the vicinity of cities and even far out in the country. By the end of the twentieth century, lawns covered an area roughly the size of the state of Iowa. These were not grasses native to North America but Eurasian grasses such as Kentucky Bluegrass (*Poa pratensis*) that grew well only if regularly watered and mowed. Lawns needed regular applications of powerful chemicals to feed the grass and kill weeds, grubs, and other pests. Such chemicals, enthusiastically applied by homeowners seeking the perfect lawn, did as much to pollute rivers and lakes as the runoff from farmers' fields.[23]

To support suburban lifestyles, all sorts of activities were built around the automobile: shopping centers, strip malls, supermarkets, and big-box stores, all of them surrounded by huge parking lots.[24] Businesses that once catered only to pedestrians turned automobile-friendly; drive-in restaurants, movie

theaters, and banks appeared on the sides of roads. At the exits of the interstate highways, hotel chains built motels for long-distance travelers.[25] Car
dealerships, filling stations, and auto repair shops opened in suburbs and
along highways. Even churches catered to the auto-bound; the first "shopping center for Jesus Christ" opened in 1954 when the Reverend Robert
Schuller began preaching in a drive-in movie theater.[26]

Intercity travel grew from 771.5 billion passenger-kilometers in 1949 to
1.8 trillion in 1969, 85 to 90 percent of which was by private car.[27] In 2006,
Americans drove almost 5 trillion kilometers.[28] Vacation travel soared, including trips to distant destinations. Meanwhile, public transportation
declined to 2.5 percent of all trips, most of them by children in school buses.[29]

Automobile historian James Flink argues that the proliferation of cars
and highways in America was driven neither by consumer choice in a free
market nor by the natural superiority of automobiles, but by the successful
lobbying of the automobile industry and automobile clubs to have gasoline
taxes earmarked exclusively for roads and highway construction.[30] However,
these factors are not mutually exclusive. Unquestionably, automobiles were
immensely popular after World War II, and suburbanization reflected cultural as well as economic choices. Long before automobiles, Americans were
a mobile people; in the late nineteenth century, the American railroad network was almost twice as long as those of the next seven countries put together.[31] What happened after World War II was the transfer of the craving
for movement from public to private transportation and the shifting taste in
housing from urban apartments to suburban houses. The consequence was
the physical manifestation of a civilization as distinctive as that of Egypt or
Rome. Once built, this civilization was irreversible. The automobile was built
into the landscape as well as into the culture.

Ford and the Amazon

The impact of automobiles was felt far beyond the nations that produced and
drove them, as rubber plantations were carved out of the old-growth forests
of Sumatra, Indochina, Malaya, and Ceylon to serve the automobile industries of America and Europe.[32] The automobile even touched Amazonia,
where there were no roads. Henry Ford, seeking an alternative to the Anglo-
Dutch monopoly on rubber supplies, decided to create a hevea plantation
in Amazonia, where such trees grew wild. In 1927–1928, he purchased a

million hectares of land along the Tapajós River. His agents planted 1.5 million trees and built a city called Fordlândia, with all the American amenities. At first, none of Ford's employees had any training in tropical agriculture or rubber planting. The climate was too dry, the land too hilly, the soil too sandy, the river too shallow for large ships during much of the year, and the workers resisted American labor practices. Worst of all, as soon as the trees had grown enough so that their canopies touched, they were devastated by the leaf blight *Microcyclus ulei*. This disease is the reason that in the wild, hevea trees are always found at a considerable distance from one another, and that the Dutch and British botanists had made such efforts to escape the blight by transplanting heveas to Southeast Asia. In 1934, at another, even larger site called Belterra, Ford's agents planted 3.6 million trees from seeds brought in from Sumatra. Once again, *Microcyclus ulei* destroyed them. In 1945, having spent $10 million on his plantations, Ford sold them to the Brazilian government for $500,000.[33]

Petroleum, Air, and Water

The history of automobiles and petroleum are so intimately intertwined that it is difficult to imagine one without the other. Per kilogram, oil contains twice as much energy as coal and is much easier to transport, store, and process. Once refined into gasoline or diesel fuel, it is the ideal fuel for motor vehicles. In the United States, oil consumption surpassed that of coal in 1948; in Europe and Japan, the same happened by 1970.[34] By 2000, liquid fuels (mainly oil) generated approximately 39 percent of the world's energy, compared with 22 percent for coal and 22 percent for natural gas; the three fossil fuels together produced over 80 percent of the world's energy.[35]

Petroleum to World War II

Most American sources attribute the drilling of the first oil well to Edwin Drake of Titusville, Pennsylvania, in 1859, but evidently James Miller Williams had a commercial oil well in Ontario, Canada, the year before.[36] For the next forty years, oil, distilled into kerosene, was used mainly for lighting in place of whale oil. Increasing demand led to a series of discoveries around the world; by 1900, there were oil wells in Azerbaijan, Romania, California, Texas, Sumatra, Oklahoma, and elsewhere.

Then came the internal combustion engine and its application to cars, aircraft, ships, and locomotives. Oil also served to fuel electric power plants, to heat buildings, and to lubricate machinery; it was the raw material for plastics, synthetics, and other petrochemicals. Soaring demand spurred the drilling of wells in Mexico, Iran, Trinidad, and Venezuela. By World War I, petroleum was attracting the attention of the great powers and led to Great Britain's seizure of Mesopotamia from the Ottoman Empire. Ever since, oil has become a strategic necessity and an element of national security.[37] It is also a major component of the world economy, with periodic gluts and shortages causing great fluctuations in stock markets, economic activities, and government policies.

The Great Boom (with an Interruption)

The period from the Allied victory in 1945 to the oil crisis of the 1970s witnessed the greatest surge in oil consumption to that time. One cause was the demand by Americans and, after 1950, by Europeans and Japanese after years of warfare. Another was the supply of oil from the Middle East. Before the war, the Persian Gulf region had shown promise of large oil deposits. Then came the discovery in 1948 of the Ghawar oil field in Saudi Arabia, the largest in the world, with proven reserves of 71 billion barrels.[38] Since it went into production in 1951, it has been producing 5 million barrels of oil a day, or 6.5 percent of the world's oil supply, as well as enormous amounts of natural gas. Other oil fields were discovered in the Gulf States, Iran, and Iraq. While the major oil companies made huge profits, independents rushed to produce as much oil as possible, driving down the price. The cost of oil declined 20 percent between 1957 and 1973 to just over $2 a barrel. With oil so cheap, people lost interest in energy efficiency or in renewable energy sources.[39]

Then came the oil crisis of 1973. In October of that year, in response to US support of Israel in the Yom Kippur War, the Arab members of the Organization of Petroleum Exporting Countries cut off oil supplies to the United States and its allies. Oil prices jumped from $2.59 a barrel in January 1973 to $11.65 a barrel in January 1974. This was followed by the Iranian Revolution of 1978–1981, which caused prices to rise even more, reaching

$35 a barrel in 1980.[40] The advanced industrial economies went into a sharp recession, with fuel shortages and long lines at the gas pumps, while the poorer nations cut back on essential services. Suddenly energy conservation was in fashion, especially in Europe and Japan. Americans switched to more fuel-efficient cars, raising average fuel efficiency for new US cars from 13.8 liters per 100 kilometers (17 miles per gallon) to 10.7 liters per 100 kilometers (22 miles per gallon).

Encouraged by the rise in the price of oil, oil geologists searched for and discovered deposits in many new sites, including the North Sea, the North Slope of Alaska, and deep waters in the Gulf of Mexico. By 1981, the world was again in an oil glut, which continues.[41] Since the nineteenth century, the world has produced and consumed close to a trillion barrels of oil. In 2011, petroleum geologists estimated that there were 1.65 trillion barrels in proven reserves. At the 2011 level of production and consumption—over 32 billion barrels per year—that amount would last 51.4 years, not counting probable (but so far unproven) reserves.[42]

As petroleum expert Daniel Yergin has explained, "The world is clearly not running out of oil. Far from it. The estimates for the world's total stock of oil keep growing." Oil shortages in the past were not caused by a decline in oil reserves underground but by wars and political crises. Oil gluts, meanwhile, are caused by new technologies that lead to new discoveries, alternate sources of energy, and greater energy efficiencies.[43]

Petroleum Technologies

There is a finite amount of oil on the planet, but nobody knows how much that is. Instead, the known deposits of recoverable oil are determined by the technology available to extract it. New technologies appear when the price of crude oil makes it worthwhile for oil companies to invest in them. One of these involves drilling ever deeper into the Earth; a few wells are more than 12 kilometers deep. Some drills, after reaching a desired depth, branch off at an angle to find oil trapped in horizontal layers.

Another important technology involves drilling at sea. Offshore drilling close to the coast began at the end of the nineteenth century. The first free-standing offshore wells were drilled after World War II in the shallow waters of the Gulf of Mexico. A half century later, under the impetus of rising oil prices, engineers devised ways to drill for oil far beyond the continental shelf, in up to 2,600 meters of water.[44] Costing several billion dollars, drilling platforms are among the largest and most complex machines ever built. Pipelines are then constructed out to productive wells. In the Gulf of Mexico alone there are some 50,000 offshore wells connected by thousands of kilometers of pipelines. As of 2010, 30 percent of American oil came from the Gulf of Mexico, half of it from wells in waters 300 to 1,500 meters deep and another third from wells in waters over 1,500 meters deep. Recent discoveries indicate 5 to 8 billion barrels of recoverable oil in deep water off the southern coast of Brazil.[45]

Tar sands (or oil sands) are another source of oil that became worth extracting and processing at the end of the twentieth century. In northwestern Canada alone, geologists estimate that there are 1.8 trillion barrels of oil, of which 175 billion are recoverable with current technology. The oil in these sands is so thick that it does not flow but has to be extracted by heating it, either above-ground in a special processing plant or underground where steam is injected directly into the deposits. As of 2011, tar sand production had used up 596 square kilometers of land and the ponds needed for toxic tailings had covered another 171 square kilometers. Extracting the oil also consumes a great deal of energy and produces 5 to 15 percent more carbon dioxide than ordinary oil wells.[46]

Hydraulic fracturing, or "fracking," involves pumping a mixture of sand and water under high pressure into rock formations. As the rocks fracture, they release trapped oil and natural gas. While this widespread technique has contributed to the abundance and low prices of oil and gas in recent years, it has also aroused concerns about its environmental impacts, especially the contamination of ground and surface waters and the possible triggering of earthquakes.[47]

Transporting oil requires yet another set of technologies. On land and over short distances underwater—as in the Gulf of Mexico or the North Sea—laying pipelines is cost effective. Most of the world's oil fields, however, are located far from the centers of consumption. Between the Persian Gulf and Europe, East Asia, or the United States, oil is shipped in tankers. Until World War II, tankers remained relatively small, in the 14,000- to 20,000-ton range. After the war, as oil consumption skyrocketed, the size of tankers increased thirtyfold. Ultra-large crude carriers with a capacity of more than 250,000 tons replaced the fleets of smaller tankers. The *Seawise Giant*, a ship that could carry over 2 million barrels of oil, was so large it could not safely pass through the English Channel.[48]

Environmental Impacts

Oil affects the environment in many different ways. The first is necessary and deliberate. Oil fields, with their thousands of derricks and kilometers of pipelines, occupy enormous amounts of land. So do refineries, vast industrial complexes that transform crude oil into gasoline, diesel fuel, and petrochemicals. Similarly, oil tanks occupy large plots on the outskirts of industrial cities. Motor vehicles require filling stations and fleets of trucks to deliver fuel to them. And at sea, tankers require special harbors to load and unload their cargo. All of these have changed the landscape of modern industrial nations.

A byproduct of petroleum, from its extraction from the Earth to its emissions from tailpipes and smokestacks, is pollution. Petroleum releases effluents in both production and consumption. When oil is pumped from the ground, it is often accompanied by methane and other hydrocarbon gases. Some of these gases leak into the atmosphere, some are burned off, and some are captured and burned as fuel in power plants, factories, and buildings.

The major pollutant released by the combustion of petroleum is carbon dioxide (CO_2). Over the past century, carbon emissions have risen from 0.6 billion tons a year in 1900 (of which almost none was from petroleum) to 7.2 billion tons in 2000 and to 9 billion tons in 2011 (around 40 percent from petroleum and most of the rest from coal, natural gas, and wood).[49]

Burning petroleum also emits carbon monoxide (a toxic gas that quickly turns to carbon dioxide), unburned hydrocarbons, nitrogen oxides, and lead. As early as the 1950s, scientists knew that unburned hydrocarbons and nitrogen oxides emitted by automobiles rose into the atmosphere where, under the influence of sunlight, they formed a layer of ozone, causing the smog that enveloped Los Angeles. In the United States, 60 to 80 percent of pollutants emitted into the atmosphere came from motor vehicles; in Orange County, California, in 1973, 97.5 percent of the air pollutants came from vehicles. For decades, the major American automobile manufacturers resisted, through advertisements and by lobbying Congress, all attempts to reduce tailpipe emissions. Only the appearance of Japanese cars that met more stringent emission standards forced them to comply.[50]

Tetraethyl lead (or "Ethyl") was introduced into gasoline in the 1920s as a means of preventing engine knocking. After studies showed that lead emissions caused mental retardation in children, it was phased out in the United States in the 1970s and banned throughout most of the world between 1988 and 2000. However, the 7 million tons of lead introduced into the atmosphere in the United States alone between 1924 and 1986 remain in the soil and water.[51] Other pollutants released by or for motor vehicles include oil refinery leaks, oil drips from vehicles, paint fumes from manufacturing and repairing vehicles, asphalt fumes, gasoline fumes from filling stations, asbestos from brake pads, salt on the roads in winter, and emissions from burning tire dumps.[52]

The United States, with the most vehicles and the highest oil consumption in the world, was at the forefront in introducing pollution controls, despite opposition from the oil and automobile industries. Positive crankcase ventilation, catalytic converters, unleaded fuels, and other technical advances reduced the amount of pollutants that each car produced by 60 to 80 percent between the 1960s and the 1990s. These measures helped clear the air in Los Angeles and other cities. But the positive benefits of pollution control have been far less than had been hoped, because they were almost canceled out by the ever-increasing number of vehicles.[53]

Other nations have followed the example of the United States, some quickly like the members of the European Union and others slowly or not at all. The worst cases are in developing countries that cannot or, for political reasons, do not want to enforce pollution control measures. As of 2004, the cities with the worst air pollution were Cairo, Delhi, Kolkata, Kanpur, and Lucknow in India; Tianjin, Chongqing, and Shenyang in China; and Jakarta in Indonesia—all in developing countries experiencing a sudden rise in motor vehicles.[54] Mexico City was notoriously polluted because of its several million old cars and because it is situated in a bowl surrounded by mountains that will not allow winds to disperse the pollutants; as a result, some 4,000 people died there of respiratory conditions every year.[55] Delhi, a city of 25 million people, suffered from the worst air pollution of all because of vehicle emissions, smoke from burning coal, cow dung, and crop residues, and dust from construction sites. The level of airborne particulates was over ten times the maximum recommended by the World Health Organization, causing 16,000 premature deaths and 6 million asthma attacks a year.[56]

Oil Disasters

More newsworthy than the drips and fumes of everyday use are the spectacular gushers that the oil industry is prone to. One of the earliest, an oil spill in Lakeview, California, in 1910–1911, was also the worst in history, losing 9 million barrels (1.4 billion liters) of oil over eighteen months. Though engineers soon learned how to plug a gusher on land, stopping one at sea was far more difficult. Among the more famous were ones off Santa Barbara, California, in 1969; the *Ixtoc I* oil well in the Bay of Campeche off Mexico in 1979; and the gusher off Scotland in 1988 that killed 169 men.[57] The most recent, and by far the worst, was the explosion of the *Deepwater Horizon* drilling rig in the Gulf of Mexico in April 2010, resulting in a gusher that lasted for eighty-eight days and spewed 4.9 million barrels (779 million liters) of oil over an area the size of Cuba.[58]

Then there were oil spills from tankers. The *Torrey Canyon*, which broke up off the coast of Cornwall in 1967, spilled 900,000 barrels (143 million liters) of crude oil into the English Channel. In 1978 the *Amoco Cadiz* broke up in heavy seas off the coast of Brittany, spilling its entire cargo of 1.6 million barrels (254 million liters). And in 1989 the *Exxon Valdez* ran aground off Alaska, spilling some 240,000 barrels (38 million liters) of crude into Prince William Sound.[59]

Almost all of these were avoidable accidents caused by human error, gross negligence, or penny-pinching at the expense of prevention. None of them, however, compare to the greatest petroleum disaster of all, the deliberate destruction of the Kuwaiti oil fields in early 1991. When the Iraq War began, the retreating Iraqi army deliberately set fire to the oil wells of Kuwait. The fires burned 6 million barrels (950 million liters) a day, and the last one was not extinguished until November 1991. The air over the Persian Gulf and nearby countries remained smoky for months.[60]

Gushers, spills, and fires have caused far more damage to the environment than to human beings. Oil spilled into the sea coated every living thing, including fish, mollusks, and sea birds. Beaches, marshes, wetlands, and rocky shores were fouled with black oil. Efforts to contain or capture the spilled oil were only moderately successful; after the *Deepwater Horizon* blowout, only one-quarter of the spilled oil was captured, while three-quarters evaporated, mixed with seawater, or turned into tar balls that washed up on beaches. Dispersants used to break up the oil slicks were toxic to fish and other aquatic creatures. In stormier waters off the Atlantic coast of Europe where containment and recovery proved impossible, attempts were made to ignite the surface slicks, thereby transferring the pollution from the water to the air. When oil is eaten by hydrocarbon-loving bacteria, as in the warm Gulf of Mexico, they also use up the oxygen in the water, causing fish to die off. Birds, seals, and sea otters coated with oil died of hypothermia, dehydration, or poisoning. Recovery is slow; it took between fifteen and twenty years for fish and mollusks to recover from the *Ixtoc I* blowout, and seven years after the *Amoco Cadiz* spill.[61]

After every spill, the oil industry introduces new technologies and procedures to avoid future spills: double-hulled ships, blowout preventers, rigorous inspections and enforcements. These are very effective; after all, of 50,000 oil wells in the Gulf of Mexico, only two have caused major spills. Yet, as the easily recoverable oil is used up, the industry turns to ever more dangerous environments: in deeper waters in the Gulf and off Brazil and Africa, and in the Arctic Ocean off Alaska and Siberia. Drilling in such areas requires very sophisticated new technologies, and complex technologies are notoriously prone to unanticipated problems with potentially devastating consequences. If a spill should occur in or near the Arctic, the recovery would be vastly more difficult and costly than in warmer areas.[62]

Industrial Agriculture

While the impacts of the automobile and oil on land, water, and air have been unprecedented, the stresses from modern industrial agriculture have arguably been just as important. Thanks to machines and chemicals, the output of industrial farming has more than kept up with the growth of population. Here too, as with autos and oil, the United States has led the way, followed by Russia, Brazil, and other countries.

Modern industrial agriculture employs monoculture, hybrid or genetically modified plants, heavy machinery, artificial irrigation, chemical fertilizers and pesticides, and enormous amounts of fossil fuels. According to historian Vaclav Smil, while the world's cultivated area increased by one-third between 1900 and 1990, harvests increased sixfold, because of an eightyfold increase in energy inputs.

The cost to the environment has been proportionately high. Twenty thousand square kilometers of formerly productive land—more than is currently devoted to farming—have been lost to deserts and badlands, or covered by buildings and pavements. In the 1990s, 60,000 to 70,000 square kilometers of productive land were lost each year. Despite the 8 million square kilometers of new land plowed up since the 1860s, the total area under cultivation began declining in the 1980s.[63] In other words, the world has been producing increasing amounts of crops on diminishing amounts of land. Nowhere is this paradox more evident than in the Great Plains of the United States.

The Great Plains

When European farmers came to plow up the Great Plains in the late nineteenth century, they found the most fertile soil on Earth, formed over millennia from wind-blown dust from crumbling rocks of the Rocky Mountains. The native grasses provided high-quality forage for bison, pronghorns, and other herbivores. Over thousands of years, these grasses created a layer of humus containing, in the top 15 centimeters of soil, over 6 tons of organic matter per hectare.

Conditions in the Plains varied a great deal, however, for average temperatures diminished from south to north and precipitation from east to west. While the eastern portion of the Plains was almost ideal for agriculture, the western half—approximately west of the 98th meridian—suffered from erratic and sometimes very deficient rainfall. Western Kansas and Oklahoma, eastern Colorado, and northern Texas were the most vulnerable parts of the Plains. Thus, in western Kansas, average rainfall was 500 millimeters, sufficient for growing wheat, but one year in five it was over 600 millimeters, producing bumper crops, and one year in six it fell below 400 millimeters, causing crop failures. For farmers who planted wheat, yields could vary from 22.5 to 1,613 liters per hectare (.3 to 21.3 bushels per acre), depending on the rainfall.[64]

The first wave of European farmers and ranchers arrived in the southern Plains after the American Civil War. The construction of railroads made settlement much easier than before. During the extraordinarily rainy decade 1878–1887, boosters, speculators, and government agents persuaded would-be farmers that "rain follows the plow." However, the climate of the 1890s proved so difficult that many farmers abandoned the land.

The rains that returned in the first decades of the twentieth century lured more settlers eager to plow up the grasslands make their fortune producing wheat. This time they were aided by machines. Reapers and combines, which both reaped and bundled the wheat, dated back to the nineteenth century. But these machines had to be powered by horses, and growing the feed for horses took up a quarter of the nation's farmland, not to mention a great deal of manpower.

Gasoline-powered tractors, introduced in the second decade of the twentieth century, freed farmers from the need to keep and feed horses. The numbers of these machines exploded, from 1,000 in 1910 to 1 million in 1932. A two-plow tractor could plow as much land in a day as eight horses and a three-plow tractor, as much as eleven horses; some were even bigger. Along with the new tractors came another innovation: the disk plow that could plow up two or three times as much as the moldboard plow and left the soil pulverized instead of in clods. For farmers who invested in machines to plow their fields and harvest their crops, the results were spectacular. While in 1830, harvesting a hectare of wheat had taken 145 man-hours, by 1930 in the Great Plains it only took 7.5 man-hours.[65]

When the First World War broke out in Europe, western Europe was cut off from Russia, its traditional source of wheat, driving up prices. To fill this need, farmers in the midwestern United States, encouraged by government agents, railroad companies, and land speculators, plowed up millions of hectares of virgin grasslands.[66] Thousands of would-be farmers came to plow, sow, and harvest, but otherwise lived in towns—"suitcase farmers" they were called. After the war, prices dropped, then rose again. With rising prices and several years of good rain in the late 1920s, farmers renewed the Great Plow-Up of the war years. From 1925 to 1930, they plowed up another 2.13 million hectares of native vegetation.[67]

Not only were they plowing, but they were also using a technique called "dry farming." This consisted of deep plowing to firmly pack the subsoil, then covering it with "dust mulch," a layer of powdered topsoil designed to conserve the scant moisture that fell on the southern Plains. As late as 1931, Walter Prescott Webb, the great historian of the Plains, could write: "Another reason for stirring the soil in the Great Plains country is to prevent the blowing of the soil by the strong winds and the consequent damage to growing plants."[68]

The Dust Bowl

Webb was wrong, for the combination of the Great Plow-Up and dry farming caused both a social crisis and an ecological disaster on the Great Plains. First came the Great Depression. The combination of good harvests and falling demand brought down the price of wheat to one-tenth what it had been in 1918, not enough for farmers to pay their debts.[69] Then came a drought that lasted from 1931 to 1940. With it came winds that sucked the "dust mulch" off the fields and carried it hundreds, even thousands of kilometers away. For the first four months of every year between 1933 and 1936, there were, on average, nine dust storms a month, each lasting for hours. The windstorm that raged on May 12, 1934, carried dust as far as New York and Washington, DC. By then, over 100 million hectares of land had been severely damaged. On April 14, 1935, a black blizzard darkened the skies from Colorado to the East Coast and even dropped dirt on ships 500 kilometers out in the Atlantic. On some farms, the wind scraped the topsoil off the land, leaving only the sterile hardpan. Drifts up to 8 meters high collected along fencerows and on the windward sides of houses and barns. By 1935, over 13 million hectares lay open to the wind, without grass or crops to protect the soil. Dust got into houses, watches, engines, and lungs. Pneumonia caused by dust became epidemic in some counties. Along with the dust came rabbits, grasshoppers, crickets, flies, and windblown Russian thistles and other weeds. People fled, three-quarters of a million of them refugees from the greatest environmental disaster in North American history.[70]

The US Department of Agriculture estimated in 1935 that, as a result of the conditions that produced the Dust Bowl, 20 million hectares of land had been ruined and abandoned. The Soil Conservation Service, established that year, reported that 43 percent of the land at the heart of the Dust Bowl had been seriously damaged and that croplands were losing 2.7 billion tons of soil each year. Ecologists and soil conservationists advised planting grasses in drier areas and on eroded land. They recommended sorghum, a more drought-resistant crop than wheat. They suggested contour tillage and terracing on slopes. The government also began to plant shelter belts of trees in moister areas—though not in the dry Dust Bowl itself—in order to slow down the winds.[71]

Then World War II broke out, grain prices shot up just as the rains returned, and all of these well-meaning reforms were forgotten. Farmers who chafed at regulations, restrictions, and shelter belts plowed up the grasslands with the same fervor as their predecessors. As the *Saturday Evening Post* explained: "The voice of two-dollar wheat is far more persuasive than scientific facts on wind, rain, sun and soil."[72]

When drought returned to the southern Plains from 1950 to 1956, it brought dust storms just like those of the 1930s. Farmers, squeezed between the cost of machinery, chemicals, and other inputs and the price of grain, plowed up as much land as they could or sold out to agribusiness

Figure 11.1. Farmer and sons walking in the face of a dust storm, Cimarron County, Oklahoma. Photograph by Arthur Rothstein. Prints and Photographs Division, Library of Congress, LC-DIG-ppmsc-00241.

corporations. Four to six million hectares were damaged each year, more than in the 1930s, for a total of 8.5 million hectares.[73]

The same situation arose again in the 1970s. In 1972, in response to massive wheat purchases by the Soviet Union, wheat prices shot up. Urged by Secretary of Agriculture Earl Butz to plant "fence-row to fence-row," farmers plowed up 16,000 square kilometers of marginal lands, much of it in Dust Bowl country. By 1981, up to 21,000 square kilometers of grasslands had been plowed under. Once again, windstorms peeled topsoil off the newly plowed but still naked land. In the 1970s, the United States lost 4 billion tons of soil a year, 1 billion tons more than in the 1930s. Crop insurance and farm-price supports prevented a repetition of the social disaster of the '30s—but not the environmental damage. Gradually, conservation has taken hold; in 1982, the nation lost 3 billion tons of soil, and in 2001, just under 2 billion tons, less than since the 1920s but still an enormous amount. Since independence, the United States had lost one-third of its topsoil.[74]

In spite of the environmental damage done to the soil in the Great Plains, the region has continued to produce huge crops, many times more than the American market can absorb. There are two reasons for this enormous productivity: chemicals and water.

Chemical Agriculture

In the early twentieth century, when midwestern farmers replaced their draft animals with tractors, they turned to chemical fertilizers to replace the manure their animals had provided for free. Chemical fertilizers have been known since the 1840s, when Justus von Liebig explained the theory of plant fertilization and John Bennett Lawes patented superphosphate fertilizer. Starting in the 1840s, farmers in Europe and North America also applied guano, seabird dung with a high concentrate of nitrogen and phosphorus, imported from the west coast of South America.[75] Potassium, the third essential fertilizer, was found in deposits of potash in Germany, South Carolina, and other places.

Of the three basic plant nutrients, nitrogen was in shortest supply. In 1913, the German chemist Fritz Haber and the engineer Carl Bosch began producing anhydrous ammonia, the form of nitrogen available to plants, on an industrial scale.[76] The process required a great deal of energy, however, and was used sparingly in agriculture until the 1950s. New technologies, using natural gas as raw material, enabled a price drop by two-thirds, and production rose exponentially, doubling in the 1960s and doubling again in the 1970s. This allowed world food production to keep pace with population growth. Over half of the fertilizers applied to fields ended up in lakes and coastal waters, rather than being taken up by the crops for which they were intended. There, they stimulated the growth of algae blooms that absorbed oxygen from the water, leaving too little for fish and other marine animals to survive.[77] Agricultural chemicals made from gas or oil used a great deal of energy in their manufacture. Modern industrial agriculture also required petroleum and natural gas to power the machinery and to transport and process the crops.

Fertilizers were not the only chemicals used in agriculture. Monoculture invited pests and weeds to multiply, and worldwide shipments of agricultural products allowed them to spread. Pesticides based on lead or arsenic did not limit themselves to killing pests and weeds but affected any other insect or animal that absorbed them, including farmworkers. Research during World War II produced several herbicides such as 2,4-D, a major ingredient of the defoliant Agent Orange used during the Vietnam War, and DDT, a potent insecticide. Later research produced chlordane, dieldrin, malathion, and other pesticides. Altogether, the output of synthetic pesticides grew from 117,500 kilograms in 1947 to 289 million kilograms in 1960.

Rachel Carson denounced these and many other agricultural and garden chemicals in her best-selling book *Silent Spring* (1962).[78] As a result, such chemicals came under scrutiny and DDT was eventually banned. Yet the outpouring of chemicals hardly slowed down. In the United States between 1964 and 1982, pesticide use increased by 170 percent, even though the cultivated acreage stayed the same. By 1990, over 50,000 different chemical compounds were in use worldwide.[79] These pesticides often failed to live up to their promises. In many cases, as insect pests developed resistance to them, chemical companies responded by introducing new toxins into the environment. In many cases, these chemicals proved to be toxic to humans, even those living far away, and were blamed for incidents of leukemia and other cancers.[80]

The Ogallala Aquifer

The other factor that allowed midwestern and especially Great Plains agriculture to flourish in spite of soil depletion was water. Water has always been in short supply on the Great Plains, especially west of the 98th meridian. An ingenious technological innovation—center-pivot irrigators—saved the Plains from becoming once again what early European visitors had called "the Great American Desert."

The water that makes green circles on the Plains every year comes not from streams and rivers but from the world's largest freshwater aquifer, the Ogallala. According to environmental historian John Opie, this aquifer stretches over 450,000 square kilometers from South Dakota to northern Texas and contains 37 trillion liters of water in a layer up to 300 meters thick. Early settlers believed that the aquifer was an underground river replenished by rain and therefore inexhaustible. It is not a river, however, but a layer of wet sand varying from 3 to 300 meters in thickness. Most of that water has been there for at least 3 million years.[81]

In the nineteenth century, farmers knew there was water under their land, but they did not suspect there could be so much. Beginning in the 1850s, they erected metal windmills to provide water for their homes, livestock, and gardens, though such simple devices could not water more than 1 hectare apiece. In the 1890s, new centrifugal pumps could irrigate up to 29 hectares per well, but they could lift water only 6 meters, leaving most of the aquifer inaccessible. Not until the 1930s, with the introduction of turbine pumps powered by automobile engines, could wells irrigate up to 56 hectares apiece with water drawn from much greater depth than older pumps. In Texas, the number of such powerful pumps rose from 170 in 1930 to 8,356 in 1948 and to 42,225 in 1957.[82]

Though powerful pumps made water available in sufficient quantities to irrigate entire farms, distributing it to the crops remained problematic. Ditch and furrow irrigation, as practiced in many dry lands, was very wasteful, and laying pipes and sprinklers was costly and required too much labor. The solution, the center-pivot irrigator, was patented in 1949 by Frank Zybach, a farmer and tinkerer from Colorado. Consisting of a 396-meter-long pipe carried on seven or more wheels that revolved around a central pivot, it could pump up to 3,785 liters per minute to irrigate up to 53.4 hectares of a standard 65 hectare (160-acre) quarter-section. Circling the field every half-day to seven days, it could spray not only water but also fertilizers, pesticides, and herbicides as needed. Unlike other irrigation systems, it required no leveling or ditch digging and could operate on sandy soil and uneven ground. Though the equipment, fuel, and chemicals cost twice as much per hectare as did dryland farming, the yields were more than three times as large.[83]

Economically, the results have been spectacular. By the early twenty-first century, 5.7 million hectares of cropland were irrigated with Ogallala water, producing over one-fifth of the United States harvest. Even thirsty crops like maize, soybeans, and cotton could profitably be grown where previously only dryland crops like wheat and sorghum survived. Land that had once been called the Great American Desert was transformed into one of the most productive agricultural regions in the world and made the United States the world's foremost exporter of agricultural products.

Like so much in American agrarian history, Ogallala irrigation has a temporary feel to it. Water not taken up by plants percolates through the soil back into the aquifer, carrying with it fertilizers, herbicides, and pesticides. Ninety-five percent of the water pumped from the Ogallala contains nitrates, sulfates, and other chemicals. Suburban developments with dispersed septic systems also contribute to polluting the aquifer. Far in excess of the maximum safe levels established by government scientists, these pollutants cause health problems for the people and animals that get their water from wells.[84]

Just as serious, from an ecological point of view, is the depletion of the Ogallala. Since widespread irrigation began in the 1950s, farmers have withdrawn ten times more water than the rate of recharge. As a result, the aquifer has suffered a net loss of between 312 and 454 cubic kilometers of water, or 9 to 11 percent of the total.[85] The loss varies enormously from one part of the region to another. In South Dakota and Wyoming, the aquifer is still thick but not usable, due to the cold climate and the presence of minerals and pollutants. In parts of Kansas, Oklahoma, and Colorado, the water table has dropped by up to 50 meters, making future irrigation problematic. And in Texas and New Mexico, accessible water has declined by 20 to 25 percent.

As the water table recedes, pumping requires increasing amounts of fuel, shrinking the economic value of irrigation; in some places it is no longer worthwhile. Many forms of life feel the squeeze. As springs dry up, streams and rivers shrink, and fish and other wildlife become scarce. Only plants with deep roots, like the mesquite that can send its roots down 20 meters into the ground, can survive.[86]

Geologists and hydrologists have been aware of the depletion of the Ogallala aquifer for a long time, and so have farmers. Yet, as historian Donald Green pointed out, farmers prefer "a super abundance for a few years and then nothing" to "a reasonable abundance for many years." Donald Worster expressed this view even more dramatically: "This, then, is the agriculture that America offers to the world: producing an incredible bounty in good seasons, using staggering quantities of machines and fossil fuels to do so, exuding confidence in man's technological mastery over the Earth, running along the thin edge of disaster."[87]

Industrial Animal Husbandry

In spite of droughts and dust storms and aquifer depletion, American agriculture continues to produce an abundance of wheat, maize, soybeans, and other crops year after year, as do farms in other modern industrial nations.

Where do all those crops go? Some are eaten by humans, but much of the world's harvest consists of grain fed to animals, providing many humans with an extraordinary and unprecedented high-protein diet of milk, eggs, and meat.[88] Farmers are able to do this by applying modern industrial and scientific methods to the raising of animals. For humans, industrial animal husbandry is a mixed blessing: more, tastier, and affordable food, though with increased risks of obesity and other health problems for consumers. For the animals involved and for the environment, the consequences are grim.

Wheat, maize, and soybeans are the crops most commonly fed to animals. In 1994–1996, the United States produced 58 million tons of wheat, 22 percent of which was fed to animals and 70 percent was eaten by humans. That same year, the United States grew 354 million tons of maize, 60 percent of which was fed to animals; most soybeans also become animal feed.[89] It is an inefficient process, for it takes 6.9 kilograms of grain to produce 1 kilogram of pork; for beef, the ratio is 4.8 to 1; for chickens, 2.8 to 1; for eggs, 2.6 to 1.

Americans have long been the world's champion meat eaters. During the twentieth century, per capita meat and fish consumption in the United States rose from 63.4 kilograms per year in the 1910s to 100.3 kilograms in the 1990s. Of that, the proportion of beef rose to 42 percent in the 1970s, then declined to 31.5 percent; the proportion of veal and lamb dropped drastically; and the proportion of chicken rose from 4.4 kilograms (3.1 percent of the total) to 69.1 kilograms (32.7 percent). In short, 80 percent of the increase in total meat and fish consumption was due to chicken.[90]

The process of turning feed into meat is not for the squeamish. Calves are born on ranches, where they spend their first six months grazing. Then they are moved to pens and fed increasingly with maize. After that, they are shipped to feedlots, crowded, filthy, and choked with dust. There, they eat a mixture of maize, hay, alfalfa, fat (often beef tallow), protein supplements, vitamins, and synthetic estrogen and urea. Some feedlots feed their animals ground up chicken feathers, the bedding and manure from chicken farms, and recycled food wastes. These diets have two consequences. One is to fatten the cattle quickly, leaving their flesh marbled with saturated fats, the kind of beef that consumers prefer.[91] The other is to make the animals sick, for cattle evolved to eat grass, not starch and animal proteins. As a result, virtually all of them suffer from pneumonia, enterotoxemia, coccidiosis, polio, and abscessed livers, and from gas trapped in their rumen. To make sure the animals gain weight until they are ready to be slaughtered, feedlot operators lace their feed with antibiotics.[92]

Industrial methods have also revolutionized the raising of chickens, especially since World War II. Of the estimated 50 billion chickens in the world in 2009—9 billion in the United States—over two-thirds were produced by industrial methods.[93] Egg-laying hens are kept in small cages, the bottoms of which consist of wire meshes that allow feces to drop out, with conveyor belts to remove the eggs. Hens lay up to 300 eggs in the year before they are slaughtered. Broilers, hens raised for meat, are kept in large houses containing 20,000 to 30,000 birds, and are fed for six to seven weeks before being slaughtered. Like cattle, their feed is optimized for rapid growth and laced with antibiotics to prevent diseases.[94]

Conditions in these poultry factories are calculated to maximize the output of eggs and meat and minimize costs. Growers trim the beaks of day-old chicks to reduce aggression. In the chicken houses, the air is laced with ammonia from the accumulated wastes, causing respiratory diseases and inflamed eyes among the animals. Egg-laying chickens lose so much calcium to their eggs that they suffer from osteoporosis, so they spend most of the day lying down in their own excrement and get blisters, sores, and often broken legs and ruptured tendons. Unlike cows, whose first six months of life correspond to their natural environment and life cycle, chickens live in pain throughout their short lives.[95]

For industrially raised pigs, conditions are slightly better. They are confined by the thousands in giant air-conditioned barns. They are fed grains, soybeans, and meat or bone meal, which, being omnivorous animals, they can digest better than cows. Like cattle, they are also given drugs to make them gain weight fast.

The diet rich in animal products that consumers enjoy weighs heavily on the rest of the planet.[96] The wastes of thousands of animals crammed together are pumped into manure lagoons, where nitrogen, phosphorus, bovine growth hormones, and heavy metal residues form a soup too toxic to use as fertilizer. Since feedlots don't all have sewage-treatment facilities, the wastes accumulate, then flow into nearby waterways or seep into groundwater. Ammonia, methane, hydrogen sulfide, and particulate matter are spread by the wind and can cause health problems among humans and animals.[97]

Antibiotics also seep into the environment. Seventy percent of antibiotics used in the United States are fed to animals to prevent infections and promote growth. Half to three-quarters of them pass through the animals and are excreted and return to the land and the water supply. Once out in the environment, they affect wild animals and fish, and human health as well. They also promote the emergence of antibiotic-resistant micro-organisms in ways that scientists are only beginning to understand.[98]

China Since Mao

Thanks to corporate advertising campaigns, the media, and travelers' reports, American-style consumerism has spread throughout the world. Wealthy people—and increasingly an emerging middle class—aspire to own a car, a home, appliances, and other accoutrements of the modern consumer society. Nowhere is this phenomenon more striking than in China.

In 1978, two years after Mao Zedong died, his successor Deng Xiaoping turned China away from Mao's idiosyncratic views on the economy toward a flexible market-driven economy. In the 1980s, under the slogan "To get rich is glorious," he devolved authority over the economy to provincial and municipal officials. Many state-owned enterprises gave way to private or locally owned companies. Agricultural communes were broken up and land returned to the peasants under the "household responsibility system." Special Economic Zones were created to encourage foreign investors to open plants or to form joint enterprises with Chinese investors. The state-centered developmentalist agenda did not disappear, but it was reduced to a few state-owned enterprises and to those areas that governments control even in free-market nations, such as building roads, railways, and major hydraulic projects.

The result of the privatization of most of the Chinese economy was a burst of entrepreneurial energy and a booming economy that is still booming thirty years later. Gross domestic product per capita leaped from $251 in 1980 to $796 in 1990, to $2,379 in 2000, and to $7,553 in 2010.[99] Suddenly, a middle class emerged, and even a spate of multi-millionaires. In 2009, the Chinese surpassed Americans as the number one buyers of cars, the largest proportion of homeowners, and the greatest polluters.[100]

At the same time, China was opened to the rest of the world. As foreign investments soared, the country began manufacturing myriad products for export. It also experienced a huge influx of foreign tourists and business-people and an even greater outflow of Chinese citizens seeking education, business opportunities, and entertainment in foreign countries. The world looked with astonishment at this unprecedented economic miracle.

On a deeper level, however, there were continuities with the recent past. The emphasis was still on producing as much as possible—albeit a different mix of products than in the Mao era. In the process, the environment took a back seat; in Robert Marks's words: "Deng Xiaoping's developmentalism conceived of nature as a vast reserve to be plundered for human ends."[101] A fast-growing economy needed ever larger amounts of farmland, water, and energy, all of which were in short supply in China. In the rush to get rich, officials paid lip service to environmental protection regulations, then ignored them.

Land

When Deng announced the "household responsibility system," farmers understood that what had once belonged to the state was now theirs—but, given the vagaries of Chinese politics, perhaps not for long. So they rushed to cut down all the trees they could as quickly as possible, before the tide turned. Regions in the south and southwest of China that still had some forests were especially vulnerable; Sichuan's forest cover dropped from 28 percent in 1970 to 8 percent in 1990. The deforestation of the mountains of Sichuan and the headwaters of the Yangzi deprived those areas of the ability to retain water after heavy rains, causing a massive flood in the Yangzi basin in 1998. Meanwhile, as China's appetite for timber continued unabated, it began importing massive quantities from Siberia, Myanmar, Indonesia, and other countries, thereby transferring the ecological burden of its economic success to others.[102]

The age-old Chinese need for more arable land also put pressure on the grasslands of the north and northwest, as farmers sought new fields to till, even where rainfall was marginal and uncertain. Grasslands were also threatened by more intensive pasturing. In the 1980s, seeking to profit from the Western fad for cashmere sweaters and scarves, herders switched from sheep to goats; unlike sheep that eat only the leaves of grass, goats eat the roots as well, destroying the plants. By 2005, China was producing three-quarters of the world's cashmere. Partly as a result, 90 percent of China's grasslands have become overgrazed, degraded, or turned into deserts. Since 1994, the Gobi Desert, to the north and west of China's most inhabited regions, has expanded by 65,000 square kilometers.[103] When the wind blows in from the northwest, as it does on average thirty-five times a year, it carries enough sand and dust to darken the skies over Beijing and block highways and railroad tracks. Dust-laden winds from China even produce haze in the United States.[104]

However, China did find an escape from its looming food crisis: chemical fertilizers. Soon after US president Richard Nixon's visit to Beijing in 1972, China began purchasing foreign-made industrial plants to produce ammonia. Later it began building its own, and by 1990 it was self-sufficient in synthetic fertilizers. Today, twice as many people live in China as under Mao; if most have enough to eat, it is thanks to agricultural chemistry. Agricultural chemical runoff has also caused algal blooms to appear on more than half of China's lakes.[105]

Coal and Air

In contrast to its shortage of water, China has abundant coal, upon which it relies for 70 percent of its energy, compared to 22 percent for the United States or 20 percent for Japan. From the late 1970s to 2008, China's coal production rose from 600 million to 2.75 billion tons per year, much of it from small illegal mines. By then, China was building one or two new coal-fired power plants every week, and coal was also used in industrial boilers and in household stoves. Most of it was inferior bituminous coal that produced soot, sulfur dioxide, and nitrogen dioxide, along with carbon dioxide. Scrubbers and other pollution-abatement methods were rare. China's CO_2 emissions have increased by 8 percent a year since 2007—much of it from burning coal—making it the number one emitter in the world. Of the world's thirty most polluted cities, twenty are in China.[106]

Coal burning and the refining of metals, especially zinc, also released poisonous mercury. In 2003, Chinese industries emitted 767 tons of mercury into the atmosphere, more than the United States, Europe, and India combined. This toxin, carried by prevailing winds over the Pacific Ocean, entered the food chain, ending up in tuna and other fish that people eat. The rest got carried by the jet stream around the world and was eventually deposited thousands of kilometers from its origin.[107]

Automobiles

Overshadowing the growth of cities and the production of coal was the astonishing rise of a Chinese automobile industry. Under Mao Zedong, the

Chinese government pursued a policy of self-sufficiency; in 1960 China produced 22,574 copies of Soviet sedans for official use. In 1978, two years after Mao Zedong's death, there were still only 1,358,400 officially registered motor vehicles in the People's Republic, about one for every thousand Chinese. Many of them were trucks. As late as 1990, China produced only 42,000 cars a year. Then came the boom. As a middle class emerged, automobile production took off, increasing from just over 2 million automobiles in 2000 to 18.4 million in 2011, more than any other country in the world. Almost all them were sold to Chinese customers; unlike the Japanese, the Chinese have not yet begun exporting cars in significant numbers.[108]

To stay ahead of the flood of new cars being bought every year, the Chinese government undertook the most extensive highway-building program in history. It began in 1989 with 147 kilometers of expressways. By January 2013, China had built 97,355 kilometers, more than the United States or all of Europe. In addition, it had built over 3.5 million kilometers of other roads.[109] It also built more rail and highway bridges than the rest of the world combined.[110]

Most of China's development has taken place in cities, especially in its eastern provinces. The result, not unexpectedly, has been gridlock. Beijing, which was already crowded with 1.5 million vehicles in 2000, had 5.4 million in 2012 and was adding a thousand new cars every day. American-style suburban development has not yet begun, but other aspects of the automobile culture have already taken hold, such as huge shopping malls with enormous parking lots, including the two largest in the world.[111]

China's oil consumption doubled between 2000 and 2010, reaching 11 percent of the world's total, but its pollution control measures have not kept up. At night, its cities are invaded by trucks whose diesel engines emit sulfur-laden smog and particulates. The pollution produced by runaway development reached an extreme case in Beijing in January 2013, when particulates soared to 886 micrograms per cubic meter of air, twenty-two times the amount that the World Health Organization considers safe. Some of its pollutants reach the upper atmosphere and are wafted over the Pacific Ocean by the jet stream and deposited in the United States.[112] The Chinese government, caught between the urge to develop its economy and a growing public dissatisfaction, has reacted with a mixture of denial and new regulations, so far with little impact.[113] China is, yet again, at a turning point. Prodded by citizens' complaints and international pressures, the Chinese government is becoming aware of the economic costs and political risks of air and water pollution, deforestation, and resource depletion. It has begun to address these problems by enforcing environmental regulations, investing in solar and wind energy, and encouraging electric vehicles and other green technologies. Whether these policies will be enough to balance the frenetic pace of development and consumerism is still to be seen.[114]

Conclusion

The economic boom that began after World War II—with occasional recessions, to be sure, but no repetition of the Great Depression—had many causes. Peace between the major powers diverted economic energies from destruction to construction. Of all the new technologies, automobiles had the largest effect on the world economy. Technological innovations in electricity, agriculture, and communication also played a part. But at the heart of the process of development was a stroke of good luck for consumers: cheap oil.

So successful was the resulting economic growth fueled by consumer demand that it trumped even the frantic efforts of totalitarian governments to generate growth on command. As a result, consumer-driven growth spread around the world, not only to command economies like the former Soviet Union but also to former Third World nations like China, Brazil, and India. In all these areas, countries around the world have followed the American example in the use of automobiles, in the consumption of oil, and in mechanized agriculture and animal husbandry. In China, consumer-based development has succeeded beyond anyone's expectations.

The growth of the world economy since World War II, with its rising living standards, better heath, personal mobility, and suburban lifestyles, has been accompanied by—and has indeed produced—corresponding environmental changes, some of them disastrous: the Dust Bowl, oil spills, the depletion of freshwater resources, and air and water pollution. The acceleration of growth and the very real prosperity that it created have come at a price, namely, the rapid transformation of the global environment from semi-natural to fully anthropocentric. The impact has been particularly visible in the extinctions of living species, the changing climate, and the depletion of ocean life.

12

Climate Change and Climate Wars

By the late twentieth century, humans had transformed large parts of the Earth. Yet the changes humans had made to the land, the air, and the water were all local or regional. Only in the last quarter of that century did scientists begin to document anthropogenic changes that affected the entire planet, and only at the very end of the century did the concept of planetary environmental changes reach the media, the general public, and political leaders. Not surprisingly, these findings have aroused controversies: how reliable they are, who is to blame for the changes, and what can be done about them? This and subsequent chapters deal with three kinds of historic and recent changes at the global level: the climate, the oceans and their inhabitants, and terrestrial species.

Climates of the Past

In the billions of years of its existence, the climate of the Earth has fluctuated many times, often dramatically, from "Snowball Earth" 650 million years ago, when the planet was covered with ice, to a period called the Paleocene-Eocene Thermal Maximum 55 million years ago, when there was no snow or ice anywhere.

The Ice Age

Some 2.4 million years ago, the Earth entered the Pleistocene, popularly known as the Ice Age. Despite this name, the climate was not consistently cold but was unstable, with periods of deep freeze lasting on average 90,000 years alternating with interglacials, or warming periods, lasting on average 10,000 years. During the deep freezes, the average global air temperature dropped to 12°C (54°F) and the average ocean surface temperatures to 7°C (45°F). Gigantic ice sheets covered most of the Northern Hemisphere, locking up much of the world's water, causing sea levels to drop 200 meters lower than today. South of the ice sheets, grasslands replaced forests, to the benefit of mammoths, wooly rhinos, and musk oxen. Then came the interglacials, often very suddenly, causing sea levels to rise by 2 meters within a hundred years. Between the warm and cold periods, the climate flickered; in the words of climatologist Richard Alley, "a crazily jumping climate has been the rule, not the exception."[1]

The Holocene

Starting 20,000 years ago, as the climate began to warm up, ice sheets melted and the sea level rose. Over a 10,000- to 12,000-year period, global temperatures rose by 8° to 11°C (14°–20°F) on average. About 12,000 years ago, a new epoch began, one that geologists call the Holocene, which witnessed the beginning of agriculture and the emergence of civilizations around the world; it continues to this day. It is also one of the most unusual in the history of the planet, for its climate has been more placid for longer than any other period in the past several million years.

Yet the ice did not give up without a few spasms. For a thousand years beginning 12,800 years ago, much of Europe, Siberia, and North America turned too cold for humans, while the Middle East turned cool and dry. A similar cold snap began 8,200 years ago but lasted only 200 years. Compared to the Ice Age and its last echoes in the early Holocene, the climate changes that occurred in historic times with such dramatic (and sometimes disastrous) consequences for people—the droughts that doomed Akkad, the Maya, and the Anasazi; the Medieval Climate Anomaly; and the Little Ice Age—were mere blips; in Richard Alley's words, they "appear as slow one-degree shifts in the ice-core record, not as abrupt ten-degree jumps."[2]

What explains the comparatively placid nature of the climate over the past 10,000 years? The best explanation is that as humans began to grow rice, raise cattle, and cut down forests, they released greenhouse gases into the atmosphere.[3] Though very little changed each year, over 10,000 years these anthropogenic gases (that is, those caused by humans) compensated for what would have been a natural cooling of the climate, leading, perhaps, to the beginning of a new ice age.

This balancing act ended over two centuries ago with the spread of industry. The Earth's climate rose 0.7°C (1.3°F); of that, 10 percent was due to an increase in solar irradiance and a decline in volcanic activity, and the other 90 percent to human activities. As in past epochs, the change was erratic. From the 1940s to the 1970s, the climate seemed to be cooling by 0.3°C (0.5°F), a change climatologists blamed on "global dimming," a decrease in solar irradiance caused by industrial pollutants in the atmosphere.[4] Then the warming returned and has accelerated ever since.

The Anthropocene

Recently, a growing number of scientists have given the name Anthropocene to a new geological epoch in which they attribute global warming to the actions of human beings. In 2002, atmospheric chemist Paul Crutzen popularized the term, arguing that it began in the late eighteenth century with an upsurge in emissions of carbon dioxide due to the burning of coal in the Industrial Revolution.[5] More recently, environmental historian John McNeill has argued that "the weight of the evidence points to a date in the middle of the twentieth century, something like 1945 or 1950."[6] Though the term has not yet received the imprimatur of the International Commission of Stratigraphy or of the International Union of Geological Sciences, its use has spread from scientific and academic circles to the mainstream media.[7]

Global Warming Today

Evidence that the climate has been getting warmer is incontrovertible. The year 1981 was the warmest on record until then; in the 1990s, three years were even warmer; since then, almost every year has been warmer than the previous one.[8] While scientists and much of the general public have accepted the reality of climate warming, two issues remain contentious. Is the warming natural or is it caused by humans? Regardless of the cause, should we be doing something about it?

The Scientific Consensus

Among climate scientists, there is an almost complete consensus on the anthropogenic causes of global warming. All 928 articles on the subject published in refereed scientific journals between 1993 and 2003 agree on this point, as do the reports of the United Nations' Intergovernmental Panel on Climate Change, the National Academy of Science, the American Meteorological Society, the American Geographical Union, and the American Association for the Advancement of Science. An analysis of 1,372 climate researchers and their publications showed that "(i) 97–98% of the climate researchers most actively published in the field . . . support the tenets of ACC [anthropogenic climate change] outlined by the Intergovernmental Panel on Climate Change, and (ii) the relative climate expertise and scientific prominence of the researchers unconvinced of ACC are substantially below that of the convinced researchers."[9]

At the heart of the consensus lies one of the largest organized scientific efforts in history to analyze and quantify a subject of such global importance. The first scientific conference on climate change took place in Austria in 1985. After a meeting of 300 scientists and policymakers from forty-eight countries in Toronto in 1988, the United Nations Environmental Programme and the World Meteorological Organization established the Intergovernmental Panel on Climate Change (IPCC).[10] Its task was to issue a periodic Assessment Report, which it did in 1990, 1996, 2001, and 2007, and 2013, along with special reports on emissions, renewable energy, climate mitigation, extreme climate events, and other related topics. These reports were the work of hundreds of scientists and vetted by delegates from more than a hundred countries. As a result, they took the most conservative, lowest-common-denominator position on climate change. Nonetheless, their findings were shocking.

The IPCC Reports

The IPCC's First Assessment Report, issued in 1990, concluded: "Emissions resulting from human activities are substantially increasing the atmospheric concentration of greenhouse gases . . . resulting on average in an additional warming of the Earth's surface." Furthermore, it warned, "long-lived gases would require immediate reductions in emissions from human activities of over 60% to stabilize their concentrations at today's levels." If humankind continued business as usual, the report predicted, the sea level would rise by 20 centimeters by 2030 and by 65 centimeters by 2100.[11]

The IPCC's Second Assessment report, issued in 1996, confirmed the conclusions of the First Report, but in a more unequivocal way:

Projections of global mean temperature change and sea level rise confirm the potential for human activities to alter the Earth's climate to an extent unprecedented in human history; and the long timescales governing both the accumulation of greenhouse gases in the atmosphere and the response of the climate system to these accumulations means that many important aspects of climate change are effectively irreversible.[12]

Besides global warming and rising sea levels, the Second Report also predicted increasingly frequent heat waves and adverse effects on agriculture. It warned that developing countries and poor people would be the most vulnerable to these changes.[13]

The Third Assessment Report, issued in 2001, repeated the conclusions of the first two reports but with more scientific certainty. Even though its conclusions were toned down at the insistence of the delegates from the United States, China, and Saudi Arabia, they were even stronger than those of previous reports. They explicitly confirmed that global warming was due to human activities and not to natural causes.[14]

The Fourth Report, issued in 2007, was even more unambiguous: "Warming of the Climate System is unequivocal, as is now evident from observations of increases in global average air and ocean temperatures, widespread melting of snow and ice and rising global average sea level." Furthermore, "impacts [of climate change] will very likely increase due to increased frequencies and intensities of some extreme weather events." "Many impacts [of climate change] can be reduced, delayed or avoided by mitigation." However, "unmitigated climate change would, in the long term, be likely to exceed the capacity of natural, managed and human systems to adapt."[15]

The Fifth Assessment Report, published in 2013, is the most adamant of all: "Warming of the climate system is unequivocal." The reason is clear: "The atmospheric concentrations of carbon dioxide, methane, and nitrous oxide have increased to levels unprecedented in at least the last 800,000 years." The reason for that increase in greenhouse gases is also clear: "It is extremely likely [i.e., with a 95 to 100 percent probability] that human influence has been the dominant cause of the observed warming since the mid-20th century."[16]

In short, the more thoroughly climate scientists studied the changing climate, the more adamant they became that human activities are endangering the Earth.

Greenhouse Gases

The IPCC reports and other research have implicated a number of causes of global warming. The most important was the increase in the proportion of greenhouse gases in the atmosphere produced by the great increase in the number of humans on the planet and economic growth.

The connection between greenhouse gases and climate goes back a long way. The French physicist Jean-Baptiste Joseph Fourier (1768–1830) was the first to suggest that the Earth's atmosphere acted as a kind of greenhouse. The English physicist John Tyndall (1820–1893) discovered that carbon dioxide in the atmosphere prevented infrared rays—that is, heat—from escaping into space, thus keeping the planet warm. And the Swedish chemist Svante Arrhenius (1859–1927) suggested that industrialization was causing the proportion of CO_2 to rise.[17]

CO_2 is not the only gas that prevents infrared rays from escaping. Methane (CH_4), emitted by livestock flatulence; by decaying vegetation in rice paddies, wetlands, and peat bogs; by melting permafrost; and from oil wells, accounts for 16 percent of the greenhouse effect; its proportion of the atmosphere has risen from 350–700 parts per billion before industrialization to 1,750 parts per billion today. Nitrous oxide (N_2O) is a byproduct of agricultural fertilizers and livestock and burning fossil fuels as well as coming from natural sources such as tropical soils; its proportion has reached 310 parts per trillion. CFCs (chlorofluorocarbons) have, per molecule, a 10,000 times more powerful greenhouse effect than CO_2 but have been phased out in most countries since 1987.

Some atmospheric changes may mitigate global warming somewhat. Clouds keep infrared radiation from returning to space like other greenhouse gases, but they also reflect sunshine back into space. Sulfate aerosols injected into the atmosphere by volcanic eruptions also reflect sunlight, producing a temporary cooling effect.[18]

Carbon Dioxide

Scientists have proved conclusively what Tyndall and Arrhenius suggested, namely, that the proportion of CO_2 in the atmosphere correlates extremely well with global temperatures and that human activities have caused it to rise.

Not all of the carbon dioxide emitted by humans has stayed in the atmosphere. Between 25 and 30 percent has been absorbed by the oceans, while 15 to 20 percent has been taken up by growing plants and trees.

Starting in 1958, the American chemist Charles David Keeling began measuring the proportion of CO_2 in the atmosphere at the Mauna Loa Observatory in Hawaii, which had the least polluted air of any observatory. After his death in 2005, his son Ralph Keeling and other scientists continued these observations at Mauna Loa and many other places around the world. Climate scientists were also able to analyze air bubbles contained in ice cores from Greenland and elsewhere, revealing the proportion of CO_2 in the atmosphere at points in the distant past.[19]

What they found is this: from 800,000 to 10,000 years ago, CO_2 fluctuated between 200 and 300 parts per million (ppm), in rhythm with global temperatures and sea levels. From then until the late nineteenth century, it ranged from 275 to 285 ppm. Then it began to rise. In 1900 it was 297 ppm. In 1958, when Keeling first published his findings, it stood at 315 ppm. By 1995 it had risen to 360 ppm, and by 2005 to 378. In May 2013 it reached 400 parts per million, a proportion not seen since 3 million years ago. Not only has this proportion grown, but its rate of increase has risen as well, from 1 ppm per year in the 1950s to 2 to 2.5 ppm per year in the early 2000s, driven largely by the 11 to 12 percent per annum increase in China's fossil fuel emissions. The consequence is not just that the Earth is getting warmer but that it is doing so faster than ever before.[20]

Before the twentieth century, most of the carbon in the atmosphere came from deforestation, as trees were burned for fuel. Since 1900, the main source has been the burning of fossil fuels, at first mostly coal but increasingly oil and natural gas. Since 2000, global emissions of greenhouse gases have been growing by 3 percent per year. In 2012, burning fossil fuels produced 35.6 billion tons of CO_2, 58 percent more than in 1990; another 3 billion tons came from deforestation, mainly in the tropics. Coal production, once surpassed by oil as the source of most of the CO_2 being emitted, has been rising again, especially in developing countries; from 1965 to 2011, the use of coal has increased sevenfold in Brazil and India, thirteenfold in Mexico, and fifteenfold in South Korea and China. The United States was long the world's foremost source of carbon emissions, with 28.5 percent of the world's total between 1850 and 2008, but it has recently been overtaken by China, with 23.6 percent of global emissions in 2009.[21]

The Impact of Global Warming

Much of the debate about global warming concerns the future. However, its effects are already being felt around the world. One such impact is the increasing number of storms since the mid-1970s.[22] As tropical seas warm up, they release more energy into the atmosphere in the form of hurricanes, cyclones, and typhoons. Between 1974 and 2005, the number of category 4 and 5 hurricanes (the most destructive) almost doubled, and the total amount of energy released by hurricanes worldwide increased by 60 percent. Among the more famous disasters are the tropical storms that pounded Haiti in 2004, leaving thousands dead, and Hurricane Katrina in August 2005, which devastated New Orleans and nearby wetlands and barrier islands.[23]

Another effect is an increase in rainfall, for warm air can hold more moisture than cold air; for every degree of warming, rainfall increases by about 1 percent. But this rainfall is not evenly distributed, with more of it in the higher latitudes and less in the tropics. The drought in the Sahel from 1965 to 2005 was one result of this shift. In Australia, rainfall has diminished by 15 percent since 1975. The American West has suffered from the driest weather in 700 years; snow cover has diminished, with less snow falling on the mountains and melting three weeks earlier than in the 1940s, with adverse effects on electricity production, tourism, and agriculture. In the Levant, the drought that lasted from 2007 to 2010, the worst on record, caused widespread crop failures and mass migration of farmers to urban centers. The civil war in Syria that followed this drought caused thousands to flee to other countries. In 2015, climatologists led by Collin Kelley concluded that "human influences on the climate system are implicated in the current Syrian conflict" and that "anthropogenic forcing [of the climate] has increased the probability of severe and persistent droughts in the region."[24] Taking a longer view, climatologists working with Aklo Kitoh wrote: "It is projected that, by the end of this century, the Fertile Crescent will lose its current shape and may disappear altogether. The annual discharge of the Euphrates River will decrease significantly (29–73%), as will the stream flow in the Jordan River."[25]

Globally, glaciers and permanent snow cover have shrunk by over 10 percent since the mid-twentieth century.[26] In Greenland, the area that undergoes summer melting increased by 30 percent from 1980 to 2010. The front edge of the Jakobshavn Glacier, the world's largest and fastest-moving, retreated 5 kilometers between 1997 and 2003. Glaciers are also retreating in the European

Alps. Glacier National Park in Montana, which had 150 glaciers in 1850, had only 30 left by 2010 and will probably be glacier-free by 2030. In Alaska, the terminus of the Columbia Glacier has retreated 600 meters per year since 1982; during 2001, it retreated by 30 meters per day. On the Pacific side of the Andes, glacial runoff is almost the only source of water and hydro-electricity. There, the shrinking of the glaciers is threatening the livelihood of nearly 80 million people. The Quelccaya Glacier in southern Peru is retreating at a rate of 61 meters per year, ten times faster than in the 1960s.[27] The Himalayan glaciers and snowpack, the main source of water for the Indus, Ganges, and Brahmaputra rivers and the Tibetan glaciers that feed the Yangzi, Mekong, Salween, and Irrawaddy rivers have shrunk by 7 percent since the 1970s.[28]

The Arctic Ocean has warmed by 2° to 3°C (4°–6°F) during the last half-century. The extent of summer sea ice has shrunk from an average of 11 million square kilometers between 1900 and 1950 to about 6 million kilometers in 2010. Summer sea ice covered 25 percent less area and was only half as thick in 2000 as in 1950; by 2007, summer sea ice was down 60 percent and its volume was down 90 percent from its twentieth-century average. In 2008, both the Northwest Passage (north of Canada and Alaska) and the Northeast Passage (north of Russia and Siberia) were open to ships for the first time in history.[29]

The oceans have changed as well. As glaciers have melted and as the oceans have expanded as they warmed, the sea level has risen by 19 centimeters since the early twentieth century. During most of the twentieth century, the level of the oceans rose by 1.4 millimeters per year, but in the 1990s it rose by 3.1 millimeters per year. A rise of a few millimeters does not make great headlines, but when the Larsen-B ice shelf in Antarctica, a piece of ice the size of Rhode Island or Luxembourg, collapsed, that made a splash. An ice shelf melting does not raise the level of the ocean because it is already part of the sea, but its collapse speeds up the flow of the glaciers behind it, and when they reach the sea, they do raise the sea level. As ice and snow melt, the land and water that are no longer covered reflect less sunlight, causing the climate to warm faster. This is the reason that the Arctic Ocean, where sea ice is disappearing faster than expected, is also warming up faster than any other part of the world.[30]

The Future of the Planet

In the case of climate change, scientists construct computer models that mimic the operations of the climate in order to come up with plausible trends in the future. The reports of the IPCC average out the best climate models to reach a consensus; but there is always the possibility of shocks, unpredictable events that overturn the best predictions.

Predictions

The first two IPCC Assessment Reports, issued in 1990 and 1995, predicted global warming of 0.3°C (0.5°F) per decade unless something was done to slow down greenhouse gas emissions. The Third Report, published in 2001, predicted that the average global temperature would rise by 1.4° to 4.5°C (2.5° to 8.1°F), depending on the amount of greenhouse gas emissions; in the worst case, if CO_2 rises by 2.5 percent per year, then global temperatures will rise by 5.8°C (10.4°F). The Fourth Assessment Report (2007), based on new data, confirmed these predictions. It contended that the climate will rise by 1.5° to 4°C (4° to 7.2°F) over pre-industrial levels by the year 2100, possibly even by 2.4° to 4.6°C (4.3° to 8.3°F) if nothing is done to mitigate the effects of human actions.[31]

According to the Fifth Assessment Report (2013), global surface temperatures will continue to rise and likely exceed 1.5°–2°C (2.7°–3.6°F) by the end of the twenty-first century and will continue to rise after that. A study in the journal *Nature* in 2014 indicated that the 2°C goal was "effectively un-achievable" and that it "allowed governments to ignore the need for massive adaptation to climate change."[32] Temperatures in the Arctic will rise the most, causing the extent and thickness of summer sea ice to continue shrinking. Differences between wet and dry regions of the world will increase, with dry regions becoming drier, and wet areas receiving more rain than today.[33]

The sea level, which rose 19 centimeters between 1900 and 2010, will continue to rise, more rapidly than before 1970. The oceans, which have absorbed about 30 percent of the anthropogenic carbon dioxide, will become progressively more acidic.

These numbers are averages for the whole globe. But what will it mean for different parts of the world? Overall, the higher latitudes will feel the effects of global warming more than the tropics. The North Pole will be 6° to 9°C (11° to 16°F) warmer than today. Much of the sea ice in the Arctic Ocean will melt; by how much will depend on whether global temperatures rise by 2°C (3.6°F), in which case much sea ice would remain year-round, or by 4°C (7.2°F), in which case it will disappear entirely during the summer months.

In either case, the Arctic Ocean will be open to shipping during part or all of the year. In Siberia, Russia, Scandinavia, Canada, and Alaska, forests will move north, replacing the permafrost, and make agriculture possible in areas never before farmed. The Greenland glaciers will begin to melt at a rapid rate.

The temperate zone will feel a variety of effects. Almost certainly, there will be more heat waves and, in places, heavy rains, and probably also more storms and high tides. Warm seasons will expand by one month at either end, with fewer very cold winters. In the United States, the Southeast, Midwest, and Southwest could experience droughts on average every other year, ten times more often than today.[34] Northern Europe will feel the effects of global warming more than southern Europe, but overall, the north and west, which are already rainy, will become even rainier, while the Mediterranean lands will become even drier than today. Crops that have been bred to give maximum yields under current climatic conditions are likely to be more seriously affected than their predecessors if temperatures and moisture move out of their optimum range.

In the tropics, temperatures will change less than in the higher latitudes, but other effects will be severe. Chinese glaciologist Yan Tangong predicted that by 2100 half of Tibet's glaciers will have melted; as a result, there will be less water in the rivers of China and Southeast Asia during the growing season, reducing the agriculture on which several billion people depend for their survival. Shifting patterns of rainfall will also affect large parts of Africa, the Middle East, and South America.[35]

As glaciers melt and the oceans become warmer, sea levels will rise. As of 2007, over 100 million people (including 3.5 million Americans) lived within a meter above sea level. By the end of the twenty-first century, the sea level could rise between 25 and 75 centimeters, and by a meter or more in the century after that.[36] Wealthy nations like the Netherlands will have the means to build sea walls but will still be affected by increasing numbers of storms. But even high levees are no guarantee of protection from storms, as Hurricane Katrina proved in New Orleans in 2005. Cities near the sea will be in trouble. Of sixteen cities with over 15 million inhabitants, eleven are on coasts or estuaries. New Orleans, Tokyo, Bangkok, and Shanghai are already sinking due to pumping water from the aquifer or to the weight of their buildings compacting the soil beneath them. There is a plan to build storm barriers to protect Venice, already suffering from more frequent floods (called *acqua alta*) than in the past, but the cost will be ever worse pollution of the lagoon on which it is built. In Bangladesh, where two-thirds of the population live within a meter of the sea level, the effects will be devastating, as they will in other river deltas in Myanmar, Thailand, and Vietnam.[37]

When sea levels rise, where will people go? Atiq Rahman, executive director of the Bangladesh Center for Advanced Studies and that nation's leading climatologist, wrote: "These migrants should have the right to move to the countries from which all these greenhouse gases are coming. Millions should be able to go to the United States."[38] Will they be welcomed?

Finally, atolls and low islands like Kiribati, Tuvalu, and the Maldives will shrink and eventually disappear. Whether their people, and those in other countries whose lands are flooded, could sue the perpetrators of global warming is a matter for lawyers to argue.[39]

Time Frames

The question many have asked is this: When will all these changes take place? Here, the uncertainty is greater than the changes themselves. The proportion of CO_2 in the atmosphere could rise anywhere from 550 ppm (if efforts are made to reduce emissions) to 1,200 ppm (under a "business as usual" scenario). According to the 2001 IPCC report, sea levels will rise between 0.1 and 1 meter in the twenty-first century but could continue to rise up to 2 meters over the next 500 years, depending, again, on how much greenhouse gas humans emit.

One aspect is worth noting, however. Even if humans ceased pouring greenhouse gases into the atmosphere, the CO_2 already there will remain for tens of thousands of years before gradually diminishing. Human actions cannot reverse the trend.[40] As the 2013 IPCC Report points out, though the impact of climate change can be moderated by vigorous actions, the effects of past and present behaviors are essentially irreversible: "Most aspects of climate change will persist for many centuries even if emissions of CO_2 are stopped."[41]

Does it matter what happens 100, 500, or 10,000 years from now? To geologists and paleontologists, these are tiny moments in the vast unfolding of time on this planet. According to William Ruddiman, if not for humans, the Earth would already be starting to cool off. Curt Steger, the author of *Deep Future: The Next 100,000 Years of Life on Earth*, has tried to predict the distant future. As he points out, without human intervention, the planet would enter a new ice age some 50,000 years from now. But, thanks to humans burning fossil fuels, that will be postponed to 130,000 years from now. Meanwhile, for several thousands of years, the world will be 2° to 4°C (3°–7°F) warmer than today, sea levels will have risen 6 to 7 meters, ice caps and glaciers will have melted, and the oceans will have acidified. But this will happen only if humans severely restrict their greenhouse gas emissions. On the other hand, if humans continue to burn fossil fuels, then the next ice age will be postponed until half a million years from now. Meanwhile, the Earth will become hot, with shrunken continents surrounded by much larger and more acidic oceans—a ghastly fate, should any humans still be around to witness it.[42]

These, then, are the trends that computer models have allowed scientists to predict with a fair degree of confidence: that Earth will get warmer, ice will melt, the sea will rise, and storms and droughts will afflict humankind. Just how much will depend on the amount of greenhouse gases that humans emit in the future. But the climate is complex and interconnected, and it is unlikely to change as smoothly as the IPCC reports predict. In particular, three possible shocks might disrupt these scenarios, because the climate system contains feedback loops that allow small events to trigger large changes.

Greenland and West Antarctica

One such feedback loop is the albedo or reflectivity of different surfaces. Ice and snow have a very high albedo, that is, they send much of the sun's radiation back into space. When they melt, they are replaced by water or land, which, being darker, absorbs the sunshine, hence get even warmer, causing more nearby ice and snow to melt, causing yet more warming. Once this process starts, it is irreversible.

The IPCC reports, based on conservative extrapolations from recent data, did not take into account the possibility of sudden and rapid changes. One such possible change is the melting of the ice sheets that cover Greenland, West Antarctica, and the Antarctic Peninsula. Should these gigantic ice sheets melt, sea levels would rise. By how much is a matter of debate. Geophysicist Henry Pollack estimates that sea levels will rise by 6 feet, or almost 2 meters.[43] Climatologists Robert DeConto and David Pollard write: "Antarctica has the potential to contribute more than a metre to sea-level rise by 2100 and more than 15 metres by 2500, if emissions continue unabated."[44]

Just as worrisome is the possibility that if the Greenland ice cap melted, it might trigger another sudden cooling of the Northern Hemisphere.

Around 12,000 years ago, the average temperature in Greenland dropped by 14° to 17°C (25° to 30°F), while glaciers advanced over Europe and North America. The cause of that sudden freeze is thought to have been the melting of the Laurentide ice sheet that covered much of Canada, causing enormous amounts of freshwater to flow down the St. Lawrence River and into Hudson Bay, shutting down the Gulf Stream.

The Gulf Stream, an ocean current one hundred times greater than the Amazon, carries warm water from the Gulf of Mexico northeastward toward Europe, warming the westerly winds. As it approaches Europe, the water cools and, being saltier, hence denser than the surrounding North Atlantic, it sinks. As it sinks, it pulls yet more warm salty tropical seawater behind it. Meanwhile, at the bottom of the Atlantic, the colder water flows south, forming what oceanographers call the Great Conveyor Belt.

If global warming were to cause the Greenland ice sheet to melt, the sudden increase in freshwater flowing into the North Atlantic would dilute the salty water of the Gulf Stream, preventing it from sinking, thereby shutting down the Conveyor Belt. Without a constant flow of warm water, the climate of Europe would become much colder and North America would also experience some cooling. Meanwhile, the rest of the world would become warmer, so that the difference between the climates of the tropics and of the higher latitudes would increase. Africa, Asia, and the Middle East might become drier, with stronger winds carrying more dust. Sea level would rise 5 or 6 meters, flooding most of the world's seaside cities. There is some evidence that this is beginning to happen, but it is not clear how significant this trend is.[45]

Could such a disaster be brought on by global warming? Among climate scientists, this possibility is a matter of debate. According to Richard Alley, "There is a significant possibility that greenhouse warming could trigger enough extra rain, snow, and ice-sheet melting to partially or completely shut down the north Atlantic conveyor circulation."[46] However, the IPCC considers such a scenario to be very unlikely, as do scientists like Curt Steger and Tim Flannery.[47]

Methane and Permafrost

Methane is many times more potent as a greenhouse gas than carbon dioxide. So far, most of it has come from rotting vegetation in the tropics and from livestock. But the possibility of a major release of methane has climate experts worried. The reason is that enormous quantities of methane are trapped in slushy sediment at the bottom of oceans and in permafrost in the Arctic, covering about one-quarter of the Northern Hemisphere. These methane bubbles formed millions of years ago when organic matter was digested by archaea able to live without oxygen.

Should the permafrost begin to melt, as is beginning to happen in Siberia and other parts of the far north, the methane it contains would be released into the atmosphere, adding between 10 and 35 percent more greenhouse gases than humans produce. This would accelerate global warming, which would melt more permafrost, and cause an unending loop. So far, this scenario is described as a "wild card" or a "ticking time bomb."[48]

Amazonia

The third wild card is the future of the Amazon rainforest. To thrive, trees require water and carbon dioxide. The water that falls on the eastern part of the forest comes from the South Atlantic Ocean. As the trees transpire, the water they release forms clouds that winds carry farther west, where it falls as rain, is transpired, forms clouds, and rains yet farther west, again and again. A climate model that takes vegetation into account has shown that rising temperatures will reduce plants' transpiration, thereby reducing rainfall over much of the forest. By 2100, the model predicts, basin-wide rainfall will drop from 5 millimeters to 2 millimeters per day, and to almost zero in northeastern Amazonia. Should that happen, the trees would die, and forest fires and the decomposition of the soil would release yet more CO_2, hastening global warming in another positive feedback loop. If the model is correct, by 2100 the Amazon forest would be reduced from its current 80 percent cover to 10 percent. What had been the world's largest and richest rainforest would turn into a dry savanna, except for areas that would become deserts.[49]

POLITICS AND CLIMATE WARS

Given the near-unanimity of the scientific consensus, the urgency of their conclusions, and the dangers to the planet that they portend, it is not surprising that they have aroused political passions. The reactions have taken place on international, national, and personal levels.[50]

The Rio Conference of 1992

The first IPCC Assessment Report led the United Nations to call a Conference on Environment and Development, popularly called the Earth Summit, that was held in June 1992 in Rio de Janeiro. It was attended by delegates of more than 170 nations, over 100 of them represented by their heads of government or heads of state, as well as thousands of activists, representatives of environmental NGOs and women's organizations, business leaders, religious figures, and journalists. It was the largest international conference in history.[51]

The goal of the conference was to discuss three major global problems—deforestation, loss of biodiversity, and climate change—but climate change dominated the proceedings and the media reports. On that subject, the world's nations were split into factions.

The delegates of the poorer countries pointed out that 80 percent of the world's resources were consumed by 20 percent of its people living in the industrialized world. They contrasted the "luxury emissions" produced by the gas-guzzling autos of the rich with the "survival emissions" of the poor. At the time, the United States emitted 4 to 5 tons of carbon per person per year, compared to 0.6 tons in China and 0.2 tons in India. The developing nations' delegates saw restrictions on greenhouse gas emissions as an obstacle to their development and asked why they should pay for the sins of the rich countries. As Dr. Mark Mwandosya, representative of Tanzania, put it: "Very many of us are struggling to attain a decent standard of living for our peoples and yet we are constantly told that we must share in the effort to reduce emissions so that industrialized countries can continue to enjoy the benefits of their wasteful life style."[52]

At first, United States president George H. W. Bush refused to attend the conference, torn between the advice of his Council of Economic Advisors and of the Environmental Protection Agency. But then he reluctantly decided to go. While European nations were eager to set timetables and targets for greenhouse gas reductions, the United States was opposed. Once in Rio, President Bush, under pressure from conservatives and the oil and gas industries, instructed the American delegation to dilute or block most diplomatic initiatives.

The document that concluded the conference, called the United Nations Framework Convention on Climate Change, was signed by 154 nations including the United States. Under the terms of that agreement, which was not legally binding, the rich countries pledged to reduce their greenhouse gas emissions to the 1990 level by the year 2000, a wildly optimistic promise. Developing countries agreed only to monitor their emissions.[53]

Kyoto and After

The next major international conference, held in Kyoto, Japan, in 1997, was called in response to the IPCC Second Assessment Report of 1996. Unlike the Rio conference, where the participants made promises they could not keep, the goals of the Kyoto conference were to agree on binding targets for greenhouse gas emissions and devise mechanisms to implement them.

Once again, the world split into warring camps. Developing countries, led by India, China, and Brazil, refused to accept binding commitments because they were about to embark on a period of extraordinary economic growth. As at Rio, they blamed the rich countries for having caused global warming by burning coal and oil and proposed that the rich should be the first to take legally binding steps to reduce their emissions. Countries like the Seychelles that were about to be swamped by rising oceans demanded compensation from those who, they claimed, had caused the problem.[54]

As at Rio, the United States stood against the demands of the poorer countries. It insisted that all countries should share the burden of reducing emissions, especially since the larger Third World countries were expected to produce more emissions than the rich countries within the next twenty years. Furthermore, restricting emissions would harm the US economy, which was heavily dependent on fossil fuels.

The result of the conference, called the Kyoto Protocol, was a very watered-down version of the expectations of the organizers and the scientists of the IPCC. China and India accepted no restrictions of their emissions. Australia and Saudi Arabia expressed reservations. The European Union agreed to reduce its emissions by 8 percent, the United States by 7 percent, and Japan and Canada by 6 percent. However, even before the conference, the United

States Senate had passed a unanimous resolution to reject any agreement that did not require developing countries to reduce their emissions too. Though President Bill Clinton signed the Kyoto Protocol, neither he nor his successor George W. Bush ever submitted it to the Senate.

The Kyoto Protocol came into force (for the signatories) in 2005 when Russia signed, but Japan refused to sign for a second term because the restrictions did not apply to its chief rival, China. Canada, which became a major fossil fuel producer thanks to the exploitation of the oil sands of Alberta, withdrew in 2011. Global carbon emissions, far from declining to 1990 levels, rose from 6 gigatons in 1990 to 8.5 gigatons in 2007, a 40 percent increase, exceeding the IPCC's worst-case scenario.[55] In short, the Kyoto Protocol was an expression of good intentions but little else.[56]

As the scientific prognosis grew darker and public concerns mounted, the governments of the world responded with ever more frequent conferences. In December 2009, a third major international conference was held in Copenhagen, attended by delegates of 193 nations. China announced that it was not responsible for the global warming that was caused by the industrialized nations but offered to cut its carbon dioxide emissions per unit of economic growth by 2020 to 40–45 percent below their 2005 level; this meant that if its economy grew (almost a certainty), its emissions could grow as well, just not as fast as before. Of the 193 nations that attended, 138 signed or pledged to sign a non-binding agreement, but no treaty was forthcoming. The media called this conference a major fiasco.[57]

A year later, another conference, this one at Cancún, Mexico, achieved a more positive result, namely a pledge to prevent average global temperatures from rising more than 2°C (3.6°F) above pre-industrial levels. But it did not require that the signatory nations adopt the technological or economic changes that scientists felt necessary to avoid dangerous climate changes.[58] At yet another conference, held in Warsaw in 2013 in response to demands by representatives of small island states threatened by a rising sea level, the US State Department representative on climate issues Todd D. Stern warned: "Lectures about compensation, reparations and the like will produce nothing but antipathy among developed country policy makers and their publics."[59]

The 2015 U.N. Climate Change Conference in Paris aimed to limit global warming to 2°C or less compared to pre-industrial levels by cutting greenhouse gas emissions to zero in the second half of the twenty-first century. Most of the governments represented at the conference accepted this goal in principle. However, since there is no binding enforcement mechanism, its implementation is far from certain. Compromising this agreement is the dramatic drop in fossil fuel prices, triggering a rise in consumption, hence emissions. Yet there is hope. In early 2015 a report by the International Energy Agency showed that emissions in 2014 had remained constant while the world economy had continued to grow—the first time such a decoupling between emissions and growth had been noted.[60] As of 2017, thanks to an increase in the use of natural gas in place of coal and a decline in the price of solar and wind energy, greenhouse gas emissions remained stable despite a growing world economy. Whether this was the beginning of a hopeful trend or only a blip remains to be seen.

The reasons for the disparity between the pious resolutions that issued from international conferences and the actual consequences of human activities are many. International agreements do not translate into national policies. National policies do not translate easily into the behavior of businesses and individuals. And the goal of reducing greenhouse gas emissions without hindering economic growth depends on technological innovations, some of which are still on the drawing board while others—for example, solar panels and electric cars—are spreading only gradually.

National Politics

The idea of global warming, its connections to fossil fuel consumption, and the scientists' predictions provoked political reactions in every country, but none so dramatic as in the United States. "An Inconvenient Truth," a film by former vice-president Al Gore, presented a strong message about the dangers of global warming. Meanwhile, Cassandras announced the coming apocalypse in the most dire terms. Thus, according to the Australian philosopher and public intellectual Clive Hamilton, "The kind of climate that has allowed civilization to flourish will be gone and humans will enter a long struggle just to survive."[61]

On the opposite side are the deniers, or, as they prefer to call themselves, the skeptics. Some agree that global warming is real but claim it is a natural phenomenon in which humans have played no part; S. Fred Singer and Dennis T. Avery's *Unstoppable Global Warming: Every 1,500 Years* is an example of this genre.[62] Then there is a large and growing literature claiming that global warming is an elaborate hoax fomented by a cabal of scientists and leftists, with titles such as *Red Hot Lies: How Global Warming Alarmists Use Threats, Fraud, and Deception to Keep You Misinformed; Power Grab: How Obama's Green Policies Will Steal Your Freedom and Bankrupt America; Eco-Tyranny: How the Left's Green Agenda Will Dismantle America*; and *The Real Global Warming Disaster: Is the Obsession with "Climate Change" Turning Out to Be the Most Costly Scientific Blunder in History?*[63]

In the United States, global warming denial is more than a literary genre; it is a political stance, with Democrats generally claiming concern for the environment and Republicans reflecting the interests of the fossil fuel industries. Under the administrations of Ronald Reagan, George H. W. Bush, and George W. Bush, scientists working for the Environmental Protection Agency and the National Oceanic and Atmospheric Administration complained that their work was interfered with and their reports altered by politically appointed administrators.[64] During the 2011 presidential campaign, Republican candidates Ron Paul and Rick Perry called climate change "a hoax," while front-runner Mitt Romney said: "My view is that we don't know what's causing climate change on this planet and the idea of spending trillions and trillions of dollars to try and reduce CO_2 emissions is not the right cause for us."[65] That year, the US House of Representatives defeated the proposed amendment that stated: "Climate change is occurring, is largely caused by human activities, and poses significant risks for public health and welfare."[66]

In 2017, President Donald Trump withdrew the United Status from the Paris agreement on climate, to the dismay of the European Union and even China, which had belatedly taken a leadership role in climate awareness.

Science on the Defensive

Things came to a head during the so-called hockey stick controversy. The expression was first used by climate researchers Michael Mann and others in scientific journals in which they argued that global temperatures, which had been rising slowly for a long time, suddenly began rising fast in the late twentieth century, like the curve of an ice-hockey stick. In fact, the same pattern—a slow rise followed by a sharp increase in the late twentieth century—is also evident in energy use, in CO_2 concentration, in sea level rise, in Arctic sea ice shrinkage, and in glacial melting.[67] A flurry of books and articles followed, pitting global warming "skeptics" against Michael Mann in particular and climate scientists in general. Conservative media denounced prominent scientists and the IPCC. Scientists, however, were ill-prepared for a public-relations battle and were woefully under-represented in Congress.[68]

The attack on climatologists was part of a larger questioning of science that had been taking place, especially in the United States, over the past decades. Science, in the public mind, had long been associated with technological innovations: vaccines and antibiotics, radar, rockets and space exploration, synthetic rubber, plastics and other useful chemicals, computers and satellites, and other "miracles." By the end of the century, however, science was being attacked from various angles. Christian fundamentalists objected to the teaching of evolution and to stem-cell research. Environmentalists denounced the chemical industries for poisoning the land and waters. Nuclear reactor accidents at Three Mile Island (1979), Chernobyl (1986), and Fukushima Daiichi (2011) provoked a revulsion against what was once believed to be a safe and environmentally benign source of energy. Genetically modified crops raised questions in the United States and were banned outright in the European Union. Meanwhile, despite years of effort and millions of dollars of funding, major scientific and medical problems—cancer, heart disease, AIDS, fusion energy—remained unsolved. In this atmosphere of skepticism and doubt, climate science, with its gloomy predictions and calls to restrict the pleasures and benefits of burning fossil fuels, was a target for science skeptics of various persuasions.[69] Climate scientist William Ruddiman complained: "I know of no precedent in science for the kind of day-to-day onslaughts and perversions of basic science now occurring in newsletters and web sites from interest groups. These attacks have more in common with the seamier side of politics than with the normal methods of science. Both the environmentalists and (especially) the industry extremists should leave the scientific process alone."[70]

The Climate Lobbies

In climate politics as in all other areas of politics, important economic interests were at play. And these interests influenced politicians through well-funded lobbies. In the United States as early as the 1980s, Peabody Energy, the world's largest coal company, backed the production of the video "The Greening of Planet Earth" that claimed that carbon dioxide would raise agricultural yields, thereby ending world hunger. The Exxon Mobil Corporation and other fossil fuel companies funded several Washington think tanks such as the George C. Marshall Institute. Coal, oil, gas, chemical, and automobile companies financed the Global Climate Coalition, the Partnership for Climate Action, the Heritage Foundation, the Competitive Enterprise Institute, and other think tanks as well as fake citizens action groups such as the Advancement of Sound Science Coalition. What these lobbyists achieved is to make people believe that scientists disagreed about global warming (when they do not), leaving the public confused.[71]

By 2006, most fossil fuel companies had stopped denying the existence of global warming but argued instead that there was no hurry and no changes would be needed for another twenty years. The spokesperson, Patrick Michaels, a senior research fellow at the libertarian Cato Institute, admitted that "human-induced climate change is indeed real, but that this will not lead to an environmental apocalypse," and "there is plenty of time—a century or so—for technological development that will be more efficient and emit far less carbon dioxide."[72]

By lobbying against climate science and supporting conservative candidates, these companies succeeded in postponing any meaningful action and casting the United States in the role of obfuscator at international conferences. In the words of Australian environmental scientist Tim Flannery: "It is impossible to overestimate the role these [American energy] industries have played over the past two decades in preventing the world from taking serious action to combat climate change."[73]

Global Warming and the Public

Think tanks, lobbyists, conservative media, and politicians do influence public opinion, but only in the direction in which the public is leaning anyway. In the United States, that direction has been toward increasing skepticism. A poll taken in 2006 showed that 92 percent of Americans had heard of global warming, 90 percent believed that the United States should reduce greenhouse gas emissions, and 70 percent believed that it should do so regardless of what other nations did. In April 2008, only 71 percent of Americans believed there was solid evidence of global warming and 47 percent believed it was due to human activities. And by October 2009, only 57 percent believed in global warming and 36 percent blamed it on humans. When asked whether global warming was a serious problem, in

2008 44 percent of the people interviewed agreed that it was, but by 2009, only 35 percent did so, and Americans ranked global warming as last in a list of twenty priorities. In other words, they were experiencing global warming fatigue.[74]

As American politics have become more polarized, so have Americans' views on global warming. In 1997, there were only slight differences between Republicans and Democrats on the subject, with 48 percent of Republicans and 52 percent of Democrats expressing concern. By 2008, however, the gap had grown by 34 percentage points, with 42 percent of Republicans and 76 percent of Democrats expressing concern.[75]

The American Way of Life

The recession of 2008–2011 did reduce demand for fossil fuels, but once the recession lifted, Americans went back to buying big cars and houses, heating and air conditioning their buildings, and otherwise consuming energy. After all, cheap energy is the very foundation of the American way of life. As Ari Fleischer, press secretary for President George W. Bush, put it: "The President believes that [energy use is] an American way of life and that it should be the goal of policy makers to protect the American way of life. The American way of life is a blessed one."[76] In a society in which people are judged by their consumption of goods and services and in which consumption correlates closely with the consumption of fossil fuels, reducing one's carbon footprint is perceived as reducing one's standard of living and even threatening one's identity.

Soft Denial

On a personal level, reducing greenhouse gas emissions is fraught with difficulties. Even people who believe that global warming is real are poorly motivated to do something about it. This attitude of "soft denial" is not a matter of ignorance. As sociologist Kari Norgaard found: "Respondents who are better informed about climate change feel less rather than more responsible for it.... People stopped paying attention to global climate change when they realized that there is no easy solution for it."[77]

In May 2014, the Yale program on Climate Change Communication released the results of a survey that showed that people were more concerned about the climate when it was referred to as "global warming" than when it was called "climate change," because the former suggests an increase in extreme weather, while the latter makes people think of natural weather fluctuations.[78]

Unsurprisingly, people have ambivalent feelings about global warming. After all, humans evolved in a tropical African environment. Most people who live in the temperate zone prefer warm weather to cold, and spring and summer to fall and winter. Many take winter vacations in warmer places, and many others move to warmer regions when they retire. Some even celebrate the blessings of global warming. As cultural historian Wolfgang Behringer put it: "The earth will continue to grow warmer even if every country behaves in model fashion and dramatically reduces its waste gases. . . . But it is not as bad as the older predictions of an imminent ice age: cooling has always resulted in major social upheavals, whereas warming has sometimes led to a blossoming of culture."[79]

Then there is the technology to which humans are committed, such as the automobile. Most Americans not only have automobiles, but they also depend on them. Outside of a few big cities, it is no easy matter to shop for groceries on foot or to take mass transit to work. It will be a decade or two before hybrid and electric vehicles could replace the millions of gasoline-powered cars on the road today, even if consumers agree to this. Nor is the United States unique in this regard. Europeans and Japanese also drive cars, albeit smaller ones. The Chinese are quickly catching up. And other countries in Asia and Latin America are joining the new modern lifestyle, with its traffic jams, air pollution, and greenhouse gas emissions.

While the United States waffles in its response to climate change, India faces an intractable dilemma. On the one hand, air pollution in Delhi and other Indian cities is among the worst in the world; furthermore, India stands to suffer more than other countries from global warming, droughts, and the shrinking of Himalayan glaciers. On the other hand, India cannot develop its economy and lift its people out of poverty without relying, for years to come, on burning its abundant and cheap coal supplies.[80]

Meanwhile, China is also playing both sides of the climate game. Its government strongly supported the 2015 Paris climate accord and its National Energy Administration plans to spend $360 billion through 2020 on solar and wind power to complement—but not to replace—coal-fired power plants. Meanwhile, Chinese companies are responsible for 700 of the 1,600 new coal-fired power plants under construction or being planned worldwide.

Once built, these plants will expand the world's coal-fired plant capacity by 43 percent, making it virtually impossible to meet the goals set in the Paris agreement.[81]

Then there is the awkward problem of timing. There is a considerable delay between the emission of greenhouse gases and the resulting global warming. The consequences of today's global gas emissions may not be fully felt for decades. Likewise, cutting back on emissions today may not reverse the global warming trend for centuries or even millennia. A well-meaning person's sacrifices for the common good will have only an infinitesimal impact on the planet, and that in the distant future. In Ruddiman's words: "The benefits from the Kyoto reductions would occur decades in the future and even then would be undetectably small in people's lives, but the costs would be felt right away in higher prices in our daily lives. Politicians tend to favor actions that work the other way: benefits that are immediate and costs that lie far off in the future."[82] So does the public.

Hope and Blame

Meanwhile, there is the hope, even the expectation, of a brighter future. Since the eighteenth century, people in the West have come to believe in progress, meaning technological innovations and a greater per capita consumption of goods and services. So have communists, even (perhaps especially) Chinese Communists. And economists, with few exceptions, have set growth as the goal of their policy recommendations. The mass consumption society that originated in the United States has spread around the world, with "development" in poorer countries as the analog of "growth" in the richer ones. Economic growth has become a fetish. Most people reconcile economic growth with the effects of greenhouse gas emissions on a limited planet by hoping for technological breakthroughs that will somehow save the world from the ill effects of previous generations of technological innovations. Hence, ideas such as "clean coal," carbon capture and storage, even geo-engineering the planet have become popular.

As the performance of governments in international conferences makes clear, there is always blame-shifting. China, India, and other developing countries can (rightly) point to the industrialized West as the first and greatest polluters of the Earth, producing at one time three-quarters of global emissions. As they pointed out at Kyoto and other conferences, why should they restrain their long-overdue economic development to compensate for the past sins of the rich countries?

Meanwhile, the vast and fast-growing Chinese middle class, finally freed from centuries of poverty, is indulging in a Western-style orgy of mass consumption that demands an equally massive production of fossil fuels. As a result, China has recently surpassed the United States in greenhouse gas emissions. Why should Americans take painful measures to save the planet if the Chinese and other "emerging markets" won't do their share?[83]

Conclusion

We live on a planet with a volatile climate system. The benign Holocene of the past 10,000 years, during which civilizations evolved and to which we have become accustomed, has been replaced by the more dangerous Anthropocene. In bringing this about, humans have had much to answer for: first farmers and herders, then industries, then a mass consumption society that devours fossil fuels.

These days, as scientists continue to find evidence of anthropogenic climate change, politicians engage in debates, attend conferences, and pass resolutions. Meanwhile people pursue, as always, a better life for themselves. Human beings, who are used to seeking immediate benefits, ignoring future problems, and blaming others, are ill-equipped to confront massive long-range problems on a planetary scale.

13

Plundering the Oceans

The seas and oceans are so huge and their denizens so numerous that humans believed they could not possibly make a dent in their abundance. In his work *Philosophie zoologique* published in 1809, the French zoologist Jean-Baptiste Lamarck wrote: "Animals living in the waters, especially the sea waters, are protected from the destruction of their species by Man. Their multiplication is so rapid and their means of evading pursuit or traps are so great that there is no likelihood of his being able to destroy the entire species of these animals."[1]

Eighty years later, in his inaugural address at a conference on fishery, the British biologist Thomas Huxley said: "I believe . . . that the cod fishery, the herring fishery, the pilchard fishery, the mackerel fishery and probably all the great sea fisheries are inexhaustible; that is to say, that nothing we do seriously affects the number of fish. And any attempt to regulate the fisheries seems consequently, from the nature of the case, to be useless."[2]

In 1912, the French marine biologist Marcel Hérubel wrote: "The sea is inexhaustible, and there can never be a general and simultaneous depopulation. The ocean fisheries will always be copious and easy, and their yield will be greater as the ocean becomes more familiar and the methods more perfect."[3]

And as late as 1950, marine biologist and ecologist Rachel Carson wrote in her book *The Sea Around Us*: "[Man] has returned to the mother sea only on her own terms. He cannot control or change the ocean as, in his brief tenancy of Earth, he has subdued and plundered the continents."[4]

Since no one knew then how many fish and whales there were in the sea, people estimated their abundance by the amount caught. As the global catch kept increasing, it seemed logical to deduce that the quantities left in the seas and oceans were "copious" and "inexhaustible." Indeed, the catch kept rising, from about 4 million tons in 1900 to roughly 20 million tons in 1940 and to 80–90 million tons in the 1980s.

Yet the amount caught reflected only the increased power of industrial technology applied to fishing and whaling. After 1980, the fish catch peaked and began to decline. By 2000 it had fallen to about70 million tons. And in the 2010s, the United Nations Food and Agriculture Organization reported that two-thirds of the species fished since the 1950s had experienced collapse, and the rate of depletion was accelerating.[5]

The variety of species in the seas and oceans of the world is enormous, and their fates vary just as enormously. This chapter will concentrate on three iconic forms of marine life: one, whales, that was nearly depleted but then given a new lease on life; another, cod, that was almost extinguished; and a third, salmon, that was a great success, but only on human terms.

Whales

Long before sailors went out to sea to hunt whales, people who lived near the coasts occasionally happened upon beached animals that provided food for many people for many weeks. Medieval Dutch and Norse literature mentions beached whales but not whale hunting. The Inuit of Greenland may have been the first to venture offshore to hunt whales; killing 120 to 130 whales a year—mostly immature yearlings—provided most of their meat and blubber but did not affect the whale population as a whole. In the Middle Ages, the Basques also practiced whaling in small rowboats in the Bay of Biscay, killing on average one whale a year per fishing village.[6]

Some whales migrated in herds that followed regular routes near coastlines. When they came to the surface to breathe, they were fairly easy to kill. Right whales (*Eubalaena glacialis*) were buoyant when dead and could be towed to shore. Whales proved very useful in many ways. Their blubber, when boiled down, produced oil for lamps. Whale oil could also be used for lubrication and to make soap, varnish, and paint. Such was the demand for whale oil in Europe that whaling became a major industry.[7]

Pre-Industrial Whaling

Deep-sea whaling and fishing evolved slowly, along with the development of oceangoing ships. After the Basques had depleted the whale population in the Bay of Biscay, they began to venture into the open ocean. In the early sixteenth century, they set out in ships of around 250 tons with a crew of fifty, first to the North Sea and off Iceland, then off Labrador. By the 1530s, Basque whalers were killing whales off Newfoundland, Labrador, and Greenland and bringing the blubber to seasonal camps in Newfoundland to render it into oil.

In the early seventeenth century, explorers searching for a northwest passage to China discovered rich whaling grounds off Svalbard in the Arctic Ocean north of Norway. In the many bays and inlets of the island, bowhead whales (*Balaena mysticetus*) sought shelter to give birth and raise their calves. Basque whalers were soon joined, then pushed out, by English, Dutch, Danish, Norwegian, and German whalers. Altogether, they killed between 300 and 450 whales a year. That bounty lasted three decades. By the 1670s, having depleted the waters off Spitsbergen, whalers moved on to the Davis Straits west of Greenland and to Baffin Bay in northern Canada. By the end of the eighteenth century, whaling in the North Atlantic and Arctic oceans had petered out.[8]

Meanwhile, the Portuguese had discovered humpback (*Megaptera novaeangliae*) and southern right (*Eubalaena australis*) whales off the coast of Bahia in Brazil, where females came close to shore to calve. When whalers on shore spotted whales, they would jump into their boats and harpoon the newborn calves. As the mothers rushed in to protect their calves, the whalers would harpoon the mothers, then tow them to shore, cut the blubber into strips and render it in cast-iron pots. During the sugarcane harvest, whale oil was used to light the mills and lubricate the cane-crushing machinery that operated day and night. Whale meat was fed to slaves and ships' crews. In 1614, the governor of Brazil declared whaling a royal monopoly and gave exclusive contracts to a few whalers. As a result, the whaling industry grew slowly. By 1770, a factory at Santa Catarina Island was processing some 500 whales a year, a sustainable harvest.[9]

By then, however, New England whalers had appeared off the shores of Brazil. A hundred years earlier, as the number of humpback and right whales off the New England coast shrank, whalers from Nantucket had begun venturing farther out to sea. In 1712, one ship encountered sperm whales (*Physeter macrocephalus*) in deep water off New England. Not only did whales of this species provide more and better oil, but they also contained spermaceti, a wax-like substance that could be used to make high-quality candles. To catch these whales that stayed far from shore, New England mariners built larger ships equipped with fireplaces and iron pots in which they could render the blubber on board and store the oil in casks. By 1774, 360 whaling ships sailed from Nantucket, New Bedford, and other New England ports. Finding too few sperm whales in the North Atlantic, they began sailing south, to the

Azores, West Africa, the Falkland Islands, and Brazil. For a time, whaling was risky because the Portuguese captured American whalers, as did the British during the American War of Independence. Once peace with Britain was signed in 1783, New England whalers returned to the South Atlantic. Between 1804 and 1817, they killed and processed 193,522 whales at sea, probably more than the Portuguese had in two centuries of whaling. In 1823, one ship returned to port with 2,600 barrels of oil taken from 170 whales.[10]

Three centuries of whaling had a dramatic impact on the whale population of the Atlantic. Gray whales (*Eschrichtius robustus*) disappeared from that ocean in the seventeenth century. In the mid-eighteenth century, humpback, right, and bowhead whale populations practically vanished from North Atlantic and Arctic waters. By the early nineteenth century, they and sperm whales were becoming rare in the South Atlantic as well. Five other species—blue, fin, sei, Bryde's, and minke whales—survived by keeping their distance from whaling ships.[11] By the late nineteenth century, after having severely reduced the whale population of the Atlantic, whalers from Europe and North America turned to the Pacific Ocean.

By targeting the most vulnerable whales—calves and females—whalers not only reduced the current whale population but also diminished its ability to reproduce. The parallel with the extermination of large land-based mammals thousands of years earlier is striking. If men in rowboats armed with hand-held harpoons could deplete an ocean of whales in 300 years, then it is not surprising that men with spears were able to empty whole continents of megafauna, one animal at a time.

Industrial Whaling

Industrial methods accelerated whale hunting dramatically. In 1865, the Norwegian Svend Foyn invented an explosive harpoon fired from a cannon that allowed whalers to hunt far-off prey. He also introduced the first steam-powered whaling ship. The newly equipped whalers' first targets were blue whales (*Balaenopterus musculus*); at up to 30 meters in length and weighing up to 170 tons, they were the largest animals that have ever lived on the planet. When blue whales were gone in a part of the ocean, whalers turned to fin, humpback, and sei whales. As these whales sank when killed, Foyn also invented a means of keeping them afloat by injecting them with compressed air until they could be towed to shore or to a nearby factory ship to be butchered. By 1896, whalers were killing 2,000 whales a year.[12]

During World War I, the British barred German whaling ships from the oceans; they also barred Norwegian ships, though ostensibly neutral, to prevent them from supplying Germany with whale oil, a key ingredient in the manufacture of dynamite and margarine. Meanwhile, German submarines made it too risky to hunt whales in the North Atlantic. As a result, British whalers could hunt only in the South Atlantic and off the coast of Antarctica. The total world catch, which had been 22,900 whales in 1913–1914, dropped to 9,468 in 1919–1920 before rising again.[13]

After the war, whalers returned, seeking oil for the manufacture of margarine, soap, and pharmaceuticals. Besides the longtime American, British, and Norwegian whalers, the Japanese brought increasing competition to the industry. Technological innovations in 1925 introduced factory ships with stern slipways that would rendezvous with diesel-powered catcher boats and winch an entire whale carcass aboard in order to process it at sea; the days of heroic whalers in rowboats hurling harpoons were over. With these ships, whalers could stay at sea for months at a time and hunt in the frigid waters surrounding Antarctica. Whalers harvested 29,649 blue whales in the 1930–1931 season alone. In 1938, 45,010 whales of all species were caught off Antarctica by Norwegian, British, and German fleets. The Japanese, as part of their mobilization for war in the 1930s, increased their proportion of the world's catch from 1 percent in 1930 to 12 percent in 1938.

As soon as the Second World War broke out, Great Britain once again prevented all other nations' ships from hunting whales in the Atlantic. In April 1940, when its troops briefly occupied Norway, they seized all the Norwegian stocks of whale oil and shipped them to Britain. Whaling ships were turned into tankers, cargo ships, minesweepers, and other naval vessels. Many were lost to submarines, and only a few returned to whaling after the war. During wartime, whaling continued in the South Pacific off Peru, the only part of the oceans free of submarines. As a result, the number of whales killed in 1942–1943 dropped to 8,390 (998 of them off Antarctica), before rebounding to 13,387 in 1945–1946.[14]

The Aftermath

After World War II, the nations that had endured the war were desperate for oils and fats for soap, margarine, and other consumer products. Encouraged by their governments, whalers from Great Britain, Norway, Japan, and the Soviet Union went hunting with renewed vigor. Japanese and Soviet whalers, newcomers to the industry, found their prey in the Pacific and Antarctic oceans.[15]

The International Convention for the Regulation of Whaling, signed in 1946, was intended not so much to protect the animals as to allocate the surviving whales to the various whaling nations. The result was to encourage whalers to catch as many whales as possible in order to get a full cargo before the international quota was reached, even whales that had not had a chance to fatten up during the feeding season.[16] In the words of whaling historian Gordon Jackson: "So long as there was a demand for oil, so long as whales were to be found in regular locations, and so long as they were caught by a system of free competitive enterprise, there was no hope whatsoever."[17]

By the 1970s, as the number of blue whales dropped to a few hundred, an anti-whaling movement arose in the Western public and press. In 1972, the United Nations passed a resolution calling for the total cessation of whaling. The International Whaling Commission, founded in 1949, voted a moratorium on whaling in 1982. Faced with worldwide condemnation and a decline in profits, the Soviet Union gave up whaling in 1987, followed by Japan a year later.

Whaling was still allowed for aboriginal whale hunters in Alaska, Canada, and Siberia, however. Another exception was made for whaling for scientific research. The Japanese government interpreted this broadly, for the Japanese are the only non-aboriginal people who regularly eat whale meat. Though a member of the International Whaling Commission, Japan objected in 1981 to a ban on the use of explosive harpoons and to a moratorium whaling. Thus in 2000, Japanese whalers killed 440 minke whales in Antarctic waters and 50 minke and 50 sei whales in the North Pacific. Norway also found "scientific research" to be a convenient loophole and harvested 552 minke whales in 2001.[18]

Altogether in the course of the twentieth century, whalers killed an estimated 2.9 million whales. As the Atlantic whale population had been decimated in the nineteenth century, 90 percent of those killed between 1900 and 1999 were found in the Pacific Ocean and around Antarctica.[19]

As a result of the moratorium, Southern Hemisphere right and humpback whales and Pacific gray whales have made a comeback. In the Atlantic, whales are struggling to recover. The right whales are the most endangered, as their numbers are declining; by a recent count, only 300 to 350 are left in the North Atlantic and 500 in the North Pacific Ocean. Those numbers are so small, and the reproduction rate of these giant animals so low, that they may not avoid eventual extinction. Other species are better off; as of 2018, there were an estimated 10,000 to 25,000 blue whales, 26,000 gray whales, 100,000 orca or killer whales (*Orcinus orca*), 60,000 humpback whales, 515,000 minke whales, 100,000 fin whales, and 100,000 belugas (*Delphinapterus lencas*).[20]

While the moratorium has been (partially) successful, another threat has appeared: the decline in the krill population in Antarctic waters. Krill (*Euphausia superba*) are tiny shrimp-like creatures that feed on the plankton that grows abundantly in those frigid waters, especially below the ice. They, in turn, are the preferred food for many animals, including baleen whales (blue, humpback, bowhead, minke, and gray) that eat them by the ton, as well as fish and the penguins and seals that eat the fish. In recent decades, as climate change has warmed the Antarctic Peninsula and sea ice has shrunk, the krill population has been in steep decline. Though penguins are the first to suffer, the whale population is also affected by the growing scarcity of food.[21]

Cod and Other Fish

Whales were not the only oceanic animal to lure humans onto the deep sea. Long before Columbus began importuning the monarchs of Spain with his schemes of sailing to the fabled lands of the East, humble fishermen were taking their small ships west into the Atlantic Ocean in pursuit of cod.

As they never stop growing, some old cod reach almost 2 meters in length and 100 kilograms in weight. In the Middle Ages, 1-meter-long cod were common. In the 1960s, the average cod weighed 4.5 to 9 kilograms, but as the larger ones were fished out, the average declined to less than 1.7 kilograms.

Cod migrate in dense schools that come together to spawn in the late winter and early spring in waters less than 55 meters deep where cold and warm currents meet, nourishing the plankton, krill, herring, capelin, and squid that cod consume. Two such spawning grounds lie between Norway and the Lofoten Islands north of the Arctic Circle, and the Dogger Bank in the North Sea between England and Denmark. The richest were the Grand Banks east of Newfoundland and the Georges Bank off of New England where the North American continental shelf extends into the Atlantic Ocean and where the Labrador Current brings upwelling nutrients.[22]

Pre-Industrial Cod Fishing

Cod are among the fish most prized by humans because their tender and flaky white flesh contains almost no fat. When dried and salted, cod keep well without spoiling. Since the Middle Ages, codfish was especially appealing to Christians forbidden to eat meat on Fridays and Saturdays and during Lent. From the twelfth century on, Norse fishing boats visited the Lofotens to gut and dry their catch in the freezing wind. Later, Basques and others also entered the trade in dried and salted cod.[23]

Signs of overfishing in European waters appeared in the late Middle Ages, encouraging fishermen to venture ever further into the Atlantic.[24] There have long been rumors that fishermen from Europe may have fished off Newfoundland before 1492, but there is no evidence to confirm this. The first recorded cargo of Newfoundland cod was brought by the English ship *Gabriel* that returned to Bristol in 1502; it had reached the Grand Banks, the richest fishing ground in the world. News spread fast, and by 1510, Bretons, Normans, and Basques were sailing to Newfoundland and the Grand Banks. At first they came in small boats carrying fifteen to twenty men who caught, dressed, and salted cod before returning to Europe, making two or three trips per season. Later, in the seventeenth century, English fleets returned with 35,000 tons of dried cod a year on average. In the eighteenth century, increasing numbers of ships brought home on average four times that many each year. By then, fishermen had established year-round settlements on the Newfoundland coast. While some sailors went out in longboats to catch cod, others gutted and dried them on shore and sold them to ships that carried them to Spain and Portugal, then returned to England with wine and salt before setting out on their next trip across the Atlantic. The French preferred "wet" cod, not dried but heavily salted to preserve it.

From the mid-sixteenth to the mid-eighteenth centuries, over half of all the fish eaten in Europe was cod. The harvest of cod by Europeans grew to enormous size, from 47,000 tons a year in the seventeenth century to 100,000 tons a year by the late eighteenth century. Yet the number of cod the fleets caught kept growing. The supply seemed inexhaustible.[25]

Industrial Fishing in British Waters

The Industrial Revolution reached the British fishing trade in the 1860s, when steamers began replacing sailing vessels in the North Sea. Rather than casting lines with hooks, steamers pulled trawls, large nets held open by beams up to 15 meters long that were dragged along the sea floor, raking up bottom-dwelling fish like cod, haddock, and plaice, even shrimp and mollusks. On board the ships, ice was used to preserve the fish and railroads quickly transported them to markets all over Britain. Fish landings in Britain doubled from 50,000 tons in 1840 to 100,000 in 1860. As supplies increased, so did the British people's demand for fish and chips.[26]

Under such pressure, the North Sea fish stocks began showing signs of depletion as early as the 1890s. As the North Sea catches declined, British trawlers moved out into the high seas around Iceland in the mid-Atlantic and Spitsbergen in the Arctic Ocean where they could haul in ten to twenty times more fish per day than in the North Sea.

During the First World War, when submarines made fishing too dangerous, North Sea fish populations experienced a dramatic recovery. In the North Atlantic, only Icelandic fishing boats still went out for cod. But fishing fleets returned soon after the war was over, and new technologies increased the pressure on the fish stocks. Disks and rollers allowed trawling along rough bottoms where juvenile fish lived. By the late 1930s, British trawler fleets were landing up to 700,000 tons of fish a year.

The Second World War brought another reprieve for marine life. As the British transformed their trawlers into minesweepers and barred other Europeans from fishing, only Icelanders kept on fishing. When the war ended, fish stocks in the eastern North Atlantic were larger and healthier than they had been in decades. Following a resurgence from 1946 to 1965, the British catch reached 600,000 tons.

In the 1960s, fishing boats in England's coastal waters switched from drift nets with large holes to purse seines with small holes that could also catch small fish. As a result, the herring catch dropped precipitously, from 1.7 million tons in 1966 to 20,000 tons in 1970.[27] Herring were eaten not only by humans but also by cod, thereby denying the remaining cod population a chance to recover. Two-thirds of the major commercial fish species were declared "outside safe biological limits." The sea floor in British waters was so depleted by trawlers and dredgers that there was not enough food left for fish to survive and reproduce. The last British North Sea trawler was tied up in 2002.[28]

The story of fishing in the Firth of Clyde illustrates the impact of modern fishing methods on a marine ecosystem. This large fjord in western Scotland once abounded with fish, shellfish, porpoises, even whales. In the late nineteenth century, after steam trawlers brought the fisheries close to collapse, the government banned bottom fishing. The ban lasted a century and allowed the fish population to recover. Then, in 1984, under the government of Margaret Thatcher, the ban was lifted and trawlers returned, this time with heavy dredges that scraped the bottom, even rocky places. First herring disappeared, then cod, plaice, and sole. The few boats still active catch scallops and prawns, for there are no fish left in the Clyde.[29]

Fishing Off North America

Since the sixteenth century, European fishermen had known that the richest fish stocks—especially cod—were to be found off the coast of northern North America, from Cape Cod to Labrador. After World War II, a host of new technologies were brought to bear upon these fishing grounds. Lightweight polymer nets and mile-long monofilament lines allowed much larger catches than in the past. Factory trawlers, much larger than before the war and powered by huge diesel engines, were guided to the fish by sonar and spotter planes, legacies of wartime. Huge trawls were hauled up their stern ramps every four hours, day and night. The first of these trawlers was the British factory ship *Fairtry*, built in 1954; at 2,800 tons displacement, it was far larger than the largest prewar trawler. Fleets of trawlers came from Eastern Europe, Japan, and Taiwan, as well as from Britain, Spain, and other western European countries. The Newfoundland fleet, closest to the Grand Banks, grew from five in 1950 to sixty-one in 1976, subsidized by a government eager to support the largest industry in the province. In the 1960s, the Soviet Union, facing declining inland fish catches, built a fleet of oceangoing trawlers.[30] Soviet trawlers of 8,000 tons displacement could catch and haul in 100 tons of fish in an hour. By 1963 almost a thousand trawlers, many of them among the largest fishing vessels in the world, were operating in Atlantic waters. Not all the fish caught were desirable; between one-quarter and one-third of the fish hauled up were discarded as by-catch, while others were injured and died in the water. On board the factory ships were machines to slice up and quick-freeze fish and turn them into fish fillets and fish sticks, as well as equipment to collect the cod-liver oil and to transform the wastes into fish meal.[31]

The results were amazing, for a short time. Until the 1950s, the catch off Newfoundland had increased slowly up to 100,000 tons a year. Then came a huge jump to 800,000 tons in 1965. Between 1960 and 1975, the trawlers brought in 8 million tons of cod, as much as had been caught in the two and a half centuries before 1750. In 1980, the catch dropped to 150,000 tons. When the yield dropped, the trawlers intensified their efforts, raising the catch to 250,000 tons a year in the mid-1980s. Then the cornucopia came to an end.[32]

Codfish Wars and Exclusive Zones

Meanwhile, coastal nations began to see fishing on the high seas, which had always been free to all, as a territorial issue. Icelanders, whose economy depended on fishing, began to resent the presence of British trawlers as little as 3 nautical miles (5.6 kilometers) off their shores, especially when their own catch began to falter. In 1950, they unilaterally extended their territorial waters from 3 to 4 nautical miles (5.6 to 7.4 kilometers) offshore, then in 1958 to 12 nautical miles (22.2 kilometers), then in 1972 to 50 nautical miles (92.6 kilometers). Thus began the "codfish wars" between Iceland and the United Kingdom, in which Icelandic coast guard vessels cut the trawls of British fishing vessels, and the British sent naval ships into Icelandic waters to protect its trawlers. In retaliation, Iceland extended its exclusive fishing zone to 200 nautical miles (370 kilometers) off its shores in 1975. Meanwhile, the United States and Canada were not thrilled to see gigantic Soviet and other foreign trawlers 3 nautical miles off their coasts. Following the example of Iceland, the United States extended its exclusive zone to 200 nautical miles in 1976, followed by Canada in 1977. Protected marine areas now cover 3.5 percent of the oceans.[33]

These restrictions not only kept foreigners out but also allowed governments to regulate the size of their fleets, the size of trawls, days at sea, total catch, and other aspects of fishing. Though done in the name of conservation, in reality it was a protectionist, not a conservationist measure. Thus the US government spent $800 million on loan guarantees and other financial incentives to build an American fishing fleet, while Canada took similar measures. Meanwhile, parts of the continental shelf remained outside these zones. The result was to increase, not decrease, the pressure on fish stocks.[34]

Collapse and Moratorium

The inevitable consequence of massive industrial fishing came in 1992, when cod stocks off North America collapsed. The biomass of northern cod is estimated to have dropped by 80 percent between 1962 and 1977. Worse, the number of larger cod, those best able to reproduce, fell by 94 percent. Tuna, haddock, menhaden, and flounder stocks fell too, as did alewives, blueback, capelin, shad, and other fish that cod ate. Within a few years, the cod population was reduced to 1 percent of its historic abundance. The Canadian government placed a moratorium on cod fishing, throwing 30,000 fishers out of work. But it was too late, as the cod stocks were so diminished that they could no longer recover, and the moratorium had to be extended indefinitely. The US government followed suit in 1994, closing Georges Bank to commercial fishing. Two years later it passed the Sustainable Fisheries Act, which stated that "fish were assumed to be inherently scarce unless proven otherwise," a complete reversal of the traditional attitude toward fish, which assumed that there were plenty of fish until there were no more.[35]

Maximum Sustainable Yield

Marine biologists and fisheries managers had long realized that the number of fish in the ocean was not infinite and that overfishing could dangerously deplete fish stocks. After World War II, they adopted a theory called Maximum Sustainable Yield (MSY) that aimed to control fishing on scientific grounds. This theory argued that there was a specific harvest of fish that corresponded exactly to the ability of the fish stocks to reproduce themselves; hence, logically, the yield would be sustainable forever. The idea was derived from the nineteenth-century German theory of scientific forestry that taught that one could keep a forest producing an optimum amount of lumber on a sustainable basis by culling old trees in order to allow young ones to grow faster. Since adult female cod produced over 7 million eggs apiece in a spawning season, it followed logically that the number of adult cod caught could not affect the number of offspring the survivors produced. Removing older, slow-growing adult fish would free up food supplies to support a larger number of faster-growing young fish.[36]

However, unlike trees, which can be counted and measured, the cod population was unknown. In lieu of an accurate census, scientists estimated the size of a fish population based on the number of fish caught. This meant that as long as trawlers were bringing in great quantities of fish, the population was assumed to be healthy. Unknown to the scientific managers, cod and many other fish were not spread out evenly across the ocean floor but huddled together when their numbers were depleted. With sonar, trawlers could locate the last remaining schools of fish and haul them in. As historian Carmel Finley explains: "The technological capacity to catch fish is so great that high catches can be obtained from stocks nearing collapse."[37] MSY was also based on the assumption that fish stocks are independent of their environment and that the only influence on their reproductive capacity was the number of breeding adults left after the harvest. Yet a level of fishing that might be harmless one year can be devastating the next, for fish reproduction is highly variable and sensitive to environmental factors such as water temperature, the availability of nutrients, and the number of rivals or predators. Despite the moratorium, by 2015, the cod population of the Gulf of Maine, once a fertile spawning ground for Atlantic cod, had dropped to 3 percent of sustainable levels, largely because of a rise in water temperature.[38] Some fish stocks, when under pressure, are vulnerable to collapse; such was the case of California sardines, Peruvian anchovies, and Atlantic cod. A few, like Atlantic herring and Pacific halibut, rebounded when the pressure eased; cod did not.[39] Marine biologists could not test their theory against reality. In short, Maximum Sustainable Yield was policy disguised as science and designed to encourage unsustainable harvests and to justify subsidies to trawlers and jobs for fishermen. In the end, the cod population was treated like a mine, not a renewable resource.

The Future of Cod

Since the 1990s, cod fishing off North America has been restricted to limited recreational fishing and scientific research. Scientists are still hoping that the cod on Georges Bank will recover by 2026, but there are reasons to think otherwise. The seabed has been so badly damaged by the trawls that the entire ecosystem has changed. The fish that cod preyed on are gone, replaced by invertebrates like crabs, prawns, lobsters, and sea urchins, while rays and small sharks have taken the place of cod. With the older cod gone, young ones cannot find the traditional spawning grounds of the species.[40]

Meanwhile, the big factory trawlers have moved on to other seas. Every year they sweep an estimated half of the world's continental shelves, leaving behind gravel, mud, and sand. The global fishing fleet is estimated to be two and a half times larger than is necessary to catch what the oceans can sustainably produce. Cod is still available in supermarket fish departments, but it comes, in limited supplies, from Alaska, Norway, or Iceland. For a decade from the mid-1980s to the mid-'90s, orange roughy (*Hoplostethus atlanticus*), caught off New Zealand, was featured in finer restaurants. Thanks to new deep-sea trawls, the capture rate shot upward from zero in 1978 to 90,000 tons in 1990. Then, overfishing caused it to collapse to 12,000 tons by 2009. Pacific pollock, with a taste similar to cod, has also been in decline. Atlantic bluefin tuna, prized by the Japanese for its delicate flavor in sushi and sashimi, has been reduced by 95 percent in the past century.[41] These and other fish are being harvested at an unsustainable rate, especially since Chinese trawlers have joined the world's fishing fleet.[42] Thus have edible ocean fish been depleted, one after another, with one exception: salmon.

Salmon

For over four centuries, cod was the least expensive fish available to Europeans and one of their main sources of protein, while salmon was an expensive luxury, except for people who lived near salmon streams. Today, cod is costly and salmon is abundant and comparatively cheap. Just as the peoples of the Fertile Crescent had domesticated sheep and goats 10,000 years ago after they had depleted the numbers of gazelles and other game, their descendants in the twentieth century turned to domestication to compensate for the growing scarcity of wild fish.

Salmon are anadromous fish, meaning they live as adults in the ocean but migrate up rivers to spawn. Their offspring, called parr, spend from six months to three years in their natal stream before heading out to sea. There, their bodies change into smolts, or adolescent salmon, as they adapt to seawater. They then spend several years feeding in the ocean. Finally, adult salmon return to spawn in the very rivers where they were born, likely guided by the chemical composition of the water. This makes them easy to catch as they migrate upriver during the spawning season. Most salmon die after they spawn, although some Atlantic salmon can spawn several times before they die.

North Atlantic salmon (*Salmo salar*) once hatched in the rivers of Scotland, Scandinavia, New England, and eastern Canada. As adults, they spent a year or more feeding off Greenland and the Faroe Islands before returning to their natal rivers. Their abundance benefited the local inhabitants but did not lead to a long-distance trade. Since the Industrial Revolution, however, pollution, dams, and agricultural runoff have reduced the wild Atlantic population to tiny numbers in a few streams in Scotland and Maine.

The Pacific Northwest

The different species of wild Pacific salmon—coho, chinook, chum, sockeye, and others—are all members of the genus *Oncorhynchus*. They are born in the rivers of the American West Coast, British Columbia, and Alaska, then spend their adult years in the Gulf of Alaska before returning to the stream of their birth. They were the main source of food and wealth of the Indians living along the northwest coast of North America. They were also an integral part of a complex ecosystem. As adult salmon came upriver to spawn and die, many were caught by bears that ate the fleshy and oily parts and left the rest scattered on the forest floor. These remains, as they decayed, fed many other creatures, including the trees. Nutrients washed into the rivers by the heavy rains in turn helped feed the newly hatched juvenile salmon.[43]

This ancient and stable ecosystem was disrupted by the flood of people who came from Europe and the East Coast after gold was discovered in California in 1848. Known as Forty-Niners for the peak year of the gold rush, they dammed the streams of California to pan for gold and polluted them by building railroads, towns, and irrigation ditches. Logging and grazing cattle eroded the stream banks, sending silt downstream that clogged the gravel beds in which the young salmon hid from predators while they grew. The Sacramento River was losing its salmon as early as 1864.

Meanwhile a new technology, canning, allowed salmon to be preserved for human consumption year-round and for export to eastern North America. Having exhausted the Sacramento and other California rivers, salmon canning enterprises moved to Oregon and Washington State. By 1883, there were thirty-nine canneries on the Columbia River alone, and seventy-eight in British Columbia by 1900. After railways reached Oregon in 1883 and British Columbia in 1885, canneries began shipping thousands of cases of canned salmon to eastern North America and to Europe. Around the turn of the century, canning peaked at 635,000 cases on the Columbia River, at over 1 million cases in British Columbia, and at 2.5 million cases on Puget Sound. Between its peak in 1911 and the 1930s, the salmon catch dropped by half. The Pacific Northwest catch, which had averaged 15.4 million kilograms a year before 1930, dropped to 4.9 million kilograms a year in the period from 1949 to 1973 and to 635,000 kilograms in 1993.[44]

By the end of the twentieth century, salmon had disappeared in four out of ten of the rivers of California, Oregon, Washington, and Idaho and were at risk in 44 percent of the remaining ones. Only 1 percent of salmon returned to their natal streams in those four states, compared with 8 to 17 percent in British Columbia and 81 to 90 percent in Alaska. The causes of the decline are not hard to find. On the Columbia and Snake rivers alone, eighteen dams bar the route of the returning fish.[45] On these and other rivers, logging, mining, irrigation, grazing animals, and urban and industrial development have so changed the quality of the water that most eggs cannot hatch nor can most juveniles survive. Those few that do are then prey to sport fishers or to commercial fleets that locate the adult feeding grounds in the ocean and capture them in almost invisible gill-nets.[46]

The Salmon Crisis

The decline in salmon stocks caused concern among sport fishers and politicians. As early as 1875, journalists, scientists, and politicians had predicted the imminent collapse of the salmon stock, each pinning the blame on someone else. In 1981, the US Congress ordered the Bonneville Power Administration to give salmon "equal consideration" when managing the Columbia River dams. In 1985, the United States and Canada signed a Pacific Salmon Treaty and set up a board of commissioners to determine who could catch the remaining salmon and how many. In the 1990s, the Canadian government imposed a moratorium on commercial salmon fishing in the Atlantic. In the Pacific Northwest of the United States, salmon are listed under the Endangered Species Act. Despite these well-meaning regulations, the political pressure to allow more fishing prevented a real resurgence of fish stocks. As historian of salmon David Montgomery explains: "The system is set up such that when fewer fish come back than anticipated the harvest proceeds at the planned rate anyway, but when more fish than anticipated come back from the sea, the harvest increases." Soil erosion, pollution, and other conditions that have led to the decline are not addressed, as they would affect the interests of farmers, ranchers, corporations, and consumers. Meanwhile, the dams remain; installing fish ladders and other measures to reverse the salmon decline on the Columbia River alone has cost $3 billion, but to little avail.[47]

Hatcheries

Another attempt to reverse the decline of wild salmon was the creation of hatcheries, where fish eggs are hatched and the juveniles are fed and protected from predators until they are big enough to survive on their own. The idea originated in 1843, when two Frenchmen, Joseph Rémy and Antoine Géhin, devised a practical method of artificially fertilizing trout eggs. Starting in 1870, the US Fish Commission tried to remedy the declining salmon catch by founding hatcheries and transporting salmon eggs between the Atlantic and Pacific oceans and from one river to another. At first, young salmon raised in one river and transferred to another could not find their way back to their natal river to spawn. By the 1960s, better understanding of salmon ecology, along with bigger hatcheries, better food, and the treatment of salmon diseases, led to a dramatic rise in salmon production.[48]

Despite such precautions, hatcheries could not prevent the decline of salmon fishing on the West Coast of the United States; in Oregon, the production of coho salmon dropped from 3.9 million in 1976 to 1 million in 1977 and to 28,000 in 1997. Over time, hatchery salmon became smaller and fewer as measures to protect their natural habitat were neglected. Operating hatcheries was also expensive; it is said to cost the Oregon Fish and Wildlife Service—and therefore the taxpayers of Oregon—$5,000 for every fish caught by sport fishermen. And it caused environmental problems. Hatchery salmon released into rivers decimated the naturally born wild salmon; as David Montgomery explains: "Hatchery salmon eat smaller wild salmon. So they are not just bullies, they are big cannibalistic bullies." They can also carry parasites and diseases that spread to wild salmon.[49]

Alaskan Salmon

Despite these problems and failures, the global catch of wild salmon has never been greater. From 1950 to 1980 it fluctuated around 400,000 tons a year. Then it rose to around 1 million tons a year, where it has remained, with fluctuations, through 2010. Only 2,500 tons came from the Atlantic seaboard, where wild salmon have almost vanished. Of the rest, 5 percent have come from the West Coast of the United States, 15 percent from Canada, and 80 percent from Alaska.

Alaskan salmon fishery is the one bright spot in an otherwise dismal story. It was not always so. In the early twentieth century, Alaskan salmon fishing followed in the footsteps of the West Coast of the United States and Canada, with fishing fleets and canneries operating without restraint. In 1953, when the Alaskan salmon harvest plummeted, the United States government declared Alaskan salmon fishery a federal disaster area. Harvests continued to decline, reaching a low point in 1972. The next year, Alaska, a state only since 1959, imposed very strict controls over salmon fishing. One aspect was the "limited entry permit" policy that restricted permits to residents of Alaska in order to prevent the influx of fishers from the West Coast. Another was the "fixed escapement policy" that empowered salmon managers to open and close fishing on a daily basis to ensure that enough adult salmon would escape capture and go on to spawn in the rivers. Yet a third aspect of Alaska's success is an active hatchery program that releases 100 million juvenile salmon a year into the rivers, producing between 27 and 63 million adult salmon per year.[50]

Salmon hatcheries have been a success in Alaska, not only because of tight regulation but also because the salmon rivers are still in a natural state that allows fish to spawn and grow successfully, whereas on the West Coast of the United States and in British Columbia the rivers have been dammed and polluted to the point where few naturally born or hatchery salmon can survive.

Aquaculture

Fish farming has a long history. There is evidence that some Australian aborigines kept eels in pens as far back as 6000 BCE. The Chinese are known to have trapped and raised carp from 2500 BCE on, while the ancient Egyptians raised tilapia. Medieval Europeans stocked streams with fish and raised fish in ponds.[51] Hawaiians also captured and kept fish in ponds. Only in the late twentieth century, however, has fish farming become a major industry and source of food worldwide. While wild fish catches have stagnated, farmed fish production has been increasing at a rate of 8 percent per year, faster than the human population or meat production.[52]

Many animals besides fish can be grown in water, including shrimp and prawns, oysters, mussels, clams, and scallops. Shrimp, grown in ponds carved out of coastal mangrove forests, have become a major product in world trade. Among farmed fish, carp are by far the most important, with over 20 million tons harvested in 2010, followed by two other freshwater fish, tilapia and cat-fish. China is by far the world's leading producer of farmed fish, with over 15,000 square kilometers of its coastline devoted to marine aquaculture, in addition to millions of fish ponds inland.

Among farmed saltwater fish, Atlantic salmon tops the list at 1.4 million tons in 2010. Fish farmers have tried to domesticate other ocean fish, such as tuna and cod, but with little success, for they are more expensive to raise than to capture wild. Only sea bass is a distant rival to salmon among farmed salt-water fish. But given the intensive research into the domestication of fish and other seafood, aquaculture will no doubt make them the food of the future.[53]

Salmon Farming

Of all the ocean fish, salmon are the easiest to raise in captivity, for they lay large eggs with yoke sacs that feed the young for several days before they can feed on their own. Salmon farming began in Norway in the late 1960s. Eggs from adult females and milt (or semen) from males are extracted by hand and allowed to fertilize, hatch, and grow into juveniles in freshwater tanks. After twelve to eighteen months, the smolts are released into floating sea cages or net pens of 1,000 to 10,000 cubic meters suspended in sheltered fjords or bays. In such cages, up to 90,000 fish are fed for twelve to twenty-four months until they are large enough to harvest. These operations, resembling feedlots for cattle, are complex and costly and have attracted investments from large multinational corporations like BP and Weyerhaeuser.

The growth of the salmon farming industry has been astonishing. Whereas the harvest of wild salmon has leveled off at about 1 million tons a year, farmed salmon production has risen from zero in 1970 to over 2 million tons a year by 2010. In 1982, wild salmon accounted for 75 percent of all salmon; by 2007, it produced, despite an increase in the catch, only 31 percent of the total, and farmed salmon the other 69 percent. Of all the farmed salmon, 33 percent comes from Norway, 31 percent from Chile (where salmon never existed in the wild), and most of the rest from Scotland and Canada.

Salmon farming requires special environmental conditions. Young parr need clean fresh running water from unpolluted rivers. Smolts grow best in unpolluted but sheltered sea water; in both cases the water must be cold but not glacial, hence the location of the industry in the fjords of Norway, Chile, and Scotland. Being carnivorous, the smolts must be fed wild-caught fish such as anchovies, herring, menhaden, sardines, and capelin. As it takes 2 to 4 kilograms of wild fish to make 1 kilogram of salmon, feeding them contributes to the depletion of the oceans. Attempts have been made to feed salmon with vegetable meal from wheat, soybeans, or algae, but so far with little success.

Salmon farms, like cattle feedlots, bring many problems. One is waste. A farm of 200,000 salmon creates more waste than a town of 60,000 people. In Scotland, where farming is big business, salmon farms produce twice as much waste as humans. Despite the antibiotics included in their feed, farmed salmon are prone to diseases such as salmon anemia or swim bladder virus, even epizootics that can wipe out all the fish in a farm or in entire regions like Norway or Maine, where millions of farmed salmon have then had to be killed. They are also susceptible to sea lice. Untreated waste water from farms can then spread diseases and parasites to wild fish.

There are also genetic issues. Each year about 10 percent of farmed salmon escape their pens during storms or when pens are damaged. It is estimated that a hundred times more fish escape from pens in Maine than there are wild salmon left in New England waters. In 2004, half a million salmon escaped from their pens in Norway and over half a million in Scotland. Once in the ocean, they mate with wild salmon. Scientists predict that someday soon there may be no true wild salmon left. Another concern is genetically modified salmon. In 2012–2013, scientists created the genetically modified "AquAdvantage" salmon that grow to market size much faster than non-GMO (genetically modified organism) salmon; as of this writing, AquAdvantage salmon has not received government approval, but the pressures of commerce and the need to feed an ever-growing human population are likely to overcome the compunctions of environmentalists concerned with the survival of natural wild salmon.[54]

Marine Environments

The stories of whales, cod, and salmon illustrate some of the ways in which humans have attempted to extract resources from the sea. However, the human impact on the oceans is not limited to fishing and whaling, or even to extracting resources. Humans have also taken the oceans for granted or seen them as convenient dumping grounds for the detritus of human activities on land. In doing so, they have transformed the very nature of the seas.

Dead Zones

Many of the chemicals dumped into lakes and rivers or released into the air by power plants and other industries end up in the sea. So do oil spills and tar balls released by tankers discharging ballast. Some pollutants drop to the bottom, only to be kicked up again by trawlers raking the seabed. [55]

Pollutants do not spread uniformly through the oceans but concentrate in particular areas. One well-known example is the formation of dead zones. These occur when synthetic fertilizers run off agricultural lands into rivers and from there into the sea, which causes phytoplankton to bloom, covering hundreds of thousands of square kilometers. When they die, these diminutive plants decay, using up the oxygen in the water. Fish flee if they can. Those that cannot, as well as shrimp, clams, snails, crabs, and worms, die from lack of oxygen.

This has happened most famously in the Gulf of Mexico at the mouth of the Mississippi River, which discharges 584 billion cubic meters of polluted water into the Gulf every year. The result is a dead zone that has recurred every spring and summer since the 1970s. In 2002, it covered 21,756 square kilometers (approximately the area of New Jersey or Israel) and stretched from the mouth of the Mississippi westward along the coasts of Louisiana, Texas, and Tamaulipas in Mexico. A similar dead zone occurred in the Baltic Sea, which receives agricultural runoff and waste waters from the cities and industries that surround it. Likewise, Chesapeake Bay, which was once teeming with oyster beds, has become hypoxic during part of every year. Off northeastern China, shrimp ponds discharge 47 billion tons of effluents, along with over 4 billion tons from cities and industries, into the Bohai Sea, creating a massive dead zone. Other dead zones have occurred along the Atlantic seaboards of Europe and North America and off the coasts of Japan and Korea. Overall, dead zones have affected over 245,000 square kilometers of seas and oceans.[56]

Garbage

The accumulation of garbage has had an astonishing impact on marine environments. Much of it comes from landfills and garbage trucks or roadside litter that is washed down rivers to the sea. Some comes from ships that sink or lose their cargo. An estimated 10,000 containers fall off container ships every year, usually during storms. Of the products lost at sea, some—such as glass or plastic bottles, six-pack rings, plastic bags, bits of nylon ropes, nets, or fishing lines, ear-swabs, syringes, and bits of foam packaging material—float to, or near, the surface. In addition to recognizable objects, tiny synthetic particles used in beauty products or to scour paint from boats and planes also end up in the sea. So do some of the millions of tons of nurdles, the tiny bits from which plastic products are made. Synthetic polymers such as nylon, polyethylene, polypropylene, and polyvinyl chloride are not biodegradable, for no bacteria have ever evolved to consume them; yet they crumble into ever smaller particles and powders that will remain in seawater for years, possibly centuries, unless they are ingested by a fish and enter the food chain.[57]

Some beaches in Hawaii are colored red and blue by tiny bits of nylon from the breakdown of nets from fishing boats as far away as Japan. Likewise, bleach bottles made of a plastic so indestructible it can resist powerful chemicals have been traced from the United States to the beaches of Portugal.[58] Most garbage, however, stays in the ocean.

In 1992, a container was swept overboard from a ship in the Pacific and broke open, releasing its cargo of plastic ducks and toys. Likewise in 2003, when a ship lost twenty-one containers in a storm, thousands of sneakers floated free. These floating objects then got carried across the ocean, allowing oceanographers to trace the ocean currents; many ended up on the Pacific coast of North America, while others got carried across the Arctic Ocean to the Atlantic.

Ocean currents tend for the most part to form great circles called gyres, of which there are two in the Atlantic, two in the Pacific, and one in the Indian Ocean. Floating detritus is thereby pushed along a circular path that eventually concentrates it in the middle of the gyre. In 1997, when Captain Charles Moore sailed from Long Beach, California, into the North Pacific Gyre, an area that sailing ships normally avoid, he found the center of it littered with human detritus, 90 percent of which was plastic. The amount was astonishing: 1 kilogram of debris for every 400 square meters of sea. An area estimated at up to 26 million square kilometers, now known as the North Pacific Garbage Patch, contains several million tons of visible plastic.

Plastic garbage is known to endanger wildlife such as birds and turtles. Sea otters have been found choked on six-pack rings. Seagulls have been strangled by nylon nets and lines. Albatrosses, which cannot distinguish plastic from food, are known to feed bits of plastic to their chicks; at Midway Island in the middle of the North Pacific Gyre, one third of the albatross chicks that die are found to have stomachs full of plastic. Of the carcasses of fulmars (a kind of petrel) found washed up on North Sea coasts, 95 percent had bits of plastic in their stomachs. Even tiny bits of plastic enter the food chain and are ingested by barnacles, sand fleas, and other small animals, and from there make their way into the stomachs of larger animals.[59]

Coral Reefs

Equally tragic is the ongoing destruction of the world's coral reefs. These are structures of calcium carbonate, the exoskeletons of tiny animals of the order *Scleractinia* that colonize mostly the warm shallow waters of the tropics. Corals are not only beautiful in their own right but also provide shelter for innumerable small fish and other creatures. Though reefs cover less than 0.1 percent of the surface of the oceans, they contain 25 percent of all marine species.

However, they are very sensitive to temperature. When temperature rises by as little as 1 or 2 degrees, they begin to die, leaving the bleached skeletons of the animals that created them. Some coral death is natural, as when El Niño killed off the corals near the Galapagos Islands in 1982. In 1998, another El Niño killed 16 percent of the world's coral reefs, including 80 percent of the reefs off northern Sumatra. Yet another in 2002 damaged the corals near the Seychelles and Maldives islands in the Indian Ocean and large parts of the Great Barrier Reef off northeastern Australia.

But human actions also threaten coral reefs. Climate change warms the tropical seas, making corals more vulnerable to weather anomalies. From August to November 2005 water temperature in the eastern Caribbean was 3°C (5.4°F) above normal, bleaching 80 percent of the corals and killing 40 percent of them. In 2010, water temperatures in the western and southern Caribbean were higher than in 2005, causing even more bleaching. During 2015–2016, record temperatures triggered coral bleaching throughout the tropics. The Great Barrier Reef was especially hard hit. Pollution of seawater near ports and estuaries and along shipping lanes also damage the reefs. Runoffs of synthetic fertilizers cause algae to bloom, suffocating corals. Dumping cyanide into the water or exploding dynamite to stun or kill fish, as has been the practice of fishers in Southeast Asia, also destroys corals. As of this writing, 10 percent of the world's corals are dead and 60 percent are at risk from human activities, especially in Southeast Asia, where 80 percent of the corals are endangered. By 2050, scientists expect all the remaining corals to be in danger.[60]

The Rising Sea Level

These issues mentioned are current—and mostly regional—problems caused by the industrialization of the world. But gradually becoming noticeable are issues that affect all seas and oceans and which are certain to become even more evident in years to come.

One of these is the rising sea level. Over the past hundred years, it has risen by 20 centimeters, or 2 millimeters per year, but in the past fifteen years, the rate has increased to 3 millimeters per year and will continued to rise throughout the twenty-first century and beyond.

The rise has two causes. One is global warming, which causes the water already in the seas and oceans to expand. The fourth IPCC report (2007) predicted a rise of between 18 and 58 centimeters by the end of the twenty-first century, depending on the level of fossil fuel emissions and the resulting global temperatures. Other scientists, however, have predicted much greater rises in sea level than those suggested by the IPCC. The reason for the discrepancy is that the IPCC report took into account only the thermal expansion of existing seawater and not the possible melting of the Greenland and West Antarctic ice sheets. By taking into account the addition of meltwater, these scientists have predicted rises of up to 2.4 meters.[61]

The impact of rising sea levels on islands and coastal areas will vary enormously depending on the amount of the rise, the physical features of the shorelines, and the measures taken by people to mitigate the damage. Steep and rocky shores like those of the West Coast of the United States will be the least affected. Most endangered will be low-lying islands and gently-sloping coasts, marshes, and river deltas. Some islands, like Tuvalu and the Maldives, will probably disappear. Sandy coasts, especially in areas prone to hurricanes and storm surges like Florida and the Gulf of Mexico, will be severely threatened. Most vulnerable of all will be the great river deltas already densely inhabited by people. Among those most at risk will be the Mississippi Delta and coastal Louisiana, the Mekong and Red River deltas in Vietnam, the Ganges-Brahmaputra Delta in Bengal and Bangladesh, the Irrawaddy Delta in Burma, and the Nile Delta in Egypt.

Much will depend on the response of the nations involved. The Netherlands, much of which is already at or below sea level, will survive thanks to its wealth and the skill and determination of its people. Bangladesh, nine-tenths of which is a delta or floodplain, cannot afford elaborate dikes, levees, and pumps like those that the Dutch are installing. There, storm surges killed 500,000 people in 1970 and another 140,000 in 1991. Even if future storms cause fewer deaths, millions, perhaps tens of millions, of people will become environmental refugees.[62]

Acidification

One of the most powerful—if only recently discovered—consequences of human actions is the acidification of the oceans. Before the Industrial Revolution, seawater was slightly alkaline, with a pH of approximately 8.2. As we burn fossil fuels, 25 to 30 percent of the carbon dioxide we produce is absorbed by the oceans. There, it reacts with water to form carbonic acid (H_2CO_3). As seawater has absorbed millions of tons of carbon dioxide since the Industrial Revolution, its pH has dropped by 0.1, meaning that its acidity has increased by 30 percent. By the end of the twenty-first century, the pH of seawater is expected to be around 7.8, a 150 percent increase in acidity since humans began burning fossil fuels. Even if humans were to stop burning fossil fuels, the oceans will continue to acidify because of the residual carbon dioxide in the atmosphere.

Seawater will taste much the same to humans, but for the creatures that live in the sea and build their shells out of calcium carbonate ($CaCO_3$), acidification is a death threat. Clams, oysters, mussels, and other mollusks will have difficulty building their shells, as will crustaceans like crabs and lobsters. Coccolithophorids, foraminifera, and pteropods, tiny creatures that are the base of the food chain, will have difficulty surviving, as will those farther up the food chain like krill that are food for many fish, seals, and whales. Acidification will weaken the calcium carbonate exoskeletons of stony corals that constitute the framework of coral reefs, adding to the damage caused by rising temperatures. Not since 55 million years ago will the oceans have been so acidic and so impoverished of life.[63]

Conclusion

It is clear that humans have made huge inroads into the populations of whales, fish, and other creatures of the sea, and they are in the process of changing the oceans themselves—their temperature, their chemical composition, and the flotsam and jetsam in their waters. The seas and oceans are not being emptied of life but are being transformed by human action.

In some instances, as in the cases of whales and cod, bans and moratoriums have prevented extinction and even allowed the recovery of some stocks, albeit at a much lower level than in the past. In other cases, such as salmon, careful regulation and hatcheries in Alaska and farming in other places have kept up the supply for human consumption, even as the number of naturally born wild salmon has shrunk. Shrimp, oysters, mussels, and clams are being raised in artificial environments in extraordinary numbers.

For humans, the future of the oceans may seem rosy. Scientists are working to make it profitable to farm other desirable fish, such as sea bass, bluefin tuna, barramundi, and swordfish. Others, meanwhile, are attempting to create genetically modified fish that can better serve the needs of humans. Yet this optimistic scenario is open to question. Writing in the journal *Science*, scientist Boris Worm and his co-authors predict "the total collapse of all taxa currently fished by the mid-21st century" and, worse, "business as usual would foreshadow serious threats to global food security, coastal water quality, ecosystem stability, affecting current and future generations."[64]

In the last few centuries, and especially in the past decades, people have increasingly turned to the oceans to supplement terrestrial food supplies. Yet ocean biota have only recently become endangered. On land, the impact of human actions started much earlier and has been much deeper and even, in some cases, irreversible.

14

Extinctions and Survivals

On September 1, 1914, a pigeon named Martha died in the zoo in Cincinnati, Ohio. She was no ordinary pigeon but the last of her species, the passenger pigeon. Just a few decades before, passenger pigeons (*Ectopistes migratorius*) had been the most numerous birds in the world. They lived throughout the eastern and midwestern North America but returned to the Appalachian forests every year to breed, where they built up to a hundred nests in a single tree. They had thrived for hundreds of thousands of years by breeding so prolifically that natural predators like cougars, eagles, and hawks hardly made a dent in their numbers.

The introduction of breech-loading shotguns in the 1870s sealed the passenger pigeon's fate. So easy were they to bring down that hunters killed them by the hundreds to provide food for hogs, or even as fertilizer. Pleas to limit or ban hunting were ignored, as hunters competed for the most birds killed in one day; in one competition, the winner produced 30,000 carcasses. By the 1890s, their numbers were dropping precipitously. Attempts to breed them in captivity failed, for they would breed only in large flocks. The last bird in the wild was killed in 1900.[1] Martha's passing symbolized one of the greatest transformations the world has witnessed in several million years: the extinction of species at the hands of humans. Balancing these extinctions were the survival of others, and the flourishing of still others.

Extinctions

Extinctions are to species as death is to individual organisms. Since the first multi-cellular organisms appeared some 530 million years ago, anywhere up to 50 billion different species have existed on Earth. Yet only an estimated 30 million species exist on Earth today, most of them insects. In other words, 99.9 percent of all the species of living beings that ever inhabited the Earth have become extinct, making this a natural process.

Background Extinctions

Before humans began migrating out of Africa, an average of one species out of every 1 to 4 million vanished every year, and the average life span of species ranged from 1 to 11 million years.[2] Invertebrate species have the longest life spans, ranging from 5 to 11 million years, while mammalian species' life spans are only 1 million years on average. This is known as the background extinction rate. These averages, of course, conceal a great diversity. At one end of the spectrum, small populations that occupy small geographic ranges or depend on a specialized diet are particularly prone to extinction, either from genetic deterioration or from misfortune, such as a spell of bad weather, a disease, or an invasion by an alien organism. Other species, such as sharks, cockroaches, horseshoe crabs, and gingko trees have survived for hundreds of millions of years.

Uniformitarians and Catastrophists

The idea that species could become extinct is less than 200 years old. Until the early nineteenth century, naturalists like Carl Linnaeus and the Count of Buffon believed that the world was static and that every species that ever lived since Creation was still alive. Thomas Jefferson, for example, thought that gigantic mammoth bones found in Kentucky meant that there were still mammoths living somewhere but had not yet been found.[3]

In the early nineteenth century, scientists observing rock formations realized that the Earth had an immensely long history, leaving evidence in the form of different rock strata formed by sedimentation, volcanic eruptions, erosion, and other forces. Paleontologists discovered and analyzed thousands of fossils and created theories to explain their appearance and disappearance in different geological strata. Foremost among them was the Frenchman Georges Cuvier, a prolific author who incorporated fossils into the Linnaean classification system and established extinction as an explanation for the existence of fossils that bore little or no resemblance to living animals. In one of his most significant works, *Discours sur les révolutions de la surface du globe* (1822), he argued that what had caused the different strata and their fossils was periodic floods that changed the land and wiped out entire species so that new species could take over. This theory is called catastrophism.

Geologists drew a different conclusion. In his *Principles of Geology; Being an Attempt to Explain the Former Changes in the Earth's Surface, by Reference to Causes Now in Operation* (1830–1833), Charles Lyell, the most famous geologist of his day, argued that the present is the key to the past and that all features of the Earth's surface can be explained by the same forces that operate today, without recourse to imaginary catastrophes. His theory, known as uniformitarianism, was convincing to Charles Darwin, Louis Agassiz, and other naturalists and became the dominant paradigm of nineteenth- and early twentieth-century geology.[4]

For a long time, the idea of catastrophic events in the history of the world was rejected by most scientists. Gradually, however, accumulating evidence convinced them that the Earth had periodically undergone violent changes in which not just individual species but entire genera and even families of plants and animals vanished from the geological record, only to be replaced by entirely new kinds of flora and fauna. By the 1980s, sufficient evidence had accumulated to overturn the uniformitarian paradigm and bring back catastrophism in a new and stronger form. What this evidence showed is that in the course of millions of years of life on Earth, extinctions happened at a fairly regular rate, interrupted by occasional minor extinction events. Five times, however, mass extinctions—defined as events in which 75 percent or more of all species disappeared—radically changed the natural world.[5]

The Big Five

Sixty-six million years ago, three-quarters of all the species of plants and animals on Earth vanished, most famously the dinosaurs. For a long time, scientists puzzled over the causes of this event, which they called the Cretaceous-Paleogene or K-Pg extinction. In 1980, physicist Luis Alvarez; his son, geologist Walter Alvarez; and two chemists, Frank Asaro and Helen Michel, announced that they had discovered, in many places around the world, a sedimentary layer containing iridium, a metal that is rare on Earth but often found in outer space. They argued that an asteroid had smashed into the Earth, spreading iridium and causing the K-Pg extinction. Their hypothesis was confirmed in 1990 by the discovery of a 150-kilometer-wide crater, called Chicxulub, in northern Yucatan and the adjacent Gulf of Mexico. Radiating out from that crater, a powerful tsunami swept through much of Mexico and the Midwest of the United States. The dust raised by the impact blotted out the sun, causing a year-round winter that may have lasted decades, preventing plants from growing and starving most animals.[6]

The Chicxulub asteroid was not the only disaster to strike the Earth's biota. About 66 million years ago, even before the asteroid struck, a flood of basalt began covering the center of the Indian sub-continent, creating the Deccan Traps, poisoning the atmosphere with hydrogen sulfide and causing global warming. According to paleontologist Peter Ward and geobiologist Joseph Kirschvink, "The Deccan Traps softened the world. The asteroid finished the job."[7]

The K-Pg event of 65 million years ago was not the first mass extinction but the fifth in the history of the Earth.[8] During the first, known as the Late-Ordovician 450 to 440 million years ago, over 80 percent of marine species died out. The second one, called Late-Devonian, took place 375 to 360 million years ago. The most catastrophic of all was the third one, the End-Permian or Permian-Triassic 245 to 250 million years ago, which wiped out up to 96 percent of marine animal species for which there is a fossil record, along with an unknown number of terrestrial species; the cause was probably a massive eruption of lava in Siberia that released methane and other toxic gases.[9] The fourth, called End-Triassic or Triassic-Jurassic, took place 210 to 200 million years ago.[10] Then came the K-Pg extinction, which wiped out the dinosaurs. After the past five mass extinctions, it took from 20 to 100 million years for the full diversity of life on Earth to recover.[11]

The Sixth Extinction

We are now entering the sixth mass extinction in the history of the Earth. As Peter Raven, past president of the American Association for the Advancement of Science, wrote in 2000: "We have driven the rate of biological extinction, the permanent loss of species, up several hundred times beyond its historical levels, and are threatened with the loss of a majority of all species by the end of the 21st century."[12]

Extinctions caused by humans are nothing new. Almost as soon as humans reached a new land, scores of large mammals and birds went extinct. As paleoanthropologist Paul Martin has argued—and has now convinced most other scientists—human hunter-gatherers caused the extinctions of many animals in Eurasia beginning 100,000 years ago, in Australia and New Guinea 30,000 to 40,000 years ago, in the Americas 13,000 years ago, in the islands of the Mediterranean 12,500 years ago, in New Zealand 1,000 years ago, and in Madagascar 200 years ago. While each case has been subject to scrutiny and controversy, the pattern that emerges is clear: only human actions, not climate change or diseases, could have led to so many extinctions at those particular times.[13]

In the last four centuries, the rate of extinctions has risen sharply. Since 1600, of the animal extinctions we know about, 115 were of mammals (35 percent of which were on islands) and 171 of birds (90 percent on islands). In addition, many species of mollusks, fishes, amphibians, insects, reptiles, crustaceans, and plants known to science have also vanished, not to mention many more species that were never discovered or classified by scientists because they were small in size, few in numbers, or found only in small areas. Overall, the rate of extinctions (depending on the kind of organism and the environment they lived in) has increased to between 100 and 10,000 times the normal or background extinction rate. Scientist are still debating whether this qualifies as a mass extinction; as biologist Anthony Barnosky explains, "The Earth could reach that extreme within just a few centuries if current threats to many species are not alleviated."[14]

Charismatic Species

Rates of extinction are dry statistics. More eye-catching are examples of charismatic species that have been lost in recent centuries. The seventeenth and eighteenth centuries saw the killing of the last aurochs (a wild bovine) in Poland in 1627 and the Steller's sea cow (a relative of the manatee and dugong) in 1768. The archetypal extinction of the early modern period is that of the dodo, a large flightless bird up to 1 meter tall that lived on the Indian Ocean island of Mauritius and was first described by Dutch sailors in 1598. German traveler J. A. de Mandeslo, who stopped in Mauritius in the 1640s, wrote: "The island is not inhabited and whence it comes the birds are so tame that a man may take them into his hand and they are commonly killed with cudgels."[15] The dodo's survival was threatened not only by hunters but also by pigs, cats, rats, and macaques brought to the island by the Dutch and by the destruction of its forest habitats. The last claimed sighting dates from 1688, less than a century after their first encounter with humans.[16]

The rate of extinctions picked up in the nineteenth century with the disappearance of the bluebuck in South Africa around 1800, the Atlas bear in

North Africa in the 1870s, the Falkland Island wolf in 1876, and the quagga (a kind of zebra) in South Africa in 1888. The great auk (*Pinguinus impennis*), a flightless bird that lived in the North Atlantic, went extinct when the last breeding pair were killed on an island off of Iceland in 1844.[17]

The twentieth century saw even more extinctions, among them the tarpan (a wild horse) in Russia in 1909, the Carolina parakeet in 1918, the thylacine or Tasmanian wolf in 1936, the Bali tiger in 1937, the Caribbean monk seal last seen in 1952, and the Guam flycatcher in 1984. The twenty-first century has so far witnessed the disappearance of the Pyrenean ibex in 2000, the ivory-billed woodpecker last reported seen in 2004, and the western black rhino and Java tiger, both last seen in 2011. These are the famous cases of (mostly) mammals and birds last seen and registered with the International Union for Conservation of Nature. For every one of these charismatics, there are numerous inconspicuous or unknown animals, not to mention insects and plants, that have vanished before scientists could discover and name them.[18]

Threatened and Doomed

Besides species already extinct, there are many others that are not yet extinct but are threatened or are in danger of extinction, or even "committed to extinction," that is, doomed. This includes one-quarter of all mammals, one-third of all amphibians, four out of every ten species of turtles and tortoises, one out of every eight species of birds, and half of all known fish species, plus one-eighth of all known species of plants.[19] At this rate, Leakey and Lewin asserted, "as many as 50 percent of the Earth's species may disappear by the end of the next [i.e., the twenty-first] century."[20]

Frogs are an example of species threatened with extinction. Throughout the world, the fungus *Batrachochytrium dendobatidis* is killing frogs. This fungus originated among African clawed frogs and was spread throughout the world from the 1930s on by obstetricians who used these frogs in pregnancy tests for their patients. When frogs escaped or were thrown out, the fungus also escaped.[21]

Birds are particularly well studied. The 2014 *State of the Birds* report issued by the North American Bird Conservation Initiative—a long-running, continent-wide project—lists 230 species on its Watch List. Some—albatrosses, petrels, plovers, sandpipers, and four species of grouse—are now threatened. Others—eastern meadowlarks, northern bobwhites, nighthawks, and black-throated sparrows—are in steep decline.[22]

Legal action can prevent the extinction of even very uncharismatic species, at least temporarily. Beginning in 1900, the United States government enacted laws and signed treaties designed to protect animals, especially migratory birds. In 1969, the US Congress passed the Endangered Species Act (revised in 1973), which granted protection to animals listed by the US Fish and Wildlife Service. By the end of the twentieth century, of the species protected by that act, fewer than 10 percent were rebounding, 27 percent were stable, and 33 percent were declining; as for the remainder, the available data are insufficient. Meanwhile, the Nature Conservancy estimated that 16 percent were "in imminent danger of extinction."[23]

Some regulations have provoked well-publicized lawsuits. The snail darter, a tiny fish that lives in the rivers of Tennessee and was only discovered in 1973, became the object of a lawsuit in 1975 to prevent its extinction by the construction of a dam on the Little Tennessee River. After the US Supreme Court ruled to stop the construction of the dam, Congress amended the law to save it. The snail darters were transferred to other rivers, but were once again listed as endangered in 1984.[24]

The Northern spotted owl has also been the object of a lawsuit, which pitted two federal agencies, the Bureau of Land Management, which wanted to open up more forest land to commercial logging, and the Fish and Wildlife Service, which argued that this would destroy the owl's habitat and drive it to extinction. As of 2013, this owl was in rapid decline in the United States and only thirty breeding pairs remained in British Columbia.[25]

Insufficient as is the protection of the law in the United States and other wealthy countries, it is almost non-existent in much of the world, where many species are losing the race to survive. Such is the case of the Sumatran rhinoceros, the Javan rhinoceros in Vietnam, the giant panda in China, the Siberian tiger, the Sundarbans tiger in India and Bangladesh, gorillas and black rhinos in Africa, orangutans in Borneo, and other species that are sought after for their pelts or body parts or are losing their habitats to loggers and farmers.[26]

Causes of Extinction

The sudden rise in extinctions and threats to wild species during the last four centuries has three causes: habitat loss, hunting and poaching, and invasive species. In some places, global warming is also beginning to threaten wildlife.

Habitat Loss

The primary cause of the current extinctions is the appropriation of natural habitats for human use, driven by the growth in population (in poor countries) and in standards of living (in rich countries). Using such technologies as power saws and bulldozers, more forest land was turned into cropland in the thirty years between 1950 and 1980 than in the hundred years before. By 1990, Africa had lost 68 percent of its forest areas; India had lost 78 percent, Mexico 66 percent, and the United States 28 percent. Much of this was new forests; if we factor in old-growth forests, which contain more species of trees and a much greater diversity of other plants and animals than secondary forests, the losses reach 99 percent in Nigeria, 95 percent in the United States, and over 80 percent in most other countries.[27] Leaving parcels of forest intact is not enough to prevent extinctions, as some species require large areas and stable communities to survive. Thus the destruction of forests in the West Indies, Mexico, and Central and South America, where migratory songbirds winter, caused a 50 percent decline in their density in the mid-Atlantic states of the United States between the 1940s and the 1980s.[28]

Habitat loss is not confined to forests. From the early Neolithic to the present, wetlands have also been transformed for human use. Marshes and mangrove swamps are among the most biologically diverse of all ecosystems—and the most threatened. Mangrove swamps in Southeast Asia have been uprooted to make room for shrimp farms. Shallow marshes have been turned into rice paddies or drained for dry farmland. And in the vicinity of cities, wetlands have become industrial zones or suburban developments.

The same is true of deserts. Though they may seem too dry to sustain life, many are home to plants such as cacti and to animals such as rattlesnakes and kangaroo rats. In the last century, deserts have been the site of mining and oil extraction, suburban sprawl, or nuclear tests.

Hunting and Poaching

Habitat destruction affects all kinds of creatures, from top predators to lowly insects and tiny plants. Hunting and poaching, in contrast, target only a few key animals. Some wild animals are killed for food—for instance, by rainforest inhabitants in Africa, New Guinea, and the Amazon or by Inuit in the Arctic. Hunting and sport fishing also take place in advanced industrial countries but generally under controlled conditions that do not deplete the animal populations.

More harmful is the trade in endangered species and their body parts. In the nineteenth century, many Hawaiian birds were hunted to extinction by collectors, to be stuffed and displayed in homes and museums.[29] Chinese cuisine features such delicacies as cobras, pangolins, ostriches, giant soft-shell turtles, giant salamanders, crocodiles, civet cats, frogs, tigers, bear paws, birds' nests, and shark fins. A survey taken in 1999 in sixteen Chinese cities found that half the inhabitants had eaten wild animals.[30] Besides food, wild animals are also killed for their fur, their skins, their feathers, or their tusks and horns, all of which enter into an extensive international wildlife trade, much of it illegal yet quite profitable. In Iran, poachers hunt gazelles and use dynamite to kill sturgeons in the Caspian Sea. And finally, ranchers kill animals they consider pests, as in the case of wolves in the western United States or kangaroos in Australia.[31]

As with extinct species, many threatened species are charismatic or at least media celebrities. Such is the wolf, which was exterminated in England in the sixteenth century, in Germany in 1904, in the United States (except in Alaska and northern Minnesota) in 1960, and in Scandinavia in the 1960s and '70s. In Russia and Siberia, wolves, though hunted, were never in danger of extinction. Since then, in Europe and the United States a resurgence of elk and deer, wolves' favorite prey, and a shift in public opinion have led to a comeback for wolves and a renewed controversy between environmentalists and ranchers. Similar stories could be told of bison and grizzly bears.

Invasive Species

Invasions by alien species are one of the prime causes of the decline or extinction of native species. Successful invaders must be opportunistic, that is, able to thrive in disturbed environments. They must also reproduce quickly, disperse easily, and have a high tolerance for a wide variety of conditions. Weeds and pests are such creatures, but the definition would also apply to humans and many of their domesticated and commensal animals, including rats, dogs, and cats, and herbivores, such as horses, cattle, sheep, pigs, and goats. Plants too invaded the Americas; some, such as sugarcane, wheat, and coffee, were deliberately introduced by humans while others came uninvited. Small creatures such as earthworms, honeybees, cockroaches, mosquitoes, and pathogens of all kinds also accompanied the invaders. Biological invasions have continued and, in some places, have accelerated.

Vulnerable Biomes

The impact of alien invasions depends not only on the nature of the invasive species but also on the indigenous plants and animals they encounter in their new habitats. Some habitats have been most dramatically transformed.

Tropical Rainforests

Tropical rainforests are exceptionally prone to extinctions. By one estimate, 550 million hectares of tropical forest were lost between 1950 and 2000.[32] Though they occupy only 6 percent of the Earth's land area, they contain over half its species of terrestrial plants, animals, and insects. The Peruvian Amazon rainforest, for instance, contains fifty times more species of trees than the much larger Canadian boreal forest.[33] By one conservative estimate, it is home to over 3 million species of birds, mammals, and plants out of a global total of 5 million, plus some 30 million insect species. Yet the huge diversity of species means that any one species may include a small number of individuals.

Deforestation has turned vast areas of rainforest into a scattering of forest islands in a sea of grasslands or fields. These islands are often too small to sustain species that are thinly spread over large areas. Some species are enmeshed in complicated relationships with prey, predators, or pollinators, without whom they cannot survive. In such situations, the demise of one species can trigger a cascade of other extinctions.[34]

In 1978, Alwyn Gentry and Calaway Dobson, field biologists from the Missouri Botanic Garden, visited a ridge in the Andean foothills of Ecuador called Centinela. This ridge, like hundreds of other in the lower Andes, was isolated from others by the rainforest that surrounded it; it was an ecological island. There the biologists discovered ninety species of animals, insects, orchids, and epiphytes that lived nowhere else. They also found farmers beginning to encroach on the ridge. By 1986, the ridge had been completely cleared for cacao and other crops, and all the unique endemic species were extinct. The extinct species of Centinela are famous only because they came to the attention of scientists before their demise. No doubt, the same scenario played out in hundreds of other cases without being noticed.

Islands

Of all the different environments on Earth, islands are particularly vulnerable to invasive species. Because many were isolated from the rest of the world until humans arrived, their plants and animals deviated enough from their continental ancestors to form original endemic species. In the absence of predators, many island birds evolved to become flightless and to build their nests on the ground. Furthermore, islands, being small, offer little refuge for endangered plants and animals. New World species were often at a disadvantage to imported Old World species that became their competitors or predators; the same phenomenon is even more true of island species. (Sea birds, able to fly long distances and colonize many islands, were much safer.) Ecologists have found that the number of species that can survive on an island or other isolated area is proportional to the size of that area; thus, fragmenting a forest into several smaller woods will reduce the carrying capacity of each area and increase the probability of local extinctions.[35]

Because islands are smaller than continents, the changes that take place there are more sudden and thorough. On islands, a combination of hunting, fire, forest clearing, and the introduction, often inadvertent, of non-native animals and plants by humans caused environmental changes and affected small as well as large animals.[36]

The islands we know the most about are those settled in relatively recent times, such as Polynesia. Between 2,100 and 1,000 years ago Polynesian navigators sailed to Fiji, then on to the Marquesas and later to Hawaii and Easter Island. They were farmers who brought with them domesticated dogs, chickens, pigs, and rats and introduced coconut, taro, yam, banana, and breadfruit plants from New Guinea or Southeast Asia, as well as sweet potatoes that originated in South America. The transformations they brought to the islands they visited and the extinctions they caused have been well documented. Far from being the tropical paradises that tourists envision, the Pacific islands were ecologically impoverished and became even more so when humans arrived.

New Zealand is the largest land mass settled by Polynesians. Before humans arrived, 85 to 90 percent of its two islands were covered with forests inhabited by some of the most distinctive fauna and flora in the world, with more unique animal species than anywhere else on Earth. In the absence of the large mammals and marsupials that inhabited the continents, their niches were occupied by birds, descendants of those that had flown there long before. The most impressive were the moas, flightless birds up to 3 meters high and weighing up to 250 kilograms, and Haast's Eagles, large avian raptors that preyed on moas. Other species included giant flightless ducks, geese, coots, swans, pelicans, and ravens.

Polynesians may have visited the islands as early as 50 CE, but the first settlers, calling themselves Maori, arrived in the twelfth or thirteenth century. Only dogs and Polynesian rats (*Rattus exulans*) survived the resettlement to New Zealand. The settlers, finding themselves in a land full of edible animals that showed no fear of humans, gave up farming and became full-time hunters. In the process, they exterminated half the indigenous bird species, with moas being easily dispatched with clubs and spears. Meat was so abundant and easy to obtain that much was wasted. Within 300 or 400 years of the arrival of humans, all the moas had been killed or died out. Their extinction had repercussions on other animals, such as the Haast's Eagles that died out when their prey disappeared. Fur seals, no doubt hunted for their pelts in a cool climate, were exterminated on the North Island, but survived in small numbers on remote coasts of the South Island. Imported dogs and rats decimated or exterminated smaller native animals. Rats decimated palms by eating their seeds.[37]

In this hunter's paradise, the human population grew fast. Once they had consumed the moas and other easy prey, the Maori were forced to turn to growing taros and sweet potatoes and eating dogs and rats. To make it easier to hunt and grow crops, they set fire to the forests. By 1769, when the first Europeans visited the islands, half the forests were gone. With food harder to come by, the growth of population also leveled off. The history of the Maori was interrupted when Europeans came to settle in the nineteenth century, decimating them with imported diseases.[38]

Other islands followed the same trajectory as New Zealand, albeit at different times. Before humans arrived some 1,300 years ago, Madagascar was home to *Arpyornis*, the world's largest bird, up to 3 meters tall and weighing up to half a ton, as well as giant tortoises, giant lemurs (the 200-kilogram *Archeolemurs*), pygmy hippopotamuses, and many other unique animals. By 1700, all the native mammals, birds, and reptiles weighing over 10 kilograms had vanished except crocodiles.[39] The same happened to the giant ground sloths, tortoises, monk seals, flightless owls, and other native animals of Cuba and the West Indies, animals that survived for 4,000 years and then vanished when humans reached the islands. As for the larger native animals of the Mediterranean islands, some were hunted to death; others were driven to extinction by human-introduced dogs, rats, goats, and diseases; and the rest found their habitats destroyed by fire or agriculture.[40]

Hawaii is the archetypal example of species extinctions. When Polynesians arrived, their rats, dogs, and pigs attacked flightless birds and ate their eggs, driving over half the native bird species to extinction. The Europeans who came after 1778 introduced cattle, goats, cats, mongooses, and yet more rats that decimated other endemic species. In the twentieth century, twelve surviving bird species became endangered, another twelve were unlikely to survive, and only eleven were safe. Hawaii now has more species of alien than of native plants, some of them deliberately introduced, others accidentally.[41] When the introduced giant African snail *Achatina fulica* turned into an agrarian pest, another snail, the carnivorous *Euglandina rosea*, was brought in from Florida to control the African snail, which it did, but it also drove several native snails to extinction.[42]

Another threat to Hawaiian birds has been avian malaria caused by the *Plasmodium relictum* and spread by the mosquito *Culex quinquefasciatus*. This disease is not a major cause of death among birds worldwide. On islands where birds have no genetic resistance, however, it is deadly. Introduced to Hawaii in 1826, probably by mosquito stowaways on a sailing ship, it spread throughout the archipelago and began decimating endemic birds living below an altitude of about 1,500 meters, above which cold temperatures prevent mosquitoes from reproducing. Since the 1980s, ten species of birds have become extinct as a result.[43] In their place, Indian mynah birds (*Acridotheres tristis*), a recent arrival, have proliferated.

A recent dramatic case of destructive invasion is the introduction of brown tree snakes (*Boiga irregularis*) to Guam. These snakes, natives of New Guinea, were accidentally imported after World War II. In Guam, they found an ideal environment, with plenty of birds' eggs, lizards, and even small mammals to eat. Within a few years, they reached a density of 100 per hectare. In less than forty years, Guam lost seven out of twenty-five species of endemic birds, with another eleven species substantially reduced. For example, the population of Guam flycatchers (*Myiagra freycineti*) dropped from 450 in 1981 to zero in 1984.[44]

Lakes

From a scientific perspective, any isolated area that is surrounded by a different environment that most plants, animals, and insects find difficult or impossible to cross is an ecological "island." Examples include mountaintops, oases, lakes, or areas surrounded by natural barriers and human development, such as pockets of rainforest in otherwise deforested areas. Species native to such "islands" are as vulnerable as those on geographical islands.

Lake Victoria in East Africa, for example, is home to countless species of cichlid fishes, many of them the varieties that aquarium owners love, but it is also a major resource to the fishing peoples who inhabit its shores. In the 1950s, in a misguided effort to improve the lot of fishing people, British colonists introduced the Nile perch (*Lates niloticus*), which can grow up to 2 meters long and weigh up to 140 kilograms. This alien fish proceeded to devour the native cichlids, driving several hundred species to extinction. As many cichlids ate algae, their disappearance resulted in algal blooms that depleted the oxygen in the water, accelerating the decline of other native species.[45]

The Great Lakes of North America have also been affected by invasive species. Some swam up the Erie and Welland canals. Such were the eel-like sea lampreys that destroyed the Great Lakes' trout population and, with it, one of the world's great commercial fisheries. Soon thereafter came alewives (or river herring) that replaced all but 10 percent of other fishes. In the summer of 1967 and several times thereafter, alewives died by the billions, fouling the shores and bottoms of the lakes with their decaying stinking corpses.

Soon after the St. Lawrence Seaway opened to oceangoing ships in 1959, other creatures appeared in the lakes, carried in the ballast water the ships discharged when they loaded cargo. One such was the zebra mussel (*Dreissena polymorpha*), a native of southern Russia that arrived in 1988; these mussels attached themselves to the walls of canals and locks and to the intake pipes of water-treatment plants. A couple of years later, quagga mussels (*Dreissena bugensis*) replaced the zebra mussels. These mollusks eat phytoplankton—the bottom of the food chain—thereby damaging what was left of the fish population. In short, the history of life in the Great Lakes replicates—at a much faster pace—the impact of invasive on native species in the Americas. Most recently, the Great Lakes are on the verge of being invaded by Asian carp that proliferate in the Mississippi watershed. Frantic efforts and a major lawsuit have attempted to delay this invasion.[46]

Continents

Invasions are not limited to islands and lakes but can spread to vast areas of continents. Ships and airplanes have sped up the pace of introductions. In recent times, the classic case is European rabbits (*Oryctolagos cuniculus*) in Australia. Rabbits accompanied the First Fleet from England in 1788, but did not thrive. This changed in 1859 when Thomas Austin, an immigrant to Australia, had twenty-four wild rabbits sent from England in order to hunt them as he was accustomed to back home. Within a few rabbit generations, their numbers had proliferated into the millions, eating up the grass that settlers had hoped to feed their sheep (another alien species) and causing soil erosion and other ravages to the environment. Efforts to control the population explosion by hunting, poison, and rabbit-proof fences failed. Finally, in 1950 the *mixoma* virus, a pathogen that kept the rabbit population in check in England, was released in Australia, causing the population to drop from about 600 million to roughly 100 million. It later rebounded and has now stabilized between 200 and 300 million.[47]

No recent invader has been quite as successful as the Australian rabbit, but others have had a serious impact nonetheless. In the Great Plains of North America, native grasses such as buffalo grass, bunchgrass, and sagegrass were plowed under to grow wheat or were destroyed by grazing cattle. In their place came Eurasian plants that had evolved to resist cattle. Cheatgrass, sharp enough to damage the mouths of cows, was first noticed in British Columbia in 1889; by 1930 it had spread to 40 million hectares. Spotted knapweed, another Eurasian grass, arrived in bags of alfalfa seeds in 1883 and spread at a rate of 27 percent per year. Like knapweed, yellow spurge and star thistle were thorny and unpalatable to cattle. Most obvious to travelers are the tumbleweeds that roll with the wind across the prairie.[48] Wallace Stegner, chronicler of the American West, describes the transformation:

> Once we were gone, the prairie should have settled back into something like its natural populations in their natural balances, except. Except that we had plowed up two hundred acres of buffalo grass, and had imported Russian thistle—tumbleweed—with our wheat seed. For a season or two, some wheat would volunteer in the fallow fields. Then the tumbleweed would take over, and begin to roll. We homesteaded a semi-arid steppe and left it nearly a desert.[49]

Several other invaders have made news periodically. In the southeastern United States, a vine called kudzu (*Pueraria lobata*), imported from Japan in the late nineteenth century, has been spreading at a rate of 61,000 hectares a year. As it grows, it covers trees, shrubbery, and utility poles. It is extremely tenacious and can be eradicated only by digging up its roots.[50] Water hyacinth (*Eichhornia crassipes*), a native of South America, was imported as an ornamental plant in the late nineteenth century. It does to waterways what kudzu does to land vegetation. It can double in two weeks and cover the entire surface of a pond, killing other plants and animals by blocking their sunlight. It once spread throughout the lakes, rivers, and wetlands of Louisiana and Florida until efforts were made, at great expense, to control its spread.

Animals are also invasive. Among the newsworthy ones are the Burmese python (*Python bivittatus*), a native of Southeast Asia and Java, which can reach a length of over 4 meters. These snakes were introduced into Florida as pets in the 1970s. When they grew too big, their owners released them into rivers, canals, or the Everglades, where they attack not just birds and small mammals but even panthers and alligators. Smaller, but no less ferocious, are the red imported fire ants *Solenopsis invicta*. Armed with a painful sting, they reproduce rapidly and can survive in a great variety of conditions, including floods and droughts. They attack insects, lizards, and birds and also eat many plants. Exported from South America in the 1930s, they have become a scourge in the United States, China, Australia, and the Philippines. These and many other less well known invaders are transforming environments around the world.[51]

Before Columbus, the Americas were an ecological "island" that only a few organisms were able to reach. As a result, indigenous plants and animals did not develop resistance to the pathogens found in the Eastern Hemisphere. Since the two hemispheres have come into contact, the forests of North America have been hit by one Asian tree disease after another. The chestnut blight, caused by the fungus *Cryphonectria parasitica*, was brought to America around 1900. Though Eurasian chestnut trees are resistant to the blight, American chestnuts, which constituted one-quarter of the trees in the eastern forests, succumbed; by 1940 very few were left. Likewise, Dutch elm disease, caused by the parasitic fungus *Ophiostoma ulmi* and spread by elm bark beetles, has infected elms throughout Europe and North America; efforts to contain the disease have been relatively successful and elms still survive in many cities willing to bear the cost of containment. Most recently,

Figure 14.1. Trees dying from pine beetle infestation in Colorado, early twentieth century. iStock/PhilAugustavo.

hemlocks in North America have become infected by the hemlock wooly adelgid (*Adelges tsugae*), a parasite that gradually poisons the tree.[52]

The Arctic

In some places, it is global warming rather than invasive species that is endangering native plants and animals. Nowhere on Earth is its impact more evident than in the Arctic, where the climate is changing twice as fast as the global average. In parts of Alaska, winters are now 2° to 3° C (4° to 5°F) warmer than they were in the 1970s. Sea ice covers less of the Arctic Ocean than it did in the twentieth century. As of this writing, the minimum extent of summer sea ice was reached in September 2012, when it covered 31.4 million square kilometers, 22 percent less than the previous minimum in 2007, or 50 percent less than the average for the period 1979–2000. The average thickness of the ice has also declined by 80 percent since 1979. Given current trends, scientists predict that the Arctic Ocean will be ice free in the summer at some point during the twenty-first century.[53]

The impact on plants and animals will be dramatic. As the climate has warmed, the tree line has gradually been moving north, while rising sea levels have encroached on the land. Squeezed from both sides, the area covered by tundra has been shrinking. This is where millions of migrating birds spend the summer and raise their young, feeding on the insects that abound in the summer months. But the arrival of birds and the appearance of insects no longer coincide, reducing the nutrition available to the birds. It is therefore expected that birds will lose half their nesting habitats during the twenty-first century.

Many other animals make their homes in the tundra and Arctic grazing areas. Small mammals like ground squirrels, voles, hares, and lemmings survive year-round on the summer vegetation and in turn provide food for predators like weasels, foxes, wolverines, wolves, bears, and birds of prey. Caribou (in the Western Hemisphere) and reindeer (in Eurasia), musk oxen, and moose also make the tundra their home during part of the year. A warming climate means that rain sometimes falls on the mosses and lichens that cover the ground, then freezes, putting them out of the reach of these grazing animals. The caribou population of western Greenland and Canada's Arctic islands dropped from 26,000 in 1961 to 1,000 in 1997, and has now become endangered, while lemmings are expected to become extinct.

Animals that depend on the Arctic Ocean are in greater danger. Polar bears are good swimmers, but they need sea ice on which to rest and to rear their young. As sea ice shrinks, so do their habitats. Already polar bears weigh less and have fewer cubs, on average, than they did in the twentieth century. When sea ice disappears completely, polar bears will become extinct as a species, although some polar-grizzly hybrids will survive. Likewise, the ringed and harp seals and walruses that bear their young on the ice will die out; with them will go one of the main sources of food for polar bears.[54]

Winners and Survivors

Just recounting extinctions and dooms would give a biased picture of the state of the natural world and its future. Compensating for the lost species will be gains among other species. The most opportunistic ones will thrive by replacing those that have gone extinct. Others will survive, in reduced numbers, by being resilient and adaptable. Still others will survive only with the help of humans.

Winners

Foremost among the invasive opportunistic species are human beings. No other species has been so successful in colonizing almost all parts of the Earth, from the wettest rainforest to the driest desert, and from the tropics to the Arctic. Humans continue to transform the shores, the continental shelves, and even the depths of the oceans. In the Arctic, global warming will allow humans to thrive. Not all humans will benefit, of course. The indigenous Arctic peoples who live off hunting, whaling, or raising reindeer will find their habitats shrinking, their prey diminishing, their cultures and ways of life subverted, and their lands taken over by peoples from the temperate zone. For immigrants from the temperate zone, however, the Arctic Ocean and its nearby lands offer tremendous opportunities. Shipping will increase and ports will be built to handle the traffic. Oil and natural gas deposits have been discovered; several, such as the North Slope of Alaska, have been exploited for some time already. Other minerals are likely to be found in these formerly inhospitable lands. Canada, Alaska, Norway, and Russia expect to benefit from global warming and the opening of the Arctic. Here, more than anywhere else on the planet, humans are winning at the expense of the rest of nature.

Domesticated animals will also be winners. Some long-standing ones like cattle, dogs, and chickens are proliferating in parallel to the increases in human numbers and wealth. Recently humans have domesticated new species, such as salmon, perch, tilapia, and other farmed fishes, as well as shrimp and mollusks.

The industrialization of agriculture and of animal husbandry has led to the reduction of natural diversity in favor of uniformity. Tens of thousands of vegetable varietals have been neglected as agriculture concentrates on a few favored species or sub-species; compared to the diversity of these plants in the world, only a small number of varieties of potatoes, rice, maize, wheat, apples, and pears, and a single variety of banana—the Cavendish, or *Musa acuminata*—dominate the grocery store shelves and human diets worldwide. Of the approximately 7,000 commercial varieties of apples cultivated in North America around 1900, fifteen now account for 90 percent of the North American crop, and three—Red Delicious, Golden Delicious, and Granny Smith—account for two-thirds. Even animals are being standardized, as half of European livestock breeds were lost during the twentieth century, and of those left, 43 percent are close to oblivion.[55]

Also among the winners are the commensals, species that have adapted to living close to humans: common birds such as sparrows, pigeons, Canada geese, and seagulls; rodents such as rats, mice, and squirrels; deer and other browsers; and insects such as cockroaches, termites, and houseflies.

The successes of humanity and the biota it favored cannot be measured solely by numbers. The quality of life depends on what ecologists call "eco-system services," that is, the quality of the environment in which we live. These services include clean air, clean water, natural (even if not really wild) landscapes, a livable climate, plants and animals (other than crops and live-stock), and even microorganisms, for everything in nature interacts and depends on everything else. As humanity grows in numbers and expands into new environments, these ecosystem services are increasingly under threat.

Survivors

Many species are surviving the Sixth Extinction, but in reduced numbers. Some respond to the pressures of a changing environment by changing their habits or their locations. Global warming has affected many species this way. In an analysis of 677 species, biologists Camille Parmesan and Gary Yohe found that 62 percent are affected by the early arrival of spring, which has taken place on average 2.3 days earlier per decade. Migrant species arrive earlier, birds nest and lay their eggs earlier, plants produce buds and flowers earlier. Likewise, in the autumn, leaves fall, migrants depart, and hibernating animals hibernate later. The common murre (*Uria aalge*), a sea bird of the North Atlantic and North Pacific, has been laying its eggs on average 2.4 days earlier every decade. Each decade, European plants have been budding and flowering 1.4 to 3.1 days earlier, and North American plants 1.2 to 2 days ear-lier. In Europe, migratory butterflies have appeared 2.8 to 3.2 days earlier and migrating birds 1.3 to 4.4 days earlier each decade.

Species that occupied a given range have shifted their range by 6.1 kilometers closer to the poles per decade. Of thirty-five non-migratory European butterfly species, two-thirds have moved northward between 35 and 240 kilometers. The hummingbird hawk-moth (*Macroglossum stellatarum*) has shifted its range from the Mediterranean region to north of the Alps.

Some animals' sex ratios are determined by the ambient temperature; thus painted turtles (*Chrysemys picta*) produce more female offspring as temperatures rise, while crocodiles and alligators produce more males. In some cases, a changing sex ratio can diminish a species' survival chances.[56]

Fish also migrate. As the North Sea has warmed by 1.25°C (2.25°F) in the twenty-five years after 1980, fifteen out of thirty-six fish species surveyed moved an average of about 300 kilometers northward. These were predominantly opportunistic species of smaller fish that reproduced early in life; larger species moved less. Fish also moved down into cooler waters, but deeper waters get less sunlight and produce less food than surface waters. Marine animals that have regular spawning and feeding areas and species that are attached to particular locations, such as oysters, mussels, corals, mangroves, and seaweeds, move slowly or not at all and can be decimated by rapid temperature changes.[57]

Trees don't move, but tree species spread when their seeds germinate and survive in new environments. In the Peruvian Amazon, tree species have been migrating uphill to cooler climates a quarter meter a year, on average; some have moved 3 meters, others not at all. Tree species once common in the northeastern United States—yellow birch, sugar maple, beech, and hemlock—are appearing in Canada, while trees of the Southeast are beginning to appear in the Northeast.[58] Species that occupy mountain areas have moved upward by 6.1 meters per decade on average. Those that live at high altitude in tropical regions, such as on the slopes of Mount Kilimanjaro in East Africa, Mount Carstenz in New Guinea, and the Atherton Tablelands in northeastern Australia, have nowhere to go but up into ever-smaller areas.

Likewise, islands and regions bounded by geographical barriers allow no escape for species that would otherwise move. In the Cape Province of South Africa, the fynbos, a region with 8,000 species found nowhere else, is the richest floral kingdom outside the tropical rainforest; global warming will shrink the area in which its plants survive by half by 2050, causing many extinctions. Even national parks will become traps for endangered species.

Hybrids and Mutants

As the climate warms, species once separated by geography are moving into one another's ranges. Sometimes they mate and give birth to hybrids. As blue-winged warblers move north into the territory of golden-winged warblers, their hybrid offspring are gradually replacing the golden-winged. In the far north, as the territories of grizzly bears and polar bears begin to overlap, they have produced hybrids—"pizzlies" or "grolars." Likewise, there have been sightings of hybrids of bowhead and right whales, belugas and narwhals, and hooded and harp seals, to name a few. In natural history, such cases have always occurred. What is new is that hybrids are hastening the extinction of the weaker of the two parent species. This seems to be happening to spotted owls, the subjects of so much litigation, as barred owls move into their range.[59]

Hunting can cause mutations rather than extermination. Such is the case of elephants in southern Africa. Until the international trade in ivory was banned in 1990, hunters killed elephants for their tusks, a very profitable business. In two nature preserves in Zambia, the elephant population declined from about 35,000 in the 1970s to 2,500 in 1988. At that point, poachers were killing ten a day, but only elephants with tusks. As a result of selective killing, the proportion of tuskless female elephants rose from 10.5 percent in 1969 to 38.2 percent in 1989.[60]

Evolution operates at all scales, from elephants to microbes. Among the survivors are species that have been targeted for extermination by humans. As older antibiotics no longer have the effect on bacteria they once had, scientists are constantly seeking new ones to replace them. The plasmodia of malaria have evolved resistance to anti-malarial drugs such as chloroquine. Mosquitoes that were expected to be eradicated by DDT have evolved resistance to pesticides. Maize, potatoes, and other crops were genetically modified to contain the *Bacillus thuringiensis* or Bt, a natural pesticide, but recently Bt-resistant pests have evolved to defeat the ingenuity of the geneticists. Meanwhile, the genes inserted into crops to make them pest-resistant sometimes jump into non-crops, such as jointed goatgrass (*Aegilops cylindrica*), a weed that can reduce the yields of winter wheat by 25 to 50 percent. In recent years, resistance to herbicides, pesticides, and immunizations has proliferated, thwarting human hopes of being rid of pests, weeds, and diseases. Instead, an arms race developed, pitting living beings against the chemists and biologists who want to exterminate them.[61]

Managing Wildlife

Finally, among the survivors are species that have been rescued from the brink of extinction and will continue to depend on humans for their survival. The first challenge of wildlife managers is to gather information about endangered animals. Scientists have long used tracking devices to follow the movements and migrations of animals. Birds are captured so that metal bands can be affixed to their legs. Plastic stickers are attached to butterfly wings. These methods provide limited information, but only if the animal is recaptured. More sophisticated are radio-tracking devices, used since the 1960s to follow animals' movements for long periods of time, with researchers using hand-held receivers or even satellites. Recent advances in radio-tracking collars equipped with GPS receivers have allowed wildlife biologists to track bighorn sheep and their mountain lion predators, as well as condors, pronghorn antelopes, bears, eagles, and other animals. With the information these devices provide, biologists can capture and transport animals to new areas and prevent, or even reverse, the threat of extinction.[62]

Tracking only provides information. To save endangered animals, scientists capture some and try to induce them to reproduce in captivity. When the last forest-dwelling Hawaiian crow died in 2002, a hundred others survived in captivity, where zoologists hoped to persuade them to reproduce.[63] Sumatran rhinos, of which only a hundred are left in the world, are being cared for in zoos and reserves in Malaysia, Sumatra, and the United States; so far, only four have been born in captivity.[64] Gradually, as more and more species become endangered, they too will survive in zoos and nature reserves. This is true not only of animals but of other creatures as well, which will be saved from extinction in aquaria, botanical gardens, seed banks, tissue cultures, and liquid nitrogen tanks.[65]

In some cases, it is possible to rescue endangered animals and then release them into nature. The California condor (*Gymnogyps californianus*) is a case in point. These scavenger birds, once found throughout North America, began to decline when the megafauna went extinct. By the time Spanish settlers reached California, their range was reduced to the West Coast of North America, where they subsisted on the carcasses of beached whales and other marine mammals. As southern California began to fill up with humans, condors died by eating poison bait set out for wolves and coyotes or by ingesting lead bullets in animals shot by hunters. By the 1940s, only 150 survived in isolated canyons, and by the 1970s, there were fewer than thirty. Between 1985 and 1987, the US Fish and Wildlife Service captured the last remaining fifteen condors and placed them in the care of wildlife veterinarians in the San Diego Wild Animal Park and in the Los Angeles Zoo. Fortunately, condors reproduced well in captivity. With successful breeding, their numbers increased enough that some could be released into thinly populated parts of the American West. As of 2014, there were 439 condors in the wild and in captivity. With numbers that low, however, condors have to be constantly tracked and monitored, at great expense.[66]

Even more dramatic is the case of the whooping crane. These large and gracious birds reproduce slowly. In 1941, there were only twenty-one wild and two captive whooping cranes left. A spokesman for the United States Wildlife Service called them "intolerant of civilization" and blamed their imminent extinction on their "lack of cooperation." Conservation methods, however, have led to their recovery, so that by 2011 there were an estimated 437 in the wild and 165 in captivity.

"In the wild" and "in captivity" are relative terms, however. Whooping crane eggs are hatched at the Patuxent Wildlife Research Center in Maryland. The young birds are then shipped to the Necedah National Wildlife Refuge in Wisconsin for flight training. Humans who handle the birds dress in white robes with crane-head puppets at the end of one arm, so that the young birds will imprint in them and not on normal-looking humans. Wild cranes are migratory birds that follow their parents on their first migration. Without parents, humans have had to play their parts. They do this by training the young birds to imprint on costumed pilots flying ultra-light aircraft. The trip from Wisconsin to Florida can take up to three months, with stop-overs in the fields of farmers sworn to stay out of sight. As of 2009, Operation Migration had trained seventy-three birds at a cost of 1.7 million dollars.[67]

These heart-warming stories do not constitute a victory for nature. With numbers so low, it would only take one disease outbreak (or one shift in government policy) to rid the world of California condors or of whooping cranes forever.[68]

Sometimes, managing wildlife raises thorny ethical dilemmas. In California, when drought reduced rivers so much that young salmon could not find their way to the sea, they were transported to the ocean in trucks or barges filled with their natal river water, in the hope that some would survive and find their way back.[69] On the Columbia River, just as the number of salmon once again began increasing—a great success for conservation—cormorants that eat salmon also began to thrive. To protect the salmon, the Army Corps of Engineers planned to shoot the cormorants. Likewise, near Bonneville Dam, sea lions have been killed in order to save salmon. For similar reasons, the Fish and Wildlife Service killed up to 3,900 barred owls that were invading the habitats of the endangered spotted owl.[70]

Figure 14.2. Training a baby whooping crane, Wisconsin, early twenty-first century. iStock.com/ Westphalia.

New Zealand, the most isolated large land mass in the world, once had a fauna composed largely of birds, many of them flightless like the moa and the kiwi. The Maori who arrived in the thirteenth century and the English who came in the eighteenth brought with them mammals that decimated the native birds and sent several species into extinction.[71] Today, in an effort to save the last surviving native bird species, private environmentalists and government conservationists are making a concerted effort to exterminate all mammals in the wild, such as rats, mice, and stoats, even feral cats.[72]

As is becoming ever more evident, it is no longer sufficient to rescue animals from the brink of extinction, breed them in captivity, and release them into the wild. The very idea of "the wild" has become problematic in much of the world. Endangered species have therefore become "conservation reliant," meaning they will always have to be monitored and protected. Maintaining viable populations will require continuing interventions. Of the species in the United States that are listed as endangered or threatened under the

Endangered Species Act of 1973, 84 percent are conservation-reliant. In the future, the challenge facing conservation managers will be overwhelming.[73]

Conclusion

Who will mourn the passing of a few more species every year? Who, other than paleontologists and schoolchildren, cares about the dinosaurs and mastodons of yesteryear? Who, other than environmentalists, has noted the passing of the moa, the dodo, the quagga, and the thylacine?

It is true that when the last polar bear or panda dies, probably in a zoo under the care of loving zookeepers as was the case of the passenger pigeon and the Carolina parakeet, there will be an outcry and heart-breaking headlines. Less celebrity-species like the nene goose and the monk seal will merit only passing mentions. The disappearance of thousands of insects, plants, frogs, toads, reptiles, and other less-known living beings may perhaps be noted in articles in scientific journals. As for the hundreds of thousands, perhaps millions, of creatures that will disappear without the benefit of scientific discovery, no human will miss them. The extinction of some—bugs, weeds, and pests—may even be celebrated by humans.

Does it matter? This only begs the question: for whom? Does nature have a value outside of the human experience? As ecologists tell us, creatures do not just live side by side with others; they interact, they eat or are eaten, they coexist or conflict, they depend on one another. And when one goes, others are weakened. Just as the extinction of giant herbivores turned Australia into a tinderbox of flammable grasses, so too the disappearance of lesser creatures impacts others and their environments. We humans form part of the natural world and depend, like all other living beings, on nature's ecosystem services. If we ignore or jeopardize these services, eventually we will feel the effects.

15

Environmentalism

On April 22, 1970, some 20 million Americans took part in peaceful demonstrations, rallies, and parades on environmental issues. Among them were students at 2,000 American colleges and universities and at 10,000 primary and secondary schools; in New York City, a million people gathered in Central Park for a one-day teach-in on the environment.

Suggestions for an environmental demonstration had been building for awhile. Peace activist John McConnell proposed the idea of an annual Earth Day to the UNESCO conference on the environment in 1969. US senator Gaylord Nelson suggested that April 22, 1970, be the first Earth Day. Everyone was surprised by the turnout and the enthusiasm of the crowds. That day marked the start of a new era in which environmentalism became a popular political movement, not just in the United States but around the world.[1]

People have long been aware of their impact on the natural environment, and some have called for restraint. That concern for the natural world, which we call environmentalism, has roots in the distant past. This chapter traces the evolution of environmental ideas from largely spiritual concerns in ancient times, to a utilitarian movement called conservationism in early modern times, to the ethical and aesthetic values of preservationism in the twentieth century, and to the political movement and policies of recent times. Finally, it looks at what difference environmentalism has made to the environment.

Pre-Industrial Environmentalism

Concern for the environment existed even among hunter-gatherers for whom the natural world was full of spirits. For religious reasons, the Telefol people of New Guinea did not hunt the long-beaked echidna, a slow-breeding, slow-moving, egg-laying mammal. In southeast New Guinea, people refused to harm sea turtles. Across Australia, local totem animals were strictly protected, while others were reserved as special foods for clan elders.[2] Aborigines had "story places" that were off-limits to humans so that important animals like tree kangaroos could survive and reproduce and re-populate adjacent hunting grounds.[3] In Micronesia, taboos forbade hunting certain birds, turtles, and sea animals for religious reasons or to reserve them for chiefs and royalty.[4] Historical geographer William Denevan has con-cluded: "There has been considerable debate over whether pre-historic in-digenous land use conserved or depleted resources. The best response is that sometimes there was intentional conservation; other times practices were destructive and not sustainable of resources and habitat."[5]

People in ancient and classical civilizations also had ambivalent attitudes toward nature. In his analysis of these attitudes among the educated elites of the Mediterranean world and of Europe from ancient times to the eight-eenth century, cultural historian Clarence Glacken found a great variety of attitudes, but their most common belief was that nature had been created to serve man's needs. "If the earth was divinely ordered for life, man's mission on earth was to improve it. Such an interpretation found room for triumphs in irrigation, drainage, mining, agriculture, plant breeding. If this interpre-tation of man serving as a partner of God overseeing the earth were correct, understanding man's place in nature was not difficult."[6]

This bifurcated view of nature as "divinely ordered" yet there for human use is found in one of the earliest written documents, *The Epic of Gilgamesh*. In it, the supreme god Enlil assigns the monster Humbaba to guard the sa-cred Cedar Forest. But Gilgamesh, the king of Uruk, kills Humbaba and fells the trees to make a door. The tension between a divinely created nature and the human demand for its resources is also found in the biblical admonition to subdue the Earth and have dominion over every living thing.

How did ancient peoples balance creation with exploitation? There is no single answer to this question. At one end of the spectrum were the Romans, who carved straight roads across hilly terrain, built aqueducts to carry water across valleys, and laid out their towns in a grid. At the other end were some Eastern religions that exhibited an extreme sensitivity toward nature. The Jains of India practice non-violence toward all forms of life, wearing masks to avoid accidentally inhaling insects and sweeping the path before them to avoid stepping on ants. In China, Taoism was a philosophy of natural-ness, non-violence, and passivity, with a spiritual view of nature. These and similar belief systems convinced certain Western humanists that Asian peo-ples had a more benign attitude toward nature than Western peoples. But deforestation, irrigation, and other forms of exploitation of nature belie these views. As cultural geographer Yi-fu Tuan explained: "The publicized environmental ethos of a culture seldom covers more than a fraction of the total range of environmental behaviour. It is misleading to derive one from the other."[7] Sinologist Mark Elvin even characterized Chinese environmental history in the title of an article as "3,000 Years of Unsustainable Growth."[8]

Besides some vague attitudes toward nature, many pre-industrial peoples followed certain very precise rules regarding particular parts of the natural world. In Nepal, Thailand, and parts of Africa, communities preserved particular woods as sacred groves for religious ceremonies and initiation rites. On a larger scale, many European kings and nobles established game preserves and employed game wardens to deter poachers. The most extensive of all hunting preserves was that of the Manchus who ruled China from 1644 to 1912. They reserved all of Manchuria, their homeland, as a hunting ground and forbade Chinese people from moving there. The Chinese-exclusion policy was not aimed at protecting nature per se but at providing the imperial court with game and other forest products imported from Manchuria and at giving young Manchu men a place to practice their military skills. By this policy, the Manchus delayed for several centuries the transformation of forests into arable land that was taking place elsewhere in China. Not until the early nineteenth century did fear of Russian penetration persuade them to allow Chinese immigration.[9]

Traditional concerns for the natural environment were a mixture of spirituality and self-interest on the part of elites. In the early modern period, a new form of environmentalism arose, more utilitarian and driven by concerns for national security. Among the main motivations was ensuring a sustainable supply of timber for urban construction and shipbuilding. In many places, forests were protected, not for the game that inhabited them but for the timber they provided. In the early fourteenth century, Portugal reforested the pine groves of Leiria and appointed mounted forest rangers to guard them.[10] In the late fifteenth century, Venice, which depended on timber for its ships and as pilings for its buildings, established state forests protected by forest guards. Its leaders also ordered the planting of trees along the streams that fed into the northern Adriatic Sea to prevent silting of the lagoon on which the city was built.[11] In the seventeenth century, the Japanese government imposed strict rules to protect the archipelago's forests and to preserve the supply of timber for houses, temples, and fortresses.[12] At the same time, the French minister Colbert's Forest Ordinance of 1669 prioritized the state's need for timber over the rights of private landowners.

Colonial Environmentalism

These various customs and policies were but precursors of modern environmentalism. The first attempts to understand and protect the environment as a whole appeared in the island colonies of the European powers in the eighteenth century. As historian Richard Grove has argued, it is in their island colonies—Madeira, St. Helena, Mauritius, St. Vincent, and others—that Europeans first learned how quickly their exploitation could lead to ecological disaster and developed the first programs to mitigate its impact.

Mauritius is a case in point. While it was under Dutch rule from 1638 to 1710, settlers harvested all the accessible hardwood trees. In 1721, the island was taken over by the French, who established slave-using sugar plantations. In 1767, when Pierre Poivre, a botanist with experience in South and East Asia, became governor, he founded the botanical garden of Pamplemousses, the first in the southern hemisphere. With the help of naturalists Philibert Commerson and Jacques-Henri Bernardin de Saint-Pierre, he undertook the first scientific survey of the island's environment. Inspired by the ideas of philosopher Jean-Jacques Rousseau as well as by Colbert's Forest Ordinance of 1669, they attempted to undo the damage done to the environment by the timber merchants and plantation owners. One-quarter of all landholdings had to remain forested to prevent erosion, and all trees within 200 meters of bodies of water were protected by law. Later ordinances established a professional forest service and forbade clearing trees more than one-third the way up the sides of mountains. These measures were aimed at preventing the complete deforestation of the island, as had happened to Madeira under Portuguese rule two centuries before.[13]

As the environmental impacts of colonialism became glaring, Europeans overseas responded in several ways. One was to engage in scientific research. In the eighteenth century, colonial administrators employed physicians or surgeons interested in natural sciences not only to care for the health of Europeans but also to develop economic resources such as spices and tropical hardwood trees. By 1838, the British East India Company employed 800 surgeons who produced a flood of information on the natural environments of India. In correspondence with scientists in Europe, they also transmitted ideas about the exploitation of tropical colonies and its impact on the colonies' environments.[14]

A persistent ecological theory was desiccation, namely, the idea that deforestation created drought. This idea had a long pedigree. Theophrastus of Eresos, known as the "father of botany," believed that the deforestation of Greece and Crete had led to a decline in rainfall. Following in his footsteps, eighteenth-century physiologist Stephen Hales established a relationship between trees and rainfall and argued that one-fifth of the Caribbean islands should be set aside as a "rain reserve."[15]

Another response to the economic development of the time was economic conservationism, which aimed to maximize the productivity of the environment in a sustainable way. In response to the increasing depredations brought on by colonialism and global trade, Europeans' concern for environmental issues spread from isolated islands and enclaves to much of the world. This interest was focused on specific environments and particular resources within those environments. Deforestation in India and the rising demand for timber for shipbuilding, railroads, and construction led the British to adopt German scientific forestry practice. The same motivation soon inspired similar policies in Australia, the Cape Colony, Indochina, and Java.

In Latin America, royal administrators had long regulated the timber trade, reserving the best trees for their countries' navies. Independence gave powerful men the opportunity to seize royal and Indian lands and sell the timber they contained for quick profits. By the 1860s, governments began intervening to mitigate the deforestation. In Mexico, President Benito Juárez introduced the first forest law in 1861. In Brazil, after the Tijuca forest overlooking Rio de Janeiro had been denuded by coffee planters, it was replanted with 72,000 seedlings in 1861 to protect the watershed on which the city relied for its water. In Peru, legislation passed in 1906 encouraged the killing of condors, gulls, and falcons to protect the cormorants that produced guano on offshore islands—a crude form of environmental management.[16]

In Africa, European hunters with powerful rifles threatened the population of wild game and led to fears of extinction for many species. In 1907, British conservationists and hunters worried about the depletion of game founded the Society for the Preservation of the Wild Fauna of the Empire.[17] Two years later, delegates of the European colonial powers signed a Convention for the Preservation of Animals, Birds and Fish in Africa. This treaty gave complete protection to gorillas, giraffes, and chimpanzees and required licenses to hunt elephants and gazelles; lions and leopards, however, were designated as "vermin" and bounties were offered for their killing. South Africa and Southern Rhodesia, both white-settler colonies, established national parks and game reserves designed to exclude Africans and to reserve big game for white hunters.[18] In the words of historian Mark Cioc, such agreements were "best understood as international hunting treaties rather than as conservation treaties."[19]

American Environmentalism Before the 1960s

To the ideas and practices of Europeans, Americans added the concept of wilderness. Of course, true wilderness did not exist in the lower forty-eight states and territories of the United States, for this part of the world had been inhabited for thousands of years by Indians who left their mark on the environment. But in the eyes of most whites, whatever was not occupied by people of European origin was a wilderness, and the inhabitants thereof were mostly "wild" as well. As the frontier between white-dominated and Indian territories moved westward, Americans debated about the "wild" environment and what to do with it.

Conservationism

One school of thought, called conservationism, argued that newly acquired environments in the West should be exploited for the benefit of humans, but in an efficient and sustainable way, either by the government or under government supervision, in order to avoid the wholesale clear-cutting that had occurred in northern Michigan and Wisconsin. Conservationists argued that collaboration between foresters, irrigation engineers, soil scientists, wildlife managers, and other experts would create managed environments that would both benefit society and protect plants, animals, and landscapes for generations to come.[20]

Foremost among the thinkers who espoused this view of nature was George Perkins Marsh, a native of Vermont who traveled widely throughout the Middle East and Europe. In his seminal book *Man and Nature; or, Physical Geography as Modified by Human Action* (1864), he noted with dismay that "man is everywhere a disturbing agent. Wherever he plants his foot, the harmonies of nature are turned to discords. The proportions and accommodations which insured the stability of existing arrangements are overthrown."[21] To prevent further damage, he argued that the power that humans possessed to transform the natural world should entail a commensurate sense of responsibility.

Preservationism

Not all people concerned with the natural environment thought in terms of resource sustainability. Some thinkers, appalled by the ugliness of industrialization, formed more ethical and aesthetic responses. In England, the back-to-the-land movement was part of the romantic movement. Its followers advocated a return to the land, or rather to a mythical rural landscape of forests and mountains and lakes and villages. Among its more famous proponents were the British poets William Wordsworth and John Clare and the writers John Ruskin and William Morris. Their influence led to the establishment of the National Trust in Britain to preserve land and old buildings. In Germany, the back-to-the-land movement came later than in Britain; its best-known advocate was the poet Rainer Maria Rilke.[22] And in the United States, Henry David Thoreau famously celebrated the blessings of nature and simple living in his book *Walden; or, Life in the Woods*.[23]

In the United States, a movement called preservationism emerged. Its apostle was the naturalist John Muir, who spent much of his life in California and the American West extolling the beauty of the natural world. He saw nature as existing not only to benefit humans but as having an intrinsic value of its own: "Nature's object in making animals and plants might possibly be first of all the happiness of each one of them, not the creation of all for the happiness of one. Why should man value himself as more than a small part of the one great unit of creation?"[24] His was an aesthetic and spiritual view of nature as an antidote to civilization "to ease the mind and sanctify the spirit." But Muir was also a practical man who founded the Sierra Club, the first environmental organization, in 1892. He befriended Theodore Roosevelt and was instrumental in the creation of national parks in the West.[25]

Another very influential figure in the American environmental movement was Aldo Leopold. At first, as professor of game management at the University of Wisconsin, he advocated the extermination of predators for the benefit of ranchers and hunters.[26] Gradually, however, he evolved from a conservationist into an environmental ethicist and helped found the Wilderness

Society. A visit to Europe led him to complain that the Germans "taught the world to plant trees like cabbages."[27] In his seminal work *A Sand County Almanac*, published in 1949 shortly after his death, Leopold wrote: "Anything is right when it tends to preserve the integrity, stability, and beauty of the biotic community. It is wrong when it tends otherwise."[28]

American Environmental Policies before the 1960s

The ideas of the environmentalists and the need to settle a continent in a more orderly manner led the United States government to implement a number of policies. One was the creation of Yellowstone National Park, the first of its kind. Covering over 4,000 square kilometers, it was carved out of the Wyoming Territory in 1872. In 1888, New York State created the Adirondack State Park covering 2,400 square kilometers. The purpose of these parks was not to preserve natural environments for their own sake but as recreation areas for an increasingly urban population and as symbols of America's uniqueness, a patrimony as distinctive as Europe's castles, cathedrals, and other historic monuments.

Federal interest in natural areas was not limited to national parks. Presidents Benjamin Harrison (1889–1893) and Grover Cleveland (1893–1897) designated 142,000 square kilometers of public land as forest reserves, making the federal government the landowner of most of the West. President Theodore Roosevelt (1901–1909), an enthusiastic hunter and outdoorsman, carried this policy much further by increasing the size of the national forests to over 700,000 square kilometers. He declared, "Forest protection is not an end in itself; it is a means to increase and sustain the resources of our country and the industries which depend upon them. The preservation of our forests is an imperative business necessity." He also established fifty-one national wildlife refuges and several more national parks and persuaded Congress to pass the Reclamation Act of 1902, which put the federal government in the business of building dams and canals and controlling water supplies in the arid West.[29]

To implement his policies, Roosevelt chose his friend Gifford Pinchot as head of the United States Forest Service. Pinchot exemplified the conservationist approach to forestry; as he wrote, "The object of our forest policy is not to preserve the forests because they are beautiful . . . or because they are refuges for the wild creatures of the wilderness. The forests are to be used by man. Every other consideration comes secondary." A student of German forestry, he believed that only the government, not private individuals and enterprises, could be entrusted with managing the forests for the long term. Yet his belief in state-managed forestry did not long survive his tenure as the nation's chief forester. Soon after his departure from office in 1910, the national forests were opened to exploitation by private lumber companies.[30]

Until 1912, conservationists and preservationists coexisted, for there was room enough in the American West for both wilderness areas and productive forests. The two sides clashed that year over the plans to dam the Tuolumne River in California in order to turn the Hetch Hetchy Valley in Yosemite National Park, one of the most scenic parts of the West, into a reservoir to supply freshwater to the San Francisco Bay area. On one side of the debate were the city of San Francisco, President Woodrow Wilson, and the US Congress; on the other were John Muir and the Sierra Club. Construction of a dam began in 1919 and the valley was flooded in 1923. The project is still controversial, with many preservationists arguing for the restoration of the valley to its pre-1912 state.[31]

Environmental policies returned to the limelight in the 1930s as part of President Franklin Roosevelt's New Deal. One incentive was the Dust Bowl. As it became obvious that monoculture under private enterprise had caused a disaster, the federal government turned to soil ecologists for advice. In 1935, Congress passed the Soil Conservation Act to help states and farmers with soil conservation measures. The government's advice was accepted as long as crop prices were low and farmers were desperate for help; when prices rose again during and after World War II, many of the practices were abandoned.[32] The New Deal also engaged in large-scale environmental transformation in the South, where the Tennessee Valley Authority built thirty-four dams and had an impact on an area of over 100,000 square kilometers.[33]

Automobile Tourism and the Wilderness

The goal of the Roosevelt administration was not to protect the environment but to revive the American economy during the Depression. One key to doing so was automobile tourism. Ever since railroads had crossed the continent, the national parks and scenic areas of the nation had attracted those who could afford first-class accommodations and services in "wilderness" areas. In fact, the railroad companies had lobbied Congress for the creation of national parks as tourist destinations. The Central Pacific Railroad passed close to Yosemite Valley in 1869, but visitors still had to face a long horseback ride into the valley until a stagecoach road was built to transport the less hardy. To make visits to the Grand Canyon easier for tourists, the Santa Fe Railroad built a branch line from its transcontinental route to the south rim in 1901, along with a hotel on the very rim of the canyon, built four years later.[34]

Hardy and wealthy tourists were few in numbers, however, and their impact on the environment was minimal. What turned the national parks into tourist destinations for middle-class Americans was the automobile. Automobiles arrived in the parks early; by 1915, a thousand cars a year were reaching Yellowstone Park. But it was the New Deal that truly brought cars into the wilderness. Federal programs built roads in scenic areas like the Blue Ridge Parkway begun in 1935 and the Natchez Trace Parkway in 1938, as well as numerous roads to and in the national parks. The Civilian Conservation Corps, whose main purpose was to give employment to young white men, was sent into the parks and scenic areas to build ranger stations, lodges, nature centers, hiking trails, and other amenities. The government also granted concessions to private companies to build hotels, motels, restaurants, souvenir stands, campgrounds, and entertainment facilities in the parks.

Despite these amenities, automobile tourism revived slowly during the Depression. After the Second World War and especially from the 1960s on, came the boom, as thousands, then millions of Americans—for the most part middle-class white suburbanites—took their vacations in the great outdoors. Not only did they come by automobile; they also brought snowmobiles, off-road vehicles, and camping and backpacking equipment, while light airplanes and helicopters offered rides over the Grand Canyon, leaving enough exhaust fumes to blur the view of the other side. Areas that had once been cattle and sheep ranches were crisscrossed with trails for hikers, horseback riders, mountain bikers, and all-terrain vehicle drivers, and dotted with ski runs, white-water rafting concessions, dude ranches, and luxury mansions and condominiums.[35]

The Birth of Political Environmentalism

As late as the 1960s, American environmentalism included two competing philosophies. Conservation was largely concerned with exploiting natural resources in an orderly and sustainable manner. Meanwhile, preservationists believed that areas of particular natural beauty should be preserved for their own sake and for the aesthetic and spiritual uplift of visitors from urban and industrial regions.

The major environmental organizations in the United States worried about wilderness areas, forests, mountains, deserts, and the Great Plains, all places far removed from where most people lived. Environmental questions were, to most Americans, of little relevance to their lives compared to suburbanization, the consumer society, the economy, race relations, and the Cold War. Then, just as the Cold War and mass consumerism were at their peak, environmentalism took on an entirely new and much more powerful form, as people throughout the world began to realize that environmental issues impacted their lives directly.

The Rise of Environmental Awareness

The first environmental movement that was concerned with human health arose in response to nuclear testing. As nuclear explosions in the atmosphere released radioactive fallout, anti-nuclear activists raised a hue and cry about their threat to human health, especially that of children. Protest movements arose not only in the United States and Great Britain but also in Japan where the memory of Hiroshima and Nagasaki were still vivid.[36]

Another factor that triggered the new environmentalism was marine biologist Rachel Carson's *Silent Spring* (1962).[37] Her book was an attack on the indiscriminate use of pesticides and herbicides, not only in agriculture but also along roadsides, in cities, and in suburban backyards. Nowhere did Carson mention John Muir, George Perkins Marsh, Aldo Leopold, or any of the classic environmental organizations, nor anything before World War II. She attacked the foundations of postwar American affluence as expressed by the slogan of the DuPont chemical company: "Better Things for Better Living . . . through Chemistry."

The manufacturers of such chemicals as DDT, 2,4-D, chlordane, and heptachlor claimed that their products affected only the insects or weeds that people wanted to be rid of and were harmless to people and their animals. Carson argued that all living beings interact with their environments, hence any new chemical has effects up and down the food chain and far beyond its intended target. "Along with the possibility of the extinction of mankind by nuclear war, the central problem of our age has therefore become the contamination of man's total environment with substances of incredible potential for harm," she wrote, calling such substances "the elixirs of death."[38]

Carson singled out DDT for special condemnation. This chemical had proved valuable during World War II for ridding the Pacific islands of malaria-infected mosquitoes, saving many American soldiers' lives. After the war, it was quickly adopted by farmers, municipalities, and homeowners. In areas where it was sprayed, not only did insects die, but so did the birds that ate them, and the animals that ate birds, and so on up the food chain. Carson implied that these chemicals, applied in unnecessarily huge doses, threatened human life as well. Her chapter on carcinogens, written as she herself was dying of cancer, proved particularly unsettling. Her argument resonated with a vast audience of bird watchers, nature enthusiasts, wildlife managers, and public health professionals.

Within weeks of publication, her book became a runaway bestseller. Carson was attacked by chemical industry spokesmen on the grounds that she did not have a PhD in chemistry and was a "communist" and a "spinster." It also caught the eye of politicians who sensed a change in the mood of the American people. And it reawakened the old environmental organizations and created new ones. From then on, the environment was not just somewhere "out West." It was everywhere.[39]

Environmental Disasters

A book alone could not have sparked a movement. Many people were shocked by a string of disasters. In 1969, an oil spill polluted the waters off Santa Barbara, California, killing thousands of birds, elephant seals, sea lions, and other wildlife, and harming the fishing and tourist businesses. That same year, the Cuyahoga River, long thick with toxic pollutants, caught fire in Cleveland, Ohio, capturing the attention of the media and the public. In 1978, the inhabitants of Love Canal, a neighborhood of Niagara Falls, New York, learned that several schools and housing developments had been built on top of a toxic waste dump left behind by the Hooker Chemical Company in the 1940s and '50s. Lois Gibbs, a resident of the neighborhood whose son had fallen ill from exposure to chemicals, organized her neighbors and attracted the attention of the national media and politicians; in 1981, she founded the Citizens' Clearinghouse for Hazardous Wastes to coordinate protests against hazardous waste dumps in other parts of the country.[40] In 1979, a nuclear power plant at Three Mile Island in Pennsylvania suffered a meltdown, releasing radioactive coolant into the area. Though no one was killed, the incident aroused worries about the long-term dangers of radioactivity.[41] In 1982, the government revealed that dioxin, a powerful carcinogenic chemical, was included in oil residues that had been sprayed on streets of Times Beach, Missouri, to keep down the dust. After a public outcry, the townspeople were evacuated, their homes were demolished, and the area was sealed off.

Other environmental problems also came to the public's attention, such as air pollution in cities, especially Los Angeles; acid rain from coal-burning power plants that was damaging forests in northeastern North America; the disappearance of wetlands and of the wildlife that depended on them; and the decline of whale populations. These American crises were amplified by headlines about terrible disasters overseas. Toxic chemicals dumped into Minamata Bay, Japan, between 1932 and 1968 sickened thousands of people and caused horrendous birth defects. A toxic gas leak at a chemical plant in Bhopal, India, in 1984 killed several thousand people and injured half a million. And the catastrophic meltdown of a nuclear power plant in Chernobyl, Ukraine, in 1986 spread radioactive particles over much of Europe.

Environmental Movements

The efflorescence of environmental literature and the series of environmental crises and disasters had two consequences: the rise of movements concerned with the entire environment, not just wild places and creatures; and the passage of laws and regulations designed to mitigate or prevent further damage to the environment.

Neither books nor crises can, by themselves, explain the enormous popularity of environmentalism in the 1970s. Polls showed that 82 percent of Americans named control of air and water pollution as the top priority for government action. This was part of a larger phenomenon, as Americans—especially young Americans—protested against racial segregation, the Vietnam War, sexual discrimination, and the consumer society.

Many others made a more lasting commitment by joining environmental organizations. The traditional organizations grew prodigiously. By 1991, the National Wildlife Federation had 5,600,000 members, the Sierra Club 650,000, and the Audubon Society 600,000; overall, the ten largest environmental organizations in the United States had 7,790,000 members. These organizations held public hearings, started public relations campaigns, and lobbied Congress on environmental issues. Thus, when the Bureau of Reclamation announced in 1966 that it planned to build two dams that would flood 240 kilometers of the Colorado River, including part of the Grand Canyon, the Sierra Club aroused public protests that led to the plan's being abandoned two years later. Likewise, the Sierra Club persuaded the Disney Corporation to stop plans to build a ski resort in Mineral King Valley in California, which later became a national park.[42]

Besides these older mainstream organizations, new ones sprang up. Some, like the Environmental Defense Fund (1967), the Natural Resources Defense Council (1971), and the Sierra Club Legal Defense Fund (1971), hired teams of lawyers, economists, and scientists and specialized in environmental litigation. Others, like Friends of the Earth, founded in 1969, and Greenpeace, founded in 1971, engaged in active campaigns that led protests against nuclear testing and whale hunting. Even more radical were groups like Earth First!, "deep ecologists" who demanded a reversal of modern industrial society and engaged in direct-action protests against bulldozers and chainsaws in forests.[43]

Environmental movements also appeared far beyond the big organizations. In the 1970s and '80s, local groups sprang up to protest local environmental issues. Many protests emanated from poor and minority communities deliberately targeted as sites for toxic waste disposal. This movement was called environmental justice because it combined protests against both racial discrimination and environmental abuses. Such was the case of a predominantly black community in Warren County, North Carolina, chosen as both the site of a government planned community called Soul City and as a place to dump polychlorinated biphenyls (PCB) wastes. Protests were led by Cora Tucker, a local resident. By the end of 1992, there were 10,000 such local environmental groups in the United States.[44] While older environmental organizations had traditionally been concerned with the wilderness, the West, and the out of doors, many of the newer ones were inspired, founded, or led by women who were more concerned with the health of their families than with plants, animals, and landscapes.[45]

Laws and Policies

Before 1970, American environmental legislation was scarce and policies were weak. In the 1970s, reflecting the public mood of Earth Day, Congress and President Nixon passed a whole series of environmental laws. The National Environmental Policy Act, signed into law on January 1, 1970, established the Environmental Protection Agency (EPA), which quickly became one of the largest regulatory bodies in the federal government.

The first major piece of legislation was the Clean Air Act of 1970. In the face of opposition by the auto, oil, and coal industries, the government banned the use of lead in gasoline and mandated scrubbers on coal-fired power plants that emitted sulfur dioxide. This was followed in 1972 by the Clean Water Act, the Marine Mammals Protection Act, and the Pesticide Control Act, among others. That year the use of DDT was banned in the United States, though it continued to be manufactured for sale in developing countries. The following year saw the passage of the Endangered Species Act. The last major environmental legislation was the Alaska National Interest Lands Conservation Act of 1980, signed by President Jimmy Carter. It created ten new national parks covering 176,880 square kilometers and added 228,240 square kilometers to the wilderness preservation system and 13,560 square kilometers to the national forests. While these acts did not immediately put a stop to the forbidden activities in their areas, they gave the big environmental organizations the opportunity to sue polluting corporations and municipalities; this was the American way to achieve progress. Altogether, between 1972 and 1988, federal expenditures for pollution abatement, regulation and monitoring, and research and development came to over $1 trillion, or $60.3 billion per year on average.[46]

The backlash was not long in coming. President Ronald Reagan (1981–1989) opposed excessive government regulations as a hindrance to business and reined in the EPA's ability to enforce the environmental laws. Later presidents—George H. W. Bush, Bill Clinton, and George W. Bush—proclaimed themselves "environment-friendly" but did little to reverse the course of American policy since 1980. Environmentalism, once a subject of bipartisan enthusiasm, became, like so much else, a contentious issue in a political system divided between Republicans and Democrats. Still, the political conflicts on environmental issues cannot wholly be blamed on the politicians. They reflected the views of the American public. Satisfied that the most crucial environmental issues had been dealt with, more worried

about jobs, the economy, and their standard of living, Americans had become bored or cynical about past crises.[47]

Environmentalism in Other Countries

Of all the nations of the world, the United States has been at the forefront in
environmental ideas and literature and in politics, laws, and regulations. That
is neither because the United States was the most polluted nation on Earth
nor because its people had the most delicate feelings toward nature. Perhaps
it was because the United States had undergone a rapid and uncontrolled industrialization next to a thinly populated "wilderness," where its effects were
dramatically obvious: forests clear-cut, bison carcasses covering the plains,
entire species of birds disappearing almost overnight, suburbs spreading
over huge areas, towns built on toxic dumps, and so much else. But there
is also the fact that Americans are opinionated and seldom intimidated by
governments or corporations. Democracy and a free press brought environmentalism, like so many other issues, out into the open. Other nations developed their own versions of environmentalism, some more and some less
inhibited than the American version.

Germany

In Germany, environmental concerns arose during the Second Empire
(1871–1918) in response to runaway industrialization and the massive deterioration of the environment that it caused. Thinkers like geographer
Friedrich Ratzel and botanist Hugo Conwentz associated the love of nature
with love of the Fatherland. Meanwhile, *Naturschutz* (nature protection)
societies arose to protect particular parts of the environment from development and deterioration. These groups included scientific societies, professional organizations of foresters and fisherman, public health officials, and
outdoor enthusiasts' clubs such as the League for Bird Protection and the Isar
Valley Society. A popular movement arose in Berlin in 1908–1915 when the
inhabitants, with the support of Kaiser Wilhelm, prevented the Prussian state
from clearing the Grünewald, a forest outside of Berlin, to make way for a
housing project.[48]

After World War I, the Nazis appropriated the jargon of the conservationist movement, stressing their love of the land, the soil, the forests, and the peasantry. After coming to power in 1933, the Nazis passed the *Reichsjagdgesetz* (Reich Hunting Law) in 1934 that banned hunting and made Hermann Goering master of the German forests. In 1935 they introduced the *Reichsnaturschutzgesetz*, or Reich Nature Protection Law. Once in power, however, the Nazi regime built superhighways and encouraged industrial enterprises, especially those concerned with rearmament. The Second World War overshadowed all environmental concerns and did considerable damage to rural as well as urban environments, especially in the east.[49]

One unfortunate consequence was to give German environmentalism a bad name. Only gradually did the nationalist and racist arguments for *Naturschutz* evaporate. In their place emerged a new environmentalism concerned with air pollution, highway construction, airport expansion, and other local issues. In the process, *Naturschutz* was replaced by *Umweltschutz*, or protection of the environment. There was also a great deal of concern regarding air pollution in industrial areas where coal was the main source of energy. Efforts to abate air pollution date back to the 1950s, starting in Rhineland-Westphalia, the state that encompasses the Ruhr, the heartland of German heavy industry. They culminated in the passage in West Germany of a Clean Air Act in 1970, followed by a series of other environmental protection laws and regulations.[50] East Germany, meanwhile, took a different path. To avoid importing fuel, industries and buildings used brown coal, which was abundant locally but which left a pall of smog over the entire country.

Most striking was the rise of a radical anti-nuclear movement in West Germany. Some of it was aimed at the proliferation of nuclear weapons in the hands of American and Soviet forces, based on a legitimate fear that the next world war would start with a massive nuclear bombardment on German soil. Much, however, was aimed at the proliferation of nuclear power plants and the fear that they might accidentally release radioactive particles or be sabotaged by terrorists. Lawsuits and local protests, some peaceful, others violent, succeeded in delaying the construction of nuclear power plants in some places and preventing them in others. This environmental movement was more concerned with the survival of humanity than with the rest of nature, but it was closely associated with more traditional environmental concerns.

What most clearly distinguished West German from American environmentalism was the Green Party that sprang up in the late 1970s as an out-
growth of radical, feminist, Marxist, and anti-nuclear groups. Formed as a
party in 1980s and led by the charismatic Petra Kelly, the Greens won seats
in the Bundestag (the German parliament) in 1983. In alliance with green
parties from Belgium and the Netherlands, it also won seats in the European
Parliament in 1984. By 2013, it held sixty-three seats (out of 630) in the
Bundestag and exercised a strong influence on the Social Democratic Party.[51]

As a result of these various movements at the local and national levels, the
government enacted environmental policies and improvements. German
industries have led the world in the development of pollution-control and
clean-energy technologies. Nuclear power plants are being phased out. Air
quality improved considerably; by 2000, particulate matter had declined by
88 percent, lead by 98 percent, sulfur dioxide by 544 percent, and carbon dioxide by 25 percent.[52]

The Soviet Union and Its Successors

Russia/USSR faced a very different environmental situation from that of the
United States and Germany. It was an enormous land with seemingly limit-
less forests, minerals, and other natural resources. Yet it had, until well into
the twentieth century, a poorly developed economy and a state that was both
tyrannical and desperate to industrialize, which led to a shocking neglect of
the environment.

Between the end of the Russian Civil War in 1921 and the beginning of
Stalin's Five-Year Plans in 1928, biology and ecology flourished briefly
among scientists with an interest in nature. In 1921, Lenin issued a decree for
"the Protection of Monuments of Nature, Gardens and Parks" and, in 1923,
a Forest Code aimed at reforestation and sustained-yield logging. By 1929,
zapovedniki or nature preserves covering almost 4 million hectares were devoted to scientific research into sustainable exploitation.[53]

The Five-Year Plans put an end to these benevolent attitudes toward nature. In its effort to industrialize at maximum speed, the Communist Party
sought the highest possible yields and the destruction of the "useless" parts
of nature. While paying lip service to environmental and human health, the
Stalin regime distrusted environmentalists and conservationists. In order to
increase timber production from 178 million to 280 million cubic meters per

year, forests were clear-cut. Protected areas were reduced from 125,000 to 15,000 square kilometers.[54]

Nonetheless, throughout the Stalin period, a small and inconspicuous conservationist movement survived. Scientists associated with the All-Russian Society for the Protection of Nature, the Moscow Society of Naturalists, the Geographical Society of the USSR, and the All-Union Botanical Society remained independent of party control. They also succeeded in keeping the remaining zapovedniki off-limits to any uses but scientific research.[55]

Until the 1980s, the regime hushed up cases of damage to the environment, especially those caused by the armed forces or military industries. The nuclear meltdown at Chernobyl in 1986, which could not be hidden, triggered a public outbreak of protests about other environmental issues, such as the paper mill that dumped toxic effluents into Lake Baikal in Siberia, the largest body of clean freshwater in the world. In the early 1990s, as the Soviet Union unraveled, environmentalists often allied themselves with pro-democracy movements. In Lithuania, Latvia, Estonia, and Armenia, "green" parties won seats in local parliaments. In Ukraine, the Rukh party that emerged after Chernobyl called for political and economic autonomy. After a secret underground nuclear test in Kazakhstan in 1989 leaked radioactive gases into the atmosphere, the Republic of Kazakhstan declared its independence in 1991 and closed the nuclear weapons test site at Semipalatinsk in 2000.[56]

Even in the Russian Republic, political protests and electoral campaigns centered around environmental issues. As the Russian economy imploded, polluting industries were shut down. Environmental non-governmental organizations (NGOs) were founded and laws were passed protecting the environment. Yet, according to Aleksey Yablokov, head of the non-governmental Center for Russian Environmental Policy, Russia continued to lose 16 million hectares of Siberian forest every year to cutting, pollution, and fires, more than were lost in Amazonia.[57] In 1990, the State Committee on Environmental Protection or *Goskompriroda* presented a plan to clean up the environment. Its chairman, Nikolai Vorontsov, wrote: "In contrast to the economy, ecology requires not 10-year planning, but 50-year planning at a minimum, long-range, extraordinary decisions." Retrofitting Soviet-era installations with anti-pollution devices would be prohibitive; just replacing defective municipal water pipes with pipes fitted with anti-corrosion linings would cost almost as much as the entire Soviet gross domestic product for 1991. According to Communist Party chairman Mikhail Gorbachev, the cost of containing and decontaminating the Chernobyl nuclear power plant—18 billion rubles (about $18 billion at the time)—virtually bankrupted the Soviet Union, not counting the loss of productive land and resources and the on-going social costs to the peoples of the affected areas.[58] To this day, Russia and the former Soviet republics contain some of the most damaged and polluted areas on Earth, especially those contaminated by radioactivity. In 2007, when President Vladimir Putin eliminated the last of the forest preserves, there was barely any protest. Environmentalism in Russia had largely subsided.[59]

China

China today faces daunting environmental problems. In the north and west, water is scarce and increasingly polluted. In the east and south, water is abundant and floods still threaten. Forests are almost gone. Grasslands are overgrazed. Forty percent of wild mammals and over 70 percent of wild plant species in China are endangered; one out of every four of the world's endangered species of wildlife is found in China. Air quality in major cities is among the worst in the world. Water quality is just as bad.

The Chinese leaders have become increasingly aware of the environmental costs of rapid development. Under Deng Xiaoping (1978–1997), they viewed environmental problems as inevitable byproducts of industrialization. A conference in 1984 declared environmental protection to be a "strategic task," yet prioritized economic development. After Deng, the rhetoric changed, perhaps in response to international pressures. The tenth Five-Year Plan (2001–2005) made environmental protection a priority. In 2002, President Hu Jintao's Scientific Development Concept included "a harmonious development between Man and Nature." In 2006, the Chinese Communist Party announced that "promoting harmony between man and nature" was an important step in building a "harmonious society" and Premier Wen Jibao stressed the need "to move from a model of growth that stresses the economy to one which balances the economy and the environment."[60]

Official lip service to environmental concerns was reflected in changes in administration. The Environmental Protection Leading Group, founded in 1974, met only twice in nine years and had little impact. In 1984 the government established an Environmental Protection Commission, upgraded it in 1989 to a National Environmental Protection Agency, in 1998 to a State Environmental Protection Administration, and in 2008 to a full-fledged Ministry of Environmental Protection.[61]

Despite the rhetoric and the bureaucracy, the Chinese environment has continued to deteriorate. The Three Gorges Dam project, designed to prevent floods as well as to produce electricity, has changed the ecology of the middle Yangzi River Valley and poses a risk of earthquakes. China's Go West Campaign has encouraged Han Chinese to resettle in Tibet and Xinjiang in order for the government to maintain its control over the Tibetan, Uighur, and other non-Han inhabitants and over the natural resources of those far western regions. In their eagerness to prosper in their new environments, the settlers have caused rampant deforestation, overgrazing, and the extinction of native plants and animals. Overgrazing has turned grasslands into semi-deserts, causing dust storms that reach the more densely populated eastern parts of China. Mines, oil wells, pipelines, railroads, and roads are transforming these once-preindustrial regions as they did the American West.[62]

A few of the central government's initiatives have been environmentally sound, if not always effective. The Three Norths Shelter Project is aimed at stabilizing the soil and stopping the southward advance of the northern desert by planting trees on 4 million square kilometers of land along a 5,000-kilometer-long belt. Commercial logging has ceased in the upper Yangzi and mid-to-upper Yellow River areas, and plantations of saplings now grow where mature forests once stood.[63]

The government has also established nature reserves. By the end of 2004, China had over 2,000 of them, covering nearly 14 percent of its area. Many, however, have no budget, staff, or boundaries, and exist only on paper. Without oversight—or with the connivance of local officials—trees are felled, tigers are hunted, and other resources are exploited by hunters, fishers, quarriers, and loggers. The Chang Tang Reserve in northwestern Tibet, established in 1993, is the largest in the world, covering 284,000 square kilometers; once the home of wild yaks, gazelles, antelopes, sheep, and other wild ungulates, it has since been occupied by over 3,000 families and a million head of livestock.[64]

The main reason the rhetoric, the bureaucracy, and the laws have not slowed down the deterioration of the environment is because decision making does not lie with the central government but with provincial and local authorities. Local Environmental Protection Bureaus are understaffed and their personnel and finances are controlled by local governments. The income of these bureaus comes from fees levied on polluters, a disincentive to combat pollution. Local officials, encouraged to develop their local economies, overlook environmental standards and cover up offenses.[65]

While local officials have ignored or concealed environmental problems, the Chinese public has become increasingly aware of them, especially industrial pollution that threatens their health. The number of rural "mass incidents" rose from 10,000 in 1994 to 74,000 in 2004, 5,000 of which were about environmental issues. In April 2005, 30,000 to 40,000 villagers in Zhejiang Province protested against thirteen chemical factories until the Ministry of Environmental Protection ordered them shut down. The number of environmental "disputes" and "complaints" has also grown. Civic environmental movements have been notoriously weak, however, for the state has repressed all forms of independent civic action.

In 1994, in an effort to keep these popular feelings under its control, the government founded Friends of Nature, the first environmental non-governmental organization. By 2008, 3,539 environmental NGOs had been founded in China. One-third are supported by the government, but all need government approval. In the words of historian Bao Maohong, they know to "practice discretion and not overstep their bounds." Nonetheless, they have had some successes, such as postponing the construction of dams on the Nu (or Salween) River in the far south, one of the last wild rivers in China. They also persuaded the central government to save the endangered snub-nosed monkey of Yunnan from loggers and the rare Tibetan antelope from poachers, but failed to save the Yangzi dolphin from extinction.[66]

Since 2000, the Chinese government, aware of the dangers of global warming and of untrammeled industrialization, has become environmentally conscious. It has signed international environmental accords and pledged to reduce its carbon emissions. It is also promoting clean energy; as of 2012, China led the world in wind farms, solar panels, and electric vehicles.[67] Over its long history, China has grown in population and production, despite periodic environmental catastrophes and demographic collapses. Recently, the rate of growth has accelerated. Is it sustainable, or must runaway growth inevitably lead to a crash?

The Environmentalism of the Poor

In industrialized nations—now including China—environmentalism focuses on two kinds of issues: damage to the natural environment and threats to human health. Some, like the United States and Germany, have managed to combine environmental protection with satisfactory economic growth. In others, like the former Soviet Union and China, environmental protection has proved to be politically divisive and largely suppressed.

A third kind of environmentalism is most often found in the developing nations of the Global South. This involves protests against the harm that peasants and hunter-gatherers suffer when the environments upon which their livelihood depends are affected by modernization. People in traditional rural occupations rely on forests, bush, and grasslands for fuelwood, game, edible plants, pastures for their livestock, and building materials. Modernization projects such as commercial logging, fishing, paper mills, dams, oil wells, and mines and smelters threaten their sources of livelihood. To many of the rural poor, certain parts of nature, such as mountains and forests, also have sacred values.[68]

Protests against modernization have erupted in India against a dam, in Thailand against eucalyptus plantations, in Nigeria against oil installations, in Brazil against agribusinesses, and in Kenya and Sarawak against deforestation, among others. Unlike the environmentalists in the United States who lobby Congress and use lawsuits to achieve their goals, those in poor countries more often use direct action, such as staging demonstrations and sit-down strikes, blockading roads, and employing other forms of protests similar to those of Mahatma Gandhi and the American civil rights movement.

The protests are often led by women who, in many societies, are the ones who gather firewood, tend animals, collect freshwater, and harvest edible plants. Such was the Green Belt Movement in Kenya, founded by Wangari Matthai in 1977, that has demonstrated for environmental conservation, women's rights, and reforestation. In many ways, their movements combine human rights and environmental protection; as such, they resemble the environmental justice movement in the United States.[69]

India

Environmentalism in India arose from a conflict over forests. Scientific forestry had long aroused sporadic protests from dispossessed rural people. During India's struggle for independence, Mahatma Gandhi's vision for his country favored rural villages over cities and industries. As he said in 1928, "God forbid that India should ever take up industrialization after the manner of the West. The economic imperialism of a single tiny island kingdom

[Britain] is today keeping the world in chains. If an entire nation of 300 million took to similar economic exploitation, it would strip the world bare like locusts."[70]

After independence, Gandhi's successors embraced industrialization, state-led developmentalism, and Western-style consumerism. They privileged mammoth projects such as steel mills, large dams, and nuclear power plants.[71] In this context, two environmental movements arose. Middle-class environmentalism, drawing its inspiration from Western models, was primarily concerned with wildlife preservation and urban air pollution. Popular environmentalism, meanwhile, aimed to protect the rural poor and marginal "tribal" peoples from the negative impacts of modernization.

As the Indian economy grew, so did the demand for railroad ties, timber, paper, plywood, even cricket bats, while agriculture encroached on forests. In reaction to logging and deforestation, protests erupted among people whose livelihood depended on access to forests. The first such protest to gain international attention was the Chipko movement that began in 1973 when a group of villagers in the foothills of the Himalayas stopped loggers from felling a stand of hornbeam trees; the movement was named Chipko, meaning "to hug," after the protestors' tactic of hugging trees to prevent logging. As word of the protest spread, similar demonstrations erupted throughout northern India's forest belt. The demonstrations effectively halted commercial forestry in the Indian Himalayas.[72] Popular environmental movements soon expanded to protest the erection of dams and the appropriation of coastal lands for commercial shrimp and prawn aquaculture. In many cases, they merged with movements for women's rights, temperance, and local forest-based industries.

Since the 1960s, India has witnessed many conflicts over the use of land, forest management, mining, fishing, dam building, and industrial pollution.[73] In the 1980s, the national government created a Ministry of the Environment and Forests, ostensibly to prevent rapacious commercial logging. During the 1990s, however, the government's interest in economic development and global trade trumped environmental concerns. The new ministry, created more to bolster India's international prestige than to enforce real reforms, became mired in disputes with other, more powerful and better funded ministries.[74] In the euphoria of a booming economy, both Gandhian ideals and peasant environmentalism have largely been sidelined.

Brazil

Almost all the nations of Latin America have environmental protection laws on their books, but enforcement ranges from non-existent (in the case of Haiti) to very effective (in Costa Rica), often at the whim of the person or group in power.[75]

In Brazil, under the military dictatorship from 1964 to 1988, building the new capital city of Brasilia, pushing highways through the Amazon forest, and developing industries overshadowed concerns for the environment.[76] In response, two distinct forms of environmentalism appeared in the 1970s and '80s. The first, led by middle-class activists inspired by developments abroad, focused on specific projects, such as plans to build airports in forest preserves or high-rise buildings near beaches. This largely urban movement received a strong impulse from the problems that arose in Cubatão, one of Brazil's fastest growing industrial cities. By the early 1980s, its air was so polluted that many inhabitants suffered from respiratory problems and birth defects. Then, in 1984, a pipeline burst, killing hundreds and destroying an entire neighborhood. This disaster awakened a new environmental consciousness among Brazil's urban citizens.[77]

The other movement was similar to that of India in some ways, but not in others. Small farmers began organizing against ranchers, loggers, and large hydroelectric projects. A Movement of Landless Rural Workers was organized in 1984. The following year, the rubber tappers and Brazil-nut collectors in Amazonia organized themselves under the leadership of Francisco "Chico" Mendes. Organizations of indigenous people, rubber tappers, and smallholders got support and considerable publicity from international organizations such as the World Wildlife Fund, the National Wildlife Federation, and the Environmental Defense Fund. Their protests against the exploitation of the poor and the destruction of the rainforest led a ranchers' association to have Mendes assassinated in 1988.[78]

After 1990, with democracy reestablished in Brazil, environmental NGOs proliferated, and political, religious, and business leaders began to espouse environmental values. Under pressure from environmentalists both within and outside of Brazil, the government began to pay more respect to the forest and its inhabitants. Indigenous peoples' reserves were established in 22 percent of Amazonia; in addition, environmental reserves and national forests covered half as much as the indigenous peoples' reserves. Unlike in the United States, where national parks are separate and distinct from Indian reservations, Brazil's indigenous reserves are a combination of the two, on the grounds that Indians tend to be good custodians of the rainforest. Yet two-thirds of Amazonia does not enjoy protected status, and to police all the reserves is impossible. Deforestation continues, but the rate of deforestation in Amazonia declined by 70 percent between 2004 and 2010.[79]

As these examples show, environmentalism has taken different forms in different countries. One major factor is the power of the state and of the ruling party. In democracies like the United States and Germany, environmental movements and organizations have been very effective in bringing about changes in the laws and improvements in environmental protection. In countries under authoritarian governments with strong developmentalist agendas, environmental movements have been largely ignored or repressed. In Brazil and India, environmental movements and organizations have competed with developmentalist policies, with mixed results.

Environmentalism and International Relations

Many environmental issues—acid rain, climate change, the ozone layer, and the precarious condition of whales and ocean fish, to name a few—are not confined to any nation. Hence the long-felt need for environmental laws and treaties that transcend national boundaries.

The Stockholm Conference of 1972

The growing interest in environmental issues after 1970 and the success of Earth Day led the United Nations to call a Conference on the Human Environment in Stockholm in 1972. Delegates from 113 nations agreed, in principle and without dissent, to protect and improve the human environment, by which they meant bettering people's health, education, and living standards. They also agreed that nations have "the sovereign right to exploit their own resources" and that environmental policies should "not adversely affect" the development of Third World nations.[80] In spite of these agreements, the conference allowed various nations to push their own agendas. The Chinese delegation blamed the world's environmental problems on capitalism and/or on Soviet revisionism and asserted that

China's goal was to "develop first and address environmental challenges one by one."[81] According to historian Shawn William Miller, "One Brazilian official argued that if anything, Brazil wanted more pollution, for this was an excellent indicator of the progress of national development."[82] And the United States tried to water down a statement on toxic substances and abstained on the question of testing nuclear weapons.[83]

Ironically, the United States, which had been at the forefront of domestic environmental policies and laws, was the most resistant to signing international treaties and conventions on environmental issues, on the grounds that they would limit its national sovereignty and its business interests. In the 1970s, it refused to sign the Law of the Seas treaty, which would have limited fishing in international waters. In 1981 it was the only nation to vote against UN Resolution 37/137, which addressed Protection against Products Harmful to Health and the Environment. In 1989, it was one of four nations that refused to sign a Convention on the Control of Transboundary Movements of Hazardous Wastes and Their Disposal. In the 1990s, it refused to sign a treaty protecting Antarctic resources.[84]

The United States did, however, agree to some environmental conventions. The most important was the Montreal Protocol on Substances that Deplete the Ozone Layer, which came into effect in 1989. It was aimed at stopping the production of chlorofluorocarbons (CFCs), a chemical used in aerosol cans and in refrigeration and air conditioning equipment that had caused a hole to appear in the ozone layer protecting the Earth from dangerous ultraviolet radiation. Though some industrialists protested, substitutions for CFCs were readily available and the ban caused no lasting harm. Similarly, the United States was a signatory of the Convention on the International Trade in Endangered Species of Wild Flora and Fauna, which became law in 1975.[85]

The Rio Conference of 1992 and After

The Stockholm conference of 1972 resulted only in vague assurances. In 1992, the United Nations organized a Conference on Environment and Development to be held in Rio de Janeiro. Most of the attention was focused on climate change, but deforestation and biodiversity were discussed as well. Developing nations were opposed to putting deforestation on the agenda, for the products of forests and of deforested land were major contributors to their economic growth. Meanwhile, they demanded that rich countries pay for the genetic information derived from their plants and animals by pharmaceutical and biotechnology companies, an initiative that the United States resisted.

In the end, the Rio Conference produced few tangible results. Although the United States signed the UN Framework on Climate Change, it refused to sign an agreement on biodiversity. In response to the demand by developing countries that the rich nations help them develop in an environmentally benign manner, Germany pledged $6.3 billion, the European Economic Community $4 billion, Japan $1.4 billion, and the United States just $25 million.[86]

Since Rio, the attention of the world's governments has focused on global warming, while deforestation, biodiversity, urban pollution, and other problems have largely been ignored. The United Nations has sponsored a series of conferences: Kyoto 1997, Cancún 1998, Johannesburg 2002, Copenhagen 2009, Warsaw 2013, and Paris 2015. Gradually, more and more governments have come to accept the need to reduce greenhouse gas emissions and their own responsibility to further this goal. Meanwhile, emissions continue and the goal of stopping the rise in global temperatures recedes further into the future.

The diplomatic history of environmentalism remains to be written. Its general outline, however, reveals a giant chasm. Most of the world's governments, influenced by the reports of scientists, are sincerely worried about the global environment, especially climate change. Yet several countries—notably the biggest contributors to environmental problems—find it politically difficult to implement the agreements reached by their diplomats. One reason is the opposition of corporations with a vested interest in the status quo, such as coal, oil, and chemical companies and automobile manufacturers (most recently Volkswagen). Another is the waxing and waning of support from a public torn between concern for the environment and worries about jobs and income. And a third—perhaps the most pervasive—is the continued belief in endless economic growth and the threat to that belief represented by the idea of living on a planet with finite resources.

Environmentalism and the Environment

There are several reasons that the environmental movement arose in the late twentieth century and not before. One was the fear of radioactivity from nuclear testing, power plant accidents, and the threat of nuclear war. Another was the growth of chemical industries and their production or accidental emissions of toxic substances. Yet another was the pollution of the air and water by vehicles, power plants, the coal and oil industries, and agricultural chemicals. Taken together, these and other changes in the power of humans over nature created a backlash both in wealthy industrial countries and in those just beginning to industrialize. The environmental movements that grew from this backlash have attempted to mitigate the effects of population growth and economic development, with varying degrees of effectiveness.

In the United States, the Endangered Species Act of 1973 has reversed the declining numbers of once endangered alligators, peregrine falcons, brown pelicans, and several other species.[87] In 1995, the US Fish and Wildlife Service reported that 10 percent of endangered species were improving, 40 percent were declining, and the rest were stable or unknown.[88] Other changes have been positive as well. DDT has been banned and other pesticides are controlled, reducing the threat to birds. The runoff of agricultural fertilizers and the dumping of industrial wastes into lakes and rivers have also been reduced; as a result, the water quality of the Cuyahoga River and Lake Erie have improved dramatically, and fish have returned to the Ohio, Hudson, and other rivers where they had once died out.

Air pollution has been reduced, or at least has leveled off, thanks to automobile emission controls and scrubbers on fossil-fueled power plants. Gasoline no longer contains lead harmful to growing children. Acid rain, once a scandal, is now seldom mentioned, perhaps because it is no longer doing as much harm to forests as it once did.

Atmospheric nuclear testing has stopped. Since the disasters at Three Mile Island and especially at Chernobyl, nuclear power plants are no longer being built in the West. In fact, since the Fukushima Daiichi meltdown in Japan in 2011, nuclear power plants in Japan and Germany are being decommissioned, as are a few aged ones in the United States. Yet, because the demand for electricity continues to grow, more fossil fuels are being consumed than would be the case with more nuclear plants; it is a trade-off between a constant but certain harm and a low probability of a major catastrophe.

As for fossil fuels, in the United States coal-fired power plants are slowly being supplanted by oil-fired and, more recently, natural gas-fired ones. That is not all for the good, for much oil comes from Alaska, where its extraction and transportation have damaged large areas, and much gas comes from underground formations newly accessible by hydraulic fracturing, with risks of their own. Recycling is now a regular feature of daily life in the United States and other wealthy countries. Yet 90 percent of American garbage still goes unrecycled, and fewer than 5 percent of the thousands of old toxic dumps have been cleaned up.[89] Overall, however, the American environment is no longer in a state of crisis.

In other parts of the world, the outlook varies considerably. In Western Europe, the environment is much better protected than it was in the immediate postwar years. Though crises occur every so often, the days of the dreaded pea-soup fog over London and the foul air over the English Midlands and the German Ruhr are no longer as common as they once were. In Eastern Europe, the legacy of the communist regimes still weighs heavily on the industrial areas. And in Belarus and Ukraine, the radioactive fallout of the Chernobyl disaster will take decades, if not centuries, to clean up. The same is true of Russia and other former Soviet republics, where large tracts will remain polluted or radioactive well into the future, harming the health of current and future generations.

Every week or two the Chinese add another coal-fired power plant, and every year another 17 million or so automobiles. Old cities are being razed to make way for high-rise buildings, while other cities are springing up in what for centuries had been farmland. Pollution of air and water is now among the worst in the world, a danger to the health of China's people. As of this writing, information coming out of China revealed that the nation was burning 17 percent more coal than the government had reported.[90] The Chinese government may be optimistic about the future, but historians are skeptical. Sinologist Robert Marks writes: "Centuries of exhortation to stem deforestation, to halt the degradation of the environment, and to maintain harmony between man and nature, have been followed by even more deforestation, environmental degradation, and loss of habitat and species."[91]

Urban pollution is not limited to China. Other large and fast-growing cities in developing countries—Mexico City, Santiago, São Paulo, Delhi, Bangkok, Cairo, and many others—are suffering from the same delay between urban growth and environmental protection that Manchester and similar Western industrial cities went through a century or more ago.

Conclusion

There is no question that the majority of humanity is better off than it was 100 years earlier. Despite a frenetic growth of population, people in Eurasia and the Americas eat better and live longer, healthier lives than at any time in the past; sub-Saharan Africa is likely to catch up in the twenty-first century. Much of this improvement is a result of industrialization and economic development; the rest is due to environmental movements and policies.

Environmentalism has scored many successes, but these successes have all been regional or local. Government developmentalism, corporate-led globalization, and public mass consumerism have succeeded in transferring environmental problems from one place to another. Today, the air in Los Angeles and London is much cleaner than it was in the 1950s, and the air in Beijing, Delhi, and Cairo is much dirtier.

Clean air and water and the protection of endangered species are necessary to maintain a planet that is livable for both humans and the rest of nature. That is why they have received the support of the public and of governments in so many countries. Yet the action so far is not sufficient. For all their successes, the environmental movements have failed to slow down, let alone reverse, the largest threat of all: global warming

Epilogue

One Past, Many Futures

For the first time in history, humans are having an impact on the entire planet, its climate, its flora and fauna, its landscapes and oceans. The human population will soon reach 10 billion and is fast spreading into the Arctic and the tropical rainforests. The global economy is also growing, lifting entire nations out of poverty. Our technology is advancing fast and making ever greater demands on the planet's natural resources. Meanwhile, areas of wilderness are shrinking, rare plants and animals are disappearing, and the climate is warming. It is time to ponder the future relations between humans and the rest of nature, and what the past can teach us.

The Past

The long shadow of our Paleolithic ancestors hangs over us. Stone Age humans were more intelligent than other creatures, and more creative artistically and technologically. They were opportunistic and adaptable and able to accumulate and share knowledge, thanks to the gift of language. But they were also competitive, voracious, and aggressive, causing the extinction of countless other species.

Farming and herding gave humans a powerful advantage over the rest of nature, turning many species into domesticated plants and animals, or into pests and weeds. These newfound powers also brought out another human trait: territoriality. As humans acquired land and other resources, each group treated others as rivals or as enemies to be killed or enslaved. Later, as farming societies coalesced into states and civilizations, many pushed their exploitation of the environments to the limit of their carrying capacity and beyond, resulting in collapses of both human populations and many species of plants and animals. Though the gains made by humans were periodically reversed by natural disasters, humans soon recovered and regained their power over the rest of nature.

The Industrial Revolution opened up an entirely new phase in the interactions between humans and nature. Access to accumulated stores of fossil fuels multiplied the energy at the disposal of humans. Capitalism helped concentrate labor and wealth to maximize the exploitation of nature. Twentieth-century developmentalism combined the power of states with that of capital and industry to further empower humans in their efforts to dominate the natural world. The result was polities and economies devoted to endless growth and societies to endless consumption. In the face of these advances in human power over nature, conservation, preservation, and environmentalism seem like rearguard actions to save what is left of the natural world. Now, with global warming, human actions are beginning to threaten the future of humankind as well as the rest of nature.

History also teaches another lesson, namely, that humans' relationship with the environment is a two-way street. In the distant past, humans lived (and died) at the mercy of the elements. Erupting volcanos, diseases, droughts and floods, sudden climate changes, earthquakes, and other natural shocks could damage—and sometimes wipe out—whole societies. In addition to completely "natural" events, there were unanticipated and unwanted natural repercussions of human actions, such as the exhaustion of game and other natural resources, the salinization and waterlogging of irrigated land, the metastasis of diseases into pandemics, and the changing climate. In short, nature is not a passive victim but has agency and will play a role in the environment of the future as it has in the past. Given our Paleolithic minds, our traditional institutions, and our modern technologies, how can we face the challenge of a changing planet?

A Triangle of Futures

A triangle of possibilities highlights where the future of the planet is likely to lie.

A Sustainable World

In one corner of the triangle would be a world in which humans have achieved a sustainable balance between their needs and the protection of the biosphere, with a climate that has leveled off and no longer represents a threat. To achieve such a world would require deep changes in our habits and in our very nature. Writers who have contemplated such a future have expressed their hopes with expressions such as "finding the right balance between power and responsibility," "humility and restraint," "wise and responsible management," and "let us hope that we will act wisely."[1] There is no question that some people are capable of exercising wisdom, restraint, and humility, but can mankind as a whole do so, given our cultures and our political and economic institutions?

Business as Usual

In a second corner of the triangle lies the possibility that we will continue along the existing path we have been on—in other words, business as usual. There will be technological advances, to be sure: solar panels, wind farms, electric cars, and so on. But overall, a growing population and a rising average standard of living will mean more resources consumed, more effluents, more greenhouse gases, and therefore a more crowded, hotter, and more polluted world.

It is not hard to imagine such a world. Where once the United States was considered the most modern country, today it is China. Already, Beijing, Shanghai, and other cities are turning into megalopolises, with glittering high-rises, high-speed trains, broad avenues filled with new cars, and tens of millions of people enjoying unprecedented prosperity. But these cities are also more crowded and polluted and subject to more extreme weather than in the past. All the while, China's remaining natural landscapes and wild plants and animals are fast disappearing.

Managing Planet Earth

At the third corner of the triangle, would be a scenario very different from the first two. Rather than denying reality or pinning their hopes on changing human nature, some thinkers have advocated changes in our institutions or new technologies. Scientists have wondered whether humankind should become responsible for managing the planet in a rational way. As early as 1989, the mainstream magazine *Scientific American* devoted an entire issue to "Managing Planet Earth."[2] More recently, Edward O. Wilson has expressed this hope: "We have entered the Century of the Environment, in which the immediate future is usefully conceived as a bottleneck. Science and technology, combined with a lack of self-understanding and a Paleolithic obstinacy, brought us to where we are today. Now science and technology, combined with foresight and moral courage, must see us through the bottleneck and out."[3]

This raises the question: How can we combine—or even imagine combining—science and technology with foresight and moral courage? We can identify two tentative answers to this question: biologists have proposed ways to mitigate the Sixth Extinction, while climate issues have attracted the attention of physicists and chemists.

Managing the Biosphere

Biologists concerned with the threats to wild plants, animals, and their habitats have come up with two proposals. One seeks to preserve entire habitats and their plants and animals from further human encroachment, or even to roll back some of the recent encroachments. In other words, they propose preservationism on a global scale. The other seeks to reverse the Sixth Extinction by resurrecting some of the recently extinct species.

Rewilding

Most living beings are adapted to specific environments; pandas, for instance, can survive only in bamboo forests and spotted owls only in old growth forests in western North America. To save such creatures "in the wild"—that is, outside of artificial human-created environments—means saving their habitats. Since "natural" environments are increasingly limited and fragmented by cities, highways, and other human constructions, a movement has recently arisen to recreate environments that are, if not natural, then as close to natural as possible. This movement, known as "rewilding," aims to reconnect patches of semi-natural areas into large swaths of land where the original native species could thrive once again.[4]

For example, in the southeastern United States, where the longleaf pine forests that once covered 60 percent of the land have been clear-cut, a project is now under way to reestablish longleaf pines on close to 9,000 hectares;

this would allow black bears, gopher tortoises, and bison to live where their ancestors once did. Other "wilderness corridors" have been suggested for Appalachia, northern Canada, a "Western Wildway" stretching from Mexico to Alaska, and even New England.[5]

Another proposal aims at rebuilding American bison herds in a natural environment. Today, of the half million bison in North America, fewer than 20,000 are managed for conservation, while the rest are raised for meat and are gradually becoming domesticated. To save the "true" bison, vast new areas of the Great Plains will have to be set aside.[6]

Another attempt to rewild involves plants in northeastern Siberia. There, ecologists Sergey and Nikita Zimov have established a 14,000-hectare Pleistocene Park and stocked it with elk, moose, reindeer, horses, and bison. By grazing the mossy tundra, these animals are expected to turn it into a form of grassland called mammoth steppe that once covered much of the circum-Arctic land before the mammoths became extinct. Since grass is a better insulator than tundra, the Zimovs expect that replacing tundra with it would prevent the methane and carbon dioxide now locked up in the permafrost under the surface from being released by global warming and accelerating global warming even more.[7]

Even in Europe, the idea of rewilding is starting to catch on. In a dry area of western Spain where cattle ranching is no longer profitable, an organization called Campanarios de Azaba has established a 522-hectare nature reserve with the intention of introducing wild horses, red deer, ibexes, wolves, European bison, and lynxes. Elsewhere a conservation foundation called Rewilding Europe plans half a dozen projects covering 970,000 hectares by the year 2020.[8]

Rewilding, as it is currently being proposed, involves native and still extant plants and animals but in vastly greater areas than they currently occupy. If that seems ambitious, consider a proposal called Pleistocene rewilding. This idea is to introduce, if not the animals that lived in an area before they became extinct, then their nearest analogues as proxies. Thus, in special game parks in North America, African lions would play the role of extinct American lions, African cheetahs of American cheetahs, and elephants of mammoths. This would create a second Serengeti in the American West. Once seemingly far-fetched, the idea is slowly gaining ground among biologists, ranchers, and government officials.[9]

The most radical solution of all is that proposed by biologist Edward O. Wilson: "The only solution to the 'Sixth Extinction' is to increase the area of inviolable natural reserves to half the surface of the Earth or greater. . . . But it also requires a fundamental shift in moral reasoning concerning our relations to the living environment."[10]

These proposals are noble and praiseworthy, but—with the exception of Wilson's half-Earth—they would affect only small areas of the world and hardly reverse centuries of human actions.

De-Extinction

Since rewilding would require radical changes in institutions and even in human nature, some thinkers have proposed protecting the environment without making so many demands on society by turning instead to technology. One such proposal is de-extinction—in other words, recovering at least a few of the creatures that have gone extinct at the hand of humans.[11]

The first step, now being carried out in many places around the world, is to collect and conserve as much genetic material as possible to allow future scientists to save endangered species and perhaps even to resurrect extinct ones. Thus the seeds of 5,000 food crops are being stored in an underground vault on the island of Svalbard, not far from the North Pole. At the San Diego Zoo and at the American Museum of Natural History in New York, living cell cultures and the sperm, eggs, and embryos of a thousand or more creatures are being preserved in liquid nitrogen. And the National Zoo in Washington maintains the world's largest collection of frozen milk from exotic animals.[12]

Geneticists believe that it may be possible to find enough DNA in the remains of a recently extinct animal species to resurrect a member of that species. A recent attempt took place in 2003 with a bucardo or Spanish mountain goat (*Capra pyrenaica pyrenaica*). The last bucardo died in captivity in 2000, but not before some of its DNA had been extracted and deep frozen. Nuclei from this bucardo's cells were then injected into goat eggs emptied of their DNA and then implanted in surrogate goat mothers. Of seven pregnancies, one resulted in a live birth, but alas the little bucardo clone died a few minutes later.[13]

If this one came close to succeeding, others cannot be far behind. Rewilding Europe has been trying to bring back the aurochs (*Bos primigenius*), a species of wild cattle once found throughout Eurasia and North Africa, the last of which was killed in 1627. By comparing heritage breeds of longhorn cattle with DNA extracted from fossilized aurochs, geneticists are hoping to

recreate if not a true aurochs, then at least a reasonable facsimile of these animals and establish herds on unused farmlands in the Balkans and in Spain
and Portugal.[14]

In 2012, geneticists Stewart Brand and Ryan Phelan founded the Revive
& Restore Project to use the tools of molecular biology to resurrect extinct
animals. Their first project is to start with band-tailed pigeons, then change
their genome to come as close as possible to that of the extinct passenger
pigeon, then breed these new quasi-passenger pigeons in captivity until
there are enough to release in the wild.[15] Meanwhile, scientists in Korea
have attempted to implant into an elephant a mammoth egg taken from a
frozen mammoth corpse in Siberia, hoping thereby to revive a long-extinct
species.[16]

Of course, what good would reviving one extinct individual do? Even
if geneticists could revive two or more animals that could then reproduce,
how would they live? Passenger pigeons, before they were killed off, lived
and reproduced in swarms of thousands. And mammoths, even if a pair were
created that was able to reproduce, would need an enormous area to live in.
If one or several members of an extinct species were revived, they would require a whole environment of their own, in an already crowded world.[17]

Managing the Climate

Managing the biosphere might slow down the Sixth Extinction, but it would
leave other environmental problems untouched. The more challenging
and urgent problem today is preventing runaway global warming without
changing either human nature or social institutions. So far in the twenty-
first century, carbon dioxide emissions have increased at three times their
rate of growth in the 1990s.[18] Since our insatiable demand for energy is
increasing, much attention is being paid to finding other, less-polluting,
sources of energy that might reverse, or at least slow down, the resulting
global warming trend.

Since the 1990s, there has been a great deal of interest in mitigating global
warming by reducing carbon dioxide emissions and replacing fossil fuels
with "clean energy." In September 2006, *Scientific American* devoted an entire issue to "Energy's Future beyond Carbon," featuring gains in efficiency,
clean coal, nuclear power, solar cells, wind turbines, and biofuels, among
others. A 2009 article in that same journal offered a plan "to determine how

100 percent of the world's energy, for *all* purposes, could be supplied by wind, water and solar resources, by as early as 2030."[19]

Such promises abound, but their realization keeps receding into the future. In most parts of the world, the potential for hydroelectric power has already been reached. Wind farms have begun to appear in large numbers in China, in North America, and along the coasts of northern Europe. Ocean barrages and wave and current generators are still at the experimental stage. Geothermal plants can be built only in a few remote areas. Solar power, the ultimate solution to mankind's energy needs, is still too costly to replace traditional sources of energy. Nuclear power plants, which once promised to make electricity "too cheap to meter," have suffered so many catastrophic accidents and so many cost over-runs that few nations are building new ones and some are even dismantling old ones.[20]

In an article entitled "The Long Slow Rise of Solar and Wind," historian Vaclav Smil compared the current transition from fossil fuels to renewable energy to three earlier transitions: from wood to coal, from coal to oil, and from oil to natural gas. In each case, it took fifty to sixty years before the new fuel displaced its predecessor as the dominant source of energy. The transition to clean energy is likely to take just as long, if not longer. In 2014, after decades of government-subsidized research and development, wind, solar, and biofuels still accounted for only 3.35 percent of the US energy supply, while "old" renewables (hydroelectric and burning wood wastes from lumbering) accounted for 10 percent. Meanwhile, in 2012, the United States derived 87 percent of its energy from fossil fuels, down from 88 percent in 1990. In short, as Smil points out, "the great hope for a quick and sweeping transition to renewable energy is wishful thinking."[21]

Geo-Engineering

Given the difficulty of changing the world's technological infrastructure and the habits of billions of consumers, a few bold thinkers have been seeking technological fixes that would be compatible with the existing global economy and with the lifestyles and aspirations of the world's peoples. Thus in the 1990s, scientists concerned with global warming began to think of ways to engineer the climate. They kept such ideas very quiet, for fear that they would provide an excuse for fossil fuel companies, climate change deniers, and conservative politicians to resist any changes. Finally in 2006,

Paul Crutzen, a Nobel Prize–winning atmospheric chemist, published an article in the journal *Climatic Change* that offered a plan for how to prevent global warming in spite of continued carbon dioxide emissions.[22] This triggered a public debate about geo-engineering, defined by the Royal Society as "the deliberate large-scale manipulation of the planetary environment to counteract anthropogenic climate change."[23]

The very idea of manipulating the entire planetary environment—as opposed to changing parts of it piecemeal—has aroused a major controversy. On one side are those who believe that tinkering with the climate would be either ineffective or dangerous, or both. Clive Hamilton, the most forceful of its critics, has called it "humankind's technological arrogance."[24] Historian of climate control James Fleming writes: "Global climate engineering is untested and untestable, and dangerous beyond belief."[25] And Doug Parr, the chief scientist at Greenpeace UK, wrote: "The scientist's focus on tinkering with the entire planetary system is not a dynamic new technology and scientific frontier, but an expression of political despair."[26]

On the other side of the controversy are the enthusiasts for geo-engineering. Some are scientists like Edward Teller, the "father of the hydrogen bomb," and Lowell Wood, an astrophysicist involved with President Ronald Reagan's "Star Wars" Strategic Defense Initiative. The United States armed forces have also expressed an interest in controlling the weather.[27] Conservative think tanks, such as the Hoover Institution, the Marshall Institute, the American Enterprise Institute, and the Heartland Institute funded by the Exxon Corporation, have also supported the idea of geo-engineering. Ironically, many supporters of geo-engineering are also known for denying the very existence of climate change to which geo-engineering would be a solution.[28]

Just how would geo-engineering slow down or prevent global warming? Some proposals are indeed wacky: cutting down all the boreal forest so that sunshine would reflect off the snow on the newly barren landscape; painting mountains white to prevent glaciers from melting; and pushing the Earth into a slightly larger orbit so that it would receive less sunshine.[29] But most fall into two categories: removing and storing carbon dioxide and diminishing solar radiation.

Carbon Capture and Storage

If carbon dioxide emissions are the cause of global warming, then reducing the amount of carbon in the atmosphere would slow down the warming effect. Several methods have been proposed. One that has attracted much attention is so-called clean coal, a proposal to extract carbon dioxide from the smokestacks of coal-fired power plants. The trouble is that despite huge amounts of government money spent on research, this technology does not yet exist and will not come online until the 2030s; when it does, it will affect only power plants, not any of the myriad other sources of CO_2.

If carbon dioxide could be removed from power-plant smokestacks, where would it go? One possibility might be to inject it into saline aquifers, giant cavities under the ocean floor. Another might be pumping it deep underground to replace natural gas and oil as they are being extracted. Yet another might be pumping it into the deepest parts of the ocean, where it would become a liquid heavier than water and, hopefully, stay there.

Even if fossil fuel-burning power plants stopped emitting CO_2, a great deal would still remain in the atmosphere. So several scientists have suggested methods of removing it from the air. One such method involved creating phytoplankton "blooms" in the ocean. To grow, these minute plants require phosphorus, nitrogen, carbon, and iron. As iron is often the limiting factor in their growth, by sprinkling iron filings over the oceans, it is argued, these plants would multiply, absorbing CO_2 in the process. Then, when they died, they would sink to the ocean bottom, taking their carbon with them. Since the 1990s, a dozen experiments have been performed, but the results have been disappointing.[30] Furthermore, adding carbon to the oceans, whether by pumping carbon dioxide down to the seabed or by stimulating phytoplankton blooms, would make the oceans more acidic. To counter that, it has been proposed to sprinkle powdered lime (calcium oxide) or limestone (calcium carbonate), alkaline substances that would neutralize the acidification. To be effective, however, all of these proposals would require massive industries and huge fleets of ships, all of them using energy and emitting greenhouse gases.[31]

Finally, there is the very sensible idea that plants, especially fast-growing trees like pines, eucalyptuses, and mangroves, absorb carbon dioxide when they grow. Unfortunately, we live in an age when entire forests are being felled and burned to make room for cattle ranches, shrimp farms, and other human uses, and there is little hope of reversing this trend as long as the growing human population demands more land and its products.

Solar Radiation Management

Reducing the amount of carbon dioxide in the atmosphere would solve one of the main causes of global warming, but doing so seems extremely difficult, outrageously expensive, and probably ineffectual. The alternative would be to ignore the cause and, instead, reduce the amount of sunshine, hence heat, entering the atmosphere. Such proposals are known as solar radiation management. Several methods have been proposed: sending trillions of mirrors or parasols into outer space to shade parts of the Earth, stirring the oceans to create bubbles that would reflect more light, spraying seawater into the air to form more light-reflecting clouds, or injecting bismuth tri-iodide into the exhaust gases of airliners to create more cirrus clouds.[32] None of these methods, however, seem remotely cost-effective or even feasible.

One method, however, does seem within reach: spraying sulfur into the stratosphere. It has long been known that volcanic eruptions, such as Laki in 1783, Tambora in 1815, and Krakatoa in 1883, caused the Earth's climate to cool for a year or more. A recent eruption, that of Mount Pinatubo in the Philippines in 1991, injected millions of tons of sulfur dioxide into the stratosphere. There, it combined with water particles to form sulfate aerosols that reflected 1 percent of incoming solar rays back into space. As a result, Northern Hemisphere temperatures declined by 0.5° to 0.6°Celsius (0.9° to 1.1°F) and global temperatures by 0.4°Celsius (0.7°F) during the following year.[33]

If volcanoes can do it, why not humans? This, in fact, was the proposal put forward by Paul Crutzen that opened the issue of geo-engineering to public discussion. Since then, it has been taken up by several other scientists, especially physicist David Keith. They propose to send a fleet of aircraft to spray sulfur dioxide, hydrogen sulfide, or sulfuric acid into the stratosphere, where they would turn into light-reflecting aerosol particles. The cost would be a few billion dollars a year, far less than any other method of cooling the Earth and well within the budget of most governments.[34]

Of course, there would be consequences, some of them unpredictable. Sulfur in the stratosphere might damage the ozone layer that protects the Earth from harmful ultraviolet light. Changing the global climate might affect different regions differently. The monsoons, on which half the world's people depend for their food, might shift. Agricultural yields might rise in the temperate zone and fall in the tropics. All scientific experiments involve failures along with successes, so experimenting with the Earth's climate might turn out to be more harmful than expected. Yet the experiment would be irreversible, for stopping it would lead to a sudden and dramatic warming since excessive carbon dioxide would still be there. Such a warming would be too rapid for plants, animals, and humans to adapt to. Small wonder that David Keith hoped this plan would take place "not anytime soon, if ever," and that Paul Crutzen wrote: "The very best would be if emissions of greenhouse gases could be reduced so much that the stratospheric sulfur release experiment would not need to take place. Currently, this looks like a pious wish."[35]

Then there are ethical and political issues. Who will decide whether to geo-engineer the planet, and how? And what if solar radiation management were successful? Would it give humanity "time to develop the political will and the technologies to achieve the needed decarbonization"? Or would it give people and nations an excuse to continue emitting greenhouse gases?[36]

Any change this momentous is bound to produce winners and losers. Low-lying nations like Tuvalu or Bangladesh would be endangered by rising sea levels, while Russia and Canada would benefit from global warming. Within nations as well, the interests of people living in low-lying or storm-prone areas would differ from those who might welcome warmer winters. Would fossil fuel companies or sulfur producers influence government decisions? How would these interests be reconciled? If no enforceable international agreement were possible, would some nations or wealthy individuals decide to take matters into their own hands by spraying the stratosphere with sulfur? And would others decide to prevent such an action, for instance by shooting down the spraying airplanes?

The Planet Machine

What if the biologists and geo-engineers were successful, and the world applied technological fixes to the problems caused by previous technologies? What would our world look like? The climate would have stabilized, thanks to carbon capture and storage and to solar radiation management. Wildlife—what was left of it—would be carefully managed in zoos, nature parks, botanical gardens, and aquaria. Endangered species would be rescued from the brink and bred in captivity or, wearing radio collars, re-introduced into special "wilderness" parks. Portions of the Earth would be devoted to carefully managed artificial "wilderness" areas. And a few extinct species—but only charismatic ones—would be brought back from the dead, to the glory of science. In other words, what was once the Planet Earth will have become the Planet Machine.

A Choice of Futures

Our future will most probably lie within this triangle, but where? Our Paleolithic minds, our territorial institutions, our industrial technologies, and our developmentalist ideologies and consumerist cultures, which were formed during our long history as a species, have so much momentum that laissez-faire will continue to characterize much of the world's approach to the environment. Yet many people will show some wisdom and exercise some restraint by recycling, buying LED lightbulbs, driving hybrid cars, and eating less meat. And planetary management will probably expand with more designated wilderness areas, more subsidies for solar and wind power, perhaps some carbon-capture and other yet-to-be-invented technologies (but probably not half-Earth or solar radiation management). Somewhere between the three points of the triangle lies the future.

But the future consists of more than trends in human behavior that we can predict, based on past performance. Nature may surprise us yet, with asteroids, pandemics, volcano eruptions, or sudden shifts in the climate. If nature intervenes, the future of the planet may well lie outside the triangle of possibilities. Whatever happens, people may well look back upon our age with envy.

Notes

Introduction

1. John A. Widtsoe, *Success on Irrigation Projects* (New York: Wiley, 1928), 138.
2. Bible, King James version, Genesis 1:28.
3. Paul J. Crutzen and Eugene F. Stoermer, 'The Anthropocene,' *Global Change Newsletter* (May 2000), 17.
4. Stephen Mosely, *The Environment and World History* (Abingdon, U.K: Routledge, 2010), 9–10.
5. Lynn White Jr., "The Historical Roots of Our Ecologic Crisis," *Science* 155 (March 10, 1967), 1003–7.
6. Erle C. Ellis et al., "Used Planet: A Global History," *Proceedings of the National Academy of Sciences* 110 no. 20 (2013), 7978–85.
7. Sing C. Chew, *World Ecological Degradation: Accumulation, Urbanization, and Deforestation, 3000 B.C.–A.D. 2000* (Walnut Creek, Cal.: AltaMira Press, 2000), 1.
8. William F. Ruddiman, "The Anthropogenic Greenhouse Era Began Thousands of Years Ago," *Climatic Change* 61 no. 3 (December 2003), 261–93, and *Plows, Plagues and Petroleum: How Humans Took Control of Climate* (Princeton, N.J.: Princeton University Press, 2005). For a similar view, see Bruce D. Smith and Melinda A. Zeder, "The Onset of the Anthropocene," *Anthropocene* 4 (December 2013), 8–13.
9. On this point, see Yi-Fu Tuan, "Discrepancies between Environmental Attitudes and Behaviour: Examples from Europe and China," *Canadian Geographer* 12 no. 3 (1968), 176–91.

Chapter 1

1. Tim Flannery, *The Eternal Frontier: An Ecological History of North America and Its Peoples* (New York: Grove Press, 2001), 221; I. G. Simmons, *Global Environmental History* (Chicago: University of Chicago Press, 2008), ch. 51; Joe Ben Wheat, "A Paleo-Indian Bison Kill," *Scientific American* (January 1967), 213–31.
2. Nina G. Jablonski, *Skin: A Natural History* (Berkeley: University of California Press, 2006), 43–55, and "The Naked Truth," *Scientific American* (February 2010), 42–49; Ann Gibbons, "Swapping Guts for Brains," *Science* 316 no. 5831 (2007), 1560; Richard W. Wrangham, *Catching Fire: How Cooking Made Us Human* (New York: Basic Books, 2009), 5–8, 97–98; Craig B. Stanford, *Upright: The Evolutionary Key to Becoming Human* (Boston: Houghton-Mifflin, 2003), 138–39; Rick Potts, "Environmental Hypotheses of Hominid Evolution," *Yearbook of Physical Anthropology* 41 (1998),

118–19; Daniel E. Lieberman, *The Story of the Human Body: Evolution, Health, and Disease* (New York: Vintage, 2014), 70–72; John L. Brooke, *Climate Change and the Course of Global History* (Cambridge: Cambridge University Press, 2014), 75–76.

3. Doug Macdougall, *Frozen Earth: The Once and Future Story of the Ice Ages* (Berkeley: University of California Press, 2004), 202–3; William H. Calvin, *A Brain for All Seasons: Human Evolution and Abrupt Climate Change* (Chicago: University of Chicago Press, 2002), 129–34.

4. Spencer Wells, *Deep Ancestry: Inside the Genographic Project* (Washington, D.C.: National Geographic, 2007), 116; Clive Finlayson, *The Humans Who Went Extinct: Why Neanderthals Died Out and We Survived* (New York: Oxford University Press, 2009), 53–54.

5. Lars Werdelin, "King of Beasts," *Scientific American* (November 2013), 35–39.

6. J. Eudald Carbonell, J. Pares, et al., "The First Hominin of Europe," *Nature* 452 (2008), 465–69; Roger Lewin, *Human Evolution: An Illustrated Introduction*, 5th ed. (Malden, Mass.: Blackwell, 2005), 159–66; Jonathan C.K. Wells and Jay T. Stock, "The Biology of the Colonizing Ape," *Yearbook of Physical Anthropology* 50 (2007), 193; Stanford, *Hunting Apes*, 130–31.

7. Naama Goren-Inbar et al., "Evidence of Hominin Control of Fire at Gesher Benot Ya'aqov Israel," *Science* 304 no. 5671 (2004), 725–27.

8. Johan Goudsblom, "Fire and Fuel in Human History," in *The Cambridge World History*, vol. 1, David Christian, ed., *Introducing World History, to 10,000 BCE* (Cambridge: Cambridge University Press, 2015), 185–207, and *Fire and Civilization* (London: Alan Lane 1992), 26–27; J. D. Clark and J. W. K. Harris, "Fire and Its Roles in Early Hominid Lifeways," *African Archaeological Review* 3 (1985), 3–27; Heather Pringle, "The Origin of Creativity: New Evidence of Ancient Ingenuity Forces Scientists to Reconsider When Our Ancestors Started Thinking Outside the Box," *Scientific American* (March 2013), 42, and "Quest for Fire Began Earlier than Thought," *ScienceNOW* (April 2, 2012); Wrangham, *Catching Fire*, 84–87; Pat Shipman, *Animal Connection: A New Perspective on What Makes Us Human* (New York: W. W. Norton, 2011), 111–15.

9. Charles K. Brain, *The Hunters or the Hunted? An Introduction to African Taphonomy* (Chicago: University of Chicago Press, 1981), 273; Lieberman, *The Story of the Human Body*, 103–4; Clark and Harris, "Fire and Its Roles," 20–21; Goudsblom, *Fire and Civilization*, 16–17.

10. Wrangham, *Catching Fire*, 14, 42, 120; see also Robert Foley, "The Evolutionary Consequences of Increased Carnivory in Hominids," in Craig B. Stanford and Henry T. Bunn, eds., *Meat Eating and Human Evolution* (Oxford: Oxford University Press, 2001), 335–37.

11. Goudsblom, *Fire and Civilization*, 12–39; Wrangham, *Catching Fire*, 99–101, 183–94.

12. The question of the origin of language has produced a vast and contentious literature. See Robin I. M. Dunbar, *Human Evolution: Our Brains and Behavior* (New York: Oxford University Press, 2016), 232–34, for an interesting explanation. See also John McWhorter, *The Power of Babel: A Natural History of Language*

(New York: HarperCollins, 2003), and Christine Kenneally, *The First Word: The Search for the Origins of Language* (New York: Viking, 2007).

13. Wrangham, *Catching Fire*, 98, 194.

14. Michael Williams, *Deforesting the Earth: From Prehistory to Global Crisis* (Chicago: University of Chicago Press, 2003), 12–13; Franz J. Broswimmer, *Ecocide: A Short History of the Mass Extinction of Species* (London: Pluto, 2002), 17–19.

15. Steven Mithin, *After the Ice: A Global Human History, 20,000–5000 BC* (Cambridge, Mass.: Harvard University Press, 2004), 10; Wrangham, *Catching Fire*, 96–97, 120–27; Wells, *Deep Ancestry*, 107–17; Wells and Stock, "Biology," 194.

16. William J. Burroughs, *Climate Change in Prehistory: The End of the Reign of Chaos* (Cambridge: Cambridge University Press, 2005), 147.

17. Steve Olson, *Mapping Human History: Race, Genes, and Our Common Origins* (Boston: Houghton Mifflin, 2002), 85; Christopher B. Stringer, "The Evolution and Distribution of Later Pleistocene Human Populations," in Elisabeth Vrba et al., eds., *Paleoclimate and Evolution, with Emphasis on Human Origins* (New Haven, Conn.: Yale University Press, 1995), 526; Finlayson, *Humans Who Went Extinct*, 125–26; Julia Lee-Thorpe and Matt Sponheimer, "Contributions of Biochemistry to Understanding Hominin Dietary Ecology," *Yearbook of Physical Anthropology* 49 (2006), 137, 143.

18. Daniel Richter et al., "The Age of the Hominin Fossils from Jebel Irhoud, Morocco, and the Origins of the Middle Stone Age," *Nature* 546 no. 7657 (June 8, 2017), 293–96.

19. Shannon L. Carto et al., "Out of Africa and into an Ice Age: On the Role of Global Climate Change in the Late Pleistocene Migration of Early Modern Humans out of Africa," *Journal of Human Evolution* 56 (2009), 139–51.

20. Gregory Cochran and Henry Harpending, *The 10,000 Year Explosion: How Civilization Accelerated Human Evolution* (New York: Basic Books, 2009), 25–64.

21. John R. Gillis, *The Human Shore: Seacoasts in History* (Chicago: University of Chicago Press, 2012), 212–22; Curtis W. Marean, "When the Sea Saved Humanity," *Scientific American* (August 2010), 54–61, and "Coastal South Africa and the Coevolution of Modern Human Lineage and the Coastal Adaptation," in Nuno F. Bicho, Jonathan A. Haws, and Loren G. Davis, eds., *Trekking the Shore: Changing Coastlines and the Antiquity of Coastal Settlements* (New York: Springer, 2011), 421–37; Kyle S. Brown et al., "An Early and Advanced Technology Originating 71,000 Years Ago in South Africa," *Nature* 491 (2012), 590–93.

22. Christopher S. Henshilwood et al., "A 100,000-Year-Old Ochre-Processing Workshop in Blombos Cave, South Africa," *Science* 34 (2011), 219–22; P. Murre et al., "Early Use of Pressure Flaking on Lithic Artifacts at Blombos Cave, South Africa," *Science* 330 no. 6004 (2010), 659–62; Ian Tattersall, "If I Had a Hammer," *Scientific American* (September 2014), 55–59; John Noble Wilford, "In African Cave, Ancient Paint Factory Pushes Human Symbolic Thought 'Far Back,'" *New York Times* (October 14, 2011); Pringle, "Origin of Creativity," 41.

23. Pierre-Jean Texier et al., "A Howiesons Poort Tradition of Engraving Ostrich Eggshell Containers Dated to 60,000 Years Ago at Diepkloof Rock Shelter, South Africa," *Proceedings of the National Academy of Sciences of the USA* 107 no. 14 (2010), 6180–85.

24. Christopher Henshilwood et al., "Emergence of Modern Human Behavior: Middle Stone Age Engravings from South Africa," *Science* 295 (2002), 1278–80; Sally McBrearty and Alison Brooks, "The Revolution that Wasn't: A New Interpretation of the Origin of Modern Human Behavior," *Journal of Human Evolution* 39 (2000), 453–563; Zenobia Jacobs and Richard G. Roberts, "Human History Written in Stone and Blood," *American Scientist* 97 (2009), 302–9; Lyn Wadley, Tamaryn Hodgskiss, and Michael Grant, "Implications for Complex Cognition from the Hafting of Tools with Compound Adhesives in the Middle Stone Age, South Africa," *Proceedings of the National Academy of Sciences of the USA*, 106 no. 24 (2009), 9590–94; Luigi Luca Cavalli-Sforza, *Genes, People, and Languages*, trans. Mark Seielstad (New York: Farrar, Straus & Giroux, 2000), 80–81; Jeffrey D. Wall and Michael F. Hammer, "Archaic Admixture in the Human Genome," *Current Opinion in Genetics and Development* 16 (2006), 606–8; Pringle, "Origin of Creativity," 43; Spencer Wells, *The Journey of Man: A Genetic Odyssey* (Princeton, N. J. : Princeton University Press, 2002), 33–56.

25. Chris Stringer, *Lone Survivors: How We Came to Be the Only Humans on Earth* (New York: Henry Holt, 2012), 219–25.

26. Brenna M. Henn, Luigi Luca Cavalli-Sforza, and Marcus M. Feldman, "The Great Human Expansion," *Proceedings of the National Academy of Sciences of the USA* 109 no. 44 (2012), 17758–64; Paul Mellars, "Why Did Modern Human Populations Disperse from Africa ca. 60,000 Years Ago?" *Proceedings of the National Academy of Sciences of the USA* 103 no. 25 (2006), 9381–86; Hugo Reyes-Centeno, "Genomic and Cranial Phenotype Data Support Multiple Modern Dispersals from Africa and a Southern Route into Asia," *Proceedings of the National Academy of Sciences of the USA* 111 no. 20 (2014), 7248–53; Carto et al., "Out of Africa," 139–5.

27. Simon Armitage et al., "The Southern Route 'Out of Africa': Evidence for an Early Expansion of Modern Humans into Arabia," *Science* 331 (2011), 453–56; Jeffrey I. Rose, "The Nubian Complex of Dhofar, Oman: An African Middle Stone Age Industry in Southern Arabia," *PLoS ONE* 6 no. 11 (2011), e28239.

28. Luca Pagani et al., "Genomic Analyses Inform on Migration Events during the Peopling of Eurasia," *Nature* 538 no. 7624 (13 October 2016), 238–42; Stanley H. Ambrose, "Late Pleistocene Human Population Bottlenecks, Volcanic Winter, and Differentiation of Modern Humans," *Journal of Human Evolution* 34 (1998), 628–30; Spencer Wells, *Pandora's Seed: The Unforeseen Cost of Civilization* (New York: Random House, 2010), 105–6, and *Journey of Man*, 98–99; Mellars, "Why Did Modern Human Populations," 9384–85; Finlayson, *Humans Who Went Extinct*, 67–71; Carto et al., "Out of Africa." Mousterian tools used by Neanderthals consisted of a stone core from which sharp flakes were struck and then retouched to make scrapers, spearheads, and blades.

29. Kate Ravilious, "Exodus on the Exploding Earth," *New Scientist* (April 17, 2010), 28; Wells and Stock, "Biology," 194; Wells, *Pandora's Seed*, 15.

30. Walter Alvarez, *T. Rex and the Crater of Doom* (Princeton, N.J.: Princeton University Press, 1997).

31. In comparison, the eruption of Mount Pinatubo in 1991 expelled 10 cubic kilometers of magma, or 1/280th as much as Toba.

32. Michael Rampino and Stanley Ambrose, "Volcanic Winter in the Garden of Eden: The Toba Super-Eruption and the Late Pleistocene Crash," in F. McCoy and W. Heiken, eds., *Volcanic Hazards and Disasters in Human Antiquity* (Boulder, Colo.: Geological Society of America, 2000), 78–80; Michael Petraglia et al., "Middle Paleolithic Assemblages from the Indian Sub-continent before and after the Toba Super-Eruption," *Science* 317 (2007), 114–16; Michael R. Rampino and Stephen Self, "Volcanic Winter and Accelerated Glaciation Following the Toba Supereruption," *Nature* 359 (1992), 50–52; M. A. J. Williams, Stephen Ambrose, et al., "Environmental Impact of the 73 ka Toba Super-Eruption in South Asia," *Paleogeography, Paleoclimatology, Paleoecology* 284 (2009), 295–314; William I. Rose and Craig A. Chesner, "Worldwide Dispersal of Ash and Gases from Earth's Largest Known Eruption: Toba, Sumatra, 75 ka," *Paleogeography, Paleoclimatology, Paleoecology* 89 (1996), 269–75; Ambrose, "Late Pleistocene Human Population Bottlenecks," 623, 632–35; Mellars, "Why Did Modern Human Populations," 9384.

33. Serena Tucci, "A Map of Human Wanderlust," *Nature* 538 no. 7624 (October 13, 2016), 179–80; Mellars, "Why Did Modern Human Populations"; Henn et al., "The Great Human Expansion"; Carto et al., "Out of Africa."

34. Vincent Macaulay et al., "Single, Rapid Coastal Settlement of Asia Revealed by Analysis of Complete Mitochondrial Genomes," *Science* 308 (2005), 1034–36; Gary Stix, "Traces of a Distant Past," *Scientific American* 299 no. 1 (July 2008), 56–63; Mellars, "Why Did Human Populations," 9383–85.

35. Patrick Manning, *Migration in World History* (New York: Routledge, 2005), 16–38; Torben C. Rick and Jon Erlandson, eds., *Human Impacts on Ancient Marine Ecosystems: A Global Perspective* (Berkeley: University of California Press, 2008), 5–20; T. R. Disotell, "Human Evolution: The Southern Route to Asia," *Current Biology* 9 (1999), R925–28; Michael D. Petraglia et al., "Out of Africa: New Hypotheses and Evidence for the Dispersal of *Homo sapiens* along the Indian Ocean Rim," *Annals of Human Biology* 37 (2010), 288–311; Wells, *Journey of Man*, 69, 95–100; Armitage, "The Southern Route"; Mellars, "Why Did Modern Human Populations Disperse"; Macauley, "Single, Rapid Coastal Settlement."

36. Mark Nathan Cohen, "History, Diet, and Hunter-Gatherers," in Kenneth F. Kiple and Kriemhild C. Ornelas, eds., *Cambridge World History of Food* (Cambridge: Cambridge University Press, 2000), 2: 65–76, and *Health and the Rise of Civilization* (New Haven, Conn.: Yale University Press, 1988), 32–37, 112–13.

37. Alfred R. Wallace, *The Geographical Distribution of Animals, with a Study of the Relationship of Living and Extinct Faunas as Elucidating Past Changes of the Earth's Surface*, vol. I (New York: Harper and Brothers, 1876), quoted in Tim Flannery, *The Future Eaters: An Ecological History of the Australasian Lands and People* (Chatswood, NSW: Reed, 1994), 181.

38. These numbers are from Paul S. Martin, "Prehistoric Overkill: The Global Model," in Paul S. Martin and Richard Klein, eds., *Quaternary Extinctions: A Prehistoric Revolution* (Tucson: University of Arizona Press, 1984), 358. See also Martin's *Twilight of the Mammoths: Ice Age Extinctions and the Rewilding of America* (Berkeley: University of California Press, 2005), fig. 1, and Martin and David

W. Steadman, "Prehistoric Extinctions on Islands and Continents," in Ross D. E. MacPhee, ed., *Extinctions in Near Time: Causes, Contexts, and Consequences* (New York: Kluwer Academic, 1999), 17–24. Other authors give different figures, but all in the same general range.

39. John Alroy, "A Multispecies Overkill Simulation of the End-Pleistocene Megafaunal Mass Extinction," *Science* 292 (2001), 1893–96; Paul S. Martin, "Prehistoric Extinctions: In the Shadow of Man," in Charles E. Kay and Randy T. Simmons, eds., *Wilderness and Political Ecology: Aboriginal Influences and the Original State of Nature* (Salt Lake City: University of Utah Press, 2002), 1–27; Elizabeth Kolbert, *The Sixth Extinction: An Unnatural History* (New York: Henry Holt, 2014), 217–35. Wolfgang Behringer [*A Cultural History of Climate*, trans. Patrick Camiller (Cambridge: Polity Press, 2010), 38–39] exonerates humans and blames climate change. For a middle position, see A. D. Barnovsky, P. L. Koch, and R. S. Feranec, "Assessing the Causes of the Late Pleistocene Extinctions on the Continents," *Science* 306 (2004), 70–75, and P. L. Koch and A. D. Barnosky, "Late Quaternary Extinctions: State of the Debate," *Annual Review of Ecology, Evolution and Systematics* 37 (2006), 215–50.

40. I. G. Simmons, *Global Environmental History* (Chicago: University of Chicago Press, 2008), ch. 43.

41. J. F. O'Connell and J. Allen, "Dating the Colonization of Sahul (Pleistocene Australia-New Guinea): A Review of Recent Research," *Journal of Archaeological Science* 31 (2004), 835–53; Giles Hamm et al., "Cultural Innovation and Megafauna Interaction in the Early Settlement of Arid Australia," *Nature* 539 (November 10, 2016), 280–83.

42. John A. Long, M. Archer, Tim Flannery, and S. Hand, *Prehistoric Mammals of Australia and New Guinea—100 Million Years of Evolution* (Baltimore: Johns Hopkins University Press, 2002).

43. Robert G. Bednarik, "Seafaring in the Pleistocene," *Cambridge Archaeological Journal* 13 no. 1 (2003), 41–66; Flannery, *The Future Eaters*, 146–53, 189; Finlayson, *Humans Who Went Extinct*, 90–101; Bryan Sykes, *The Seven Daughters of Eve: The Science that Reveals Our Genetic Ancestry* (New York: W. W. Norton, 2001), 285; Cavalli-Sforza, *Genes*, 60, 93, 170–71; Wells, *Journey of Man*, 64–66, 76.

44. Flannery, *Future Eaters*, 190–92.

45. Flannery, *Future Eaters*, 187–92. Today, the penguins of Antarctica are the only remaining large undomesticated animals that are not afraid of humans.

46. For examples of differing views, see Gifford H. Miller et al. ["Ecosystem Collapse in Pleistocene Australia and a Human Role in Megafaunal Extinction," *Science* 309 no. 5732 (2005), 287–90], who blame humans, and Stephen Wroe and Judith Field ["A Review of Evidence for a Human Role in the Extinction of Australian Megafauna and an Alternative Explanation," *Quaternary Science Reviews* 25 no. 21–22 (2006), 2,692–703] who exonerate them. See also Peter Hiscock, "The Pleistocene Colonization and Occupation of Australia," in *The Cambridge World History*, vol. I: *Introducing World History, to 10,000 BCE*, ed. by David Christian (Cambridge: Cambridge University Press, 2015), 440–43.

47. Flannery, *Future Eaters*, 207.

48. Donald Grayson, "The Archeological Record of Human Impact on Animals," *Journal of World Prehistory* 15 no. 1 (2001), 41–42.

49. Susan Rule et al., "The Aftermath of Megafauna Extinction: Ecosystem Transformation in Pleistocene Australia," *Science* 335 (2012), 1483–86; R. G. Roberts et al., "New Ages for the Last Australian Megafauna: Continent-Wide Extinction about 46,000 Years Ago," *Science* 292 (2001), 1888–92; Miller, "Ecosystem Collapse," 287–90.

50. Irina Pugash, "Genome-Wide Date Substantiate Holocene Gene Flow from India to Australia," *Proceedings of the National Academy of Sciences of the USA* 110 no. 5 (2013), 1803–8; Geoffrey Blainey, *Triumph of the Nomads: A History of Ancient Australia*, rev. ed. (Australia: Sun, 1994), 58–61; Xiaoming Wang and Richard H. Tedford, *Dogs: Their Fossil Relatives and Evolutionary History* (New York: Columbia University Press, 2008), 162–63; C. N. Johnson and S. Wroe, "Causes of Extinction of Vertebrates during the Holocene of Mainland Australia: Arrival of the Dingo, or Human Impact?," *Holocene* 13 (2003), 941–48; Flannery, *Future Eaters*, 208–16.

51. Rule, "The Aftermath"; Miller, "Ecosystem Collapse"; Roberts et al., "New Ages"; Hiscock, "Pleistocene Colonization," 443.

52. Flannery, *Future Eaters*, 233; see also 225–232.

53. Goudsblom, *Fire and Civilization*, 31–32; Blainey, *Triumph*, 67–83; Flannery, *Future Eaters*, 180, 217–24, 236–41.

54. Nicholas Wade, *Before the Dawn: Recovering the Lost History of Our Ancestors* (New York: Penguin, 2006), 96; Flannery, *Future Eaters*, 228–29, 284–85.

55. Sykes, *Seven Daughters of Eve*, 129; Wells, *Journey of Man*, 76–78, 99, 108–33, and *Deep Ancestry* 25–141.

56. William F. Ruddiman, *Earth's Climate, Past and Future* (New York: W. H. Freeman, 2008), 193–203; Tim Flannery, *The Weather Makers: How Man Is Changing the Climate and What It Means for Life on Earth* (New York: Grove Press, 2009), 59; Brooke, *Climate Change*, 130–31; Burroughs, *Climate Change*, 43–45.

57. Burroughs, *Climate Change*, 116–17. Other sources give somewhat different dates; see François Djindjian, Janusz Kozlowski, and Marcel Otte, *Le Paléolitique supérieur en Europe* (Paris: Armand Colin, 1999), 144–55; Juan Luis de Arsuaga, *The Neanderthal's Necklace: In Search of the First Thinkers*, trans. Andy Klatt (New York: Four Walls Eight Windows, 2002), 192–93; Wade, *Before the Dawn*, 30, 103–5; and Finlayson, *Humans Who Went Extinct*, 128–29, 156–60, 186.

58. Dunbar, *Human Evolution*, 252–54.

59. Burroughs, *Climate Change*, 123; Finlayson, *Humans Who Went Extinct*, 181.

60. Rachel Caspari, "The Evolution of Grandparents," *Scientific American* (August 2011), 45–49; quotation on p. 49.

61. Martin and Steadman, "Prehistoric Extinctions," 24–26. See also Julien Louys, "Quaternary Extinctions in Southeast Asia," in Ashraf M. T. Elewa, ed., *Mass Extinctions* (Berlin: Springer, 2008), 159–89, and Pat Shipman, *The Invaders: How Humans and Their Dogs Drove Neanderthals to Extinction* (Cambridge, Mass.: Harvard University Press, 2015), 156–57, 226–28.

62. Ann Gibbons, "Close Encounters of the Prehistoric Kind," *Science*, 328 (2010), 680–84; Christopher Stringer, "Evolution: What Makes a Modern Human," *Nature* 485 (2012), 33–35; Kate Wong, "Our Inner Neanderthal," *Scientific American* 22 no. 1 (Winter 2013), 82–83; Johannes Krause, "The Derived FOXP2 Variant of Modern Humans Was Shared with Neanderthals," *Current Biology* 17 no. 21 (2007), 1908–12.

63. Douglas P. Fry and Patrick Söderberg, "Lethal Aggression in Mobile Forager Bands and Implications for the Origin of War," *Science* 341 (2013), 270–73; Marta M. Lahr, Robert A. Foley et al., "Inter-group Violence among Early Holocene Hunter-Gatherers of West Turkana, Kenya," *Nature* 529 no. 7588 (2016), 394–98.

64. Recent evidence shows that Neanderthals did occasionally eat small animals, fish, mollusks, and plant foods; see Kate Wong, "Neandertal Minds," *Scientific American* (February 2015), 38, 43.

65. Virginia Morell, "From Wolf to Dog," *Scientific American* 313 (July 2015), 60–67; Shipman, *The Invaders*, 167–90, 212–13, 230–32.

66. Paul Mellars, "Neanderthals and the Modern Colonization of Europe," *Nature* 432 (2004), 461–65; B. Hockett and J. A. Haws, "Nutritional Ecology and the Human Demography of Neanderthal Extinction," *Quaternary International* 137 (2005), 21–34; Kate Wong, "Twilight of the Neanderthals," *Scientific American* (August 2009), 33–37; Finlayson, *Humans Who Went Extinct*, 78, 102, 116–28, 140–45; Dunbar, *Human Evolution*, 249–57.

67. Morten Rasmussen et al., "The Genome of a Late Pleistocene Human from a Clovis Burial Site in Western Montana," *Nature* 506 no. 7487 (2014), 225–29; Sykes, *Seven Daughters of Eve*, 280–81.

68. Burroughs, *Climate Change*, 130–32; Shipman, *Animal Connection*, 211–19; M. V. Sablin and G. A. Khlopachev, "The Earliest Ice Age Dogs: Evidence from Eliseevichi I," *Current Anthropology* 43 (2002), 795–99.

69. Maanasa Raghavan et al., "Upper Paleolithic Siberian Genome Reveals Dual Ancestry of Native Americans," *Nature* 505 no. 7481 (2014), 87–91; Ted Goebel, "Pleistocene Colonization of Siberia and Peopling of the Americas: An Ecological Approach," *Evolutionary Anthropology* 8 (1999), 208–27; Heather Pringle, "The First Americans," *Scientific American* 305 (November 2011), 38–39.

70. T. D. Dillehay, "Monte Verde: Seaweed, Food, Medicine, and the Peopling of South America," *Science* 320 (2008), 784–86; Heather Pringle, "Texas Site Confirms Pre-Clovis Settlement of the Americas," *Science* 331 (2011), 1512, and "First Americans," 38, 45; Michael Waters et al., "The Buttermilk Creek Complex and the Origin of Clovis at the Debra L. Friedkin Site, Texas," *Science* 331 (2011), 1599–1603; David J. Meltzer, *First Peoples in a New World: Colonizing Ice Age America* (Berkeley: University of California Press, 2009), 129–30; James M. Adovasio and Jack Page, *The First Americans: In Pursuit of Archaeology's Greatest Mystery* (New York: Random House, 2002), ch. 7; Nicole M. Waguespack, "The Pleistocene Colonization and Occupation of the Americas," in *Cambridge World History*, 1: 473; Burroughs, *Climate Change* 207–17.

71. Jon M. Erlandson, "Anatomically Modern Humans, Maritime Voyaging, and the Pleistocene Colonization of the Americas," in Nina G. Jablonski, ed., *The First Americans: The Pleistocene Colonization of the New World* (San Francisco: University

of California Press, 2002), 59–92, and "Paleoindian Seafaring, Maritime Technologies, and Coastal Foraging on California's Channel Islands," *Science* 331 (2011), 1181–85; Jon M. Erlandson et al., "The Kept Highway Hypothesis: Marine Ecology, the Coastal Migration theory, and the Peopling of the Americas," *Journal of Island and Coastal Archaeology* 2 (2007), 161–74; Torben C. Rick and Jon M. Erlandson, eds., *Human Impacts on Ancient Marine Ecosystems: A Global Perspective* (Berkeley: University of California Press, 2008), 77–79; Roberta Hall, Diana Roy, and David Boling, "Pleistocene Migration Routes into the Americas: Human Biological Adaptations and Environmental Constraints," *Evolutionary Anthropology* 13 (2004), 132–44; Carole A. S. Mandryk et al., "Late Quaternary Paleoenvironments of Northwestern North America: Implications for Inland versus Coastal Migration Routes," *Quaternary Science Reviews* 20 (2001), 301–14; Ted Goebel, M. R. Waters, and D. H. O'Rourke, "The Late Pleistocene Dispersal of Modern Humans in the Americas," *Science* 319 (2008), 1497–1502.

72. Peter Ward and Joe Kirschvink, *A New History of Life: The Radical Discoveries about the Origins and Evolution of Life on Earth* (London: Bloomsbury Press, 2015), 339–41.

73. Charles Q. Choi, "Lost Giants: Did Mammoths Vanish before, during, *and* after Humans Arrived?" *Scientific American* (February 2010), 21–22; Jacquelyn L. Gill et al., "Pleistocene Megafaunal Collapse, Novel Plant Communities, and Enhanced Fire Regimes in North America," *Science* 326 (2009), 1101–3; J. Tyler Faith and Todd Surovell, "Synchronous Extinction of North America's Pleistocene Mammals," *Proceedings of the National Academy of Science of the USA* 106 (2009), 20641–45; Waguespack, "Pleistocene Colonization," 469–70.

74. Wallace S. Broeker et al., "Putting the Younger Dryas Event into Context," *Quaternary Science Reviews* 29 (2010), 1078–81; Douglas J. Kennett et al., "Shock-Synthesized Hexagonal Diamonds in Younger Dryas Boundary Sediments," *Proceedings of the National Academy of Sciences* 106 no. 31 (2009), 12,623–28; R. B. Firestone et al., "Evidence for an Extraterrestrial Impact 12,900 Years Ago that Contributed to the Megafaunal Extinctions and the Younger Dryas Cooling," *Proceedings of the National Academy of Sciences of the USA* 104 (2007), 16,016–21; and Todd A. Surovell et al., "An Independent Evaluation of the Younger Dryas Extraterritorial Impact Hypothesis," *Proceedings of the National Academy of Sciences of the USA* 106 no. 43 (2009), 18, 155–58.

75. Sharon Levy, *Once and Future Giants: What Ice Age Extinctions Tell Us about the Fate of Earth's Largest Animals* (New York: Oxford University Press, 2011), 16; Gill et al., "Pleistocene Megafaunal Collapse," 1100–103; Donald Grayson, "The Archaeological Record of Human Impact on Animals," *Journal of World Prehistory* 15 no. 1 (2001), 35; Flannery, *Eternal Frontier*, 187–88, 194–95.

76. Richard A. Kerr, "Megafauna Died from Big Kill, Not Big Chill," *Science*, 300 (2003), 585; Gill et al., "Pleistocene Megafaunal Collapse," 1102; Faith and Surovell, "Synchronous Extinction."

77. Paul S. Martin and H. E. Wright Jr., eds., *Pleistocene Extinctions: The Search for a Cause* (New Haven, Conn.: Yale University Press, 1967), and Martin, "Prehistoric Overkill," 354–403.

78. Edward O. Wilson, *The Future of Life* (New York: Alfred A. Knopf, 2002), 56.

79. Doug Peacock, *In the Shadow of the Sabertooth* (Oakland, Cal.: AK Press, 2013).

80. Edmund Russell, *Evolutionary History: Uniting History and Biology to Understand Life on Earth* (Cambridge: Cambridge University Press, 2011), 21–23; Flannery, *Eternal Frontier*, 212–28.

81. Gill, "Pleistocene Megafaunal Collapse," 1100–103.

82. Wheat, "Paleo-Indian Bison Kill, 213–31; Simmons, *Global Environmental History*, ch. 51.

83. Tamara Whited et al., *Northern Europe: An Environmental History* (Santa Barbara, Cal.: ABC CLIO, 2005), 15–16; Brooke, *Climate Change*, 131–32; Burroughs, *Climate Change*, 43–45; Shaun A. Marcott et al., "A Reconstruction of Regional and Global Temperature for the Past 11,300 Years," *Science* 339 (2013), 1198–201; Peter de Menocal et al., "Coherent High- and Low-Latitude Climate Variability during the Holocene Warm Period," *Science* 288 (2000), 2198–202; Richard B. Alley, "Abrupt Climate Change," *Scientific American* (November 2004), 64; Ruddiman, *Earth's Climate*, 230.

84. Andrew Sherratt, "The Secondary Exploitation of Animals in the Old World," *World Archaeology* 15 no. 1 (1983), 90–102.

85. Jean-Pierre Boquet-Appel, "Estimate of Upper Paleolithic Meta-Population Size in Europe from Archaeological Data," *Journal of Archaeological Science* 32 (2005), 1656–68, cited in Brooke, *Climate Change*, 137.

86. Brigitte M. Holt and Vincenzo Formicola, "Hunters of the Ice Age: The Biology of Upper Paleolithic People," *Yearbook of Physical Anthropology* 51 (2008), 86–87; J.-G. Rozoy, "The Revolution of the Bowmen in Europe," in Clive Bonsall, ed., *The Mesolithic in Europe* (Edinburgh: John Donald, 1989), 14–27; Neil Roberts, *The Holocene: An Environmental History* (Oxford: Blackwell, 1998), 125–26; Williams, *Deforesting the Earth*, 21.

87. William Cronon, *Changes in the Land: Indians, Colonists, and the Ecology of New England* (New York: Hill and Wang, 1983), 49–51; Williams, *Deforesting the Earth*, 25–60. See also Shepard Krech III, *The Ecological Indian: Myth and History* (New York: W. W. Norton, 1999), 103–13; and Gordon G. Whitney, *From Coastal Wilderness to Fruited Plain: A History of Environmental Change in Temperate North America, 1500 to the Present* (Cambridge: Cambridge University Press, 1994), 107–20.

88. Goudsblom, *Fire and Civilization*, 30–31; Bruce D. Smith, *The Emergence of Agriculture* (New York: W. H. Freeman, 1995), 16–18; Charles L. Redman, *Human Impact on Ancient Environments* (Tucson: University of Arizona Press, 1999), 99; Cronon, *Changes in the Land*, 13, 28; Whited et al., *Northern Europe*, 16.

89. Wade, *Before the Dawn*, 125–31; Steven Mithen, *After the Ice: A Global Human History, 20,000–5000 BC* (Cambridge, Mass.: Harvard University Press, 2004), 160; Joseph E. Taylor, *Making Salmon: An Environmental History of the Northwest Fisheries Crisis* (Seattle: University of Washington Press, 1999), 13–38; Roberts, *Holocene*, 151–52; Flannery, *Eternal Frontier*, 239–41; Finlayson, *Humans Who Went Extinct*, 200–201.

90. Arlene Miller Rosen, *Civilizing Climate: Social Responses to Climate Change in the Ancient Near East* (Lanham, Md.: Rowman & Littlefield, 2007), 68.

91. Steven J. Mithen, *Thirst: Water and Power in the Ancient World* (London: Weidenfeld and Nicholson 2012), 18–20, and *After the Ice*, 43–44; Jared Diamond, *Guns, Germs, and Steel: The Fates of Human Societies* (New York: W. W. Norton, 1997), 136–46; Rosen, *Civilizing Climate*, 34–36, 111; Redman, *Human Impact*, 96–105; Roberts, *Holocene*, 13–48, 130, 147–48; Burroughs, *Climate Change*, 193–94.

92. Julian B. Murton et al., "Identification of Younger Dryas Outburst Flood Path from Lake Agassiz to the Arctic Ocean," *Nature* 464 (2010), 740–43; Robley Matthews, Douglas Anderson, Robert S. Chen, and Thompson Webb, "Global Climate and the Origin of Agriculture," in Lucile F. Newman, ed., *Hunger in History: Food Shortages, Poverty and Deprivation* (Oxford: Blackwell, 1990), 27–34; Macdougall, *Frozen Earth*, 197–99; Richard B. Alley, *The Two-Mile Time Machine: Ice Cores, Abrupt Climate Change, and Our Future* (Princeton, N.J.: Princeton University Press, 2000), 110–18, and "Abrupt Climate Change," 66; Brooke, *Climate Change*, 132–33; Burroughs, *Climate Change*, 190, 220.

93. Peter Bellwood, *First Farmers: The Origins of Agricultural Societies* (Oxford: Blackwell, 2005), 20–21, 51–52; Graeme Barker, *The Agricultural Revolution in Prehistory: Why Did Foragers Become Farmers?* (Oxford: Oxford University Press, 2006), 104–48, 382–85; David R. Montgomery, *Dirt: The Erosion of Civilizations* (Berkeley: University of California Press, 2007), 36; Ruddiman, *Earth's Climate*, 281; Alley, "Abrupt Climate Change," 62–69; Rosen, *Civilizing Climate*, 45, 68, 105, 124–26; Mithen, *After the Ice*, 41–53; Roberts, *Holocene*, 130–31, 143; Finlayson, *Humans Who Went Extinct*, 196–97.

94. I am indebted to Tim Webster of Yale University for this insight.

Chapter 2

1. Neil Roberts, *The Holocene: An Environmental History* (Oxford: Blackwell, 1998), 141.

2. Ester Boserup, *Population and Technological Change: A Study in Long-Term Trends* (Chicago: University of Chicago Press, 1981), 40–45; Peter Bellwood, *First Farmers: The Origins of Agricultural Societies* (Oxford: Blackwell, 2005), 21–24; Graeme Barker, *The Agricultural Revolution in Prehistory: Why Did Foragers Become Farmers?* (Oxford: Oxford University Press, 2006), 392–93, 413; Arlene Miller Rosen, *Civilizing Climate: Social Responses to Climate Change in the Ancient Near East* (Lanham, Md.: Rowman & Littlefield, 2007), 103–17; Jared Diamond, *Guns, Germs, and Steel: The Fates of Human Societies* (New York: W. W. Norton, 1997), 105–8.

3. Jared Diamond, "Evolution, Consequences and Future of Plant and Animal Domestication," *Nature*, 418 no. 6898 (2002), 700–707.

4. Michael Pollan, *The Omnivore's Dilemma: A Natural History of Four Meals* (New York: Penguin, 2006), 23–24.

5. Juliet Clutton-Brock, *Animals as Domesticates: A World View through History* (East Lansing: Michigan State University Press, 2012), 19–34, and *A Natural History*

of Domesticated Mammals, 2nd ed. (Cambridge: Cambridge University Press, 1999), 29–39; Stephen Budiansky, *The Covenant of the Wild: Why Animals Chose Domestication* (New York: William Morrow, 1992), 22–23, 60–61.

6. Carlos A. Driscoll, David W. MacDonald, and Stephen J. O'Brien, "From Wild Animals to Domestic Pets: An Evolutionary View of Domestication," *Proceedings of the National Academy of Sciences USA* 16 suppl. 1 (June 16, 2009), 9971–78; Abigail Tucker, *The Lion in the Living Room: How House Cats Tamed Us and Took over the World* (New York: Simon and Schuster, 2016), 39; Budiansky, *Covenant of the Wild,* 77–80, 97–107; Diamond, *Guns, Germs, and Steel,* chs. 7, 8, and 9.

7. Melinda A. Zeder, "Pathways to Animal Domestication," in Paul Gepts et al., eds., *Biodiversity in Agriculture: Domestication, Evolution and Sustainability* (Cambridge: Cambridge University Press, 2012), 227–59.

8. Carlos A. Driscoll, Juliet Clutton-Brock, Andrew C. Kitchener, and Stephen J. O'Brien, "The Taming of the Cat," *Scientific American* (June 2009), 68–75; Carlos A. Driscoll et al., "The Near Eastern Origin of Cat Domestication," *Science* 317 (July 2007), 519–23; Edward O. Price, *Animal Domestication and Behavior* (Wallingford, U.K.: CABI, 2002), 24; Jaromir Malek, *The Cat in Ancient Egypt* (London: British Museum, 2006), 45–72.

9. Sam White, "From Globalized Pig Breeds to Capitalist Pig: A Study of Animal Cultures and Evolutionary History," *Environmental History* 16 no. 1 (2011), 94–120; Umberto Albarella et al., eds., *Pigs and Humans: 10,000 Years of Interaction* (Oxford: Oxford University Press, 2007), 30–40, 57; Clutton-Brock, *Natural History,* 72–76, 91–99.

10. Pat Shipman, *The Invaders: How Humans and Their Dogs Drove Neanderthals to Extinction* (Cambridge, Mass.: Harvard University Press, 2015), 167–91; Virginia Morell, "From Wolf to Dog," *Scientific American* 313 (July 2015), 60–67; S. Tacon and Colin Pardoe, "Dogs Make Us Human," *Nature Australia* 27 no. 4 (Autumn 2002), 52–61; Mark Derr, *How the Dog Became the Dog: From Wolves to Our Best Friends* (New York: Overlook Duckworth, 2011), 113–29; Laura M. Shannon et al., "Genetic Structure in Village Dogs Reveals a Central Asian Domestication Origin," *Proceedings of the National Academy of Sciences of the USA* 112 no. 44 (2015), 13639–44; Clutton-Brock, *Natural History,* 49–51; Driscoll et al., "From Wild Animals."

11. Raymond Coppinger and Lorna Coppinger, *Dogs: A Startling New Understanding of Canine Origins, Behavior, and Evolution* (New York: Scribner's, 2001), 57–71, 283. See also Price, *Animal Domestication,* 25–26, 89–96, 102–3; Derr, *How the Dog,* 217; and James Gorman, "Dogs, Decoded: Tracing the Origins of Man's Best Friend," *New York Times* (December 27, 2016).

12. William J. Burroughs, *Climate Change in Prehistory: The End of the Reign of Chaos* (Cambridge: Cambridge University Press, 2005), 228.

13. P. J. Richerson, R. Boyd, and R. L. Bettinger, "Was Agriculture Impossible during the Pleistocene but Mandatory during the Holocene? A Climate Change Hypothesis," *American Antiquity* 663 no. 3 (2001), 387–411.

14. Diamond, *Guns, Germs, and Steel,* chs. 5, 8, and 10.

15. Megan Sweeney and Susan McCouch, "The Complex History of the Domestication of Rice," *Annals of Botany* 100 no. 5 (2007), 951–57; Huyuan Lu et al., "Rice

Domestication and Climatic Change: Phytolith Evidence from East China," *Boreas* 31 no. 4 (December 2002), 378–85; Zhijun Zhao and Dolores R. Piperno, "Late Pleistocene/Holocene Environments in the Middle Yangtze Valley, China, and Rice (*Oryza sativa* L.) Domestication: The Phytolith Evidence," *Geoarchaeology* 15 no. 2 (February 2000), 203–22; Charles Higham and Tracy L. D. Lu, "The Origin and Dispersal of Rice Cultivation," *Antiquity* 72 no. 278 (1998), 867–77.

16. Spencer Wells, *The Journey of Man: A Genetic Odyssey* (Princeton. N.J.: Princeton University Press, 2002), 156–57; Peter Boomgaard, *Southeast Asia: An Environmental History* (Santa Barbara, Cal.: ABC-CLIO, 2007), 36–39; Peter Bellwood, "The Austronesian Dispersal and the Origin of Languages," *Scientific American* (July 1991), 88–93; Yoshan Moodley, "The Peopling of the Pacific from a Bacterial Perspective," *Science* 323 no. 5913 (2009), 527–30; Barker, *Agricultural Revolution*, 182–230.

17. Robert Marks, *China: Its Environment and History* (Lanham, Md.: Rowman & Littlefield, 2012), 12; I am grateful to Professor Marks for allowing me to read his manuscript in press. See also Qu Geping and Li Jinchang, *Population and Environment in China* (Boulder, Colo.: L. Rienner, 1994), 21; and Hsu Cho-yuan, "Environmental History—Ancient," in *Berkshire Encyclopedia of China* (Great Barrington, Mass: Berkshire, 2009), 2:732–33.

18. Francesca Bray, *Agriculture*, vol. 6 part 2 of Joseph Needham, ed., *Science and Civilisation in China* (Cambridge: Cambridge University Press, 1984), 39–40; Marks, *China*, 24–29.

19. Richard MacNeish, *The Origins of Agriculture and Settled Life* (Norman: University of Oklahoma Press, 1991), 127; Diamond, *Guns, Germs, and Steel*, 117–21, 137–46.

20. Pat Shipman, *The Animal Connection: A New Perspective on What Makes Us Human* (New York: W. W. Norton, 2011), 224–36.

21. Jonathan Silvertown, *An Orchard Invisible: A Natural History of Seeds* (Chicago: University of Chicago Press, 2009),146–49; Joy McCorriston and Frank Hole, "The Ecology of Seasonal Stress and the Origins of Agriculture in the Near East," *American Anthropologist* 93 (1991), 46–69; Rosen, *Civilizing Climate*, 36–37, 97–99, 107–15, 20; Richard B. Alley, "Abrupt Climate Change," *Scientific American* (November 2004), 64–66; Charles L. Redman, *The Human Impact on Ancient Environments* (Tucson: University of Arizona Press, 1999), 96, 105–7; Bellwood, *First Farmers*, 44–66; Steven Mithen, *After the Ice: A Global Human History, 20,000–5000 BC* (Cambridge, Mass.: Harvard University Press, 2004), 41–78; John L. Brooke, *Climate Change and the Course of Human History* (Cambridge: Cambridge University Press, 2014), 143–49; Roberts, *Holocene*, 130–31, 149.

22. Andrew M. T. Moore et al., *Village on the Euphrates: From Foraging to Farming at Abu Hureya* (New York: Oxford University Press, 2000); Carlos E. Cordova, *Millennial Landscape Change in Jordan: Geoarchaeology and Cultural Ecology* (Tucson: University of Arizona Press, 2007), 165–75. See also Bellwood, *First Farmers*, 20, 51–52; Rosen, *Civilizing Climate*, 105; Burroughs, *Climate Change*, 186–91; Shipman, *Animal Connection*, 224, 231–36; and Roberts, *Holocene*, 130–31.

23. Bert de Vries and Robert Marchant, "Environment and the Great Transition: Agrarianization," in Bert de Vries and Johan Goudsblom, eds., *Mappae*

Mundi: Humans and Their Habitats in a Long-Term Socio-Ecological Perspective (Amsterdam: Amsterdam University Press, 2003), 89; Robert Rodden, "An Early Neolithic Village in Greece," *Scientific American* (April 1965), 95–103; Mithen, *After the Ice*, 162–63; Redman, *Human Impact*, 110–15; Roberts, *Holocene*, 189–92.

24. William B. F. Ryan and Walter Pitman [*Noah's Flood: The New Scientific Discoveries about the Event that Changed History* (New York: Simon and Schuster, 1998)] and Mithen [*After the Ice*, 153] advance this thesis. For a discussion, see Quirin Schiermeier, "Oceanography: Noah's Flood," *Nature* 430 no. 7001 (August 12, 2004), 718–19; Liviu Giosan, Florin Filip, and Stefan Constantinescu, "Was the Black Sea Catastrophically Flooded in the Early Holocene?" *Quaternary Science Reviews* 28 (2009), 1–6; and Burroughs, *Climate Change*, 221–22.

25. Michael Williams, *Deforesting the Earth: From Prehistory to Global Crisis* (Chicago: University of Chicago Press, 2003), 36–46; Tamara L. Whited, *Northern Europe: An Environmental History* (Santa Barbara, Cal.: ABC-CLIO, 2005), 23–26.

26. Amy Bogaard, " 'Garden Agriculture' and the Nature of Early Farming in Europe and the Near East," *World Archaeology* 37 no. 2 (June 2005), 177–96; Amy Bogaard et al., "Crop Manuring and Intensive Land Management by Europe's First Farmers," *Proceedings of the National Academy of Sciences USA* 110 no. 31 (July 30, 2013), 12,589–94; R. Rowley-Conwy, "Slash and Burn in the Temperate European Neolithic," in Roger Mercer, ed., *Farming Practices in British Prehistory* (Edinburgh: University Press, 1981), 85–96; Williams, *Deforesting the Earth*, 22, 35, 43–46; Bellwood, *First Farmers*, 68–84; Roberts, *Holocene*, 153–58, 196; de Vries and Marchant, "Environment," 90–92; Andrew Sherratt, "The Secondary Exploitation of Animals in the Old World," *World Archeology* 15 no. 1 (1983), 91–92; Whited, *Northern Europe*, 23–32.

27. David W. Anthony, *The Horse, the Wheel, and Language: How Bronze-Age Riders from the Eurasian Steppe Shaped the Modern World* (Princeton, N.J.: Princeton University Press, 2008), 119–20, 138.

28. Anthony, *Horse, Wheel, and Language*, 133, 200–21, 282, 300–22, 460.

29. Ewen Callaway, "DNA Deluge Reveals Bronze Age Secrets," 140–41; John Novembre, "Ancient DNA Steps into Language Debate, 164–65; Morten E. Allentoft et al., "Population Genomics of Bronze Age Eurasia," 167–71; and Wolfgang Haak et al., "Massive Migration from the Steppe Was a Source for Indo-European Languages in Europe," 207–11, all in *Nature* 522 no. 7555 (June 11, 2015). See also Anthony, *Horse, Wheel, and Language*, 328, 361.

30. Gregory Cochran and Henry Harpending, *The 10,000 Year Explosion: How Civilization Accelerated Human Evolution* (New York: Basic Books, 2009), 179–85; Peter Golden, "Nomads and Sedentary Societies in Medieval Eurasia," in Michael Adas, ed., *Agricultural and Pastoral Societies in Ancient and Classical History* (Philadelphia: Temple University Press, 2001), 75–87; Callaway, "DNA Deluge."

31. David Christian, "Silk Roads or Steppe Roads? The Silk Roads in World History," *Journal of World History* 11 no. 1 (2000), 1–26.

32. Jack Golson, "The Origins and Development of New Guinea Agriculture," 175–84, and Tim Denham, "Update: New Research in New Guinea," 184–85, in Tim Denham and

Peter White, eds., *The Emergence of Agriculture: A Global View* (London: Routledge, 2007); Tim Flannery, *The Future Eaters: An Ecological History of the Australasian Lands and People* (New York: Grove Press, 1994), 292–96; Mithen, *After the Ice*, 342–45; Diamond, *Guns, Germs, and Steel*, 106–7, 147–49.

33. Roy A. Rappaport, "The Flow of Energy in an Agricultural Society," *Scientific American* 225 no. 3 (September 1971), 116–32. According to anthropologist Axel Steensberg [*New Guinea Gardens: A Study of Husbandry with Parallels in Prehistoric Europe* (London: Academic Press, 1980), 56–65], the agricultural methods practiced in New Guinea resembled those of the first Neolithic farmers in Europe.

34. Gregory H. Maddox, *Sub-Saharan Africa: An Environmental History* (Santa Barbara: ABC-CLIO, 2006), 39; James L. Newman, "Africa South from the Sahara," in Kenneth F. Kiple and Kriemhild Conee Ornelas, eds., *Cambridge World History of Food* (Cambridge: Cambridge University Press, 2000), 2: 1332; Alley, "Abrupt Climate Change," 65–69; Mithen, *After the Ice*, 489–502; Roberts, *Holocene*, 116, 162–63; Brooke, *Climate Change*, 155–56; Burroughs, *Climate Change*, 226–28.

35. Jack R. Harlan, "Indigenous African Agriculture," in C. Wesley Cowan and Patty Jo Watson, eds., *The Origins of Agriculture: An International Perspective* (Washington, D.C.: Smithsonian Institution Press, 1992), 60–69; Fiona Marshall and Elizabeth Hildebrand, "Cattle before Crops: The Beginnings of Food Production in Africa," *Journal of World Prehistory* 16 (2002), 99–143; John R. McNeill, "Biological Exchanges in World History," in Jerry H. Bentley, ed., *The Oxford Handbook of World History* (New York: Oxford University Press, 2011), 325–42.

36. Burroughs, *Climate Change in Prehistory*, 232.

37. Jan Vansina, *Paths in the Rainforests: Toward a History of Political Tradition in Equatorial Africa* (Madison: University of Wisconsin Press, 1990), 49–68; James L. A. Webb Jr., *Humanity's Burden: A Global History of Malaria* (New York: Cambridge University Press, 2009), 34–35; D. W. Phillipson, "The Spread of the Bantu Language," *Scientific American* 236 no. 4 (April 1977), 106–14; Maddox, *Sub-Saharan Africa*, 52–56; J. G. Sutton, "The Interior of East Africa," in P. L. Shinnie, ed., *The African Iron Age* (Oxford: Clarendon, 1971), 144–48; Thomas R. de Gregori, *Technology and the Economic Development of the Tropical African Frontier* (Cleveland, Ohio: Press of Case Western Reserve University, 1969), 99–100.

38. Christopher Ehret, *An African Classical Age: Eastern and Southern Africa in World History, 1000 B.C. to A.D. 400* (Charlottesville: University of Virginia Press, 1998), 188–89, 245; Jack Goody, *Technology, Tradition and the State in Africa* (London: Oxford University Press, 1971), 30–47.

39. Richard W. Bulliet, *The Camel and the Wheel* (New York: Columbia University Press, 1990), 38–45, 132–35; Ehret, *African Classical Age*, 302–3.

40. Brooke, *Climate Change*, 152–56.

41. Clutton-Brock, *Animals as Domesticates*, 121–32.

42. Shawn William Miller, *An Environmental History of Latin America* (Cambridge: Cambridge University Press, 2007), 38; Diamond, *Guns, Germs, and Steel*, 178–79, 187–88.

43. Logan Kistler et al., "Transoceanic Drift and the Domestication of African Bottle Gourds in the Americas," *Proceedings of the National Academy of Sciences of the USA* 111 no. 8 (February 25, 2014), 2937–41.

44. Dolores R. Piperno, "The Origins of Plant Cultivation and Domestication in the New World: Patterns, Processes, and New Developments," *Current Anthropology* 52 (2011), S453–70.

45. Hugh H. Iltis, "From Teosinte to Maize: The Catastrophic Sexual Mutation," *Science* 222 (1983), 886–94; Michael D. Coe and Rex Koontz, *Mexico: From the Olmecs to the Aztecs* (London: Thames and Hudson, 2013), 38–42; Bellwood, *First Farmers*, 155–57; Roberts, *Holocene*, 131–35; Diamond, *Guns, Germs, and Steel*, 96, 137; Miller, *Environmental History*, 37–38; Mithen, *After the Ice*, 274–81.

46. S. Christopher Caran and James A. Neely, "Hydraulic Engineering in Prehistoric Mexico," *Scientific American* (October 2006), 78–85.

47. M. Edward Moseley, "Organizational Preadaptation to Irrigation: The Evolution of Early Water-Management Systems in Coastal Peru," in Theodore E. Downing and McGuire Gibson, eds., *Irrigation's Impact on Society* (Tucson: University of Arizona Press, 1976), 77–82; MacNeish, *Origins of Agriculture*, 39–45; Bellwood, *First Farmers*, 159–64; Mithen, *After the Ice*, 267–69.

48. Charles C. Mann, "Ancient Earthmovers of the Amazon," *Science* 321 no. 5893 (2008), 1148–52, and *1491: New Revelations of the Americas before Columbus* (New York: Random House, 2005), 315–20; John Hemming, *Tree of Rivers: The Story of the Amazon* (London: Thames and Hudson, 2008), 20–34; Miller, *Environmental History*, 9–17.

49. Michael Heckenberger, "Lost Cities of the Amazon," *Scientific American* (October 2009), 64–71; William M. Denevan, *Cultivated Landscapes of Native Amazonia and the Andes* (Oxford: Oxford University Press, 2001), 102–14, and "Pre-European Forest Cultivation in Amazonia," in William Balée and Clark L. Erickson, eds., *Time and Complexity in Historical Ecology* (New York: Columbia University Press, 2006), 153–63; Anna Roosevelt, *Moundbuilders of the Amazon: Geophysical Archaeology on Marajó Island, Brazil* (San Diego: Academic Press, 1991) and "The Lower Amazon: A Dynamic Human Habitat," in David L. Lentz, ed., *Imperfect Balance: Landscape Transformations in the Precolumbian Americas* (New York: Columbia University Press, 2000), 455–79; David R. Montgomery, *Dirt: The Erosion of Civilization* (Berkeley: University of California Press, 2007), 142–44; Mann, "Ancient Earthmovers of the Amazon," and *1491*, 336–46; Clark L. Erickson, "Amazonia: The Historical Ecology of a Domesticated Landscape," 157–83, and other articles in Helaine Silverman and William H. Isbell, eds., *Handbook of South American Archaeology* (New York: Springer, 2008); Hemming, *Tree of Rivers*, 17, 271–88.

50. MacNeish, *Origins of Agriculture*, 179; de Vries and Marchant, "Environment," 93–96.

51. Tim Flannery, *The Eternal Frontier: An Ecological History of North America and Its Peoples* (New York: Grove Press, 2001), 246–47; Ellen Messer, "Maize," in *The Cambridge World History of Food*, 100; Bellwood, *First Farmers*, 174–79; Diamond, *Guns, Germs, and Steel*, 151–59; Barker, *Agricultural Revolution*, 231–72.

52. Thomas W. Neumann, "The Role of Prehistoric Peoples in Shaping Ecosystems in the Eastern United States," in Charles E. Kay and Randy T. Simmons, eds., *Wilderness and Political Ecology: Aboriginal Influences and the Original State of Nature* (Salt Lake City: University of Utah Press, 2002), 141–78; Shepard Krech III, *The Ecological Indian: Myth and History* (New York: W. W. Norton, 1999), 103–13; Gordon G. Whitney, *From Coastal Wilderness to Fruited Plain: A History of Environmental Change in Temperate North America, 1500 to the Present* (Cambridge: Cambridge University Press, 1994), 107–20; William Cronon, *Changes in the Land: Indians, Colonists, and the Ecology of New England* (New York: Hill and Wang, 1983), 48–51; Williams, *Deforesting the Earth*, 25–30, 55–60.

53. Cronon, *Changes in the Land*, 37–47.

54. Tim Flannery, *The Eternal Frontier*, 227; Williams, *Deforesting the Earth*, 29, 53.

55. Genesis 3:23 and 3:19.

56. Bryan Sykes, *The Seven Daughters of Eve: The Science that Reveals Our Genetic Ancestry* (New York: W. W. Norton, 2001), 267.

57. Mark Nathan Cohen, *Health and the Rise of Civilization* (New Haven, Conn.: Yale University Press, 1988), 127–31.

58. Cochran and Harpending, *10,000 Year Explosion*, 69; Spencer Wells, *Pandora's Seed: The Unforeseen Cost of Civilization* (New York: Random House, 2010), 53.

59. William H. McNeill, *Plagues and Peoples* (Garden City, N.Y.: Doubleday, 1976), 22.

60. Helen M. Leach, "Human Domestication Reconsidered," *Current Anthropology* 44 no. 3 (2003), 349–68; Lawrence Angel, "Health as a Crucial Factor in the Changes from Hunting to Developed Farming in the Eastern Mediterranean," and Mark Nathan Cohen and George J. Armelagos, "Paleopathology at the Origins of Agriculture: Editors' Summation," in Mark Nathan Cohen and George J. Armelagos, eds., *Paleopathology at the Origins of Agriculture* (Orlando, Fla.: Academic Press, 1984), 54–56 and 585–601.

61. Cochran and Harpending, *The 10,000 Year Explosion*, 112.

62. Maciej Henneberg, "Decrease of Human Skull Size in the Holocene," *Human Biology* 60 (1988), 395–405; John P. Rushton, "Cranial Capacity Related to Sex, Rank, and Race in a Stratified Random Sample of 6,325 U.S. Military Personnel," *Intelligence* 16 no. 3–4 (1992), 401–13. For a discussion, see Kathleen McAuliffe, "If Modern Humans Are So Smart, Why Are Our Brains Shrinking?" *Discover* (September 2010).

63. Daniel Lieberman, *The Story of the Human Body: Evolution, Health, and Disease* (New York: Vintage, 2014), 190–95; Anthony, *Horse, Wheel, and Language*, 405, 439; Brooke, *Climate Change*, 221; Cochran and Harpending, *10,000 Year Explosion*, 76–80; Cohen, *Health and the Rise of Civilization*, 118–20.

64. Edmund Russell, *Evolutionary History: Uniting History and Biology to Understand Life on Earth* (Cambridge: Cambridge University Press, 2011), 89–93; John R. McNeill, "Biological Exchanges," 325–42; Nabil Sabri Enettah et al., "Independent Introduction of Two Lactase-Resistance Alleles into Human Populations Reflects Different History of Adaptation to Milk Culture," *American Journal of Human Genetics* 82 no. 1 (2008), 57–72; Sarah A. Tishkoff et al., "Convergent Adaptation of Human Lactase

Persistence in Africa and Europe," *Nature Genetics* 32 no. 1 (2007), 31–40; Cochran and Harpending, *10,000 Year Explosion*, 65–67, 74–84.

65. James L. A. Webb Jr., "Malaria and the Peopling of Early Tropical Africa," *Journal of World History* 16 no. 3 (2005), 273–84, and *Humanity's Burden: A Global History of Malaria* (New York: Cambridge University Press, 2009), 20–33; Randall M. Packard, *The Making of a Tropical Disease: A Short History of Malaria* (Baltimore: Johns Hopkins University Press, 2007), 22–31; Dorothy H. Crawford, *Deadly Companions: How Microbes Shaped Our History* (Oxford: Oxford University Press, 2007), 59–62; Deirdre Joy et al., "Early Origin and Recent Expansion of Plasmodium Falciparum," *Science* 300 no. 56 (2003), 318–21.

66. Vered Eshed et al., "Paleopathology and the Origin of Agriculture in the Levant," *American Journal of Physical Anthropology* 143 no. 1 (2010), 121–33; Jessica M. C. Pearce-Duvet, "The Origin of Human Pathogens: Evaluating the Role of Agriculture and Domestic Animals in the Evolution of Human Disease," *Biological Reviews* 81 no. 3 (2006), 369–82; Clark S. Larsen, "The Agricultural Revolution as Environmental Catastrophe: Implications for Health and Lifestyle in the Holocene," *Quaternary International* 150 (2006), 12–20; Nathan Wolfe, "Preventing the Next Pandemic," *Scientific American* 300 no. 4 (2009), 76–81; Cohen, *Health and the Rise of Civilization*, 33–47, 116–18; McNeill, *Plagues and Peoples*, 42, 51–56; Lieberman, *The Story of the Human Body*, 201–202.

67. Alfred Crosby, *Ecological Imperialism: The Biological Expansion of Europe, 900–1900* (Cambridge: Cambridge University Press, 1986), 197–98; Cohen, *Health and the Rise of Civilization*, 114–15; Cochran and Harpending, *10,000 Year Explosion*, 159–61; McNeill, *Plagues and Peoples*, 32–33; Crawford, *Deadly Companions*, 112. See also Suzanne Austin Alchon, *A Pest in the Land: New World Epidemics in a Global Perspective* (Albuquerque: University of New Mexico Press, 2003), 15.

68. Williams, *Deforesting the Earth*, 35–58; Roberts, *Holocene*, 155–201; Cordova, *Millennial Landscape Change*, 165–75.

69. Johan Goudsblom, *Fire and Civilization* (New York: Penguin, 1992), 48–53; Boserup, *Population and Technological Change*, 47–48.

70. John R. McNeill, *The Mountains of the Mediterranean World: An Environmental History* (Cambridge: Cambridge University Press, 1992), 276–80; Redman, *Human Impact*, 100–102; Roberts, *Holocene*, 186–88; de Vries and Marchant, "Environment," 89–90.

71. Gary O. Rollefson and Ilse Kohler-Rollefson, "Early Neolithic Exploitation Patterns in the Levant: Cultural Impact on the Environment," *Population and Environment: A Journal of Interdisciplinary Studies* 13 no. 4 (1992), 243–54.

72. William H. Calvin, *A Brain for All Seasons: Human Evolution and Abrupt Climate Change* (Chicago: University of Chicago Press, 2002), 77, 213–14; William Ruddiman, *Earth's Climate: Past and Future* (New York: W. H. Freeman, 2001), 277–80; Burroughs, *Climate Change*, 63; Brooke, *Climate Change*, 67–78.

73. Ruddiman, *Earth's Climate*, 277–80.

74. This section is based on William F. Ruddiman, "The Anthropogenic Greenhouse Era Began Thousands of Years Ago," *Climatic Change* 61 (2003), 261–93, and

Plows, Plagues, and Petroleum: How Humans Took Control of Climate (Princeton, N.J.: Princeton University Press, 2005), chs. 3–10. See also M. James Salinger, "Agriculture's Influence on Climate during the Holocene," *Agricultural and Forest Meteorology* 142 (2007), 96–102.

75. Ruddiman, *Plows, Plagues, and Petroleum*, 63–64.

76. See the graphs in Ruddiman, *Plows, Plagues, and Petroleum*, 77, 87.

77. Ruddiman, *Plows, Plagues, and Petroleum*, 88.

78. Ruddiman, *Plows, Plagues, and Petroleum*, 95.

79. Recent studies supporting the Ruddiman thesis include Logan Mitchell, "Constraints on the Late Holocene Anthropogenic Contribution to the Atmospheric Methane Budget," *Science* 342 (2013), 964–66, and Jed O. Kaplan, "Holocene Carbon Cycle: Climate or Humans?" *Nature Geoscience* 8 (2015), 3235–36; Michael Crucifix, "Earth's Narrow Escape from a Big Freeze," *Nature* 529 no. 7585 (January 14, 2016), 162–63; A. Ganopolkski et al., "Critical Insolation–CO_2 Relation for Diagnosing Past and Future Glacial Inception," *Nature* 529 no. 7585 (January 14, 2016), 200–203. For a contrary view, see Jörgen Olofsson and Thomas Hickler, "Effects of Human Land-Use on the Global Carbon Cycle during the Last 6,000 Years," *Vegetation History and Archaeobotany* 17 no. 5 (September 2008), 605–15.

Chapter 3

1. D. T. Potts, *Mesopotamian Civilization: The Material Foundations* (London: Athlon Press, 1997); Edmund Burke III, "The Transformation of the Middle Eastern Environment, 1500 B.C.E.–2000 C.E.," in Edmund Burke III and Kenneth Pomeranz, eds., *The Environment and World History* (Berkeley: University of California Press, 2009), 82–84.

2. Steven Mithen, *Thirst: Water and Power in the Ancient World* (Cambridge, Mass.: Harvard University Press, 2012), 48–60; Robert McC. Adams, "The Origin of Cities," in Scientific American, *Old World Archaeology: Foundations of Civilization* (San Francisco: W. H. Freeman, 1972), 137–44, and "Historic Patterns of Mesopotamian Irrigation Agriculture," in Theodore E. Downing and McGuire Gibson, eds., *Irrigation's Impact on Society* (Tucson: University of Arizona Press, 1976), 1–2; Harvey Weiss, "Beyond the Younger Dryas: Collapse as Adaptation to Abrupt Climate Change in Ancient West Asia and the Eastern Mediterranean," in Garth Bawden and Richard Martin Reycraft, eds., *Environmental Disasters and the Archaeology of Human Response* (Albuquerque, N.M.: Maxwell Museum of Anthropology, 2000), 75–77; Edward Goldsmith, *Social and Environmental Effects of Large Dams* (San Francisco: Sierra Club Books, 1986), 304–11; Potts, *Mesopotamian Civilization*, 52–55.

3. Karl Wittfogel, *Oriental Despotism: A Comparative Study of Total Power* (New Haven, Conn.: Yale University Press, 1957).

4. Denise Schmandt-Bessarat, *How Writing Came About* (Austin: University of Texas Press, 1996); C. Leonard Wooley, *The Sumerians* (New York: W. W. Norton, 1965),

16–17; Samuel Noah Kramer, "The Sumerians," in Scientific American, *Old World Archaeology*, 149; Adams, "Origin," 138–39, and "Historic Patterns," 2–4.

5. H. M. Cullen et al., "Climate Change and the Collapse of the Akkadian Empire: Evidence from the Deep Sea," *Geology* 28 (April 2000), 379–82; Harvey Weiss, "Late Third Millennium Abrupt Climate Change and Social Collapse in West Asia and Egypt," in H. Nüzhet Dalfes, George Kukla and Harvey Weiss, eds., *Third Millennium BC Climate Change and Old World Collapse* (Berlin: Springer Verlag, 1997), 711–24; Weiss, "Genesis and Collapse of Third Millennium North Mesopotamian Civilization," *Science* 261 (1993), 995–1004, and "Beyond the Younger Dryas," 75–98; Peter B. de Menocal, "Cultural Responses to Climate Change during the Late Holocene," *Science* 292 (2001), 667–73; Marie-Agnès Courty et al., "The Genesis and Collapse of Third Millennium North Mesopotamian Civilization," 995–1004, and Ann Gibbons et al., "How the Akkadian Empire Was Hung Out to Dry," 985, both in *Science* 261 (1993).

6. Thorkild Jacobsen and Robert McC. Adams, "Salt and Silt in Ancient Mesopotamian Agriculture," *Science* 128 (1958), 1251–58. The salinization scenario has come under attack by M. Powell in "Salt, Silt, and Yields in Sumerian Agriculture: A Critique of the Theory of Progressive Salinization," *Zeitschrift für Assyriologie* 75 (1985), 7–38. See also McGuire Gibson, "Violation of Fallow and Engineered Disaster in Mesopotamian Civilization," in Dowling and Gibson, *Irrigation's Impact on Society*, 7–19; Charles L. Redman, *Human Impact on Ancient Environments* (Tucson: University of Arizona Press, 1999), 128–34; Weiss, "Beyond the Younger Dryas," 91; and Mithen, *Thirst*, 69–74.

7. Karl W. Butzer, *Early Hydraulic Civilization in Egypt: A Study in Cultural Ecology* (Chicago: University of Chicago Press, 1976), 17–23; Brian Fagan, *Floods, Famines, and Emperors: El Niño and the Fate of Civilizations* (New York: Basic Books, 1999), 103–6.

8. J. Donald Hughes, *An Environmental History of the World: Humankind's Changing Role in the Community of Life* (London: Routledge, 2001), 38.

9. J. H. Breasted, *Records of Ancient Egypt* (Chicago: University of Chicago Press, 1906), 1, 188–89, quoted in Hughes, *Environmental History*, 39.

10. Barry Kemp, *Ancient Egypt: The Anatomy of a Civilization*, 2nd ed. (London: Routledge, 2005), 10–12; Butzer, *Early Hydraulic Civilization*, 4–6, 19–21, 43–51, 89–90.

11. John L. Brooke, *Climate Change and the Course of Global History: A Rough Journey* (Cambridge: Cambridge University Press, 2014), 292–93.

12. Robert C. Allen, "Agriculture and the Origin of the State in Ancient Egypt," *Explorations in Economic History* 34 (1997), 135–54; William J. Burroughs, *Climate Change in Prehistory: The End of the Reign of Chaos* (Cambridge: Cambridge University Press, 2005), 230; Rushdi Said, *The River Nile: Geology, Hydrology, ad Utilization* (Oxford: Pergamon Press, 1993), 191–93; Butzer, *Early Hydraulic Civilization*, 46–55; Kemp, *Ancient Egypt*, 10–12; Fagan, *Floods*, 99–117.

13. Robert O. Collins, *The Nile* (New Haven, Conn.: Yale University Press, 2002), 14–19; Barbara Bell, "The Oldest Records of the Nile Floods," *Geographical Journal* 136 (1970), 569–73; "Climate and the History of Egypt, the Middle Kingdom," *American Journal of Archaeology* 79 no. 3 (1975), 223–69; and "The Dark Ages in Ancient

History, I: The First Dark Age in Egypt," *American Journal of Archaeology* 79 no. 1 (1975), 1–26; Said, *River Nile*, 138–50.

14. Jane R. McIntosh, *The Ancient Indus Valley: New Perspectives* (Santa Barbara, Cal.: ABC-CLIO, 2008), 67–91, 109–22, 356; Bridget Allchin, "Early Man and Environment in South Asia, 10,000 BC–AD 500," in Richard Grove, Vinita Damodaran, and Satpal Sangwan, eds., *Nature and the Orient* (Delhi: Oxford University Press, 1998), 41; V. N. Misra, "Climate, a Factor in the Rise and Fall of the Indus Civilization: Evidence from Rajasthan and Beyond," in Mahesh Rangarajan and K. Sivaramakrishnan, eds., *India's Environmental History*, vol. 1: *From Ancient Times to the Colonial Period* (Ranikhet: Permanent Black, 2012), 53–64; Gregory L. Possehl, "Climate and the Eclipse of the Ancient Cities of the Indus," in Dalfes, Kukla, and Weiss, *Third Millennium BC*, 193–244.

15. N. V. Misra, "Climate, a Factor in the Rise and Fall of the Indus Civilization: Evidence from Rajasthan and Beyond," in Mahesh Rangarajan and Kalyanakrishnan Sivaramakrishnan, eds., *India's Environmental History* (Ranikhet: Permanent Black, 2012), 1: 53–64.

16. Gwen Robbins Schug et al., "Infection, Disease, and Biosocial Processes at the End of the Indus Civilization, *PLoS ONE* 8 no. 12 (2013); Jane R. McIntosh, *Peaceful Realm: The Rise and Fall of the Indus Civilization* (Santa Barbara, Cal.: ABC-CLIO, 2008), 173, 186–89, and *Ancient Indus Valley*, 91–94, 396–98; Nayajot Lahiri, *Decline and Fall of the Indus Civilization* (Bangalore: Orient Longman, 2000), 1–33; Jonathan Mark Kenoyer, *Ancient Cities of the Indus Valley Civilization* (Karachi: Oxford University Press, 1998), 173; Allchin, "Early Man," 43–48.

17. Kenoyer, *Ancient Cities*, 173; Allchin, "Early Man," 46–47; McIntosh, *Peaceful Realm*, 190–92, and *Ancient Indus Valley*, 119–20.

18. McIntosh, *Peaceful Realm*, 193.

19. William M. Denevan, *Cultivated Landscapes of Native Amazonia and the Andes* (Oxford: Oxford University Press, 2001), 135–37; Michael E. Moseley, *The Incas and Their Ancestors* (New York: Thames and Hudson, 2001), 26–29, 223; Brian Fagan, *The Great Warming: Climate Change and the Rise and Fall of Civilizations* (New York: Macmillan, 2008), 159–64, and *Floods*, 129–30.

20. César N. Caviedes, *El Niño in History: Storming through the Ages* (Gainesville: University of Florida Press, 2001), chs. 2 and 3.

21. Mithen, *Thirst*, 258–60.

22. Stuart J. Fiedel, *Prehistory of the Americas*, 2nd ed. (Cambridge: Cambridge University Press, 1993), 321–23; Izumi Shimada et al., "Cultural Impacts of Severe Droughts in the Prehistoric Andes: Application of a 1,500-Year Ice Core Precipitation Record," *World Archaeology* 22 no. 3 (1991), 247–70; Moseley, *Incas*, 26–33, 41–49, 223–25; de Menocal, "Cultural Responses"; Fagan, *Floods*, 124–34.

23. Denevan, *Cultivated Landscapes*, 168–69; see also 149–57; Moseley, *Incas*, 134–35; Fiedel, *Prehistory*, 330–31; Mithen, *Thirst*, 260–63; Fagan, *Great Warming*, 165–72, and *Floods*, 119–35.

24. Michael W. Binford et al., "Climate Variations and the Rise and Fall of Andean Civilizations," *Quaternary Research* 47 no. 2 (1997), 235–48; Fiedel, *Prehistory*,

330–37; Denevan, *Cultivated Landscapes*, chs. 9 and 11; Moseley, *Incas*, 27–44, 223–24; de Menocal, "Cultural Responses."

25. Terrence N. D'Altroy, *The Incas* (Oxford: Blackwell, 2002); Ann Kendall, *Everyday Life of the Incas* (London: Batsford, 1973), 140–44; Moseley, *The Incas*, 43–46; Fiedel, *Prehistory*, 327–39; Mithen, *Thirst*, 263–65.

26. Antonio Elio Brailovsky, *Historia ecológica de Iberoamérica*, vol. 1: *De los Mayas al Quijote* (Buenos Aires: Ediciones Le Monde Diplomatique, 2006), 91–93; Michael Coe, "The Chinampas of Mexico," in Scientific American, *New World Archaeology: Theoretical and Cultural Transformations* (San Francisco: W. H. Freeman, 1974), 231–34.

27. Angel Palerm, *Obras hidráulicas prehispánicas en el sistema lacustre del Valle de México* (Mexico City: Instituto Nacional de Antropología e Historia, 1973), 18–19; Shawn William Miller, *An Environmental History of Latin America* (Cambridge: Cambridge University Press, 2007), 18–22; Thomas M. Whitmore and B. L. Turner II, *Cultivated Landscapes of Middle America on the Eve of the Conquest* (Oxford: Oxford University Press, 2002), 220–24; Michael D. Coe and Rex Koontz, *Mexico: From the Olmecs to the Aztecs* (London: Thames and Hudson, 2013), 163–64, and "Chinampas," 231–37.

28. Emily McClung de Tapia, "Prehispanic Agricultural Systems in the Basin of Mexico," in David Lewis Lentz, ed., *Imperfect Balance: Landscape Transformations in the Precolumbian Americas* (New York: Columbia University Press, 2000), 135–40; René Millon, "Teotihuacán," in Scientific American, *Avenues to Antiquity: Readings from Scientific American* (San Francisco: W. H. Freeman, 1976), 221–25.

29. Mark Nathan Cohen, *Health and the Rise of Civilization* (New Haven, Conn.: Yale University Press, 1988), 54, 122–26.

30. Coe and Koontz, *Mexico*, 105–6.

31. Whitmore and Turner, *Cultivated Landscapes*, 112–14; Coe, "Chinampas," 237–38; Miller, *Environmental History*, 10–11, 22; Palerm, *Obras hidráulicas prehispánicas*, 22.

32. David Stuart, *Anasazi America: Seventeen Centuries on the Road from Center Place* (Albuquerque: University of New Mexico Press, 2000), 35–38; Bruce D. Smith, *The Emergence of Agriculture* (New York: W. H. Freeman, 1995), 156–60, 203–5; Jared Diamond, *Collapse: How Societies Choose to Fail or Succeed* (New York: Penguin, 2005), 140–44; Fiedel, *Prehistory*, 207–15.

33. Stephen Plog, *Ancient Peoples of the American Southwest*, 2nd ed. (London: Thames and Hudson, 2008), 107.

34. Plog, *Ancient Peoples*, 102–5; Christopher H. Guiterman, "Eleventh-Century Shift in Timber Procurement Areas for the Great Houses of Chaco Canyon," *Proceedings of the National Academy of Science of the USA* 113 no. 5 (February 2, 2016), 1186–190; "Sources of Chaco Wood," *Nature* 529 no. 7584 (January 7, 2016), 31–32, and Diamond, *Collapse*, 143–49; Redman, *Human Impact*, 120.

35. Larry Benson, Kenneth Petersen, and John Stein, "Anasazi (Pre-Columbian Native American) Migrations during the Middle-12th and Late-13th Centuries—Were They Drought-Induced?" *Climatic Change* 83 (2007), 187–213; Terry L. Jones et al., "Environmental Imperatives Reconsidered: Demographic Crises in Western North America during the Medieval Climate Anomaly," *Current Anthropology* 40 no. 2

(1999), 137–70; Stuart, *Anasazi America*, 119–21; Redman, *Human Impact*, 118–24; Diamond, *Collapse*, 137–55; Fiedel, *Prehistory*, 207–28; Fagan, *Floods*, 160–77. For a critique of the Chaco Canyon collapse scenario, see Michael Wilcox, "Marketing Conquest and the Vanishing Indian," in Patricia A. McAnany and Norman Yoffee, eds., *Questioning Collapse: Human Resilience, Ecological Vulnerability, and the Aftermath of Empire* (New York: Cambridge University Press, 2010), 113–41, and Plog, *Ancient Peoples*, 115.

36. Richardson Benedict Gill, *The Great Maya Drought: Water, Life, and Death* (Albuquerque: University of New Mexico Press, 2000), ch. 9: "Geology, Hydrology, and Water"; Anabel Ford, "Critical Resource Control and the Rise of the Classic Period Maya," in Scott L. Fedick, ed., *The Managed Mosaic: Ancient Maya Agriculture and Resource Use* (Salt Lake City: University of Utah Press, 1996), 299–300; Michael Coe, *The Maya*, 8th ed. (London: Thames and Hudson, 2011), 13–16; Mithen, *Thirst*, 229–30.

37. Paul W. Richards, "The Tropical Rain Forest," *Scientific American* 299 no. 6 (December 1973), 58–67.

38. Coe, *Maya*, 16–17; David Webster, *The Fall of the Ancient Maya: Solving the Mystery of the Maya Collapse* (New York: Thames and Hudson, 2002), 252–53; Fiedel, *Prehistory*, 287; Redman, *Human Impact*, 141–42.

39. Michael Williams, *Deforesting the Earth: From Prehistory to Global Crisis* (Chicago: University of Chicago Press, 2003), 49–50; Diamond, *Collapse*, 163–68; Coe, *Maya*, 14–20, 35; Webster, *Fall of the Ancient Maya*, 332–34; Mithen, *Thirst*, 226, 232–33.

40. Lisa Lucero, "The Collapse of the Classic Maya: A Case for the Role of Water Control," *American Anthropologist* 104 (2002), 815–19; Webster, *Fall of the Ancient Maya*, 255–55, 330–35; Redman, *Human Impact*, 140–45; Coe, *Maya*, 14–20, 166–67; Fagan, *Floods*, 147–55; Williams, *Deforesting the Earth*, 49–51; Gill, *Great Maya Drought*, 318–20, 371; Diamond, *Collapse*, 164–70. For a contrary view, see Scott L. Fedick, "The Maya Forest: Destroyed or Cultivated by the Ancient Maya?," *Proceedings of the National Academy of Science* 17 no. 3 (2010), 953–54.

41. William Saturno, "Sistine Chapel of the Early Maya," *National Geographic* 204 no. 6 (2003), 72–76.

42. Vernon L. Scarborough, "The Flow of Power: Water Reservoirs Controlled the Rise and Fall of the Ancient Maya," *Sciences* 32 no. 2 (March 1992), 39–43; Gill, *Great Maya Drought*, 263–66; Ford, "Critical Resource Control," 300–302; Lucero, "Collapse," 818; Mithen, *Thirst*, 233–36.

43. Lucero, "Collapse," 815; Coe, *Maya*, 21; Scarborough, "Flow of Power," 42–43; Ford, "Critical Resources Control," 301–2; Diamond, *Collapse*, 167–69; Webster, *Fall of the Ancient Maya*, 335; Mithen, *Thirst*, 230–31, 246–47.

44. Gill, *Great Maya Drought*, 313–14, 360; Diamond, *Collapse*, 170–75.

45. Brailovsky, *Historia ecológica*, 1: 90; Rebecca Storey et al., "Social Disruption and the Maya Civilization of Mesoamerica: A Study of Health and Economy in the Last Thousand Years," in Richard H. Steckel and Jerome C. Rose, eds., *The Backbone of History: Health and Nutrition in the Western Hemisphere* (Cambridge: Cambridge University Press, 2005), 290–96.

46. Coe, *Maya*, 128.

47. Eliot M. Abrams, Ann Corfinne Freter, David J. Rue, and John D. Wingard, "The Role of Deforestation in the Collapse of the Late Classic Copan Maya State," in Leslie E. Sponsel, Thomas N. Headland, and Robert C. Bailey, eds., *Tropical Deforestation: The Human Dimension* (New York: Columbia University Press, 1996).

48. Brailovksy, *Historia ecológica*, 89–91.

49. Andrew Scherer and Charles Golden, "Water in the West: Chronology and Collapse of the Western Maya River Kingdoms," in Gyles Iannone, ed., *The Great Maya Droughts in Cultural Context* (Boulder: University of Colorado Press, 2014), 211–27; Coe, *Maya*, 128.

50. Webster, *Fall of the Ancient Maya*, 254–55. See also T. Patrick Culbert, "The Collapse of the Classic Maya Civilization," in Norman Yaffee and George Cowgill, eds., *The Collapse of Ancient States and Civilizations* (Tucson: University of Arizona Press, 1988), 69–101.

51. Gill, *Great Maya Droughts*, 386.

52. Pete D. Akers, "An Extended and Higher-Resolution Record of Climate and Land Use from Stalagmite MC01 from Macal Chasm, Belize, Revealing Connections between Major Dry Events, Overall Climate Variability, and Maya Sociopolitical Changes," *Palaeogeography, Palaeoclimatology, Palaeoecology* 459 (2016), 268–88; Webster, *Fall of the Ancient Maya*, 239–44; Gill, *Great Maya Droughts*, 313–20, 331, 360–65. On the possible cause of the drought, see Gergana Yancheva et al., "Influence of the Intertropical Convergence Zone on the East Asian Monsoon," *Nature* 345 (2007), 74–77.

53. Gerald Haug et al., "Climate and the Collapse of Maya Civilization," *Science* 299 (2003), 1731. See also Diamond, *Collapse*, 173–74; Gill, *Great Maya Drought*, 280, 363; David A. Hodell, Jason H. Curtis, and Mark Brenner, "Possible Role of Climate in the Collapse of Classic Maya Civilization," *Nature* 375 (1995), 391–94; David Stahle et al., "Major Mesoamerican Droughts of the Past Millennium," *Geophysical Research Letters* 38 no. 5 (March 2011). For a contrary view, see Arthur Demarest, *Ancient Maya: The Rise and Fall of a Rainforest Civilization* (Cambridge: Cambridge University Press, 2004).

54. Benjamin Cook et al., "Pre-Columbian Deforestation as an Amplifier of Drought in Mesoamerica," *Geophysical Research Letters* 39 no. 16 (2012); Mithen, *Thirst*, 250–52.

55. Patricia A. McAnany and Tomás Gallareta Negrón, "Bellicose Rulers and Climatological Peril? Retrofitting Twenty-First-Century Woes on Eighth-Century Maya Society," in McAnany and Yoffee, *Questioning Collapse*, 154–58.

Chapter 4

1. Jean-Noël Biraben, "Essai sur l'évolution du nombre des hommes," *Population* 34 no. 1 (1979), 13–25; Massimo Livi-Bacci, *A Concise History of World Population*, 3rd ed. (Malden, Mass.: Blackwell, 2001), 26.

2. Vaclav Smil, *General Energetics* (Chichester: Wiley, 1991), 239.

3. Tertius Chandler and Gerald Fox, *Four Thousand Years of Urban Growth: An Historical Census* (New York: Academic Press, 1974), 300–317.

4. Rachel Laudan, *Cuisine and Empire: Cooking in World History* (Berkeley: University of California Press, 2013), 35–36.

5. Michael Williams, *Deforesting the Earth: From Prehistory to Global Crisis* (Chicago: University of Chicago Press, 2003), 74–77, 124.

6. R. J. Wenke, "Western Iran in the Partho-Sasanian Period: The Imperial Transformation," in F. Hole, ed., *The Archaeology of Western Iran: Settlement and Society from Prehistory to the Islamic Conquest* (Washington: Smithsonian Institution Press, 1987), 251–81.

7. Edmund Burke III, "The Transformation of the Middle Eastern Environment, 1500 B.C.E.–2000 C.E.," in Edmund Burke III and Kenneth Pomeranz, eds., *The Environment and World History* (Berkeley: University of California Press, 2009), 84.

8. Robert McC. Adams, *Land behind Baghdad: A History of Settlement on the Diyala Plains* (Chicago: University of Chicago Press, 1965), 69. See also Adams, "Historic Patterns of Mesopotamian Irrigation Agriculture," in Theodore E. Downing and McGuire Gibson, eds., *Irrigation's Impact on Society* (Tucson: University of Arizona Press, 1974), 4; and Peter Christensen, *Decline of Iranshahr: Irrigation and Environments in the History of the Middle East, 500 B.C. to A.D. 1500* (Copenhagen: Museum Tusculanum Press, 1993), 68–72.

9. Adams, *Land behind Baghdad*, 80–81. See also William H. McNeill, *Plagues and Peoples* (Garden City, N.Y.: Doubleday, 1976), 144; Adams, "Historic Patterns," 5; and Christensen, *Decline of Iranshahr*, 73–104.

10. Lynn White Jr., *Medieval Technology and Social Change* (Oxford: Oxford University Press, 1962), 41; J. Donald Hughes, *Pan's Travails: Environmental Problems of the Greeks and Romans* (Baltimore: Johns Hopkins University Press, 1994), 136–39, and *The Mediterranean: An Environmental History* (Santa Barbara, Cal.: ABC-CLIO, 2005), 33; Lukas Thommen, *An Environmental History of Ancient Greece and Rome*, trans. Philip Hill (Cambridge: Cambridge University Press, 2012), 33–36.

11. Rushdi Said, *The River Nile: Geology, Hydrology, and Utilization* (Oxford: Pergamon Press, 1993), 204.

12. Kyle Harper, *The Fate of Rome: Climate, Disease, and the End of an Empire* (Princeton, N.J.: Princeton University Press, 2017), chs. 3 and 4.

13. Russell Meiggs, *Trees and Timber in the Ancient World* (Oxford: Clarendon Press, 1982), 373; Hughes, *Pan's Travails*, 145–46; Williams, *Deforesting the Earth*, 63, 70–71.

14. *The Epic of Gilgamesh*, trans. and ed. Benjamin R. Foster (New York: W. W. Norton, 2001), 44–45.

15. Quoted in Meiggs, *Trees and Timber*, 62.

16. 2 Chronicles 2.

17. Meiggs, *Trees and Timber*, 378.

18. J. V. Thirgood, *Man and the Mediterranean Forest: A History of Resource Depletion* (New York: Academic Press, 1981), 27; Theodore A. Werteim, "The Furnace versus the Goat: The Pyrotechnologic Industries and Mediterranean Deforestation in

Antiquity," *Journal of Field Archaeology* 10 no. 4 (1983), 445–55; Williams, *Deforesting the Earth*, 62, 79 table 4.1, 445–46; Hughes, *Pan's Travails*, 75–78, 81, 86–88, 132–33.

19. Marvin W. Mikesell, "The Deforestation of Mt. Lebanon," *Geographical Review* 59 (1969), 1–28; Thirgood, *Man and the Mediterranean Forest*, 58–59, 96–101.

20. Xavier de Planhol, "Le déboisement de l'Iran," *Annales de géographie* 78 no. 430 (1969), 627–29.

21. Wertime, "The Furnace versus the Goat," 445–55.

22. Williams, *Deforesting the Earth*, 68.

23. Thirgood, *Man and the Mediterranean Forest*, 55–58; Hughes, *Mediterranean*, 41; Williams, *Deforesting the Earth*, 74–75.

24. Ian Gordon Simmons, *An Environmental History of Great Britain: From 10,000 Years Ago to the Present* (Edinburgh: Edinburgh University Press, 2001), 95; John R. McNeill, "Woods and Warfare in World History," *Environmental History* 9 (2004), 392; Thirgood, *Man and the Mediterranean Forest*, 58; Williams, *Deforesting the Earth*, 75.

25. Thirgood, *Man and the Mediterranean Forest*, 35; McNeill, "Woods and Warfare," 395–96.

26. Williams, *Deforesting the Earth*, 72.

27. Thirgood, *Man and the Mediterranean Forest*, 37.

28. Sheldon Judson, "Erosion Rates Near Rome, Italy," *Science* 160 (1968), 1444–46.

29. David R. Montgomery, *Dirt: The Erosion of Civilizations* (Berkeley: University of California Press, 2007), 62–63; Hughes, *Mediterranean*, 43.

30. John R. McNeill, *The Mountains of the Mediterranean World: An Environmental History* (Cambridge: Cambridge University Press, 1992), 74; John L. Brooke, *Climate Change and the Course of Global History: A Rough Journey* (Cambridge: Cambridge University Press, 2014), 341; Hughes, *Pan's Travails*, 82–85; Montgomery, *Dirt*, 58.

31. John McNeill, *Mountains of the Mediterranean*, 72; Thirgood, *Man and the Mediterranean Forest*, 36–40, 46, 59–60. See also Williams, *Deforesting the Earth*, 64–75.

32. George Erdosy, "Deforestation in Pre- and Protohistoric South Asia," in Richard H. Grove, Vinita Damodaran, and Satpal Sangwan, eds., *Nature and the Orient: The Environmental History of South and Southeast Asia* (Delhi: Oxford University Press, 1998), 58–61.

33. Ranabir Chakravarti, "The Creation and Expansion of Settlements and Management of Hydraulic Resources in Ancient India," in Grove, Damodaran, and Sangwan, *Nature and the Orient*, 89; Bridget and F. R. Allchin, *The Rise of Civilization in India and Pakistan* (Cambridge: Cambridge University Press, 1982), 318; Erdosy, "Deforestation," 62–65. For a contrary view, see Makhan Lal, "Iron Tools, Forest Clearance, and Urbanization in the Gangetic Plain," in Mahesh Rangarajan and Kalyanakrishnan Sivaramakrishnan, eds., *India's Environmental History*, 2 vols. (Ranikhet: Permanent Black, 2012), 65–79.

34. Madhav Gadgil and Ramachandra Guha, *This Fissured Land: An Ecological History of India* (Berkeley: University of California Press, 1993), 77–106; Chakravarti, "Creation and Expansion of Settlements," 91–97.

35. John R. McNeill, "China's Environmental History in World Perspective," in Mark Elvin and Liu Ts'ui-jung, eds., *Sediments of Time: Environment and Society in Chinese History* (Cambridge: Cambridge University Press, 1998), 35–37.

36. Mark Edward Lewis, *The Early Chinese Empires: Qin and Han* (Cambridge, Mass.: Harvard University Press, 2007), 9, 105–13; Francesca Bray, "Swords into Plowshares: A Study of Agricultural Technology and Society in Early China," *Technology and Culture* 19 no. 1 (1978), 1–13.

37. Mencius quotation in Lester J. Bilsky, "Ecological Crisis and Response in Ancient China," in Lester J. Bilsky, ed., *Historical Ecology: Essays on Environment and Social Change* (Port Washington, N.Y.: Kennikat Press, 1980), 61. See also Bao Maohong, "Environmental Resources and china's Historical Development, in John R. McNeill, José Augusto Pádua, and Mahesh Rangarajan, eds., *Environmental History: As If Nature Existed* (New Delhi: Oxford University Press, 2010), 88–90.

38. Robert B. Marks, *China: Its Environment and History* (Lanham, Md.: Rowman & Littlefield, 2012), 32–34.

39. Mark Elvin, *Retreat of the Elephants: An Environmental History of China* (New Haven, Conn.: Yale University Press, 2004), 42.

40. Francesca Bray, *Agriculture*, vol. 6 part 2 of Joseph Needham, *Science and Civilisation in China* (Cambridge: Cambridge University Press, 1984), 566; Brooke, *Climate Change*, 318–19; Marks, *China*, 58–76.

41. Quoted in Elvin, *Retreat of the Elephants*, 11, and in Marks, *China*, 60.

42. Marks, *China*, 37–50, 79–88.

43. Hsu Cho-yun, *Han Agriculture: The Formation of Early Chinese Agrarian Economy, 206 B.C.–A.D. 220* (Seattle: University of Washington Press, 1980).

44. Hsu Cho-yuan, "Environmental History—Ancient," in *Berkshire Encyclopedia of China* (Great Barrington, Mass.: Berkshire, 2009), 732–33; Marks, *China*, 101; Williams, *Deforesting the Earth*, 121–22; Elvin, *Retreat of the Elephants*, chs. 3 and 4.

45. Walter H. Mallory, *China: Land of Famine* (New York: American Geographical Society, 1926), 37.

46. Randall Dodgen, *Controlling the Dragon: Confucian Engineers and the Yellow River in Late Imperial China* (Honolulu: University of Hawaii Press, 2001), 11–12.

47. Kenneth Pomeranz, "The Transformation of China's Environment," in Burke and Pomeranz, *Environment and World History*, 129; Marks, *China*, 41, 78; Dodgen, *Controlling the Dragon*, 14.

48. Mark Edward Lewis, *China's Cosmopolitan Empire: The Tang Dynasty* (Cambridge, Mass.: Harvard University Press, 2009), 12, and *Early Chinese Empires*, 9; Mark Elvin, "3,000 Years of Unsustainable Growth: China's Environment from Archaic Times to the Present," *East Asian History* 6 (1993), 30, and *Retreat of the Elephants*, 24–25; Hsu, "Environmental History—Ancient," 2: 731; Qu Geping and Li Jinchang, *Population and Environment in China* (Boulder, Colo.: L. Rienner, 1994), 16–17; Mallory, *China*, 45–46.

49. Lewis, *China's Cosmopolitan Empire*, 9; Marks, *China*, 79.

50. Hsu, *Han Agriculture*, 28–29; Marks, *China*, 53–55, 66–71.

51. J. E. Spencer, "Water-Control in Terraced Rice-Field Agriculture in Southeastern Asia," in Downing and Gibson, *Irrigation's Impact*, 59–65.

52. Charles Higham and Tracey L. D. Lu, "The Origin and Dispersal of Rice Cultivation," *Antiquity* 72 (1998): 867–77; Steven Mithen, *After the Ice: A Global Human History, 20,000–5000 BC* (Cambridge, Mass.: Harvard University Press, 2004), 361–69; Francesca Bray, *The Rice Economies: Technology and Development in Asian Societies* (Berkeley: University of California Press, 1986), 9–34.

53. Steven Mithen, *Thirst: Water and Power in the Ancient World* (Cambridge, Mass.: Harvard University Press, 2012), 150–64; Lyman P. Van Slyke, *Yangtze: Nature, History, and the River* (Reading, Mass.: Addison-Wesley, 1988), 56–58; Hsu Cho-yuan, "China, Ancient," in Shepard Krech III, John R. McNeill, and Carolyn Merchant, eds., *Encyclopedia of World Environmental History* (New York: Routledge, 2004), 224–27; Hsu, *Han Agriculture*, 100; Marks, *China*, 77.

54. McNeill, *Plagues and Peoples*, 88.

55. Elvin, *Retreat of the Elephants*, 262–65.

56. Lewis, *Early Chinese Empires*, 24, 105–6; quotation on p. 9.

57. Hsu, "Environmental History—Ancient," 733, and "China, Ancient," 225–26.

58. Elvin, *Retreat of the Elephants*, 11–14.

59. Mark Nathan Cohen, *Health and the Rise of Civilization* (New Haven, Conn.: Yale University Press, 1989), 124–26, 133.

60. Barry W. Cunliffe, *Europe between the Oceans, 9000 BC to AD 1000* (New Haven, Conn.: Yale University Press, 2011), 394; Cohen, *Health and the Rise of Civilization*, 51, 137.

61. Brooke, *Climate Change*, 338.

62. Frederick Cartwright, *Disease and History* (New York: Mentor, 1972), 15–16.

63. Vaclav Smil, *Why America Is Not a New Rome* (Cambridge, Mass.: MIT Press, 2010), 124–35.

64. McNeill, *Plagues and Peoples*, 77–82, 107–114.

65. Gregory Cochran and Henry Harpending, *The 10,000 Year Explosion: How Civilization Accelerated Human Evolution* (New York: Basic Books, 2009), 86; Cohen, *Health and the Rise of Civilization*, 48–49; McNeill, *Plagues and Peoples*, 60.

66. McNeill, *Plagues and Peoples*, 130.

67. Robert Sallares, *Malaria and Rome: A History of Malaria in Ancient Italy* (Oxford: Oxford University Press, 2002); Robert Sallares, A. Bouwman, and C. Anderung, "The Spread of Malaria to Southern Europe in Antiquity: New Approaches to Old Problems," *Medical History* 48 (2004), 311–28; John McNeill, *Mountains*, 74, 86–87, 176–77; Hughes, *Mediterranean*, 49–50.

68. See, for example, Exodus 9:9, I Samuel 5:6 and 5:9, and Isaiah 37:36.

69. R. S. Bray, *Armies of Pestilence: The Effects of Pandemics on History* (Cambridge: Lutterworth, 1996), 9.

70. Burke A. Cunha, "The Cause of the Plague of Athens: Plague, Typhoid, Typhus, Smallpox, or Measles?," *Infectious Disease Clinics of North America* 18 no. 1 (2004), 29–42; M. J. Papagrigorakis et al., "DNA Examination of Ancient Dental Pulp Incriminates Typhoid Fever as a Probable Cause of the Plague of Athens," 206–14, and B. Shapiro et al., "No Proof That Typhoid Fever Caused the Plague of Athens," 334–35, both in *International Journal of Infectious Diseases* 10 no. 3 (2006). See also Deborah

Crawford, *Deadly Companions: How Microbes Shaped Our History* (Oxford: Oxford University Press, 2007), 75–77.

71. Harper, *The Fate of Rome*, 65–118; J. Rufus Fears, "The Plague under Marcus Aurelius and the Decline and Fall of the Roman Empire," *Infectious Disease Clinics of North America* 18 no. 1 (2004), 65–77.

72. Harper, *The Fate of Rome*, 136–45; Cunliffe, *Europe between the Oceans*, 393–94; Bray, *Armies of Pestilence*, 12–13; McNeill, *Plagues and Peoples*, 115–17, 135; Hughes, *Pan's Travails*, 187–88, and *Mediterranean*, 49; Cartwright, *Disease and History*, 17–21.

73. Bray, *Armies of Pestilence*, 13–17; McNeill, *Plagues and Peoples*, 114–19.

74. Denis Twitchett, "Population and Pestilence in T'ang China," in Wolfgang Bauer, ed., *Studia Sino-Mongolica: Festschrift für Herbert Franke* (Wiesbaden: Fritz Steiner, 1979), 42; McNeill, *Plagues and Peoples*, 132, 293–302.

75. Hans Zinsser, *Rats, Lice and History* (New York: Bantam Books, 1971), 106–7.

76. Procopius of Caesaria, *History of the Wars*, IV. xiv.4–10 in *Procopius*, trans. H. B. Dewing (London: Heinemann, 1916), 2: 329

77. John of Ephesus, *The Third Part of the Ecclesiastical History of John of Ephesus*, quoted in Bailey K. Young, "Climate and Crisis in Sixth-Century Italy and Gaul," in Joel D. Gunn, ed., *Years without Summer: Tracing A.D. 536 and Its Aftermath* (Oxford: Archaeopress, 2000), 37.

78. A. T. Grove and Oliver Rackham, *The Nature of Mediterranean Europe: An Ecological History* (New Haven, Conn.: Yale University Press), 143.

79. Cassiodorus Senator, *The Letters of Cassiodorus being a condensed translation of the Variae Epistolae of Magnus Aurelius Cassiodorus Senator by Thomas Hodgkin* (London: Henry Frowde, 1886), 518–19.

80. Margaret Snow Houston, "Chinese Climate, History, and State Stability in A.D. 536," in Joel Gunn, *Years without Summer*, 73–74.

81. Elizabeth Jones, "Climate, Archaeology, History, and the Arthurian Tradition: A Multiple Source Study of Two Dark-Age Puzzles," in Gunn, *Years without Summer*, 27; M. G. L. Baillie, "Dendrochronology Raises Questions about the Nature of the A.D. 536 Dust-Veil Event," *The Holocene* 4 no. 2 (1994), 212; Francis Ludlow et al., "Medieval Irish Chronicles Reveal Persistent Volcanic Forcing of Severe Winter Cold Events, 431–1649 CE," *Environmental Research Letters* 8 (2013), 024035 (I am grateful to Professor Ludlow for this citation); Houston, "Chinese Climate," 72; Ulf Büntgen et al., "Cooling and Societal Change during the Late Antique Little Ice Age from 536 to around 600 AD," *Nature Geoscience* 9 (February 8, 2016), 231–36.

82. Robert Dull et al., "Did the Ilopango TBJ Eruption Cause the AD 536 Event?" (paper presented at the American Geophysical Union conference, San Francisco, 2010, and the Association of American Geographers conference, New York, 2012). I am grateful to Dr. Dull for this information.

83. M. Sigl et al., "Timing and Climate Forcing of Volcanic Eruptions for the Past 2,500 Years," *Nature* 523 (July 8, 2015), 543–49. See also M. G. L. Baillie, "Proposed Re-dating of the European Ice Core Chronology by Seven Years Prior to the 7th Century AD," *Geophysical Research Letters* 35 no. 15 (2008), L15813.

84. Dallas Abbott, P. Biscaye, J. Cole-Dai, and D. Breger, "Magnetite and Silicate Spherules from the GISP2 Core at the 536 A.D. Horizon," *American Geophysical Union, Fall Meeting 2008* in http://adsabs.harvard.edu/abs/2008AGUFMPP41B1454A (accessed December 29, 2009).

85. Dallas Abbott et al., "What Caused Terrestrial Dust Loading and Climate Downturns between AD 533 and 540?" *Geological Society of America Special Papers* 505 (2014), 421–38.

86. Harper, *The Fate of Rome*, 218–35; William Rosen, *Justinian's Flea: Plague, Empire, and the Birth of Europe* (New York: Viking, 2007); Dionysios Stathakopoulos, *Famine and Pestilence in the Late Roman and Early Byzantine Empire: A Systematic Survey of Subsistence Crises and Epidemics* (Burlington, Vt.: Ashgate, 2004), 110–54, 277–94.

87. In 1984, physician Graham Twigg argued that this was not the bubonic plague; see *The Black Death: A Biomedical Reappraisal* (London: Batsford, 1984), 32–34. However, geneticist Peregrine Horden analyzed the arguments and concluded that it was certainly a new and unparalleled disease, probably plague; see her "Mediterranean Plague in the Age of Justinian," in Michael Maas, ed., *The Cambridge Companion to the Age of Justinian* (Cambridge: Cambridge University Press, 2005), 134–60. More recently, M. Harbeck et al. confirmed that it was indeed the bubonic plague; see "*Yersinia pestis* DNA from Skeletal Remains from the 6th Century AD Reveals Insights into Justinianic Plague," *PLoS Pathogens* 9 n. 4 (2013), e1003349.

88. Some dispute the role of *Xenophylla cheopsis* and suggest that the disease may have been transmitted by the human flea *Pulex irritans* or the rat flea *Nosophyllus fasciatus*; see Michael McCormick, "Toward a Molecular History of the Justinianic Plague," in Michael McCormick, ed., *The Long Morning of Medieval Europe: New Directions in Early Medieval Studies* (Aldershot: Ashgate, 2008), 83–97.

89. Jean-Noël Biraben and Jacques Le Goff, "The Plague in the Early Middle Ages," trans. Elborg Forster and Patricia M. Ranum, in Robert Forster and Orest Ranum, eds., *Biology of Man in History: Selections from the Annales: Economies, Sociétés, Civilisations* (Baltimore: Johns Hopkins University Press, 1975), 50–55; Bray, *Armies of Pestilence*, 19–20, 34.

90. Procopius of Caesaria, *History*, 2: 457–59; Horden, "Mediterranean Plague," 142; M. Meier, *Das andere Zeitalters Justinians* (Göttingen: Vanderhoeck & Ruprecht, 2003), 334, quoted in James J. O'Donnell, *The Ruin of the Roman Empire: A New History* (New York: HarperCollins, 2008), 286.

91. Procopius of Caesaria, *History*, 2: xxii–xxiii

92. John of Ephesus, *The Third Part*, in Sussman, "Scientists Doing History: Central Africa and the Origins of the First Plague Pandemic," *Journal of World History* 26 no. 2 (June 2015), 328 n. 7

93. Evagrius Scholasticus, *The Ecclesiastical History of Evagrius Scholasticus*, trans. Michael Whitby (Liverpool: Liverpool University Press, 2000), 229.

94. McNeill, *Plagues and Peoples*, 124–27.

95. Biraben and LeGoff, "The Plague," 50, 58; Michael Dols, *The Black Death in the Middle East* (Princeton, N.J.: Princeton University Press, 1977), 13, 49–50;

David Keys, *Catastrophe: An Investigation into the Origins of the Modern World* (New York: Ballantine, 1999), 17–24; Ole J. Benedictow, *The Black Death, 1346-1353: A Complete History* (Woodbridge, U.K.: Boydell Press, 2004), 39; Ionnis Antoniou and Anastasio K. Sinakos, " The Sixth-Century Plague, Its Repeated Appearance until 746 AD and the Explosion of the Rabaul Volcano," *Byzantinische Zeitschrift* 98 no. 1 (2005), 2–3; Horden, "Mediterranean Plague," 135; Dionysios Stathakopoulos, "Crime and Punishment: The Plague in the Byzantine Empire, 541–749," in Lester K. Little, ed., *Plague and the End of Antiquity: The Pandemic of 541-750* (Cambridge: Cambridge University Press, 2007), 534; Peter Sarris, "Bubonic Plague in Byzantium: The Evidence from Non-Literary Sources," in Little, *Plague and the End of Antiquity*, 120–23; Robert Sallares, "Ecology, Evolution, and Epidemiology of Plague," in Little, *Plague and the End of Antiquity*, 246–47; Rosen, *Justinian's Flea*, 194–200.

96. Giovanna Morelli et al., "*Yersinia pestis* Genome Sequencing Identifies Patterns of Global Phylogenetic Diversity," *Nature Genetics* 42 (online October 31, 2010), 140–43; Kirsten I. Bos et al., "*Yersinia pestis*: New Evidence for an Old Infection," *PLoS ONE* 7 no. 11 (2012), 1–3; Yujun Cui et al., "Historical Variations in Mutation Rate in an Epidemic Pathogen, *Yersinia pestis*," *Proceedings of the National Academy of Sciences of the USA* 110 (2013), 577–82; Harbeck, "*Yersinia pestis* DNA"; M. Thomas P. Gilbert, "*Yersinia pestis*: One Pandemic, Two Pandemics, three Pandemics, More?" *Lancet Infectious Diseases* 14 no. 4 (2014), 264–65. On the differences between historians and scientists on this question, see Sussman, "Scientists Doing History, 325–54.

97. Rosen, *Justinian's Flea*, 210; Biraben and Le Goff, "Plague," 58; Bray, *Armies of Pestilence*, 22–24; McNeill, *Plagues and Peoples*, 123.

98. Jean-Noël Biraben, *Les hommes et la peste en France et dans les pays méditerranéens*, 2 vols. (Paris: Mouton, 1975–76), 1: 25–48; McCormick, "Toward a Molecular History," 94; Christensen, *Decline of Iranshahr*, 81–82; Rosen, *Justinian's Flea*, 196–97, 220; Bray, *Armies of Pestilence*, 47; Biraben and Le Goff, "Plague," 59–62; Jones, "Climate, Archaeology, History," 28–31; McNeill, *Plagues and Peoples*, 128–29.

99. Michael W. Dols, "Plague in Early Islamic History," *Journal of the American Oriental Society* 94 (1974), 371–83, and *The Black Death*, 13–26; Dionysios Stathakopoulos, "Plague of Justinian. First Pandemic," in Joseph Byrne, ed., *Encyclopedia of Pestilence, Pandemics, and Plagues*, 2 vols. (Westport, Conn.: Greenwood, 2008), 2: 532–35, and Selma Tibi-Harb, "Plague in the Islamic World, 1500–1850," in Byrne, ed., *Encyclopedia of Pestilence, Pandemics, and Plagues*, 2: 516–19; William McNeill, *Plagues and Peoples*, 159; Bray, *Armies of Pestilence*, 31–32; Twitchett, "Population and Pestilence," 42–60.

100. Joel D. Gunn, "A.D. 536 and its 300-Year Aftermath," in Gunn, *Years without Summer*, 22; Young, "Climate and Crisis," 38–39; Bray, *Armies of Pestilence*, 24, 31.

101. Dols, "Plague," 372, and *The Black Death*, 16.

102. Houston, "Chinese Climate," 74–75; Marks, *China*, 110–11.

Chapter 5

1. The expression "Middle Ages" is a misnomer in European history, for the period 600–1500 was not the middle of anything but a new beginning. It is even less appropriate for East Asian, Middle Eastern, and African history. Yet, like so many expressions in history, it is a traditional, and therefore useful, shorthand.

2. Andrew Watson, "The Arab Agricultural Revolution and Its Diffusion, 700–1100, *Journal of Economic History* 34 (1974), 8–35, and *Agricultural Innovation in the Early Islamic World: The Diffusion of Crops and Farming Techniques, 700–1100* (Cambridge: Cambridge University Press, 1983); quotation on p. 111. See also Rushdi Said, *The River Nile: Geology, Hydrology, and Utilization* (Oxford: Pergamon Press, 1993), 207.

3. John R. McNeill, *The Mountains of the Mediterranean World: An Environmental History* (Cambridge: Cambridge University Press, 1992), 87–88; Norman Alfred Fisher Smith, *Man and Water: A History of Hydro-Technology* (New York: Scribner's, 1975), 16–17; Edmund Burke III, "The Transformation of the Middle Eastern Environment, 1500 B.C.E.–2000 C.E.," in Edmund Burke III and Kenneth Pomeranz, eds., *The Environment and World History* (Berkeley: University of California Press, 2009), 86–88; Peter Christensen, *The Decline of Iranshahr: Irrigation and Environments in the History of the Middle East, 500 B.C. to A.D. 1500* (Copenhagen: Museum Tusculanum Press, 1993), 87–94.

4. Robert McC. Adams, *Land behind Baghdad: A History of Settlement on the Diyala Plains* (Chicago: University of Chicago Press, 1965), 84.

5. Richard Bulliet, *Cotton, Climate, and Camels in Early Islamic Iran: A Moment in World History* (New York: Columbia University Press, 2009), chs. 1 and 2, 131–32.

6. Stuart J. Borsch, *Black Death in Egypt and England: A Comparative Study* (Austin: University of Texas Press, 2005), 34–38.

7. José Rodríguez Molina, *Regadío medieval andaluz* (Jaén: Diputación provincial de Jaén, 1991), 167–79; Thomas F. Glick, *Regadío y sociedad en la Valencia medieval* (Valencia: Generalitat Valenciana, 2003), 324–38; Richard C. Hoffman, *An Environmental History of Medieval Europe* (Cambridge: Cambridge University Press, 2014), 142–44.

8. María Rosa Menocal, *The Ornament of the World* (Boston: Little Brown, 2002).

9. Watson, *Agricultural Innovation*, 140.

10. Robert McC. Adams, *Heartland of Cities: Surveys of Ancient Settlement and Land Use on the Central Floodplain of the Euphrates* (Chicago: University of Chicago Press, 1981), 215–18; and *Land behind Baghdad*, 84, 115 table 25; Peter Christensen, "Middle Eastern Irrigation: Legacies and Lessons" in Jeff Albert et al., eds., *Transformations of Middle Eastern Natural Environments: Legacies and Lessons* (New Haven, Conn.: Yale University Press, 1998), 19–20; McGuire Gibson, "Violation of Fallow and Engineered Disaster in Mesopotamian Civilization," in Theodore E. Downing and McGuire Gibson, eds., *Irrigation's Impact on Society* (Tucson: University of Arizona Press, 1974), 15; Burke, "Transformation," 85.

11. Ronnie Ellenblum, *The Collapse of the Eastern Mediterranean: Climate Change and the Decline of the East, 950–1072* (Cambridge: Cambridge University Press, 2012), 23–29, 41–51.

12. Ellenblum, *Collapse of the Eastern Mediterranean*, 30–33, 49.

13. John L. Brooke, *Climate Change and the Course of Global History: A Rough Journey* (Cambridge: Cambridge University Press, 2014), 351–59; Bulliet, *Cotton, Climate, and Camels*, 69–77, 84–89, 96; Ellenblum, *Collapse of the Eastern Mediterranean*, 4–6, 61–64, 89.

14. Christensen, *Decline of Iranshahr*, 100. See also Thorkild Jacobsen and Robert McC. Adams, "Salt and Silt in Ancient Mesopotamian Agriculture," *Science* 128 no. 3334 (November 21, 1958), 1256–58.

15. Brooke, *Climate Change*, 364; Hoffman, *Environmental History*, 321.

16. Brian Fagan, *The Long Summer: How Climate Changed Civilization* (New York: Basic Books, 2004), 206–7.

17. Hubert H. Lamb, *Climate, History, and the Modern World*, 2nd ed. (London: Routledge, 1995), 159; Hoffman, *Environmental History*, 116; Brian Fagan, *The Great Warming: Climate Change and the Rise and Fall of Civilizations* (New York: Bloomsbury Press, 2008), 12–35; Brooke, *Climate Change*, 372.

18. Norman F. Cantor, *In the Wake of the Plague: The Black Death and the World It Made* (New York: Free Press, 2001), 193.

19. Bruce M. S. Campbell, *The Great Transition: Climate, Disease, and Society in the Late Medieval World* (Cambridge: Cambridge University Press, 2016), 36–37, 45–47; I am grateful to Professor Campbell for letting me read and cite this book before its publication. See also Jared Diamond, *Collapse: How Societies Choose to Fail or Succeed* (Harmondsworth, U.K.: Penguin, 1996), 211–47, and Fagan, *The Great Warming*, 88–94.

20. Lynn White Jr., *Medieval Technology and Social Change* (Oxford: Oxford University Press, 1962), 40–44.

21. White, *Medieval Technology*, 42–53; Hoffman, *Environmental History*, 122–26.

22. Roland Bechmann, *Trees and Man: The Forest in the Middle Ages*, trans. Katharyn Dunham (New York: Paragon House, 1990), 49, 65.

23. White, *Medieval Technology*, 71–73.

24. William S. Cooter, "Ecological Dimensions of Medieval Agrarian Systems," *Agricultural History* 52 no. 4 (October 1978), 458–77; R. S. Loomis, "Ecological Dimensions of Medieval Agrarian Systems: An Ecologist Responds," *Agricultural History* 52 no. 4 (October 1978), 478–83; Tamara Whited et al., *Northern Europe: An Environmental History* (Santa Barbara, Cal.: ABC-CLIO, 2005), 63–64; Hoffman, *Environmental History*, 148–49.

25. Ian Simmons, *An Environmental History of Great Britain: From 10,000 Years Ago to the Present* (Edinburgh: Edinburgh University Press, 2001), 71–72; White, *Medieval Technology*, 57–74.

26. White, *Medieval Technology*, 76. See also 56, 69–75; B. H. Slicher van Bath, *The Agrarian History of Western Europe, A.D. 500–1850*, trans. Olive Ordish (London: Edward Arnold, 1963), 54–65, and Whited, *Northern Europe*, 51–52.

27. White, *Medieval Technology*, 56–57. See also White, "The Historical Roots of Our Ecological Crisis," *Science* 155 (1967), 1203–207.

28. William TeBrake, *Medieval Frontier: Culture and Ecology in Rijnland* (College Station: Texas A&M University Press, 1985), 141–42, 190–98, 228–32, and "Land Drainage

and Public Environmental Policy in Medieval Holland," *Environmental Review* (Fall 1988): 75–93; G. P. Van der Ven, *Man-Made Lowlands: History of Water Management and Land Reclamation in the Netherlands* (Utrecht: Uitgeverij Matrijs, 2004), 52–73, 105

29. Peter Hoppenbrouwers, "Agricultural Production and Technology in the Netherlands, c. 1000–1500," in Grenville Astill and John Langdon, eds., *Medieval Farming and Technology: The Impact of Agricultural Change in Northwest Europe* (Leiden: Brill, 1997), 92–98 (quotation on p. 98).

30. Clarence Glacken, *Traces on the Rhodian Shore: Nature and Culture in Western Thought from Ancient Times to the End of the Eighteenth Century* (Berkeley: University of California Press, 1967), 310.

31. Michael Williams, *Deforesting the Earth: From Prehistory to Global Crisis* (Chicago: University of Chicago Press, 2003), 92–106; Whited, *Northern Europe*, 34–35, 63–67.

32. H. C. Darby, "The Clearing of Woodland in Europe," in William L. Thomas Jr., ed., *Man's Role in Changing the Face of the Earth* (Chicago: University of Chicago Press, 1956), 196; Slicher van Bath, *Agrarian History*, 155.

33. John McNeill, *Mountains*, 92; Cantor, *In the Wake of the Plague*, 63–65.

34. Bechmann, *Trees and Man*, 53–62.

35. Thomas T. Allsen, *The Royal Hunt in Eurasian History* (Philadelphia: University of Pennsylvania Press, 2006), 40–41, 189.

36. Bechmann, *Trees and Man*, ch. 2; Whited, *Northern Europe*, 55; Slicher van Bath, *Agrarian History*, 72–73.

37. Bechmann, *Trees and Man*, 68–70.

38. Bechmann, *Trees and Man*, 87–91; White, *Medieval Technology*, 82, 118.

39. J. V. Thirgood, *Man and the Mediterranean Forest: A History of Resource Depletion* (London: Academic Press, 1981), 49; Russell Meiggs, *Trees and Timber in the Ancient World* (Oxford: Clarendon Press, 1982), 383; Williams, *Deforesting the Earth*, 106–10; Whited, *Northern Europe*, 55, 63; Hoffman, *Environmental History*, 121.

40. Peter Brimblecombe, *The Big Smoke: A History of Air Pollution in London since Medieval Times* (London: Methuen, 1987), 26–34.

41. Karl Appuhn, *A Forest on the Sea: Environmental Expertise in Renaissance Florence* (Baltimore: Johns Hopkins University Press, 2009), 94–97; Thirgood, *Man and the Mediterranean Forest*, 47–48; Whited, *Northern Europe*, 43, 55–56, 64–68; Cantor, *In the Wake of the Plague*, 67; Williams, *Deforesting the Earth*, 90, 116.

42. S. C. Gilfillan, "The Coldward Course of Progress," *Political Science Quarterly* 35 no. 3 (September 1920), 393–410.

43. Walter H. Mallory, *China: Land of Famine* (New York: American Geographical Society, 1926), 38–40; Eric Jones, *The European Miracle: Environments, Economies, and Geopolitics in the History of Europe and Asia* (Cambridge: Cambridge University Press, 1981), 28.

44. Qu Geping and Li Jinchang, *Population and Environment in China* (Boulder, Colo.: L. Rienner, 1994), 21; Mark Elvin, *Retreat of the Elephants: An Environmental History of China* (New Haven, Conn.: Yale University Press, 2004), 25–26.

45. Ling Zhang, *The River, the Plain, and the State: An Environmental Drama in Northern song China, 1048–1128* (Cambridge: Cambridge University Press, 2016), and "Changing with the Yellow River: An Environmental History of Hebei, 1048–1128," *Harvard Journal of Asiatic Studies* 69 no. 1 (2009), 1–36.

46. Christian Lamouroux, "From the Yellow River to the Huai: New Representations of a River Network and the Hydraulic Crisis of 1128," in Mark Elvin and Liu Ts'ui-jung, eds., *Sediments of Time: Environment and Society in Chinese History* (Cambridge: Cambridge University Press, 1998), 545–56; Zhang, "Changing with the Yellow River."

47. Lyman P. Van Slyke, *Yangtze: Nature, History and the River* (Reading, Mass.: Addison-Wesley, 1988), 13; Jones, *European Miracle*, 27–28.

48. Robert Marks, *China: Its Environment and History* (Lanham, Md.: Rowman & Littlefield, 2011), 96–97.

49. Mark Elvin, *The Pattern of the Chinese Past* (Stanford, Cal.: Stanford University Press, 1973), 55; Mark Edward Lewis, *China's Cosmopolitan Empire: The Tang Dynasty* (Cambridge, Mass.: Harvard University Press, 2009), 10; Margaret Snow Houston, "Chinese Climate, History, and State Stability in A.D. 536," in Joel Gunn, ed., *Years without Summer: Tracing A.D. 536 and Its Aftermath* (Oxford: Archaeopress, 2000), 75.

50. Qu and Li, *Population and Environment*, 20.

51. Francesca Bray, *The Rice Economies: Technology and Development in Asian Societies* (Berkeley: University of California Press, 1986), 9–34.

52. Asaf Goldschmidt, *The Evolution of Chinese Medicine: Song Dynasty, 960–1200* (London: Routledge, 2009), 75.

53. Lewis, *China's Cosmopolitan Empire*, 135.

54. Lewis, *China's Cosmopolitan Empire*, 10–20, 129–34.

55. Francesca Bray, *Agriculture*, vol. 6 part 2 of Joseph Needham, *Science and Civilisation in China* (Cambridge: Cambridge University Press, 1984), 598–601.

56. Bao Maohong, "Environmental Resources and China's Historical Development," in John R. McNeill, José Augusto Pádua, and Mahesh Rangarajan, eds., *Environmental History: As if Nature Existed* (New Delhi: Oxford University Press, 2010), 90–91; J. E. Spencer, "Water Control in Terraced Rice-Field Agriculture in Southeastern Asia," in Downing and Gibson, *Irrigation's Impact on Society*, 59–65; Elvin, *Pattern*, 82, 113–26, and *Retreat of the Elephants*, 24–26; Lewis, *China's Cosmopolitan Empire*, 21–22, 132–36; Bray, *Rice Economies*, 68–91, 204–5.

57. Christian Lamouroux, "Crise politique et développement rizicole en Chine: La région de Jiang-Huai (VIII–Xe siècle)," *Bulletin de l'Ecole française d'Extrême-Orient* 82 (1995), 145–83; Michel Cartier, "Aux origines de l'agriculture intensive du Bas Yangzi (note critique)," *Annales E.S.C.* 46 no. 5 (September–October 1991), 1013–17; Elvin, *Pattern*, 113–26; Bray, *Rice Economies*, 81–83, 203, and *Agriculture*, 598–601; Marks, *China*, 150.

58. Ping-ti Ho, "Early Ripening Rice in Chinese History," *Economic History Review*, 2nd ser., no. 9 (1956–57): 200–18; Lewis, *China's Cosmopolitan Empire*, 129–33; Bray, *Agriculture*, 598–99, and *Rice Economies*, 76–78, 90–91, 203–4; Elvin, *Pattern*, 82, 114–23.

59. Williams, *Deforesting the Earth,* 117–21; Lewis, *China's Cosmopolitan Empire,* 17.

60. Mark Elvin, "3,000 Years of Unsustainable Growth: China's Environment from Archaic Times to the Present," *East Asian History* 6 (1993), 44.

61. This is the phenomenon that Clifford Geertz described, in the context of Java, in *Agricultural Involution: The Process of Ecological Change in Indonesia* (Berkeley: University of California Press, 1963).

62. Allsen, *Royal Hunt,* 215–16.

63. Jin-Qi Fang and Guo Liu, "Relationship between Climatic Change and the Nomadic Southward Migrations in Eastern Asia during Historical Times," *Climatic Change* 22 (1992), 151–69; Campbell, *The Great Transition,* 48; Brooke, *Climate Change,* 369–70.

   For a critique of climatic determinism, see William B. Meyer, "Climate and Migration," in Andrew Bell-Fialkoff, ed., *The Role of Migration in the History of the Eurasian Steppes: Sedentary Civilization vs. Barbarian and Nomad* (London: Macmillan 2000), 287–94.

64. Aaron E. Putnam et al., "Little Ice Age Wetting of Interior Asian Deserts and the Rise of the Mongol Empire," *Quaternary Science Reviews* 131 (January 2016); Neil Pederson et al., "Pluvials, Droughts, the Mongol Empire, and Modern Mongolia," *Proceedings of the National Academy of Science of the USA* 111 no. 12 (March 25, 2014), 4375–79.

65. Owen Lattimore, *Mongols of Manchuria: Their Tribal Divisions, Geographical Distribution, Historical Relations with Manchus and Chinese, and Present Political Problems* (New York: John Day, 1934), 65.

66. René Grousset, *Conqueror of the World,* trans. Denis Sinor and Marian Mackellar (London: Oliver and Boyd, 1967), 281.

67. Thomas Allsen, "The Rise of the Mongolian Empire and Mongolian Rule in North China," ch. 4 in *The Cambridge History of China,* vol. 6: Herbert Franke and Denis Twitchett, eds., *Alien Regimes and Border States, 907–1368* (New York: Cambridge University Press, 1994), 362–64; Marks, *China,* 157; Brooke, *Climate Change,* 370.

68. Marks, *China,* 112–18, 141–45, 158.

69. Roderick J. McIntosh, *Peoples of the Middle Niger: The Island of Gold* (Oxford: Blackwell, 1998), 241–45, and *The Ancient Middle Niger: Urbanism and the Self-Organizing Landscape* (Cambridge: Cambridge University Press, 2005), ch. 2; Gregory H. Maddox, *Sub-Saharan Africa: An Environmental History* (Santa Barbara, Cal.: ABC-CLIO, 2006), 64–67; Ralph A. Austen, *Trans-Saharan Africa in World History* (New York: Oxford University Press, 2010), 6–14; Christopher Ehret, *The Civilizations of Africa: A History to 1800* (Charlottesville: University Press of Virginia, 2002), 309–10; Susan Keech McIntosh, "Reconceptualizing Early Ghana," *Canadian Journal of African Studies* 42 no. 2/3 (2008), 347–73.

70. Richard W. Bulliet, *The Camel and the Wheel* (New York: Columbia University Press, 1990), 138; Pekka Masonen, "Trans-Saharan Trade and the West African Discovery of the Mediterranean World," in *Third Nordic Conference on Middle Eastern Studies* (Joensuu, Finland, June 19–22, 1995), 116–42; Austen, *Trans-Saharan Africa,* 15–17.

71. Austen, *Trans-Saharan Africa,* 14–40; Maddox, *Sub-Saharan Africa,* 66; Masonen, "Trans-Saharan Trade."

72. William C. Jordan, *The Great Famine: Northern Europe in the Early 14th Century* (Princeton, N.J.: Princeton University Press, 1996); William Rosen, *The Third Horseman: Climate Change and the Great Famine of the 14th Century* (New York: Viking 2014); Emmanuel Le Roy Ladurie, Daniel Rousseau, and Anouchka Vasak, *Les fluctuations du climat de l'an mil à aujourd'hui* (Paris: Fayard, 2011), 22–24; Jan Esper, Fritz H. Schweingruber, and Matthias Winiger, "1,300 Years of Climatic History for Western Central Asia Inferred from Tree-Rings," *Holocene* 12 no. 3 (2002), 267–77; Lamb, *Climate, History*, 165, 181–82; Cantor, *In the Wake of the Plague*, 74; Hoffman, *Environmental History*, 323–25.

73. Jordan, *Great Famine*, 147.

74. Rosen, *The Third Horseman*, 150–52.

75. Bruce M. S. Campbell, "Panzootics, Pandemics and Climate Anomalies in the Fourteenth Century," in *Beiträge zum Göttinger Umwelthistorischen Kolloquium, 2010–2011* (Göttingen: Universitätsverlag Göttingen, 2011), 177–215; "Physical Shocks, Biological Hazards, and Human Impacts: The Crisis of the Fourteenth Century Revisited," in S. Cavaciocchi, ed., *Le interazioni fra economia e ambiente biologico nell'Europa preindustriale. Seccoli XIII–XVIII* (Prato, 2010), 13–32; and *The Great Transition*, 211–27; Le Roy Ladurie, Rousseau, and Vasak, *Les fluctuations du climat*, 24; Brooke, *Climate Change*, 384; Rosen, *The Third Horseman*, 184–87.

76. Jordan, *Great Famine*, 7–20; Wolfgang Behringer, *A Cultural History of Climate*, trans. Patrick Camiller (Cambridge: Polity Press, 2010), 103–9. For a short narrative of these events, see Brooke, *Climate Change*, 375, and Fagan, *Little Ice Age*, ch. 2.

77. John R. McNeill, "Woods and Warfare in World History," *Environmental History* 9 (2004), 401; Slicher van Bath, *Agrarian History*, 142; Jordan, *Great Famine*, 7–8, 24–38, 185–86; Lamb, *Climate, History*, 195–203; Whited, *Northern Europe*, 59; Cantor, *In the Wake of the Plague*, 8.

78. Diamond, *Collapse*, ch. 8; Fagan, *Great Warming*, ch. 5.

79. Lamb, *Climate, History*, 200. See also Arsenio Peter Martinez, "Institutional Development, Revenues and Trade," in Nicola Di Cosmo, Allen J. Frank, and Peter B. Golden, *The Cambridge History of Inner Asia: The Chinggisid Age* (Cambridge: Cambridge University Press, 2009), 91.

80. Timothy Brook, *Troubled Empire: China in the Yuan and Ming Dynasties* (Cambridge, Mass.: Harvard University Press, 2010), 53–68.

81. John Dardess, "Shun-ti and the End of Yüan Rule in China," in Franke and Twitchett, eds., *Cambridge History of China*, vol. 6, 585; Brook, *Troubled Empire*, 53–54.

82. Brook, *Troubled Empire*, 53–72 (quotation on p. 72).

83. Ole J. Benedictow, *The Black Death, 1346–1353: A Complete History* (Woodbridge, U.K.: Boydell Press, 2004), 44–61; Robert S. Gottfried, *Black Death: Natural and Human Disaster in Medieval Europe* (New York: Free Press, 1983), 34–35; Robert Pollitzer, *Plague* (Geneva: World Health Organization, 1954), 13–14; Paul D. Buell, *Historical Dictionary of the Mongol World Empire* (Lanham, Md.: Scarecrow Press, 2003), 121–22;.

84. William H. McNeill, *Plagues and Peoples* (Garden City, N.Y.: Doubleday, 1976), 165–66.

85. Benedictow, *The Black Death*, 383; Campbell, *The Great Transition*, 319.

86. Campbell, *The Great Transition*, 397–402. See also Boris V. Schmid et al., "Climate-Driven Introduction of the Black Death and Successive Plague Reintroductions into Europe," *Proceedings of the National Academy of Sciences (USA)* 112 no. 10 (February 23, 2015), 3020–25; Borsch, *Black Death in Egypt and England*, 24, 55–66; Williams, *Deforesting the Earth*, 117; and Slicher van Bath, *Agrarian History*, 144.

87. Williams, *Deforesting the Earth*, 117, 124; Campbell, *Great Transition*, 395–97.

88. Ross E. Dunn, *The Adventures of Ibn Battuta* (Berkeley: University of California Press, 2005), 269.

89. Michael Dols, *The Black Death in the Middle East* (Princeton, N.J.: Princeton University Press, 1977), 43, 154–84, 215–23; Selma Tibi-Harb, "Plague in the Islamic World, 1500–1850," in Joseph P. Byrne, ed., *Encyclopedia of Pestilence, Pandemics, and Plagues*, 2 vols. (Westport, Conn.: Greenwood Press, 2008), 2: 517–18; George C. Kohn, *Encyclopedia of Plague and Pestilence: From Ancient Times to the Present* (New York: Facts on File, 2008), 225–26; Borsch, *Black Death in Egypt and England*, 24; Benedictow, *The Black Death*, 61–66; Christensen, *Decline of Iranshar*, 74, 100–103.

90. Said, *River Nile*, 166, 209.

91. Borsch, *Black Death in Egypt and England*, 25–53; Christensen, *Decline of Iranshahr*, 100–101.

92. McIntosh, *Peoples of the Middle Niger*, 244.

93. McIntosh, *Peoples of the Middle Niger*, 248.

94. Gérard L. Chouin and Christopher R. Decorse, "Prelude to the Atlantic Trade: New Perspectives on Southern Ghana's Pre-Atlantic History," *Journal of African History* 51 no. 2 (July 2010), 143.

95. For a good summary of the etiology of bubonic plague, see Benedictow, *The Black Death*, 17–20, 33–37.

96. Susan Scott and Christopher J. Duncan, *Biology of Plagues: Evidence from Historical Populations* (New York: Cambridge University Press, 2001), 107–8, 356–62 (quotation on p. 107).

97. Samuel J. Cohn, "The Black Death: The End of a Paradigm," *American Historical Review* 107 no. 3 (2002), 703–38.

98. Didier Raoult et al., "Molecular Identification by 'Suicide PCR' of *Yersinia pestis* as the Agent of Medieval Black Death," *Proceedings of the National Academy of Sciences (USA)* 97 no. 23 (November 7, 2000), 12,800–803.

99. Stephanie Haensch et al., "Distinct Clones of *Yersinia pestis* Caused the Black Death," *PloS Pathogens* (October 7, 2010); Michel Drancourt et al., "*Yersinia pestis* Orientalis in Remains of Ancient Plague Patients," *Emerging Infectious Diseases* 13 (2007), 332–33.

100. Campbell, "Panzootics, Pandemics and Climate Anomalies"; Pollitzer, *Plague*, 13–14.

101. Benedictow, *The Black Death*, 44–61; Hoffman, *Environmental History*, 289–90; Buell, *Historical Dictionary*, 121–22.

102. Justus Friedrich Karl Hecker, *Der schwarze Tod im vierzehnten Jahrhundert: Nach den Quellen für Ärzte und gebildete Nichtärzte bearbeitet* (Berlin: Herbig, 1832).

103. Jean-Noël Biraben, *Les hommes et la peste en France et dans les pays européens et méditerranéens* (Paris: Mouton, 1975–1976), 46. For similar, though not identical, numbers, see McNeill, *Plagues and Peoples*, 163, and Brook, *Troubled Empire*, 42–44.

104. McNeill, *Plagues and Peoples*, 149–65 (quotations on pp. 161–63).

105. Denis Twitchett, "Population and Pestilence in T'ang China," in Wolfgang Bauer, ed., *Studia Sino-Mongolica: Festschrift für Herbert Franke* (Wiesbaden: Fritz Steiner, 1979), 47; Marks, *China*, 119–20.

106. Elvin, *Pattern*, 175. See also Goldschmidt, *Evolution of Chinese Medicine*, 69–81.

107. Dardess, "Shun-ti and the End of Yüan Rule," 585; Brook, *Troubled Empire*, 64–65.

108. Carol Benedict, *The Bubonic Plague in Nineteenth-Century China* (Stanford, Cal.: Stanford University Press, 1996), 8–9.

109. Goldschmidt, *Evolution of Chinese Medicine*, 70–87; Shigehisa Kuriyama, "Epidemics, Weather and Contagion in Traditional Chinese Medicine," in Lawrence I. Conrad and Dominik Wujastyk, eds., *Contagion: Perspectives from Pre-Modern Societies* (Aldershot: Ashgate, 2000), 4.

110. Benedict, *Bubonic Plague in Nineteenth-Century China*, 9; Twitchett, "Population and Pestilence," 42–52, 62–63. Historian of Japan Ann Jannetta, however, denies that plague ever reached Japan before modern times; see *Epidemics and Mortality in Early Modern Japan* (Princeton. N.J.: Princeton University Press, 1987), 192.

111. Benedict, *Bubonic Plague*, 9–10.

112. Marks, *China*, 158.

113. Personal communication from Professor Bridie Minehan, July 17, 2010, and from Professor Richard von Glahn, July 17, 2010.

114. Paul D. Buell, "Qubilai and the Rats: Plague, Biology and History," *Südhoffs Archiv: Zeitschrift für Wissenschaftsgeschichte* 96 no. 2 (2012). I am grateful to Professor Buell for this citation. See also George D. Sussman, "Was the Black Death in India and China?" *Bulletin of the History of Medicine* 85 no. 3 (2011), 319–55.

115. Buell, *Historical Dictionary*, 122–22; Schmid, "Climate-Driven Introduction"; Dols, *The Black Death*, 35; Benedictow, *The Black Death*, 44–51.

116. Yujun Cui et al., "Historical Variations in Mutation Rate in an Epidemic Pathogen *Yersinia pestis*," *Proceedings of the National Academy of Sciences (USA)* 110 (2013), 577–82. See also Campbell, *The Great Transition*, 320–25, 358–408; Pederson, "Pluvials, Droughts, the Mongol Empire"; and P. C. Stenseth et al., "Plague Dynamics Are Driven by Climate Variation," *Proceedings of the National Academy of Sciences (USA)* 103 (2006), 13110–15.

117. George D. Sussman, "Scientists Doing History: Central Africa and the Origins of the First Plague Pandemic," *Journal of World History* 26 no. 2 (June 2016), 325–54; Mark Achtman et al., "Insights from Genomic Comparisons of Genetically Monomorphic Bacterial Pathogens," *Philosophical Transactions of the Royal Society B: Biological Sciences* 367 no. 1590 (2012), 860–67; Giovanna Morelli et al., "*Yersinia pestis* Genome Sequencing Identifies Patterns of Global Phylogenetic Diversity," *Nature Genetics* 42 (online October 31, 2010), 1140–43; Cui et al., "Historical

Variations"; M. Thomas P. Gilbert, "*Yersinia pestis*: One Pandemic. Two Pandemics, Three Pandemics, More?" *Lancet Infectious Diseases* 14 no. 4 (2014), 264–65.

Chapter 6

1. Franck Prugnolle et al., "A Fresh Look at the Origin of *Plasmodium falciparum*, the Most Malignant Malaria Agent," *PLoS Pathogens* 7 no. 2 (2011), e1001283.

2. William McNeill, *Plagues and Peoples* (Garden City, N.Y.: Doubleday, 1976), ch. 3: "Confluence of Disease Pools, 500 B.C.–A.D. 1200"; Alfred Crosby, *Germs, Seeds, and Animals: Studies in Ecological History* (Armonk, N.Y.: M.E. Sharpe, 1994), 97.

3. David S. Jones, "Virgin Soil Revisited," *William and Mary Quarterly* 60 no. 4 (October 2003), 703–42; Mark Nathan Cohen, *Health and the Rise of Civilization* (New Haven, Conn.: Yale University Press, 1988), 138; William McNeill, *Plagues and Peoples*, 69.

4. According to William M. Denevan, ed., *The Native Population of the Americas in 1492*, 2nd ed. (Madison: University of Wisconsin Press, 1992), xxviii–xxiv, the range was 43 to 65 million; for John Verano and Douglas H. Ueberlaker, eds., *Disease and Demography in the Americas* (Washington, D.C.: Smithsonian Institution Press, 1992), 171–74, it was 43–72 million; and according to Russell Thornton, *American Indian Holocaust and Survival* (Norman: University of Oklahoma Press, 1987), 22–25, it was at least 72 million. Other estimates can be found in Massimo Livi-Bacci, *A Concise History of World Population*, 3rd ed. (Malden, Mass.: Blackwell, 2001), 50–56, and Nicolás Sánchez-Albornoz, *La población de América Latina, desde los tiempos precolombianos al año 2000* (Madrid: Alianza Editorial, 1973), 54–71.

5. Massimo Livi-Bacci, *Conquest: The Destruction of the American Indios* (Cambridge: Polity, 2008), 6.

6. William M. Denevan, "The Pristine Myth: The Landscape of the Americas in 1492," *Annals of the Association of American Geographers* 82 no. 3 (September 1992), 369. This issue of the *Annals* also contains other articles about the Americas, both before and after 1492.

7. *The Book of Chilam Balam of Chumayel*, trans. Ralph L. Roy (Washington, D.C.: Carnegie Institution of Washington, 1933), 83, quoted in Alfred W. Crosby, *The Columbian Exchange: Biological and Cultural Consequences of 1492* (Westport, Conn.: Greenwood Press, 1972), 36.

8. Gregory Cochran and Henry Harpending, *The 10,000 Year Explosion: How Civilization Accelerated Human Evolution* (New York: Basic Books, 2009), 159; Jared Diamond, *Guns, Germs, and Steel: The Fates of Human Societies* (New York: W. W. Norton, 1997), 212–13; McNeill, *Plagues and Peoples*, 201; Thornton, *American Indian Holocaust*, 40–41.

9. Alfred Crosby, *Ecological Imperialism: The Biological Expansion of Europe, 900–1900* (Cambridge: Cambridge University Press, 1986), 197–98; Suzanne Austin Alchon, *A Pest in the Land: New World Epidemics in a Global Perspective* (Albuquerque: University of New Mexico Press, 2003), 15; James Daschuk, *Clearing the Plains: Disease, Politics of Starvation, and the Loss of Aboriginal Life*

(Regina, Sask., Canada: University of Regina Press, 2013), 2; Paul Kelton, *Epidemics and Enslavement: Biological Catastrophe in the Native Southeast, 1492-1715* (Lincoln: University of Nebraska Press, 2007), 14–28; Jane E. Buikstra, ed., *Prehistoric Tuberculosis in the Americas* (Evanston, Ill.: Northwestern University Archaeological Program, 1981); Lourdes Marquez Morfín, Robert McCaa, Rebecca Storey, and Andrés del Angel, "Health and Nutrition in Pre-Hispanic Mesoamerica," in Richard Steckel and Jerome C. Rose, eds., *The Backbone of History: Health and Nutrition in the Western Hemisphere* (Cambridge: Cambridge University Press, 2002), 230–45.

10. Elizabeth Fenn, *Pox Americana: The Great Smallpox Epidemic of 1775-82* (New York: Hill & Wang, 2001), 25–27; Cochran and Harpending, *10,000 Year Explosion*, 160–61.

11. Kelton, *Epidemics and Enslavement*, 12–14; Daschuk, *Clearing the Plains*, 2–9.

12. David Watts, *The West Indies: Patterns of Development, Culture and Environmental Change since 1492* (Cambridge: Cambridge University Press, 1987), 53–71.

13. Noble David Cook, *Born to Die: Disease and New World Conquest, 1492-1650* (Cambridge: Cambridge University Press, 1998), 23–24, and "Sickness, Starvation and Death in Early Hispaniola," *Journal of Interdisciplinary History* 32 no. 3 (Winter 2002), 349–86; Massimo Livi-Bacci, "Return to Hispaniola: Reassessing a Demographic Catastrophe," *Hispanic American Historical Review* 83 no. 1 (February 2003), 49; Angel Rosenblat, "The Population of Hispaniola at the Time of Columbus," in Denevan, *Native Population*, ch. 2; Watts, *West Indies*, 103.

14. Cook, *Born to Die*, 23–24; quotation in Crosby, *Columbian Exchange*, 45.

15. Kenneth F. Kiple and Brian T. Higgins, "Yellow Fever and the Africanization of the Caribbean," in John W. Verano and Douglas H. Ueberlaker, eds., *Disease and Demography in the Americas* (Washington: Smithsonian Institution Press, 1992), 237; Donald R. Hopkins, *The Greatest Killer: Smallpox in History* (Chicago: University of Chicago Press, 2002), 204; Cook, *Born to Die*, 230; Livi-Bacci, "Return to Hispaniola," 50–51.

16. Crosby, *Columbian Exchange*, 44–49; Robert McCaa, "Spanish and Nahuatl Views on Smallpox and Demographic Catastrophe in Mexico," in Robert I. Rotberg, ed., *Health and Disease in Human History* (Cambridge, Mass.: MIT Press, 2000), 187–88; Fenn, *Pox Americana*, 27–28.

17. William Shawn Miller, *An Environmental History of Latin America* (Cambridge: Cambridge University Press, 2007), 53; John F. Richards, *The Unending Frontier: An Environmental History of the Early Modern World* (Berkeley: University of California Press, 2001), 477–79, 503; Steckel and Rose, *Backbone of History*, 230–45; Crosby, *Columbian Exchange*, 43–47, and *Ecological Imperialism*, 201–2; Hopkins, *Greatest Killer*, 205.

18. Quoted in Miguel León-Portilla, *The Broken Spears: The Aztec Account of the Conquest of Mexico*, rev. ed. (Boston: Beacon Press, 1992), 93.

19. Crosby, *Columbian Exchange*, 48.

20. Rodolfo Acuna-Soto et al., "Megadrought and Megadeath in 16th Century Mexico," *Emerging Infectious Diseases* 8 no. 4 (April 2002), 360–62, and "When Half the Population Died: The Epidemic of Hemorrhagic Fevers of 1576 in Mexico," *FEMS*

Microbiology Letters 240 (2004), 1–5. See also Daniel Reff, *Disease, Depopulation, and Cultural Change in Northwestern New Spain* (Salt Lake City: University of Utah Press, 1990), 1127.

21. David W. Stahle et al., "The Mexican Drought Atlas: Tree-Ring Reconstructions of the Soil Moisture Balance during the Late Pre-Hispanic, Colonial, and Modern Eras," *Quaternary Science Reviews* 149 (October 1, 2016), 34–60, and "Tree-Ring Data Document 16th-Century Megadrought over North America," *Eos: Transactions, American Geophysical Union* 81 no. 12 (March 21, 2000), 121–25; John L. Brooke, *Climate Change and the Course of Global History: A Rough Journey* (Cambridge: Cambridge University Press, 2014), 432.

22. Ross Hassig, *Mexico and the Spanish Conquest* (London: Longman, 1994), 101–7; Hanns Prem, "Disease Outbreaks in Central Mexico during the Sixteenth Century," in Noble David Cook and W. George Lovell, eds., *"Secret Judgments of God": Old World Disease in Colonial Spanish America* (Norman: University of Oklahoma Press, 1992), 24–43; McCaa, "Spanish and Nahuatl Views," 169–97; Crosby, *Columbian Exchange*, 42–43; Reff, *Disease*, 236 figure 30.

23. Noble David Cook, *Demographic Collapse, Indian Peru, 1520–1620* (Cambridge: Cambridge University Press, 1981). See also Hopkins, *Greatest Killer*, 208–23; and Crosby, *Columbian Exchange*, 38–55.

24. Thomas Falkner, *A Description of Patagonia and the Adjoining Parts of South America* (London: T. Lewis, 1774), 97–98. See also Alfred J. Tapson, "Indian Warfare on the Pampa during the Colonial Period," *Hispanic American Historical Review* 42 (February 1962), 4; Rómulo Muñiz, *Los indios pampas* (Buenos Aires: Editorial Bragado, 1966), 20–21; and Crosby, *Ecological Imperialism*, 203–5.

25. Dauril Alden and Joseph C. Miller, "Out of Africa; the Slave Trade and the Transmission of Smallpox to Brazil, 1560–1831," in Rotberg, *Health and Disease*, 203–30; Warren Dean, *With Broadaxe and Firebrand: The Destruction of the Brazilian Atlantic Forest* (Berkeley: University of California Press, 1995), 64; Hopkins, *Greatest Killer*, 215–19; Geoffrey Parker, *The Global Crisis: War, Climate, and Catastrophe in the Seventeenth-Century* (New Haven, Conn.: Yale University Press, 2013), 845; I am very grateful to Professor Parker for allowing me to read his manuscript before its publication.

26. Crosby, *Ecological Imperialism*, 209–15.

27. Alan C. Swedlund, "Contagion, Conflict, and Captivity in Interior New England: Native American and European Contacts in the Middle Connecticut Valley of Massachusetts, 1616–2004," in Catherine M. Cameron, Paul Kelton, and Alan C. Swedlund, eds., *Beyond Germs: Native Depopulation in North America* (Tucson: University of Arizona Press, 2015), 146–73; Richard W. Judd, *Second Nature: An Environmental History of New England* (Amherst: University of Massachusetts Press, 2014), 48; Alfred W. Crosby, "God . . . Would Destroy Them, and Give Their Country to Another People . . .," in Crosby, *Germs, Seeds and Animals*, 109–19; Sherburne F. Cook, "The Significance of Disease in the Extinction of the New England Indians," in Kenneth F. Kiple and Stephen V. Beck, *Biological Consequences of European Expansion, 1450–1800* (Aldershot, U.K.: Variorum Ashgate, 1997), 253–74.

28. William Cronon, *Changes in the Land: Indians, Colonists, and the Ecology of New England* (New York: Hill and Wang, 1983), 88; Parker, *Global Crisis*, 822.

29. Crosby, *Ecological Imperialism*, 208.

30. Fenn, *Pox Americana*, 88–89, and "Biological Warfare in Eighteenth-Century North America: Beyond Jeffrey Amherst," *Journal of American History* 86 no. 4 (March 2000), 1552–80.

31. Quoted in Cronon, *Changes in the Land*, 90.

32. Daschuk, *Clearing the Plains*, 12–19.

33. Kelton, *Epidemics and Enslavement*,143–58; Brooke, *Climate Change*, 432–33.

34. Adam R. Hodge, "'In Want of Nourishment for to Keep Them Alive': Climate Fluctuations, Bison Scarcity, and the Smallpox Epidemic of 1780–82 on the Northern Great Plains," *Environmental History* 17 (2012), 365–403.

35. Colin G. Calloway, "The Inter-tribal Balance of Power on the Great Plains, 1760–1850," *Journal of American Studies* 16 (April 1982), 25–48; Fenn, *Pox Americana*, 88–89, 210–23; Crosby, *Germs, Seeds, and Animals*, 98; Thornton, *American Indian Holocaust*, 91–94; Daschuk, *Clearing the Plains*, 22.

36. Esther W. Stearn and Allan E. Stearn, *The Effect of Smallpox on the Destiny of the Amerindian* (Boston: Bruce Humphreys, 1945), 89–90.

37. Miller, *Environmental History*, 50.

38. Denevan, "The Pristine Myth."

39. Judith Carney, *Black Rice: The African Origins of Rice Cultivation in the Americas* (Cambridge, Mass.: Harvard University Press, 2001), passim.

40. Miller, *Environmental History*, 59–60; Crosby, *Ecological Imperialism*, ch. 7; Darwin quote on p. 160.

41. John R. McNeill, *Mosquito Empires: Ecology and War in the Greater Caribbean, 1620–1914* (New York: Cambridge University Press, 2010), 23.

42. Sidney W. Mintz, *Sweetness and Power: The Place of Sugar in Modern History* (New York: Viking Press, 1986), 23–32; Richards, *Unending Frontier*, 414–15.

43. John R. McNeill, "Agriculture, Forests, and Ecological History: Brazil, 1500–1984," *Environmental Review* (Summer 1986), 124; Michael Williams, *Deforesting the Earth: From Prehistory to Global Crisis* (Chicago: University of Chicago Press, 2003), 198–99; Miller, *Environmental History*, 81–84; Richards, *Unending Frontier*, 388–92.

44. Richards, *Unending Frontier*, 418–38.

45. Richards, *Unending Frontier*, 438–51.

46. Williams, *Deforesting the Earth*, 197–98; Crosby, *Ecological Imperialism*, 75.

47. John McNeill, "Agriculture," 124; Williams, *Deforesting the Earth*, 199–200.

48. Dean, *With Broadaxe and Firebrand*, 144–90; Richards, *Unending Frontier*, 392–93, 413, 459; Williams, *Deforesting the Earth*, 199–200; Miller, *Environmental History*, 79–82; John McNeill, "Agriculture," 125.

49. Richard H. Grove, *Ecology, Climate and Empire: Colonialism and Global Environmental History, 1400–1940* (Knapwell, U.K.: Whitehorse Press, 1997), ch. 7; Richards, *Unending Frontier*, 421–25, 433, 446–47; McNeill, *Mosquito Empires*, 27–31.

50. Randall M. Packard, *The Making of a Tropical Disease: A Short History of Malaria* (Baltimore: Johns Hopkins University Press, 2007), 53–66; Charles Mann,

1493: Uncovering the New World Columbus Created (New York: Alfred A. Knopf, 2011), 82–87, 102; Stephen M. Rich and Francisco J. Ayala, "Evolutionary Origins of Human Malaria Parasites," in Krishna R. Dronamraju and Paolo Arese, eds., *Malaria: Genetic and Evolutionary Aspects* (New York: Springer, 2006), 132; Darrett B. Rutman and Anita H. Rutman, "Of Agues and Fevers: Malaria in the Early Chesapeake," *William and Mary Quarterly* 33 no. 1 (1976), 40; John McNeill, *Mosquito Empires*, 52–57, 62–67; Curtin, *Rise and Fall*, 79–81.

51. John McNeill, *Mosquito Empires*, 33–67, and "Ecology, Epidemics and Empires: Environmental Change and the Geopolitics of Tropical America, 1600–1825," *Environment and History* 5 no. 2 (1999), 177–79; Philip Curtin, *Death by Migration: Europe's Encounter with the Tropical World in the Nineteenth Century* (Cambridge: Cambridge University Press, 1989), 130; Kiple and Higgins, "Yellow Fever," 239.

52. James D. Goodyear, "The Sugar Connection: A New Perspective on the History of Yellow Fever," *Bulletin of the History of Medicine* 52 no. 1 (Spring 1978), 5–21; John McNeill, *Mosquito Empires*, 43–44, 64, 91–97, and "Ecology," 177–78; Kiple and Higgins, "Yellow Fever," 239–45.

53. On the Philadelphia epidemic, see J. H. Powell, *Bring Out Your Dead: The Great Plague of Yellow Fever in Philadelphia in 1793* (Philadelphia: University of Pennsylvania Press, 1949).

54. John McNeill, *Mosquito Empires*, 4–5.

55. John McNeill, *Mosquito Empires*, 100.

56. All of these cases are described in detail in John McNeill, *Mosquito Empires* and "Ecology."

57. John R. McNeill, "Biological Exchange in World History," in Jerry H. Bentley, ed., *The Oxford Handbook of World History* (New York: Oxford University Press, 2011), 338; Kiple and Higgins, "Yellow Fever"; Curtin, *Rise and Fall*, 78–81.

58. William Beinart and Lotte Hughes, *Environment and Empire* (Oxford: Oxford University Press, 2007), 41–55; Eric Jay Dolin, *Fur, Fortune, and Empire: The Epic History of the Fur Trade in America* (New York: W. W. Norton, 2010), 13–22, 45–46, 283; Daschuk, *Clearing the Plains*, 7; Richards, *Unending Frontier*, 463–79, 509–11.

59. Richards, *Unending Frontier*, 463–515; Beinart and Hughes, *Environment and Empire*, 52; Cronon, *Changes in the Land*, 99–107.

60. B. L. Turner II and Karl W. Butzer, "The Columbian Encounter and Land-Use Change," *Environment* 34 no. 8 (October 1992), 16–20; Watts, *West Indies*, 104; Richards, *Unending Frontier*, 325–31, 515–18; Crosby, *Columbian Exchange*, 75–78, 111–13, and *Ecological Imperialism*, 173–77; quotation on p. 175.

61. Watts, *West Indies*, 118–19; Turner and Butzer, "Columbian Encounter"; Richards, *Unending Frontier*, 325–31, 415–18; Crosby, *Ecological Imperialism*, 190–92.

62. Edward O. Wilson, "Ant Plagues: A Centuries-Old Mystery Solved," in E. O. Wilson, *Nature Revealed: Selected Writings, 1949–2006* (Baltimore: Johns Hopkins University Press, 2006), 349; see also Wilson, "Early Ant Plagues in the New World," *Nature* 433 (January 6, 2006), 32.

63. Eva Crane, *The World History of Beekeeping and Honey Hunting* (New York: Routledge, 1999), 358–59; John B. Free, *Bees and Mankind* (London: Allen & Unwin, 1982), 115; Crosby, *Ecological Imperialism*, 188.

64. L. E. Frelich et al., "Earthworm Invasion into Previously Earthworm-Free Temperate and Boreal Forests," *Biological Invasions* 8 no. 6 (September 2006), 1235–45; David A. Wardle, "Belowground Phenomena," in Daniel Simberloff and Marcel Rejmánek, eds., *Encyclopedia of Biological Invasions* (Berkeley: University of California Press, 2011), 54–58.

65. Bernardino de Sahagún, *Historia general de las cosas de Nueva España* [1576], quoted in Alberto Mario Salas, *Las armas de la conquista* (Buenos Aires: Emecé, 1950), 159.

66. John G. Varner and Jeanette J. Varner, *Dogs of the Conquest* (Norman: University of Oklahoma Press, 1983), 61–66; Salas, *Armas de la conquista*, 159–66.

67. John J. Johnson, "The Introduction of the Horse into the Western Hemisphere," *Hispanic American Historical Review* 23 (November 1942), 587–610.

68. John Hemming, *The Conquest of the Incas* (New York: Harcourt Brace Jovanovich, 1973), 112; Salas, *Armas de la conquista,* 138. See also Daniel Headrick, *Power over Peoples: Technology, Environments, and Western Imperialism, 1400 to the Present* (Princeton, N.J.: Princeton University Press, 2010), 101–18.

69. Inca Garcilaso de la Vega, *Comentarios reales de los Incas* [first published in 1609], ed. Carlos Araníbar (Lima: Fondo de Cultura Económica, 1991), 1: 158. The idea that Indians thought Europeans were gods, propagated by Garcilaso de la Vega and others, has been debunked by Camilla Townsend, "Burying the White Gods: New Perspectives in the Conquest of Mexico," *American Historical Review* 108 no. 3 (June 2003), 659–87.

70. Clark Wissler, "The Influence of the Horse in the Development of Plains Culture," *American Anthropologist* 16 no. 1 (1914), 1–25; Francis Haines, "Where Did the Plains Indians Get Their Horses?" *American Anthropologist* 40 no. 1 (1938), 112–17; Pekka Hämäläinen, *Comanche Empire* (New Haven, Conn.: Yale University Press, 2008), 37–38; Bradley Smith, *The Horse in the West* (New York: World, 1969), 14–16; Frank Gilbert Roe, *The Indian and the Horse* (Norman: University of Oklahoma, 1955), 72–122; Theodore Binnema, *Common and Contested Ground: A Human and Environmental History of the Northwest Plains* (Norman: University of Oklahoma Press, 2001), 86–106; Robert M. Denhardt, *The Horse of the Americas*, rev. ed. (Norman: University of Oklahoma Press, 1975), 92–111.

71. On horses in Argentina, see Antonio Elio Brailovsky, *Historia ecológica de Iberamérica*, vol. 1: *De los Mayas al Quijote* (Buenos Aires: Ediciones Capital Intelectual, 2006), 146–49; Prudencio de la C. Mendoza, *Historia de la ganadería argentina* (Buenos Aires: Ministerio de Agricultura, 1928), 11–14; Felix de Azara, *The Natural History of the Quadrupeds of Paraguay and the River La Plata*, trans. W. Percival Hunter (Edinburgh: A & C Black, 1838), 5; Madeline W. Nichols, "The Spanish Horse of the Pampas," *American Anthropologist* 41 no. 1 (1939), 119–29; Carlos Villafuerte, *Indios y gauchos en las pampas del sur* (Buenos Aires: Corregidor, 1989), 16–19; Muñiz, *Indios pampas*, 36–38; and Denhardt, *The Horse of the Americas*, 171–75.

72. Antonio Vázquez de Espinosa, *Compendium and Description of the West Indies*, trans. Charles Upson Clark (Washington: Smithsonian Institution Press, 1948), 675, 694, quoted in Crosby, *Columbian Exchange*, 84–85.

73. Tapson, "Indian Warfare," 5 n. 21.

74. Elinor G. K. Melville, *A Plague of Sheep: Environmental Consequences of the Conquest of Mexico* (Cambridge: Cambridge University Press, 1994); Richards, *Unending Frontier*, 362–65.

75. Quoted in Crosby, *Columbian Exchange*, 99.

76. Crosby, *Columbian Exchange*, 85–92, and *Ecological Imperialism*, 177–79; Richards, *Unending Frontier*, 346–49, 395–401.

77. Nicholas A. Robins, *Mercury, Mining, and Empire: The Human and Ecological Cost of Colonial Silver Mining in the Andes* (Bloomington: Indiana University Press, 21011), 101–43; Brailovsky, *Historia ecológica*, 193–95; Miller, *Environmental History*, 87–91.

78. Daviken Studniki-Gizbert and David Schecter, "The Environmental Dynamics of a Fuel-Rush: Silver Mining and Deforestation in New Spain, 1522 to 1810," *Environmental History* 15 (January 2010), 94–119.

79. Dean, *With Broadax and Firebrand*, 90–97; Miller, *Environmental History*, 91–99; Richards, *Unending Frontier*, 387–405; McNeill, "Agriculture," 124–25.

80. Miller, *Environmental History*, 95–104; quotations on 102, 104. See also Dean, *With Broadax and Firebrand*, 61–65; John McNeill, "Agriculture," 123–26; and Richards, *Unending Frontier*, 377–423.

81. Cronon, *Changes in the Land*, 108–21; Williams, *Deforesting the Earth*, 203–15.

82. Cronon, *Changes in the Land*, 90–91, 122–26.

83. This is also the opinion of Miller, *Environmental History*, 57; Williams, *Deforesting the Earth*, 195; and Richards, *Unending Frontier*, 458.

Chapter 7

1. For a scientific analysis of the Little Ice Age, see Jean Grove, *The Little Ice Age* (London: Routledge, 1990). A more popular account can be found in Brian Fagan, *The Little Ice Age: How Climate Made History, 1300–1850* (New York: Basic Books, 2000). See also the series of articles in the *Journal of Interdisciplinary History* 44 no. 3 (Winter 2014).

2. William F. Ruddiman, *Earth's Climate, Past and Future* (New York: W. H. Freeman, 2008), 302–3; John F. Richards, *The Unending Frontier: An Environmental History of the Early Modern World* (Berkeley: University of California Press, 2001), 59–61; Raymond S. Bradley and Philip D. Jones, eds., *Climate since AD 1500* (London: Routledge, 1992), 659.

3. Hubert H. Lamb, *Climate, History, and the Modern World*, 2nd ed. (London: Routledge, 1995), 212.

4. Sam White, "Rethinking Disease in Ottoman History," *International Journal of Middle Eastern Studies* 42 (2010), 559, and *The Climate of Rebellion in the Early*

Modern Ottoman Empire (New York: Cambridge University Press, 2011), 123, 136–43, 175, 181.

5. Geoffrey Parker, *The Global Crisis: War, Climate, and Catastrophe in the Seventeenth-Century World* (New Haven, Conn.: Yale University Press, 2013), 816–20; Robert B. Marks, *Tigers, Rice, Silk, and Silt: Environment and Economy in Late Imperial South China* (Cambridge: Cambridge University Press, 1998), 26–27, 138–39, 196–201; Martin Heijdra, "The Socio-Economic Development of Rural China during the Ming," in *The Cambridge History of* China, vol. 8: Denis Twitchett and Frederick W. Mote, eds. *The Ming Dynasty, 1398-1644* (Cambridge: Cambridge University Press, 2008), 423–27.

6. Wolfgang Behringer, *A Cultural History of Climate*, trans. Patrick Camiller (Cambridge: Polity, 2010), 89–92.

7. James L. A. Webb, *Desert Frontier: Ecological and Economic Change along the Western Sahel, 1600-1850* (Madison: University of Wisconsin Press, 1995), 5, 15–16; Neil Roberts, *The Holocene: An Environmental History* (Oxford: Blackwell, 1998), 216; Parker, *Global Crisis*, 865.

8. Parker, *Global Crisis*, 3–6, lists these and many other extreme weather events. See also John L. Brooke, *Climate Change and the Course of Global History: A Rough Journey* (Cambridge: Cambridge University Press, 2014), 439.

9. Geoffrey Parker, *Europe in Crisis, 1598-1648*, 2nd ed. (Oxford: Blackwell, 2001), 4–5; "Crisis and Catastrophe: The Global Crisis of the Seventeenth Century Reconsidered," *American Historical Review* 113 (October 2008), 1053–79; and "States Make War but Wars also Break States," *Journal of Military History* 74 no. 1 (January 2010), 17–19; Robert B. Marks, "'It Never Used to Snow': Climatic Variability and Harvest Yields in Late-Imperial South China, 1650-1850," in Mark Elvin and Liu Ts'ui-jung, eds., *Sediments of Time: Environment and Society in Chinese History* (Cambridge: Cambridge University Press, 1998), 411–18; Lamb, *Climate*, 211–12, 237–41; Heijdra, "Socio-Economic Development," 427–28. On North America, see, for example, Sam White, "Cold, Drought, and Disaster: The Little Ice Age and the Spanish Conquest of New Mexico," *New Mexico Historical Review* 89 no. 4 (Fall 2014), 425–58, and Dee C. Pederson et al., "Medieval Warming, Little Ice Age, and European Impact on the Environment during the Last Millennium in the Lower Hudson Valley, New York, USA," *Quaternary Research* 63 (2005), 238–49.

10. Parker, *Europe in Crisis*, 4–5; Lamb, *Climate*, 227–28.

11. Lamb, *Climate*, 213–40; Grove, *Little Ice Age*, 18; Parker, "Crisis and Catastrophe," 1065–73, and *Europe in Crisis*, 21.

12. Parker, *Global Crisis*, 8–13; Behringer, *Cultural History*, 115–46.

13. Ruddiman, *Earth's Climate*, 116–36.

14. Parker, *Global Crisis*, 13.

15. Quoted in Parker, *Europe in Crisis*, 19–20.

16. John A. Eddy, "The 'Maunder Minimum': Sunspots and Climate in the Reign of Louis XIV," in Geoffrey Parker and Lesley M. Smith, eds., *The General Crisis of the Seventeenth Century* (London: Routledge, 1978), 226–55, "The Maunder Minimum," *Science* 192 (1976), 1189–203, and "Climate and the Role of the Sun,"

in Robert I. Rotberg and Theodore K. Rabb, eds., *Climate and History: Studies in Interdisciplinary History* (Princeton, N.J.: Princeton University Press, 1981), 145–67; D. T. Shindell et al., "Solar Forcing of Regional Climate Change during the Maunder Minimum," *Science* 294 (December 7, 2001), 2149–52; J. Luterbacher et al., "The Late Maunder Minimum," *Climatic Change* 49 no. 4 (2001), 441–62. See also Emmanuel Le Roy Ladurie, *Histoire humaine et comparée du climat*, vol. 1 (Paris: Fayard, 2004), 409–33, and Brooke, *Climate Change,* 381–82.

17. William S. Atwell, "Volcanism and Short-Term Climatic Change in East Asian and World History, c. 1200–1699," *Journal of World History* 12 no. 1 (September 2001), 29–98; Shanaka L. DeSilva and Gregory A. Zielinski, "Global Influence of Huaynaputina, Peru," *Nature* 393 no. 6684 (June 4, 1998), 455–58; K. R. Briffa et al., "Influence of Volcanic Eruptions on Northern Hemisphere Summer Temperature over the Past 600 Years," *Nature* 393 no. 6684 (June 4, 1998): 450–55; Luke Oman, Alan Robock, Gregory L. Stenchikov, and Thorvaldur Thordarson, "High-Latitude Eruptions Cast Shadow over the African Monsoon and the Flow of the Nile," *Geophysical Research letters* 33 (September 2006): L18711; Bertram Schwarzschild, "The Triggering and Persistence of the Little Ice Age," *Physics Today* 65 no. 4 (2012); Gifford H. Miller, "Abrupt Onset of the Little Ice Age Triggered by Volcanism and Sustained by Sea-Ice/Ocean Feedbacks," *Geophysical Research Letters* 39 (January 31, 2012), L02708; Parker, *Global Crisis*, 13–14.

18. Ruddiman, *Earth's Climate*, 303–6.

19. Ruddiman, *Plows, Plagues, and Petroleum: How Humans Took Control of Climate* (Princeton, N.J.: Princeton University Press, 2005), 119–24, and *Earth's Climate*, 306.

20. Ruddiman, *Plows, Plagues, and Petroleum*, 126–41. See also Robert Hartwell, "A Revolution in the Chinese Iron and Coal Industries during the Northern Sung, 960–1126 A.D.," *Journal of Asian Studies* 21 no. 2 (February 1962), 153–62.

21. Franz X. Faust et al., "Evidence for the Postconquest Demographic Collapse of the Americas in Historical CO_2 Levels," *Earth Interactions* 10 no. 1 (2006): 1–14; Richard J. Nevle and Dennis K. Bird, "Effects of Syn-pandemic Fire Reduction and Reforestation in the Tropical Americas on Atmospheric CO_2 during European Conquest," *Palaeogeography, Palaeoclimatology, Palaeoecology* 264 (July 2008): 25–38; R. A. Dull et al., "The Columbian Encounter and the Little Ice Age: Abrupt Land Use Change, Fire, and Greenhouse Forcing," *Annals of the Association of American Geographers* 100 (2010): 755–71; Richard Nevle et al., "Neotropical Human-Landscape Interactions, Fire, and Atmospheric CO_2 during European Conquest," *Holocene* 21 no. 5 (August 2011); Simon L. Lewis and Mark A. Maslin, "Defining the Anthropocene," *Nature* 519 (March 12, 2015), 171–809.

22. Brooke, *Climate Change*, 440–42.

23. David Stocker et al., "Holocene Peatland and Ice-Core Data Constraints on the Timing and Magnitude of CO2 emissions from Peat Land Use," *Proceedings of the National Academy of Sciences of the USA* 114 no. 7 (February 14, 2017), 1492–97; Julia Pongratz et al., "Coupled Climate-Carbon Simulations Indicate Minor Global Effects of Wars and Epidemics on Atmospheric CO_2 between AD 800 and 1850," *Holocene* 21 no. 5 (2011), 843–51; Jörgen Olofsson and Thomas Hickler, "Effects of Human

Land-Use on the Global Carbon Cycle during the Last 6,000 Years," *Vegetation History and Archaeobotany* 17 no. 5 (September 2008): 605–15.

24. Parker, "Crisis and Catastrophe"; quotation on pp. 1053–54.

25. Emmanuel LeRoy Ladurie, Daniel Rousseau, and Anouchka Vasak, *Les Fluctuations du climat de l'an mil à aujourd'hui* (Paris: Fayard, 2011), 16; Parker, *Europe in Crisis*, 5–11, and *Global Crisis*, 17–22; Behringer, *Cultural History*, 93–99, 113.

26. Jaime Vicens Vices, *An Economic History of Spain*, trans. Frances M. Lopez-Morillas (Princeton, N.J.: Princeton University Press, 1969), 416. See also Carlos Álvarez-Nogal et al., "Spanish Agriculture in the Little Divergence," *European Review of Economic History* 20 no. 4 (November 11, 2016), 452–77, and Henry Kamen, "Climate and Crisis in the Mediterranean: A Perspective," in Wolfgang Behringer, Hartmut Lehmann, and Christian Pfister, eds., *Kulturelle Konsequenzen der "Kleinen Eiszei": Cultural Consequences of the "Little Ice Age"* (Göttingen: Vanderhoeck & Ruprecht, 2005), 371.

27. Andrew P. Appleby, "Epidemics and Famine during the Little Ice Age," *Journal of Interdisciplinary History* 10 (1980), 643–63; Lamb, *Climate*, 220–24.

28. White, *Climate of Rebellion*, 123–25, 140–50, 157, 187–88, 198ff, 217, 240–48, 268; quotation on p. 222; White, "Rethinking Disease," 559–61, and "The Little Ice Age Crisis of the Ottoman Empire: A Conjuncture in Middle Eastern Environmental History," in Alan Mikhail, ed., *Water on Sand: Environmental Histories of the Middle East and North Africa* (New York: Oxford University Press, 2013), 71–90; Michael Williams, *Deforesting the Earth: From Prehistory to Global Crisis* (Chicago: University of Chicago Press, 2003), 151–57, 174, 181–83.

29. White, "Rethinking Disease," 558–61.

30. Nükhet Varlik, "Plague in the Islamic World, 1500–1850," in Joseph P. Byrne, ed., *Encyclopedia of Pestilence, Pandemics, and Plagues* (Westport, Conn.: Greenwood Press, 2008), 519–20; Michael W. Dols, "The Second Plague Pandemic and Its Recurrences in the Middle East, 1347–1894," *Journal of the Economic and Social History of the Orient* 22 no. 2 (May 1979), 168–76; Alan Mikhail, *Nature and Empire in Ottoman Egypt: An Environmental History* (New York: Cambridge University Press, 2011), 215–16, and "The Nature of Plague in Late Eighteenth-Century Egypt," *Bulletin of the History of Medicine* 82 no. 2 (2008), 250; Rushdi Said, *The River Nile: Geology, Hydrology, and Utilization* (Oxford: Pergamon Press, 1993), 213.

31. Alan Mikhail, *Under Osman's Tree: The Ottoman Empire, Egypt, and Environmental History* (Chicago: University of Chicago Press, 2017), 169–83, *Nature and Empire*, 219–21, and "The Nature of Plague," 254–66; Daniel Panzac, *La peste dans l'Empire Ottoman* (Louvain: Peeters, 1985), 58–77.

32. Panzac, *Peste*, 105–8; Varlik, "Plague," 519–21; Dols, "Second Plague Pandemic," 178–79.

33. Mikhail, *Nature and Empire*, 218–29, "The Nature of Plague," 261–69, and "Plague and Environment in Late Ottoman Egypt," in Mikhail, ed., *Water on Sand*, 111–32.

34. Mikhail, *Nature and Empire*, 201, "The Nature of Plague," 249–50, and "Plague and Environment"; White, "Rethinking Disease," 554; Varlik, "Plague," 521–22 and "From *'Bête Noire'* to *'le Mal de Constantinople'*: Plagues, Medicine, and the Early Modern

Ottoman State," *Journal of World History* 24 no. 4 (December 2013), 765–67; Panzac, *Peste*, 281–85, 295.

35. Panzac, *Peste*, 279, 295–315, 340–41.

36. Panzac, *Peste*, 327–33; Mikhail, *Nature and Empire*, 230–36.

37. Jin-Qi Fang and Guo Liu, "Relationship between Climatic Change and the Nomadic Southward Migrations in Eastern Asia during Historical Times," *Climatic Change* 22 (1992), 151–69.

38. Robert Marks, *China: Its Environment and History* (Lanham, Md.: Rowman & Littlefield, 2012), 173–74. See also his " 'It Never Used to Snow,' " 411; Timothy Brook, *The Troubled Empire: China in the Yuan and Ming Dynasties* (Cambridge, Mass.: Harvard University Press, 2010), 243–50; William S. Atwell, "A Seventeenth-Century 'General Crisis' in East Asia?" *Modern Asian Studies* 24 no. 4 (1990), 661–82; and Brooke, *Climate Change*, 445–46.

39. Helen Dunstan, "The Late Ming Epidemics: A Preliminary Survey," *Ch'ing-shih wen-t'i / Late Imperial China* 3 no. 3 (1974–76), 17. See also Marks, *China*, 173–74, and Brook, *Troubled Empire*, 65–66, 250–51.

40. Dunstan, "Late Ming Epidemics," 19.

41. Mark Elvin, *The Pattern of the Chinese Past* (Stanford, Cal.: Stanford University Press, 1973), 310–11.

42. Carol Benedict, *Bubonic Plague in Nineteenth-Century China* (Stanford, Cal.: Stanford University Press, 1996), 18–24; *China*, 172.

43. Brook, *Troubled Empire*, 244–52.

44. Marks, *Tigers, Rice, Silk, and Silt*, 144–45; quotation on p. 135.

45. Parker, "States Make War," 19–20; see also "Crisis and Catastrophe," 9–10.

46. I. Schöffer, "Did Holland's Golden Age Coincide with a Period of Crisis?," in Parker and Smith, *General Crisis*, 87–107.

47. William S. Atwell, "Some Observations on the 17th-Century Crisis in China and Japan," *Journal of Asian Studies* 45 no. 2 (February 1986), 223–44.

48. Brooke, *Climate Change*, 414, 436, 451. For a compendium of demographers' estimates, see "World Population Estimates," in Wikipedia (accessed August 2014).

49. Kenneth F. Kiple, *A Moveable Feast: Ten Millennia of Food Globalization* (Cambridge: Cambridge University Press, 2007), 135–49.

50. Duccio Bonavia, *Maize: Origin, Domestication, and Its Role in the Development of Culture* (Cambridge: Cambridge University Press, 2013)..

51. "Sweet Potato," in Wikipedia (accessed August 28, 2011).

52. William O. Jones, *Manioc in Africa* (Stanford, Cal.: Stanford University Press, 1959), 4–6, 15–16, 22–23.

53. One exception: potatoes freeze-dried by Indians in the Andes could last for years. See Charles C. Mann, *1493: Uncovering the New World Columbus Created* (New York: Alfred A. Knopf, 2011), 201–9.

54. Different authors give different figures; see, for example, Parker, *Europe in Crisis*, 22–23; Richards, *Unending Frontier*, 205; and E. A. Wrigley, *Population and History* (New York: McGraw-Hill, 1969), 79, 153.

55. Andrew B. Appleby, "Epidemics and Famine in the Little Ice Age," in Rotberg and Rabb, *Climate and History*, 63–83; Michael Flinn, "The Stabilization of Mortality in

Preindustrial Western Europe," *Journal of European Economic History* 3 (1974): 285–318; Eric L. Jones, *The European Miracle: Environments, Economies, and Geopolitics in the History of Europe and Asia* (Cambridge: Cambridge University Press, 1981), 32, 181.

56. Richards, *Unending Frontier*, 194, 205–13; Appleby, "Epidemics and Famine," 78–83; Flinn, "Stabilization of Mortality," 121–56.

57. Audrey M. Lambert, *The Making of the Dutch Landscape: An Historical Geography of the Netherlands* (London: Academic Press, 1985), 208–13; Richards, *Unending Frontier*, 214–20; Mann, *1493*, 82–83.

58. Joachim Radkau, *Nature and Power: A Global History of the Environment*, trans. Thomas Dunlap (Cambridge: Cambridge University Press, 2008), 212–21; John U. Nef, "An Early Energy Crisis and Its Consequences," *Scientific American*, 237 no. 5 (November 1977), 141; Williams, *Deforesting the Earth*, 151–85; Richards, *Unending Frontier*, 221–26.

59. Many authors, even experts on Venetian history, have claimed far larger numbers. Thus John J. Norwich [*A History of Venice* (New York: Knopf, 1972), 17 n.1] says that there were 1,156,627 pilings; Norbert Huse [*Venedig: Von der Kunst, eine Stadt im Wasser zu bauen* (Munich: C. H. Beck, 2005), 14] gives the number as 1,156,650; Joachim Radkau [*Nature and Power*, 122] gives the number as 1,150,657; Andrew Hopkins [*Santa Maria della Salute: Architecture and Ceremony in Baroque Venice* (Cambridge: Cambridge University Press, 2000), 53] says "eventually 1,156,567 wooden piles were used for the foundations"; and the author of the article "Santa Maria della Salute," in Wikipedia (English version), says the basilica was "built on a platform of 1,000,000 wooden piles."

The source of these preposterous numbers is a 1663 edition by D. Giustiniano Martinioni of a book by Francesco Sansovino, *Venetia, città nobilissima, et singolare. . . . Con aggivunta di tutte les cose notabili della città, fatte, & occore dall'anno 1580. fino al presente 1663. da D. Givstiniano Martinioni* (Venice: S.Curti, 1663), 278, in which he gives the number "un milione, cento cinquanta sei mille, e sei cento cinquanta sette Pali"; in other words, 1,156,657 pilings.

According to the author of the Italian Wikipedia article "Basilica di Santa Maria della Salute," "assuming the diameter of the pilings as 25 centimeters, it is not possible to come up with more than 16 pilings per square meter. . . . Even extending the surface to the entire open area, we come up with a figure of approximately 100,000 pilings used." The basilica measure 70 meters long by 47 meters wide, or 3,290 square meters; at sixteen pilings per square meter, that gives 52,640 pilings. After adding the plaza in front of the church, 100,000 seems a reasonable number.

60. Frederic C. Lane, *Venetian Ships and Shipbuilders of the Renaissance* (Baltimore: Johns Hopkins University Press, 1934), 219–33.

61. Williams, *Deforesting the Earth*, 183–86.

62. Richard H. Grove, *Ecology, Climate, and Empire: Colonialism and Global Environmental History, 1400–1940* (Knapwell, U.K.: White Horse Press, 1997), 49–63.

63. Alfred W. Crosby, "The Demographic Effect of American Crops in Europe," in Alfred W. Crosby, ed., *Germs, Seeds & Animals: Studies in Ecological History* (Armonk, N.Y.: M.E. Sharpe, 1994), 162–63.

64. John R. McNeill, *The Mountains of the Mediterranean World: An Environmental History* (Cambridge: Cambridge University Press, 1992), 89–91; Lamb, *Climate*, 245; Alfred Crosby, *The Columbian Exchange: Biological and Cultural Consequences of 1492* (Westport, Conn.: Greenwood, 1972), 178–81, and "The Demographic Effect," 150–52.

65. William H. McNeill, "How the Potato Changed the World's History," *Social Research* 66 (1999): 69–83, and "American Food Crops in the Old World," in Herman J. Viola and Carolyn Margolis, eds., *Seeds of Change: A Quincentennial Commemoration* (Washington: Smithsonian Institution Press, 1991) 43–59; William L. Langer, "American Foods and Europe's Population Growth, 1750–1850," *Journal of Social History* 8 no. 2 (Winter 1975), 51–66; John Reader, *Potato: A History of the Propitious Esculent* (New Haven, Conn.: Yale University Press, 2009), 22–23, 85–91; Crosby, "Demographic Effect," 153–57; John Komlos, "The New World's Contribution to Food Consumption during the Industrial Revolution," *Journal of European Economic History* 27 no. 1 (1998), 67–82.

66. Stanley Alpern, "The European Introduction of Crops into West Africa in Precolonial Times," *History in Africa* 19 (1992), 14–30; Christopher Ehret, *The Civilizations of Africa: A History to 1800* (Charlottesville: University of Virginia Press, 2002), 354; Gregory H. Maddox, *Sub-Saharan Africa: An Environmental History* (Santa Barbara, Cal.: ABC-CLIO, 2006), 85.

67. James McCann, *Maize and Grace: Africa's Encounter with a New World Crop, 1500–2000* (Cambridge, Mass.: Harvard University Press, 2005), 23–29; Brooke, *Climate Change*, 435; Alpern, "European Introduction," 24–25.

68. Jan Vansina, "Histoire du manioc en Afrique centrale avant 1850," *Paideuma* 43 (1997), 255–79; William McNeill, "American Food Crops," 57; Jones, *Manioc in Africa*, 32, 62–73; Alpern, "European Introduction," 25.

69. Marks, *China*, 172–74; Atwell, "Seventeenth-Century 'General Crisis,'" 661–82; and Heijdra, "Socio-Economic Development," 435–38.

70. Marks, *China*, 205–6.

71. Thomas T. Allsen, *The Royal Hunt in Eurasian History* (Philadelphia: University of Pennsylvania Press, 2006), 45–46.

72. Kenneth Pomeranz, "The Transformation of China's Environment," in Edmund Burke III and Kenneth Pomeranz, eds., *The Environment and World History* (Berkeley: University of California Press, 2009), 130; Marks, *Tigers, Rice, Silk, and Silt*, 161; Richards, *Unending Frontier*, 138–42.

73. Ping-ti Ho (He Bingdi), *Studies on the Population of China, 1368–1953* (Cambridge, Mass.: Harvard University Press, 1959), 139, 283; Benedict, *Bubonic Plague*, 25–26; Andre Gunder Frank, *ReOrient: Global Economy in the Asian Age* (Berkeley: University of California Press, 1998), 109–10, 168; Marks, *China*, 199–201.

74. Peter Perdue, *China Marches West: The Qing Conquest of Central Eurasia* (Cambridge, Mass.: Harvard University Press, 2005); Marks, *China*, 161–86.

75. Parker, "Crisis and Catastrophe," 1059; Ho, "Studies in the Population," 101; Williams, *Deforesting the Earth*, 216–18.

76. Randall A. Dodgen, *Controlling the Dragon: Confucian Engineers and the Yellow River in Late Imperial China* (Honolulu: University of Hawaii Press, 2001), 11–26; Pomeranz, "Transformation," 123–30.

77. Anne Osborne, "Highlands and Lowlands: Economic and Ecological Interactions in the Lower Yangzi Region under the Qing," in Elvin and Liu, *Sediments of Time*, 230; Peter Perdue, *Exhausting the Earth: State and Peasant in Hunan, 1500–1850* (Cambridge, Mass.: Harvard University Press, 1987), 113–30, 175–77; Marks, *China*, 191–96.

78. Osborne, "Highlands and Lowlands," 205–8; Pomeranz, "Transformation," 127–29.

79. Marks, *Tigers, Rice, Silk, and Silt*, 281–85, 312–14, 334.

80. Marks, *China*, 189–91.

81. Sucheta Mazumdar, "The Impact of New World Food Crops on the Diet and Economy of China and India, 1600–1900," in Raymond Grew, ed., *Food in Global History* (Boulder, Colo.: Westview Press, 1999), 62–64; Mann, *1493*, 182; Perdue, *Exhausting the Earth*, 59–75; Marks, *Tigers, Rice, Silk, and Silt*, 307–8; Ho, *Studies*, 145–46, 183–92; Richards, *Unending Frontier*, 112–28.

82. Ping-ti Ho (He Bingdi), "The Introduction of American Food Plants into China," *American Anthropologist* 57 no. 2 (April 1955), 193–94; Mazumdar, "Impact," 66–70.

83. Ho, "Introduction," 194–97; Mazumdar, "Impact," 68–69.

84. Ho, "Introduction," 191–92, and *Studies*, 185–86; Mazumdar, "Impact," 69–70.

85. Ho, *Studies*, 150.

86. Mark Elvin, "3,000 Years of Unsustainable Growth: China's Environment from Archaic Times to the Present," *East Asian History* 6 (1993), 36; Ho, *Studies*, 147–48; Perdue, *Exhausting the Land*, 88; Richards, *Unending Frontier*, 130–31; Marks, *China*, 197–99.

87. Marks, *China*, 206.

88. Brook, *Troubled Empire*, 133.

89. Marks, *China*, 174–75.

90. Williams, *Deforesting the Earth*, 220. See also Brook, *Troubled Empire*, 131–33; Marks, *Tigers, Rice, Silk, and Silt*, 320–21; Osborne, "Highlands and Lowlands," 209–10; Richards, *Unending Frontier*, 142.

91. Quotation in Marks, *Tigers, Rice, Silk, and Silt*, 161. See also Brook, *Troubled Empire*, 131–32.

92. Conrad D. Totman, *The Green Archipelago: Forestry in Preindustrial Japan* (Berkeley: University of California Press, 1989), 35; Edwin O. Reischauer, *Japan: The Story of a Nation*, 4th ed. (New York: McGraw Hill, 1990), 79–81; Richards, *Unending Frontier*, 154, 167–75.

93. Totman, *Green Archipelago*, 11–45.

94. Williams, *Deforesting the Earth*, 221–22; Totman, *Green Archipelago*, 62.

95. Richards, *Unending Frontier*, 162–63, 180–81.

96. Richards, *Unending Frontier*, 173; Williams, *Deforesting the Earth*, 223–24; "Great Fire of Meireki," Wikipedia (accessed September 2011).

97. Ramachandra Guha, *Environmentalism: A Global History* (New York: Longman, 2000), 42–43.

98. Masako M. Osako, "Forest Preservation in Tokugawa Japan," in Richard Tucker and John F. Richards, eds., *Global Deforestation and the Nineteenth-Century World Economy* (Durham, N.C.: Duke University Press, 1983), 129–45; Conrad D. Totman, *Early Modern Japan* (Berkeley: University of California Press, 1993),

226–30, 268–69, and *Green Archipelago,* 57–68; Richards, *Unending Frontier,* 173; Williams, *Deforesting the Earth,* 224.

99. Susan B. Hanley, *Everyday Things in Premodern Japan* (Berkeley: University of California Press, 1997), 54–62; Totman, *Green Archipelago,* 88–93, 124, 271–72; Williams, *Deforesting the Earth,* 149, 224–25; Richards, *Unending Frontier,* 149, 178–79.

100. Williams, *Deforesting the Earth,* 220.

Chapter 8

1. John F. Richards, *The Unending Frontier: An Environmental History of the Early Modern World* (Berkeley: University of California Press, 2001), 194–95, 235; Peter Thorsheim, *Inventing Pollution: Coal, Smoke, and Culture in Britain since 1800* (Athens: Ohio University Press, 2006), 3; B. W. Clapp, *An Environmental History of Britain since the Industrial Revolution* (London: Longman, 1994), 15–16; Rolf Peter Sieferle, *The Subterranean Forest: Energy Systems and the Industrial Revolution* (Cambridge: White Horse Press, 2001), 40–41, 102–3; Stephen Mosley, *Chimney of the World: A History of Smoke Pollution in Victorian and Edwardian Manchester* (Cambridge: White Horse Press, 2001), 16; Richard G. Wilkinson, *Poverty and Progress: An Ecological Model of Economic Development* (London: Methuen, 1973), 112–14.

2. Richards, *Unending Frontier,* 237–39.

3. Prassanan Parthasarathi, *Why Europe Grew Rich and Asia Did Not* (Cambridge: Cambridge University Press, 2011), 93–94.

4. Edmund Russell, *Evolutionary History: Uniting History and Biology to Understand Life on Earth* (Cambridge: Cambridge University Press, 2011), 66–67, 104–22. Today, *Gossypium hirsutum* accounts for 90 percent of all cotton grown worldwide, and *G. barbadense* for another 8 percent; the other two kinds of cotton have almost disappeared; see "Cotton," in Wikipedia (accessed May 2012).

5. Ted Steinberg, *Down to Earth: Nature's Role in American History,* 2nd ed. (New York: Oxford University Press, 2009), 83; Russell, *Evolutionary Biology,* 104–5; 113–19.

6. For a competent summary, see Abbott Payton Usher, "The Textile Industry, 1750–1830," in Melvin Kranzberg and Carroll W. Pursell Jr., eds., *Technology in Western Civilization,* vol. 1 (New York: Oxford University Press, 1967), 230–44. See also Joel Mokyr, *The Lever of Riches: Technological Creativity and Economic Progress* (New York: Oxford University Press, 1990), 96–100.

7. Edward Carpenter, *Towards Democracy,* 3rd ed. (London: T. F. Unwin, 1892), 452.

8. Mokyr, *The Lever of Riches,* 92–96; I. G. Simmons, *Global Environmental History* (Chicago: University of Chicago Press, 2008), 110; Sieferle, *Subterranean Forest,* 111–22; Wilkinson, *Poverty and Progress,* 116–27; Thorsheim, *Inventing Pollution,* 3.

9. Michael W. Flinn, *The History of the British Coal Industry,* vol. 2: *1700–1830, The Industrial Revolution* (Oxford: Clarendon Press, 1984), 252–53; Smil, *Energy,* 160.

10. Sieferle, *Subterranean Forest*, 127–31; Wilkinson, *Poverty and Progress*, 118–20; Pomeranz, *Great Divergence*, 67–68.

11. Daniel R. Headrick, *Technology: A World History* (New York: Oxford University Press, 2009), 99–100.

12. Michael Williams, *Deforesting the Earth: From Prehistory to Global Crisis* (Chicago: University of Chicago Press, 2003), 243, 244.

13. The first writings on air pollution was *Fumifugium* by John Evelyn, published in 1661.

14. Peter Brimblecombe, *The Big Smoke: A History of Air Pollution in London since Medieval Times* (London: Methuen, 1987), 26–34, 63; Sieferle, *Subterranean Forest*, 81–86; Thorsheim, *Inventing Pollution*, 5–6; Richards, *Unending Frontier*, 234.

15. B. R. Mitchell, *British Historical Statistics* (Cambridge: Cambridge University Press, 1988), 25–28.

16. Peter Stearns, *European Society in Upheaval: Social History since 1800* (New York: Macmillan, 1967), 112.

17. Nassau Senior, *Letters on the Factory Act, as it Affects the Cotton Manufacture, Addressed to the Right Honourable the President of the Board of Trade*, 2nd ed. (London: Fellowes, 1844), 20.

18. Stephen Mosley, *Environment in World History* (New York: Routledge, 2010), 94–97; Joel A. Tarr, *The Search for the Ultimate Sink: Urban Pollution in Historical Perspective* (Akron, Ohio: University of Akron Press, 1996), 113–18, 323–27; Clapp, *Environmental History*, 28–31, 72–75, 85–87.

19. Clapp, *Environmental History*, 14.

20. Tamara Whited et al., *Northern Europe: An Environmental History* (Santa Barbara, Cal.: ABC-CLIO, 2005), 112–13; Clapp, *Environmental History*, 19–22; Tarr, *Search for the Ultimate Sink*, 264–67; Brimblecombe, *The Big Smoke*, 96.

21. Mosley, *Chimney of the World*, 20; Brimblecombe, *The Big Smoke*, 63–67; Clapp, *Environmental History*, 26–27.

22. Harold L. Platt, *Shock Cities: The Environmental Transformation and Reform of Manchester and Chicago* (Chicago: University of Chicago Press, 2005), 448.

23. Clapp, *Environmental History*, 76–77; Brimblecombe, *The Big Smoke*, 101–6, and "Air Pollution in York 1850–1900," in Peter Brimblecombe and Christian Pfister, eds., *The Silent Countdown: Essays in European Environmental History* (Berlin: Springer Verlag, 1990).

24. Anthony N. Penna, *The Human Footprint: A Global Environmental History* (Armonk, N.Y.: M.E. Sharpe, 1999), 183.

25. Platt, *Shock Cities*, 217, 452.

26. Steven Johnson, *The Ghost Map: The Story of London's Most Terrifying Epidemic—and How It Changed Science, Cities, and the Modern World* (New York: Penguin, 2006).

27. George Rosen, *History of Public Health* (Baltimore: Johns Hopkins University Press, 1993), 205–55; David P. Clark, *Germs, Genes, and Civilization: How Epidemics Shaped Who We Are Today* (Upper Saddle River, N.J.: FT Press, 2010), 72–73; Robert Pollitzer, *Cholera* (Geneva: World Health Organization, 1959), 21–30.

28. Charles E. Rosenberg, *The Cholera Years: The United States in 1832, 1849 and 1866* (Chicago: University of Chicago Press, 1962), 2.

29. Mosley, *Chimney of the World*, 60–61.

30. W. F. Loomis, "Rickets," *Scientific American* (December 1970), 112–20.

31. Mosley, *Chimney of the World*, 17–18; Platt, *Shock Cities*, 31–32.

32. Platt, *Shock Cities*, 196–97, 205–7, 211–14, 220–27.

33. Hugh Miller, *The Old Red Sandstone* (Edinburgh, 1873) quoted in Lewis Mumford, *The City in History: Its Origins, Its Transformations, and Its Prospects* (New York: Harcourt, Brace and World, 1961), 459–60.

34. Mosley, *Chimney of the World*, 31; Platt, *Shock Cities*, 203–11, 228–29.

35. Friedrich Engels, *Condition of the Working Class in England*, trans. and ed. W. O. Henderson and W. H. Chaloner (Oxford: Blackwell, 1958), 312.

36. Theodore Steinberg, *Nature Incorporated: Industrialization and the Waters of New England* (Cambridge: Cambridge University Press, 1991), 22–23, 50–57, 79, 146, 166–67, 205–11; John Opie, *Nature's Nation: An Environmental History of the United States* (Fort Worth, Tex.: Harcourt Brace, 1998), 146–47. On fish in New England rivers, see William Cronon, *Changes in the Land: Indians, Colonists, and the Ecology of New England* (New York: Hill and Wang, 1983), 22–23.

37. Albert E. Cowdrey, *This Land, This South: An Environmental History* (Lexington: University Press of Kentucky, 1996), 71; Russell, *Evolutionary History*, 119–20.

38. Philip D. Curtin, *The Rise and Fall of the Plantation Complex: Essays in Atlantic History* (Cambridge: Cambridge University Press, 1990).

39. Cowdrey, *This Land*, 77.

40. Curtis Nettels, *The Emergence of a National Economy* (White Plains, N.Y.: M.E. Sharpe, 1977), 191.

41. On the guano rush, see Charles C. Mann, *1493: Uncovering the New World Columbus Created* (New York: Knopf, 2011), 212–15; David R. Montgomery, *Dirt: The Erosion of Civilization* (Berkeley: University of California Press, 2007), 185–88; and Jimmy M. Skaggs, *The Great Guano Rush: Entrepreneurs and American Overseas Expansion* (New York: St. Martin's Press, 1994), ch. 2

42. Cowdrey, *This Land*, 76.

43. Carville Earle, "The Myth of the Southern Soil Miner: Macrohistory, Agricultural Innovation, and Environmental Change," in Donald Worster, ed., *The Ends of the Earth: Perspectives on Modern Environmental History* (Cambridge: Cambridge University Press, 1988), 175–210; Tony Hiss, "The Wildest Idea on Earth," *Smithsonian* 45 no. 5 (September 2014), 69; Montgomery, *Dirt*, 127–41; Steinberg, *Down to Earth*, 85–87, 100–103.

44. Richard W. Judd, *Second Nature: An Environmental History of New England* (Amherst: University of Massachusetts Press, 2014), 80–94.

45. Lord Byron, "Darkness," in www.poets.org.

46. Gillen D'Arcy Wood, *Tambora: The Eruption That Changed the World* (Princeton, N.J.: Princeton University Press, 2014), 12–27; Henry Stommel and Elizabeth Stommel, *Volcano Weather: The Story of 1816, the Year without a Summer* (Newport, R.I.: Seven Seas Press, 1983), 92–100; William P. Baron, "1816 in Perspective: The View from the Northeastern United States," in C. R. Harrington, ed., *The Year without a Summer? World Climate in 1816* (Ottawa: Canadian Museum of Nature,

1992), 124–31; Hubert H. Lamb, *Climate, History, and the Modern World*, 2nd ed. (London: Routledge, 1995), 243–27, and "Volcanic Dust in the Atmosphere: With a Chronology and an Assessment of its Meteorological Significance," *Philosophical Transactions of the Royal Society*, Series A, 166 (1970), 425–533.

Harsh weather was not limited to North America. On China, see Huang Jiayou, "Was There a Colder Summer in China in 1816?," in Harrington, *The Year without a Summer*, 448–52, and Stommel and Stommel, *Volcano Weather*, 44–51. On Europe, see Wolfgang Behringer, *Tambora und des Jahr ohne Sommer: Wie ein Vulkan die Welt in die Kriese stürtzte* (Munich: C. H. Beck, 2015). See also Clive Oppenheimer, "Climatic, Environmental and Human Consequences of the Largest Known Historic Eruption: Tambora Volcano (Indonesia) 1815," *Progress in Physical Geography* 27 no. 2 (June 2003), 230–59; and Richard B. Stothers, "The Great Tambora Eruption of 1815 and Its Aftermath," *Science* 224 no. 4654 (June 15, 1984), 1191–98.

47. On Fulton, see Kirkpatrick Sale, *The Fire of His Genius: Robert Fulton and the American Dream* (New York: Free Press, 2001).

48. Louis C. Hunter, *Steamboats on the Western Rivers: An Economic and Technological History* (Cambridge, Mass.: Harvard University Press, 1949), 8–13, 62, 122–33; Carl Daniel Lane, *American Paddle Steamboats* (New York: Coward-McCann, 1943), 30–33; James T. Flexner, *Steamboats Come True: American Inventors in Action* (New York: Viking, 1944), 344–45; Sale, *Fire of His Genius*, 188.

49. Gordon F. Whitney, *From Coastal Wilderness to Fruited Plain: A History of Environmental Change in Temperate North America, 1500 to the Present* (Cambridge: Cambridge University Press, 1994), 247; Tim Flannery, *The Eternal Frontier: An Ecological History of North America and Its Peoples* (New York: Grove Press, 2001), 326; Opie, *Nature's Nation*, 131–32.

50. Hunter, *Steamboats on the Western Rivers*, 61, 269–70; Sale, *Fire of His Genius*, 188–94;

51. On the Midwestern railroad boom, see William Cronon, *Nature's Metropolis: Chicago and the Great West* (New York: W. W. Norton, 1991), 64–68.

52. Statistics from Williams, *Deforesting the Earth*, 243, Table 9.4.

53. Edmund Russell, "The Nature of Power: Synthesizing the History of Technology and Environmental History," *Technology and Culture*, 52 no. 2 (April 2011), 256–58; Cronon, *Nature's Metropolis*, 99–100, 239; Opie, *Nature's Nation*, 219, 237–39.

54. Alexis de Tocqueville, *Democracy in America*, 3rd ed., trans. H. Reeves (London: Saunders & Otley, 1838), 2: 74, quoted in Williams, *Deforesting the Earth*, 254.

55. Francis Parkman, "The Forest and the Census," *Atlantic Monthly* 55 (1885), quoted in Williams, *Deforesting the Earth*, 255.

56. Opie, *Nature's Nation*, 143.

57. William Beinart and Peter Coates, *Environment and History: The Taming of Nature in the USA and South Africa* (New York: Routledge, 1995), 39; Cronon, *Nature's Metropolis*, 151–52, 178–79; Opie, *Nature's Nation*, 142–44; Wilkinson, *Poverty and Progress*, 153–55.

58. Williams, *Deforesting the Earth*, 231–36, 251.

59. Williams, *Deforesting the Earth*, 231–49; Opie, *Nature's Nation*, 141–43.

60. Michael Williams, *Americans and Their Forests: A Historical Geography* (Cambridge: Cambridge University Press, 1989), 228.

61. Williams, *Americans and Their Forests*, 232.

62. Dan Louie Flores, *Natural West: Environmental History of the Great Plains and Rocky Mountains* (Norman: University of Oklahoma Press, 2001), 57–58. But see also Harold P. Danz, *Of Bison and Man* (Niwot: University Press of Colorado, 1997), 92–93, 113; R. G. Robertson, *Rotting Face: Smallpox and the American Indian* (Caldwell, Idaho: Caxton Press, 2001), 242–46; and Colonel Richard Irving Dodge, *Our Wild Indians: Thirty-Three Years' Personal Experience among the Red Men of the Great West* (Hartford, Conn.: Worthington, 1882; reprint New York: Archer House, 1960), 295–96.

63. Donald Worster, *Dust Bowl: The Southern Plains in the 1930s* (New York: Oxford University Press, 1979), 77.

64. Elliott West, *The Way of the West: Essays on the Central Plains* (Albuquerque: University of New Mexico Press, 1995), 15–17, 58–59, 62–65.

65. James E. Sherow, "Workings of the Geodialectic: High Plains Indians and Their Horses in the Region of the Arkansas River Valley, 1800–1870," *Environmental History Review* 16 (Summer 1992), 61–85; Pekka Hämäläinen, *Comanche Empire* (New Haven, Conn.: Yale University Press, 2008); West, *Way of the West*, 21–35, 74–76.

66. Andrew C. Isenberg, *The Destruction of the Bison: An Environmental History, 1750–1920* (Cambridge: Cambridge University Press, 2000), 93–111; Hämäläinen, *Comanche Empire*, 296–97; West, *Way of the West*, 38–40, 53–54, 72, 79–81; Danz, *Of Bison and Man*, 97–99; Flores, *Natural West*, 58–68.

67. Daniel R. Headrick, *Power over Peoples: Technology, Environments, and Western Imperialism, 1400 to the Present* (Princeton, N.J.: Princeton University Press, 2010), 180–86; Carlos A. Schwantes, *Long Day's Journey: The Steamboat and Stagecoach Era in the Northern West* (Seattle: University of Washington Press, 1999), chs. 1–4; Robertson, *Rotting Face*, 240–42.

68. Cronon, *Nature's Metropolis*, 216–17.

69. On the firearms revolution of the mid-nineteenth century, see Headrick, *Power over Peoples*, 257–65.

70. Headrick, *Power over Peoples*, 276–84; Flores, *Natural West*, 68; Isenberg, *Destruction of the Bison*, 113–19.

71. West, *Way of the West*, 18–19, 30–37, 45–47, 77–78.

72. James H. Shaw, "How Many Bison Originally Populated Western Rangelands," *Rangelands* 17 no. 5 (1995), 148–50.

73. Isenberg, *Destruction of the Bison*, 130–32. See also Cronon, *Nature's Metropolis*, 216.

74. Valerius Geist, *Buffalo Nation: History and Legend of the North American Bison* (Stillwater, Minn.: Voyageur Press, 1996), 75, 84.

75. Quoted in Geist, *Buffalo Nation*, 91.

76. Danz, *Of Bison and Man*, 100–108; Isenberg, *Destruction of the Bison*, 134; Cronon, *Nature's Metropolis*, 216.

77. Danz, *Of Bison and Man*, 104; Flannery, *Eternal Frontier*, 321.

78. Dodge, *Our Wild Indians*, 295–96.

79. Mark V. Barrow Jr., *Nature's Ghosts: Confronting Extinction from the Age of Jefferson to the Age of Ecology* (Chicago: University of Chicago Press, 2009), 108–12; Isenberg, *Destruction of the Bison*, 143; Flannery, *Eternal Frontier*, 322; Danz, *Of Bison and Man*, 109–12; Geist, *Buffalo Nation*, 91–107.

80. Walter Prescott Webb, *The Great Plains* (New York: Grosset & Dunlap, 1931), 207–37; Donald Worster, *Under Western Skies: Nature and History in the American West* (New York: Oxford University Press, 1992), 40–41; Cronon, *Nature's Metropolis*, 218–20; Steinberg, *Down to Earth*, 129–33; Danz, *Of Bison and Man*, 84.

81. Webb, *The Great Plains*, 280–95, 312–17.

82. Worster, *Under Western Skies*, 45–47, and *Dust Bowl*, 83. See also Webb, *The Great Plains*, 236–43, 280, 317; Steinberg, *Down to Earth*, 127–31; and Cronon, *Nature's Metropolis*, 220–21.

83. Donald Worster, *Rivers of Empire: Water, Aridity, and the Growth of the American West* (New York: Oxford University Press, 1985), 76–82.

84. Jessica Teisch, *Engineering Nature: Water, Development, and the Global Spread of American Environmental Expertise* (Chapel Hill: University of North Carolina Press, 2011), 20–21; Steinberg, *Down to Earth*, 174–75.

85. Andrew C. Isenberg, *Mining California: An Ecological History* (New York: Hill and Wang, 2005), 23–28; Steinberg, *Down to Earth*, 117–21.

86. Duane Smith, *Mining America: The Industry and the Environment, 1800-1980* (Lawrence: University of Kansas Press, 1987), 119; Isenberg, *Mining California*, 23–41.

87. Andrew C. Isenberg, "The Industrial Alchemy of Hydraulic Mining: Law, Technology, and Resource-Intensive Industrialization," in Jeffrey M. Diefendorf and Kurk Dorsey, eds., *City, Country, Empire: Landscapes in Environmental History* (Pittsburgh: University of Pittsburgh Press, 2005), 122–37, and *Mining California*, 38.

88. L. P. Brockett, *Our Western Empire: or, The New West beyond the Mississippi* (Philadelphia: Bradley, 1881), 106–7, quoted in Smith, *Mining America*, 6.

89. Robert L. Kelley, *Gold vs. Grain: The Hydraulic Mining Controversy in California's Sacramento Valley: A Chapter in the Decline of Laissez Faire* (Glendale, Cal.: A. H. Clark, 1959), 14, 21–56.

90. Isenberg, *Mining California*, 42–51; Smith, *Mining America*, 68–70.

91. Isenberg, *Mining California*, 25, 46–50.

92. Kelley, *Gold vs. Grain*, 57–84.

93. Kelley, *Gold vs. Grain*, 13.

94. Teisch, *Engineering Nature*, 35–41, 50; Isenberg, *Mining California*, 165–77; Smith, *Mining America*, 68–72.

95. John F. Richards, "Land Transformation," in B. L. Turner II, *The Earth as Transformed by Human Action: Global and Regional Changes in the Biosphere over the Past 300 Years* (Cambridge: Cambridge University Press, 1990), 164, Table 10-1.

96. Mark I. L'Vovich et al., "Use and Transformation of Terrestrial Water Systems," in Turner, *The Earth as Transformed*, 236, Table 14-1.

97. Vaclav Smil, *Energy in World History* (Boulder, Colo.: Westview Press, 1994), 185–87; William F. Ruddiman, *Plows, Plagues, and Petroleum: How Humans Took Control*

of Climate (Princeton, N.J.: Princeton University Press, 2005), 96, 156. See also Paul J. Crutzen, "Geology of Mankind: The Anthropocene," *Nature* 415 (January 3, 2002), 23.

Chapter 9

1. John Gallagher and Ronald Robinson, "The Imperialism of Free Trade," *Economic History Review* 6 no.1 (1953), 1–15.
2. Sir Charles Eliot, *The East Africa Protectorate* (London: E. Arnold, 1905), 4–5, quoted in John M. MacKenzie, "Empire and the Ecological Apocalypse: The Historiography of the Imperial Environment," in Tom Griffiths and Libby Robin, eds., *Ecology and Empire: Environmental History of Settler Societies* (Seattle: University of Washington Press, 1997), 216–17.
3. A. J. H. Latham, *The International Economy and the Underdeveloped World, 1865–1914* (Totowa, N.J.: Rowman & Littlefield, 1978), 71; W. Arthur Lewis, "The Export Stimulus," in W. Arthur Lewis, ed., *Tropical Development, 1880–1913* (London: Allen & Unwin, 1970), 14.
4. See Daniel R. Headrick, *The Tentacles of Progress: Technology Transfer in the Age of Imperialism* (New York: Oxford University Press, 1988).
5. John Christopher Willis, *Agriculture in the Tropics: An Elementary Treatise* (Cambridge: Cambridge University Press, 1909), 38–39.
6. David Mackay, *In the Wake of Cook: Exploration, Science and Empire, 1780–1801* (London: Croom Helm, 1985), 123–40, 168–88.
7. Lucile Brockway, *Science and Colonial Expansion: The Role of the Royal British Botanic Gardens* (New York: Academic Press, 1979); Richard Drayton, *Nature's Government: Science, Imperial Britain and the 'Improvement' of the World* (New Haven, Conn.: Yale University Press, 2000), 171–211; Headrick, *Tentacles of Progress*, 212–15.
8. Camille Limoges, "The Development of the Muséum d'Histoire Naturelle of Paris, c. 1800–1914," in Robert Fox and George Weisz, eds., *The Organization of Science and Technology in France, 1808–1914* (Cambridge: Cambridge University Press 1980), 21–40; Adrien Davy de Virville, ed., *Histoire de la botanique en France* (Paris: Société d'édition d'enseignement supérieur, 1954); Headrick, *Tentacles of Progress*, 222–31.
9. Melchior Treub, "Kurze Geschichte des botanischen Gartens zu Buitenzorg," in *Der botanische Garten "'s Lands Plantentuin" zu Buitenzorg auf Java. Festschrift zur Feier seines 75jährigen Gestehens (1817–1892)* (Leipzig: W. Engelmann, 1893), 23–78; Headrick, *Tentacles of Progress*, 219–22.
10. David Hershey, "Doctor Ward's Accidental Terrarium," *American Biology Teacher* 58 (1996), 276–81; Brockway, *Science and Colonial Expansion*, 86–87.
11. Stuart McCook, *States of Nature: Science, Agriculture, and Environment in the Spanish Caribbean, 1760–1940* (Austin: University of Texas Press, 2002), 83–84.
12. Daniel R. Headrick, "Botany, Chemistry, and Tropical Development," *Journal of World History* 7 no. 1 (Spring 1996), 5–6, and *Tentacles of Progress*, 218–22, 231–50.
13. Lewis, "Export Stimulus," 17–19, 24.

14. Roy Moxham, *Tea: Addiction, Exploitation, and Empire* (London: Constable, 2003), 57; Henry Hobhouse, *Seeds of Change: Five Plants that Transformed Mankind* (New York: Harper & Row, 1985), 95–115; Robert B. Marks, *The Origins of the Modern World: A Global and Ecological Narrative* (Lanham, Md.: Rowman & Littlefield, 2002), 113; Drayton, *Nature's Government*, 249.

15. Jane Pettigrew, *A Social History of Tea* (London: National Trust, 2001), 89; Kalipada Biswas, ed., *150th Anniversary Volume of the Calcutta Royal Botanic Garden* (Alipore: Bengal Government Press, 1942), 56; Brockway, *Science and Colonial Expansion*, 27; Moxham, *Tea*, 91–111.

16. James L. A. Webb, *Tropical Pioneers: Human Agency and Ecological Change in the Highlands of Sri Lanka, 1800–1900* (Athens: Ohio University Press, 2002), 134 Table 5.4; Victor H. Mair and Erling Hoh, *The True History of Tea* (New York: Thames and Hudson, 2009), 219–23; Drayton, *Nature's Government*, 195, 249; Brockway, *Science and Colonial Expansion*, 27–28.

17. Richard P. Tucker, "The Depletion of India's Forests under British Imperialism: Planters, Foresters, and Peasants in Assam and Kerala," in Donald Worster, ed., *The Ends of the Earth: Perspectives on Modern Environmental History* (Cambridge: Cambridge University Press, 1988),120–25,133–36; Michael Williams, *Deforesting the Earth: From Prehistory to Global Crisis* (Chicago: University of Chicago Press, 2003), 340; Brockway, *Science and Colonial Expansion*, 27–28; Moxham, *Tea*, 101–13.

18. Mark Prendergrast, *Uncommon Grounds: The History of Coffee and How It Transformed Our World* (London: Texere, 2001), 15–25.

19. Stanley J. Stein, *Vassouras: A Brazilian Coffee County, 1850–1900* (Cambridge, Mass.: Harvard University Press, 1957); quotation on p. 289. See also Frédéric Mauro, *Histoire du café* (Paris: Desjonquères, 1991), 48–73.

20. Shawn William Miller, *An Environmental History of Latin America* (Cambridge: Cambridge University Press, 2007), 131.

21. William Gervase Clarence-Smith, "Coffee Crisis in Asia, Africa, and the Pacific," in William Gervase Clarence-Smith and Steven Topik, eds., *The Global Coffee Economy in Africa, Asia, and Latin America, 1500–1989* (Cambridge: Cambridge University Press, 2003), 102.

22. Warren Dean, "Deforestation in Southeastern Brazil," in Richard Tucker and John Richards, eds., *Global Deforestation and the Nineteenth-Century World Economy* (Durham, N.C.: Duke University Press, 1983), 62–63.

23. Figures from Barbara Weinstein, *The Amazon Rubber Boom, 1880–1913* (Stanford, Cal.: Stanford University Press, 1983), 9, 218. See also Warren Dean, *Brazil and the Struggle for Rubber: A Study in Environmental History* (Cambridge: Cambridge University Press, 1987), 9, 36–38; Charles C. Mann, *1493: Uncovering the New World Columbus Created* (New York: Alfred A. Knopf, 2011), 249–62; Charles C. Stover, "Tropical Exports," in Lewis, *Tropical Development*, 58.

24. Adam Hochschild, *King Leopold's Ghost: A Story of Greed, Terror, and Heroism in Colonial Africa* (Boston: Houghton Mifflin, 1999), 160–66; Weinstein, *Amazon Rubber Boom*, 26; Mann, *1493*, 256; Drayton, *Nature's Government*, 249.

25. On the extraction—legal or otherwise—of hevea seeds from Brazil, see Dean, *Brazil and the Struggle for Rubber*, 7–23.

26. J. H. Drabble, *Rubber in Malaya: The Genesis of an Industry* (Kuala Lumpur: Oxford University Press, 1973), 117, 205, 215–19; Colin Barlow, *The Natural Rubber Industry: Its Development, Technology, and Economy in Malaysia* (Kuala Lumpur: Oxford University Press, 1978), 26.

27. William Beinart and Lotte Hughes, *Environment and Empire* (Oxford: Oxford University Press, 2007), 234–44; James C. Jackson, *Planters and Speculators: Chinese and European Agricultural Enterprise in Malaya, 1786–1921* (Kuala Lumpur: University of Malaya Press, 1968); Headrick, *Tentacles of Progress*, 245–46; Brockway, *Science and Colonial Expansion*, 158–65.

28. Brockway, *Science and Colonial Expansion*, 108–12.

29. Webb, *Tropical Pioneers*, 134 Table 5.4.

30. Headrick, *Tentacles of Progress*, 232–35; Brockway, *Science and Colonial Expansion*, 114–22; Drayton, *Nature's Government*, 206–11.

31. Daniel R. Headrick, *The Invisible Weapon: Telecommunications and International Politics, 1851–1945* (New York: Oxford University Press, 1991), 28–116.

32. *India Rubber and Gutta Percha and Electrical Trades Journal* (December 8, 1892), 156.

33. Daniel R. Headrick, "Gutta-Percha: A Case of Resource Depletion and International Rivalry," *IEEE Technology and Society Magazine* 6 no. 4 (December 1987), 12–16; William T. Brandt, *India Rubber, Gutta Percha and Balata* (London: Sampson Low, Marston, 1900), 230–36; John Tully, "A Victorian Ecological Disaster: Imperialism, the Telegraph, and Gutta-Percha," *Journal of World History* 20 no. 4 (December 2009): 559–81.

34. Headrick, "Gutta-Percha," 12–14; Tully, "Victorian Ecological Disaster," 577–78.

35. Mahesh Rangarajan, "Imperial Agenda and Indian Forests: The Early History of Indian Forestry, 1800–1878," *Indian Economic and Social History Review* 31 no. 2 (June 1994), 147–67; Satpal Sangwan, "Making of a Popular Debate: The *Indian Forester* and the Emerging Agenda of State Forestry in India," *Indian Economic and Social History Review* 36 no. 2 (1999), 189–92; Christopher Bayly, *Indian Society and the Making of the British Empire* (vol. 2.1 of *The New Cambridge History of India*) (Cambridge: Cambridge University Press, 1988), 138–40.

36. Madhav Gadgil and Ramachandra Guha, *This Fissured Land: An Ecological History of India* (Berkeley: University of California Press, 1993), 120–23; Ramachandra Guha and Madhav Gadgil, "State Forestry and Social Conflicts in British India," *Past and Present* 123 (May 1989), 145; Williams, *Deforesting the Earth*, 337–38; Richard P. Tucker, "The British Colonial System and the Forests of the Western Himalayas," in Tucker and Richards, *Global Deforestation*, 158–59.

37. John M. Hurd, "Railways," in *The Cambridge Economic History of India*, vol. 2, Dharma Kumar and Meghnad Desai, eds., *c. 1757–c. 1970* (Cambridge: Cambridge University Press, 1983), 2: 737–61; Beinart and Hughes, *Environment and Empire*, 112–15.

38. S. Ravi Rajan, *Modernizing Nature: Forestry and Imperial Eco-development, 1800–1950* (Oxford: Oxford University Press, 2006), 7–8; Michael Mann, "Ecological

Change in North India: Deforestation and Agrarian Distress in the Ganga-Yamuna Doab, 1800–1850," in Richard H. Grove, Vinita Damodara, and Satpal Sangwan, eds., *Nature and the Orient: The Environmental History of South and Southeast Asia* (Delhi: Oxford University Press, 1998), 408–14; Bayly, *Indian Society*, 138–39; Tucker, "Depletion of India's Forests," 128–33 and "British Colonial System," 156–59.

39. Ramachandra Guha, *Environmentalism: A Global History* (New York: Longman, 2000), 33–34; Tamara L. Whited, *Northern Europe: An Environmental History* (Santa Barbara, Cal.: ABC-CLIO, 2005), 122–23; Rajan, *Modernizing Nature*, 21, 198, and "Imperial Environmentalism or Environmental Imperialism? European Forestry, Colonial Foresters and the Agendas of Forest Management in British India, 1800–1900," in Grove, Damodara, and Sangwan, *Nature and the Orient*, 326–40; Raymond L. Bryant, "Rationalizing Forest Use in British Burma, 1856–1942," in Grove, Damodara, and Sangwan, *Nature and the Orient*, 829; Williams, *Deforesting the Earth*, 260–62.

40. Gregory A. Barton, *Empire Forestry and the Origins of Environmentalism* (Cambridge: Cambridge University Press, 2002), 38–39, 58–67; Benjamin Weil, "Conservation, Exploitation, and Cultural Change in the Indian Forest Service, 1875–1927," *Environmental History* 11 (2006), 320–32; Gadgil and Guha, *This Fissured Land*, 129–33; Sangwan, "Making of a Popular Debate," 193–99; Beinart and Hughes, *Environment and Empire*, 116–17.

41. Barton, *Empire Forestry*, 73–93; Rangrajan, "Imperial Agenda," 16–65; Weil, "Conservation, Exploitation, and Cultural Change," 320–29; Gadgil and Guha, *This Fissured Land*, 123–33.

42. Weil, "Conservation, Exploitation, and Cultural Change," 333.

43. Guha and Gadgil, "State Forestry and Social Conflicts," 146–58, and Guha, *Environmentalism*, 38–41; Bayly, *Indian Society*, 141–44; Sangwan, "Making of a Popular Debate," 201–3; Beinart and Hughes, *Environment and Empire*, 117–27.

44. John F. Richards and Michelle B. McAlpin, "Cotton Cultivation and Land Clearing in the Bombay Deccan and Karnatak: 1818–1920," in Tucker and Richards, *Global Deforestation*, 84–94; David Arnold, *The Problem of Nature: Environment, Culture and European Expansion* (Oxford: Blackwell, 1996), 183–84; Rajan, *Modernizing Nature*, 10–11, 82–90, 198–99, and "Imperial Environmentalism," 343–58; Richard P. Tucker, "The Depletion of India's Forests under British Imperialism: Planters, Foresters, and Peasants in Assam and Kerala," in Worster, *The Ends of the Earth*, 118–40; Bryant, "Rationalizing Forest Use," 828–45; Weil, "Conservation, Exploitation, and Cultural Change," 337.

45. Guha and Gadgil, "State Forestry and Social Conflicts," 147–50.

46. On the Mughals, see Thomas T. Allsen, *The Royal Hunt in Eurasian History* (Philadelphia: University of Pennsylvania Press, 2006), 188–89, 192–99; on the Manchus, see 45–46.

47. John M. MacKenzie, *The Empire of Nature: Hunting, Conservation and British Imperialism* (Manchester: Manchester University Press, 1988), 148–49; Gregory H. Maddox, *Sub-Saharan Africa: An Environmental History* (Santa Barbara, Cal.: ABC-CLIO, 2006), 108–11, 128.

48. Mahesh Rangarajan, "The Raj and the Natural World: The Campaign against 'Dangerous Beasts' in Colonial India, 1875–1925," in Mahesh Rangarajan and K. Sivaramakrishnan, eds., *India's Environmental History*, vol. 2: *Colonialism, Modernity and the Nation* (Ranikhet: Permanent Black, 2012), 95–142; Beinart and Hughes, *Environment and Empire*, 122.

49. MacKenzie, *Empire of Nature*, 171. See also 169–70 and Mahesh Rangarajan, *Fencing the Forest: Conservation and Ecological Change in India's Central Provinces, 1860–1914* (Delhi: Oxford University Press, 1996), 139–51, 167.

50. Rangarajan, *Fencing the Forest*, 139–40, 152, 184; MacKenzie, *Empire of Nature*, 172; Beinart and Hughes, *Environment and Empire*, 122.

51. D. G. Harris, *Irrigation in India* (London: Oxford University Press, 1923), 71–75, 90–92; Elizabeth Whitcombe, "Irrigation," in Kumar and Desai, *The Cambridge Economic History of India*, vol. 2, 677–737; Aloys Arthur Michel, *The Indus River: A Study of the Effects of Partition* (New Haven, Conn.: Yale University Press, 1967), 84–93, 104–26, 445–54; Alfred Deakin, *Irrigated India: An Australian View of India and Ceylon, Their Irrigation and Agriculture* (London: W. Thacker, 1893); George Walter Macgeorge, *Ways and Works in India: Being an Account of the Public Works in That Country from the Earliest Times up to the Present Day* (Westminster: A. Constable, 1894). For a summary of this period, see Headrick, *Tentacles of Progress*, 171–96.

52. Indu Agnihotri, "Ecology, Land Use, and Colonization: The Canal Colonies of Punjab," in Rangarajan and Sivaramakrishnan, *Colonialism, Modernity, and the Nation*, 37–63; Beinart and Hughes, *Environment and Empire*, 137–38.

53. On indigo, see Prakash Kumar, *Indigo Plantations and Science in Colonial India* (Cambridge: Cambridge University Press, 2012). On palm oil and cacao, see James C. McCann, *Green Land, Brown Land, Black Land: An Environmental History of Africa, 1800–1990* (Portsmouth, N.H.: Heinemann, 1999), 128–30.

54. Clifford Geertz, *Agricultural Involution: The Processes of Ecological Change in Indonesia* (Berkeley: University of California Press, 1963), 69, chs. 4 and 5.

55. Michael Adas, "Colonization, Commercial Agriculture, and the Destruction of the Deltaic Rainforests of British Burma in the Late Nineteenth Century," in Tucker and Richards, *Global Deforestation*, 95–110; Williams, *Deforesting the Earth*, 346–47; Lewis, "Export Stimulus," 21.

56. Yves Henry, *Economie agricole de l'Indochine* (Hanoi: Imprimerie d'Extrême-Orient, 1932), 673–79; Auguste Chevalier, *L'organisation de l'agriculture coloniale en Indochine et dans la Métropole* (Saigon: C. Ardin et fils, 1918), 20–57.

57. Shigeharu Tanabe, "Land Reclamation in the Chao Phraya Delta," in Yoneo Ishii, ed., *Thailand: A Rice-Growing Society* (Honolulu: University of Hawaii Press, 1978), 40–83.

58. "Egyptian Cotton: Its Modern Origin and the Importance of the Supply," *New York Times* (June 26, 1864). Jumal cotton is a form of *Gossypium barbadense* or sea-island cotton; see Edmund Russell, *Evolutionary History: Uniting History and Biology to Understand Life on Earth* (Cambridge: Cambridge University Press, 2011), 119.

59. Robert Pollitzer, *Cholera* (Geneva: World Health Organization, 1959), 11–19; David Arnold, *Colonizing the Body: State Medicine and Epidemic Disease in*

Nineteenth-Century India (Berkeley: University of California Press, 1993), 162; Kerrie L. MacPherson, "Cholera in China 1820–1830: An Aspect of the Internationalization of Infectious Disease," in Mark Elvin and Liu Ts'ui-jung, eds., *Sediments of Time: Environment and Society in Chinese History* (Cambridge: Cambridge University Press, 1998), 488–514; David P. Clark, *Germs, Genes, and Civilization* (Upper Saddle River, N.J.: FT Press, 2010), 71–72.

60. LaVerne Kuhnke, *Lives at Risk: Public Health in Nineteenth-Century Egypt* (Berkeley: University of California Press, 1990), 49–57; John Aberth, *Plagues in World History* (Lanham, Md.: Rowman & Littlefield, 2011), 101–3; Pollitzer, *Cholera*, 21–30; Ann Bowman Jannetta, *Epidemics and Mortality in Early Modern Japan* (Princeton, N.J.: Princeton University Press, 1987), 147–62; MacPherson, "Cholera in China," 494.

61. Arnold, *Colonizing the Body*, 161–66.

62. "Haiti's Latest Misery," *New York Times* (October 26, 2010).

63. Carol Ann Benedict, *Bubonic Plague in Nineteenth-Century China* (Stanford, Cal.: Stanford University Press, 1996), 18–23, 37–38, 49–52; quotation on p. 141. See also Robert Marks, *China: Its Environment and History* (Lanham, Md.: Rowman & Littlefield, 2012), 227–28.

64. Arnold, *Colonizing the Body*, 201–6.

65. Myron J. Echenberg, *Plague Ports: The Global Impact of Bubonic Plague, 1894–1901* (New York: New York University Press, 2007); Aberth, *Plagues*, 61–70; Beinart and Hughes, *Environment and Empire*, 168–75; Headrick, *Tentacles of Progress*, 159–67.

66. Ernest C. Large, *The Advance of the Fungi* (New York: Henry Holt, 1940), 147–58, 196–207.

67. Some readers may object that Ireland is neither a tropical nor a non-Western country. That is true, but in the nineteenth century, it was just as much a colony of Great Britain as Jamaica or Ceylon, and its inhabitants were treated just as harshly as those of any tropical colony.

68. Christine Kinealy, *A Death-Dealing Famine: The Great Hunger in Ireland* (London: Pluto Press, 1997), Cormac Ó'Gráda, *Black '47 and Beyond: The Great Irish Famine in History, Economy, and Memory* (Princeton, N.J.: Princeton University Press, 1999); Joel Mokyr, *Why Ireland Starved: A Quantitative and Analytical History of the Irish Economy, 1800–1850* (Boston: Allen & Unwin, 1983). On the potato blight, see John Reader, *Potato: A History of the Propitious Esculent* (New Haven, Conn.: Yale University Press, 2009), 191–212, and Large, *Advance of the Fungi*, 13–33.

69. Thomas Barrett, Paul-Pierre Pastoret, and William P. Taylor, eds., *Rinderpest and Peste des Petits Ruminants: Virus Plagues of Large and Small Ruminants* (Amsterdam: Academic, 2006), 100; Clive A. Spinage, *Cattle Plague: A History* (New York: Kluwer, 2003), 497–525; John Ford, *The Role of Trypanosomiasis in African Ecology: A Study of the Tsetse Fly Problem* (Oxford: Clarendon Press, 1971), 134–40; Maddox, *Sub-Saharan Africa*, 121–22; MacKenzie, *Empire of Nature*, 158.

70. César N. Caviedes, *El Niño in History: Storming through the Ages* (Gainesville: University of Florida Press, 2001), 122; Peter H. Whetton and Ian Rutherfurd, "Historical ENSO Teleconnections in the Eastern Hemisphere," *Climatic Change* 28 no. 3 (1994), 221–53.

71. Cao Shuji, Yushang Li, and Bin Yang, "Mt. Tambora, Climatic Changes, and China's Decline in the Nineteenth Century," *Journal of World History* 23 no. 3 (2012), 587–607.

72. Marks, *China*, 250.

73. Williams, *Deforesting the Earth*, 237–38. Marks, *Origins of the Modern World*, 103, gives different figures: 225 million in 1750 and 380–400 million in 1850.

74. Kenneth Pomeranz, "The Transformation of the China's Environment," in Edmund Burke III and Kenneth Pomeranz, eds., *The Environment and World History* (Berkeley: University of California Press, 2009), 125. See also Lilian Li, *Fighting Famine in North China: State, Market, and Environmental Decline, 1690s–1990s* (Stanford, Cal.: Stanford University Press, 2007), 19.

75. Kenneth Pomeranz, "Calamities without Collapse: Environment, Economy, and Society in China, ca. 1800–1949," in Patricia A. McAnany and Norman Yoffee, eds., *Questioning Collapse: Human Resilience, Ecological Vulnerability, and the Aftermath of Empire* (Cambridge: Cambridge University Press, 2010), 85–86; Marks, *China*, 230–57.

76. David A. Pietz, *The Yellow River: The Problem of Water in Modern China* (Cambridge, Mass.: Harvard University Press, 2015), 64–69.

77. Rhoads Murphey, "Deforestation in Modern China," in Tucker and Richards, *Global Deforestation*, 118–20.

78. Randall Dodgen, *Controlling the Dragon: Confucian Engineers and the Yellow River in Late Imperial China* (Honolulu: University of Hawaii Press, 2001), 144–46; Pomeranz, "Transformation," 129–32, and "Calamities," 87–89; Marks, *China*, 236–42; Li, *Fighting Famine*, 284.

79. Mike Davis, *Late Victorian Holocausts: El Niño Famines and the Making of the Third World* (London: Verso, 2001).

80. Marks, *China*, 255–57; Pietz, *The Yellow River*, 74–75.

81. Caviedes, *El Niño in History*, 199.

82. Davis, *Late Victorian Holocausts*, 25–29.

83. Angus M. Gunn, *Encyclopedia of Disasters: Environmental Catastrophes and Human Tragedies* (Westport, Conn.: Greenwood Press, 2008), 141–44; James Cornell, *The Great International Disaster Book* (New York: Scribner's, 1976), 133–42; Caviedes, *El Niño in History*, 122–24.

84. Compare the 100,000 people who died outright and the 100,000 who died of disease and famine in the cyclone that hit Chittagong in eastern Bengal on October 31, 1876, with the 6,000–12,000 persons who died in the hurricane that devastated Galveston on September 8, 1900, the worst natural disaster in US history.

Chapter 10

1. Several authors have calculated the world's population. Their estimates vary somewhat, but the differences decrease as they approach the present. Here I have consulted Paul Demeny, "Population," in B. L. Turner et al., eds., *The Earth as Transformed by Human Action: Global and Regional Changes in the Biosphere over the Past 300 Years*

(Cambridge: Cambridge University Press, 1990), 42–44; and Angus Maddison, *The World Economy*, vol 2: *Historical Statistics* (Paris: OECD, 2003), 256–57.

2. Maddison, *World Economy*, 2: 233–34. Readers will object that "average" consumption covers huge disparities among people. That is true; when an average American uses a hundred times more fossil fuel than an average African, that matters a great deal. But from the point of view of the impact on the global environment, what matters is the total, not the disparities.

3. Richard P. Tucker, "The Impact of Warfare on the Natural World: A Historical Survey," in Richard P. Tucker and Edmund Russell, eds., *Natural Enemy, Natural Ally: Toward an Environmental History of War* (Corvallis: Oregon University Press, 2004), 15–41; and "War and the Environment," in John R. McNeill and Erin Stewart Mauldin, eds., *A Companion to Global Environmental History* (Chichester, U.K.: Wiley-Blackwell, 2012), 320–27; John R. McNeill, "Woods and Warfare in World History," *Environmental History* 9 no. 3 (2004), 401.

4. There have been many descriptions of the Western Front. For this one, see Dorothee Brantz, "Environments of Death: Trench Warfare on the Western Front, 1914–1918," in Charles Closmann, ed., *War and the Environment: Military Destruction in the Modern Age* (College Station: Texas A&M University Press, 2009), 68–91.

5. Edmund Russell, " 'Speaking of Annihilation': Mobilizing for War against Human and Insect Enemies, 1914–1945," in Tucker and Russell, *Natural Enemy, Natural Ally*, 146–48; Olivier Lepick, *Grande Guerre chimique, 1914–1918* (Paris: Presses Universitaires de France, 1998).

6. Michael Clodfelter, *Warfare and Armed Conflicts—Statistical Reference to Casualty and Other Figures, 1500–2000*, 2nd ed. (Jefferson, N.C.: McFarland, 2002).

7. Many authors have repeated the estimate of global flu mortality as 50 to 100 million; see, for example John M. Barry, *The Great Influenza: The Story of the Deadliest Pandemic in History* (London: Penguin, 2005), 397, 452; and Andrew T. Price-Smith, *Contagion and Chaos: Disease, Ecology, and National Security in the Era of Globalization* (Cambridge, Mass.: MIT Press, 2009), 57–58. The most careful compilation of mortality statistics is in K. David Patterson and Gerald F. Pyle, "The Geography and Mortality of the 1918 Influenza Pandemic," *Bulletin of the History of Medicine* 65 no. 1 (1991), 4–21.

8. Carol R. Byerly, *The Fever of War: The Influenza Epidemic in the U.S. Army during World War I* (New York: New York University Press 2005), 74–80, 99; Price-Smith, *Contagion and Chaos*, 62–64.

9. Price-Smith, *Contagion and Chaos*, 70–76.

10. Alfred Crosby, *America's Forgotten Pandemic* (Cambridge: Cambridge University Press, 1989), 11, 203–7, 296–97, quotation on p. 311; Tom Quinn, *Flu: A Social History of Influenza* (London: New Holland, 2008), 123–35, 147–50. On East Asia, see Wataru Ijima, "Spanish Influenza in China, 1918–20," in Howard Phillips and David Killingray, eds., *The Spanish Influenza Pandemic of 1918–19: New Perspectives* (New York: Rutledge, 2003), 102–9. On Africa, see Rita Headrick, *Colonialism, Health and Illness in French Equatorial Africa, 1885–1935* (Atlanta: African Studies Association Press, 1994), 170–80.

11. Martin Gilbert, *Winston S. Churchill*, vol. 4: *1916–1922* (London: Heinemann, 1975), 494.

12. Charles Townshend, *Britain's Civil Wars: Counterinsurgency in the Twentieth Century* (Boston: Faber & Faber, 1986), 147–48.

13. David E. Omissi, *Air Power and Colonial Control: The Royal Air Force, 1919–1939* (Manchester: Manchester University Press, 1990), 160; Townshend, *Britain's Civil Wars*, 147–48.

14. Rudibert Kunz and Rolf-Dieter Müller, *Giftgas gegen Abd el Krim: Deutschland, Spanien und der Gaskrieg in Spanisch-Morokko, 1922–1927* (Freiburg: Verlag Rombach, 1990), 58–59, 74–90.

15. Sebastian Balfour, *Deadly Embrace: Morocco and the Road to the Spanish Civil War* (New York: Oxford University Press, 2002), 124–56; David S. Woolman, *Rebels in the Rif: Abd el Krim and the Rif Rebellion* (Stanford, Cal.: Stanford University Press, 1968), 196, 204–5; James S. Corum and Wray R. Johnson, *Airpower in Small Wars: Fighting Insurgents and Terrorists* (Lawrence: University of Kansas Press, 2003), 72–77.

16. Sven Lindqvist, *A History of Bombing*, trans. Linda Haverty Rugg (New York: New Press, 2001), 70; Giorgio Rochat, *Guerre italiane in Libia e in Etiopia: Studi militari, 1921–1939* (Treviso: Pagus, 1991), 143–76.

17. Diana Lary, "The Waters Covered the Earth: China's War-Induced Natural Disaster," in Mark Selden and Alvin Y. So, eds., *War and State Terrorism: The United States, Japan, and the Asia-Pacific in the Long Twentieth Century* (Lanham, Md.: Rowman & Littlefield, 2004), 143–70; Robert Marks, *China: Its Environment and History* (Lanham, Md.: Rowman & Littlefield, 2012), 261–62; David A. Pietz, *Engineering the State: The Huai River and Reconstruction in Nationalist China, 1927–1937* (London: Routledge, 2002), 106; Micah Muscolino, "Conceptualizing Wartime Flood and Famine in China," in Simo Laakkonen, Richard Tucker, and Timo Vuorisalo, eds., *The Long Shadows: A Global Environmental History of the Second World War* (Corvallis: Oregon State University Press, 2017), 97–115; quote on 97.

18. Peter Cotgreave and Irwin Forseth, *Introductory Ecology* (Oxford: Blackwell Science, 2002), 247–51; Edward O. Wilson, *The Diversity of Life* (Cambridge, Mass.: Harvard University Press, 1992), 226–28.

19. Susan D. Lanier-Graham, *The Ecology of War: Environmental Impacts of Weaponry and Warfare* (New York: Walker, 1993), 26–31.

20. Paul Josephson, "The Costs of War for the Soviet Union," in Laakkonen, Tucker, and Vuorisalo, *The Long Shadows*, 75–96.

21. Richard P. Tucker, "The World Wars and the Globalization of Timber Cutting," 110–37, and "The Impact of Warfare," 30–32, in Tucker and Russell, *Natural Enemy, Natural Ally;* "War and the Environment," in McNeill and Mauldin, *Companion to Global Environmental History,* 329–30; and "The Depletion of India's Forests under British Imperialism: Planters, Foresters, and Peasants in Assam and Kerala," in Donald Worster, ed., *The Ends of the Earth: Perspectives on Modern Environmental History* (Cambridge: Cambridge University Press, 1988), 126–27; Westing, *Warfare in a Fragile World,* 53–58; Michael Williams, *Deforesting the Earth: From Prehistory to Global Crisis* (Chicago: University of Chicago Press, 2003), 388; William M. Tsutsui,

"Landscapes in the Dark Valley: Toward an Environmental History of Wartime Japan," in Tucker and Russell, *Natural Enemy, Natural Ally*, 195–216.

22. Arthur Westing, *Warfare in a Fragile World: Military Impact on the Human Environment* (London: Taylor and Francis, 1980), 4–5, 193, and *Ecological Consequences of the Second Indochina War* (Stockholm: Almqvist and Wiksell, 1976), 12–13, 46–49.

23. Arthur W. Westing, "Herbicidal Damage to Cambodia," in *Harvest of Death: Chemical Warfare in Vietnam and Cambodia* (New York: Free Press, 1972), 188–96, and *Ecological Consequences*, 24–38; Jeann Moger Stellman et al., "The Extent and Pattern of Usage of Agent Orange and Other Herbicides in Vietnam," 681–85, and Declan Butler, "Flight Records Reveal Full Extent of Agent Orange Contamination," 649, both in *Nature* 422 no. 6933 (April 17, 2003); A. K. Orians and E. W. Pfeiffer, "Ecological Effects of the War in Vietnam," *Science* 168 no. 3931 (1970), 545–54; Ronald M. Nowak, "Wildlife of Indochina: Tragedy or Opportunity?" *National Parks and Conservation Magazine* 50 no. 6 (1976), 14; J. A. McNeely, "Biodiversity, War, and Tropical Forests," *Journal of Sustainable Forestry* 16 no. 3 (2003), 9; Philip Jones Griffiths, *Agent Orange: 'Collateral Damage' in Viet Nam* (London: Trolley, 2005), 16–20, 164–69.

24. John R. McNeill and Peter Engelke, *The Great Acceleration: An Environmental History of the Anthropocene since 1945* (Cambridge: Cambridge University Press, 2014), 161–65.

25. Merrill Eisenbud, "The Ionizing Radiation," in Turner et al., *The Earth as Transformed*, 456.

26. Kate Brown, *Plutopia: Nuclear Families, Atomic Cities, and the Great Soviet and American Plutonium Disasters* (Oxford: Oxford University Press, 2013), 4–7; John M. Findlay and Bruce Hevly, *Atomic Frontier Days: Hanford and the American West* (Seattle: University of Washington Press, 2011).

27. Jacob Darwin Hamblin, *Poison in the Well: Radioactive Waste in the Oceans at the Dawn of the Nuclear Age* (New Brunswick, N.J.: Rutgers University Press, 2008), 177–79; Paul Josephson, "Technology and the Environment," in McNeill and Mauldin, *Companion to Global Environmental History*, 353; McNeill, *Something New under the Sun: An Environmental History of the 20th-Century World* (New York: W.W. Norton, 2000), 342–44; Brown, *Plutopia*, 150–57.

28. Brown, *Plutopia*, 1–7.

29. Nowak, "Wildlife of Indochina," 18.

30. Nowak, "Wildlife of Indochina," 15–18; Orians and Pfeiffer, "Ecological Effects," 548–53; Westing, "Herbicidal Damage," 196–97, and *Ecological Consequences*, 18–19, 32–36.

31. Orians and Pfeiffer, "Ecological Effects," 553.

32. "Rocky Flats National Wildlife Refuge," "Rocky Flats Plant," and "Vieques, Puerto Rico," Wikipedia (accessed December 2012).

33. Gordon L. Rottman, *The Berlin Wall and the Intra-German Border, 1961–89* (Oxford: Osprey, 2008), 23–29.

34. Lisa M. Brady, "Life in the DMZ: Turning Diplomatic Failure into an Environmental Success," *Diplomatic History* 32 no. 4 (September 2008), 585–88. I thank Jenifer Van Vleck for bringing this article to my attention.

35. Or, as Kenneth Boulding put it: "Anyone who believes exponential growth can go on forever in a finite world is either a madman or an economist," in "The Economics of the Coming Spaceship Earth," in Victor D. Lippit, ed., *Radical Political Economy: Exploration in Alternative Economic Analysis* (Armonk, N.Y.: M.E. Sharpe, 1996), 362. Among the few economists who question the idea of infinite growth is Herman E. Daly; see his "Economics in a Full World," *Scientific American* (September 2005), 100–107, and *Beyond Growth: The Economics of Sustainable Development* (Boston: Beacon Press, 1996).

36. McNeill, *Something New under the Sun*, 336.

37. Richard J. Samuels, *"Rich Nation, Strong Army": National Security and the Technological Transformation of Japan* (Ithaca, N.Y.: Cornell University Press, 1994).

38. I owe the concept of "developmentalism" to Robert Marks, *China*, 71.

39. McNeill and Engelke, *The Great Acceleration*, 193–96.

40. Quoted in C. Wright Mills, *The Marxists* (Harmondsworth: Penguin Books, 1963), 278–79.

41. Quoted in Murray Feshbach and Alfred Friendly Jr., *Ecocide in the USSR: Health and Nature under Siege* (New York: Basic Books, 1992), 43.

42. Paul Josephson, *Industrialized Nature: Brute Force Technology and the Transformation of the Natural World* (Washington: Island Press, 2002), 18–36.

43. Feshbach and Friendly, *Ecocide in the USSR*; Boris Komarov, *The Destruction of Nature in the Soviet Union* (White Plains, N.Y.: M. E. Sharpe, 1980).

44. Feshbach and Friendly, *Ecocide in the USSR*, 57–59.

45. Feshbach and Friendly, *Ecocide in the USSR*, 73–76. On inland fish and fishing during the Soviet era, see Paul R. Josephson, "When Stalin Learned to Fish: Natural Resources, Technology, and Industry under Socialism," in Jeffrey M. Diefendorf and Kurk Dorsey, eds., *City, Country, Empire: Landscapes in Environmental History* (Pittsburgh, Pa.: University of Pittsburgh Press, 2005), 162–92.

46. Komarov, *The Destruction of Nature*, 3–19.

47. Feshbach and Friendly, *Ecocide in the USSR*, 116–24; see also Komarov, *The Destruction of Nature*, 3–16, 37–38.

48. Komarov, *The Destruction of Nature*, 20–45; Glenn E. Curtis, *Russia: A Country Study* (Washington: Library of Congress, 1998), 139.

49. Feshbach and Friendly, *Ecocide in the USSR*, 99–100; Komarov, *The Destruction of Nature*, 20–31.

50. Douglas R. Weiner, *Models of Nature: Ecology, Conservation and Cultural Revolution in Soviet Russia* (Bloomington: Indiana University Press, 1988), 248; Feshbach and Friendly, *Ecocide in the USSR*, 157–58; Josephson, "The Costs of War."

51. McNeill and Engelke, *The Great Acceleration*, 168–72.

52. Elizabeth Economy, *The River Runs Black: The Environmental Challenge to China's Future*, 2nd ed. (Ithaca, N.Y.: Cornell University Press, 2010), 49.

53. Economy, *The River Runs Black*, 52.

54. Judith Shapiro, *Mao's War against Nature: Politics and Environment in Revolutionary China* (Cambridge: Cambridge University Press, 2001), 75–85; Marks, *China*, 285; Economy, *The River Runs Black*, 52.

55. Bao Maohong, "Environmentalism and Environmental Movements in China since 1949," in McNeill and Mauldin, *Companion to Global Environmental History*, 476–77.

56. Frank Dikötter, *Mao's Great Famine: The History of China's Most Devastating Catastrophe, 1958–62* (New York: Walker, 2010).

57. Shapiro, *Mao's War against Nature*, 76–91; Marks, *China*, 270–74; Economy, *The River Runs Black*, 50–52.

58. Shapiro, *Mao's War against Nature*, 5, 115; Marks, *China*, 285–86.

59. Shapiro, *Mao's War against Nature*, 13; Marks, *China*, 267–68, 290–91; Edward B. Vermeer, "Environmental History—Modern," *Berkshire Encyclopedia of China* 2: 736–46.

60. Williams, *Deforesting the Earth*, 435–37.

61. Paul W. Richards, "The Tropical Rain Forest," *Scientific American* 229 (1973): 58–67; Wilson, *Diversity of Life*, 273.

62. Philip M. Fearnside, "Deforestation in Brazilian Amazonia: History, Rates, and Consequences," *Conservation Biology* 19 no. 3 (2005), 680–88.

63. Williams, *Deforesting the Earth*, 434–62; Susanna Hecht and Alexander Cockburn, *The Fate of the Forest: Developers, Destroyers, and Defenders of the Amazon* (London: Verso, 1989), 100, 110–22, 133, 141; Hemming, *Tree of Rivers*, 289–96, 306–12; Josephson, *Industrialized Nature*, 141–71.

64. The figures vary somewhat. Kenneth Pomeranz, "Advanced Agriculture," in Jerry H. Bentley, ed., *The Oxford Handbook of World History* (Oxford: Oxford University Press, 2011), 257–58, gives the figures of 40 million hectares in 1900 and 280 million in 2000; Boris G. Rozanov, Viktor Targulian, and B. S. Orlov, "Soils," in Turner et al., *The Earth as Transformed*, 210, say 480,000 square kilometers (48 million hectares) in 1900 and 2.2 million square kilometers (220 million hectares) in 1984; Mark I. L'Vovich and Gilbert F. White, "Use and Transformation of Terrestrial Water Systems," in Turner et al., *The Earth as Transformed*, 242, say 2,509,000 square kilometers (250.9 million hectares) in 1985.

65. D. G. Harris, *Irrigation in India* (London: Oxford University Press, 1923), 71–75, 90–92; Elizabeth Whitcombe, "Irrigation," in Dharma Kumar, ed., *The Cambridge Economic History of India*, vol. 2: *c. 1857–c.1970* (Cambridge: Cambridge University Press, 2008), 677–737; Aloys Arthur Michel, *The Indus River: A Study of the Effects of Partition* (New Haven, Conn.: Yale University Press, 1967), 84–93, 104–26, 445–54; Alfred Deakin, *Irrigated India: An Australian View of India and Ceylon, Their Irrigation and Agriculture* (London: W. Thacker, 1893); George Walter Macgeorge, *Ways and Works in India: Being an Account of the Public Works in That Country from the Earliest Times up to the Present Day* (Westminster: A. Constable, 1894). For a summary of this period, see Daniel R. Headrick, *The Tentacles of Progress: Technology Transfer in the Age of Imperialism, 1850–1940* (New York: Oxford University Press, 1988), 171–96.

66. "Narmada River," Wikipedia (accessed April 2013).

67. Elizabeth Whitcombe, "The Environmental Costs of Irrigation in British India: Waterlogging, Salinity, Malaria," in David Arnold and Ramachandra Guha,

eds., *Nature, Culture, Imperialism: Essays on the Environmental History of South Asia* (Delhi: Oxford University Press, 1995), 237–59.

68. Rohan D'Souza, "Water in British India: The Making of a 'Colonial Hydrology,'" *History Compass* 4 (July 2006), 625.

69. Edward Goldsmith, *Social and Environmental Effects of Large Dams* (San Francisco: Sierra Club Books, 1986), 246.

70. Rushdi Said, *The Nile: Geology, Hydrology, and Utilization* (Oxford: Pergamon Press, 1993), 168, 213–18; Robert O. Collins, *The Nile* (New Haven, Conn.: Yale University Press, 2002), 140–81; Harold E. Hurst, *The Nile: A General Account of the River and the Utilization of Its Waters*, rev. ed. (London: Constable,1957), 28–54; John Waterbury, *Hydropolitics of the Nile Valley* (Syracuse, N.Y.: Syracuse University Press, 1976), 26–33; Edmund Burke III, "The Transformation of the Middle Eastern Environment, 1500 B.C.E.–2000 C.E.," in Edmund Burke III and Kenneth Pomeranz, eds., *The Environment and World History* (Berkeley: University of California Press, 2009), 102; Ian Douglas, "Sediment Transfer and Siltation," in Turner et al., *The Earth as Transformed*, 229; Headrick, *Tentacles of Progress*, 196–206.

71. Donald Worster, *Rivers of Empire: Water, Aridity, and the Growth of the American West* (New York: Oxford University Press, 1985), 160–70.

72. Marc Reisner, *Cadillac Desert: The American West and Its Disappearing Water*, 2nd ed. (New York: Penguin, 1993), 165–66; Donald Worster, *Under Western Skies: Nature and History in the American West* (New York: Oxford University Press, 1992), 56.

73. Reisner, *Cadillac Desert*, 126–44.

74. Reisner, *Cadillac Desert*, 151–52; Worster, *Under Western Skies*, 61; John Opie, *Nature's Nation: An Environmental History of the United States* (Fort Worth, Tex.: Harcourt Brace, 1998), 328–29.

75. Josephson, *Industrialized Nature*, 41–53; Reisner, *Cadillac Desert*, 156–65.

76. Reisner, *Cadillac Desert*, 168; see also Worster, *Under Western Skies*, 56.

77. United States Bureau of Reclamation, "Hoover Dam FAQs," at www.usbr.gov/lc/hooverdam/faqs/lakefaqs.html, accessed April 2013; Opie, *Nature's Nation*, 324; Reisner, *Cadillac Desert*, 120, 473–75.

78. Reisner, *Cadillac Desert*, 121; Worster, *Rivers of Empire*, 321–23; Opie, *Nature's Nation*, 330–37.

79. Reisner, *Cadillac Desert*, 158, 485, 501–11; Worster, *Under Western Skies*, 46.

80. Kenneth Pomeranz, "The Transformation of China's Environment, 1500–2000," in Burke and Pomeranz, *The Environment and World History*, 141–47; Economy, *The River Runs Black*, 52; Shui Fu, " A Profile of Dams in China," in Qing Dai, comp., and John G. Thibodeau and Philip B. Williams, eds., *The Red Dragon Has Come! The Three Gorges Dam and the Fate of China's Yangtze River and Its People* (Armonk, N.Y.: M. E. Sharpe, 1998), 18–22; Economy, *The River Runs Black*, 52.

81. Luna B. Leopold, "Sediment Problems at the Three Gorges Dam," in Dai Qing, *Red Dragon*, 194–99; Douglas, "Sediment Transfer and Siltation," in Turner et al., *The Earth as Transformed*, 217–28; Zuo Dakang and Zhang Peiyuan, "The Huang-Huai-Hai Plain," in Turner et al., *The Earth as Transformed*, 475–76; Marks, *China*, 300–304; Shang Wei, "A Lamentation for the Yellow River: The Three Gate Gorge Dam

(Sanmenxia)," in Dai Qing, *Red Dragon*, 143–59; Shapiro, *Mao's War against Nature*, 62–63; Pietz, *Engineering the State*, 183–93, 211–26.

82. Yi Si, "The World's Most Catastrophic Dam Failures: The August 1975 Collapse of the Banqiao and Shimantan Dams," in Dai Qing, *Red Dragon*, 25–38; Marks, *China*, 299–302; Shapiro, *Mao's War against Nature*, 63–64; Fu Shui, "A Profile of Dams," in Dai Qing, *Red Dragon*, 22–23.

83. Economy, *The River Runs Black*, 69–71; Pomeranz, "Transformation," 140; Marks, *China*, 303–4; Vaclav Smil, *Energy in World History* (Boulder, Colo.: Westview Press, 1994), 190.

84. Shapiro, *Mao's War against Nature*, 204–5.

85. Unlike American politicians, who are all lawyers, the recent leaders of China— Premiers Li Peng (1988–1998), Zhu Rongji (1998–2003), and Wen Jiabao (2003–2013) and General Secretaries Jiang Zemin (1998–2002) and Hu Jintao (2002–2012)—were either engineers or geologists.

86. Dai Qing, "The Three Gorges Project: A Symbol of Uncontrolled Development in the Late Twentieth Century," in Dai Qing, *Red Dragon*, 3–24; Economy, *The River Runs Black*, 67–68; Audrey Ronning Topping, "Foreword" to Dai Qing, *Red Dragon*, xvii– xx; Gorild Heggelund, *Environment and Resettlement Politics in China: The Three Gorges Project* (Burlington, Vt.: Ashgate, 2004); Marks, *China*, 304–308; Pomeranz, "Transformation," 143–46.

87. Edward Wong, "Plans for China's Water Crisis Spur Concern," *New York Times* (June 1, 2011); Marks, *China*, 304.

88. Economy, *The River Runs Black*, 68–72; Marks, *China*, 302–12.

Chapter 11

1. The best social and cultural history of consumerism is Frank Trentmann, *Empire of Things: How We Became a World of Consumers, from the Fifteenth Century to the Twenty-First* (New York: HarperCollins, 2016), but it neglects the environmental impacts.

2. Henry Ford, *My Life and Work* (1922), ch. 4, quoted in "Henry Ford," in Wikiquotes (January 2013).

3. John B. Rae, *The Road and the Car in American Life* (Cambridge, Mass.: MIT Press, 1971), 40–44, 50; James C. Flink, *The Car Culture* (Cambridge, Mass.: MIT Press, 1975), 141.

4. Bernhard Rieger, "The Automobile," in John R. McNeill and Kenneth Pomeranz, eds., in *The Cambridge World History*, vol. 7: *Production, Destruction, and Connection, 1750–Present*, part 2: *Shared Transformations?* (Cambridge: Cambridge University Press, 2015), 478.

5. Flink, *The Car Culture*, 147–49.

6. Robert S. Lynd and Helen M. Lynd, *Middletown: A Study in American Culture* (New York: Harcourt Brace, 1929), 254–56.

7. Flink, *The Car Culture*, 155–60; Rae, *The Road and the Car*, 73–74, 134–35.

8. Official statistics since 1960 are found in US Department of Transportation, Federal Highway Administration, Highway Statistics Series, 2010, Chart DV-1C in www.fhwa.dot.gov and US Department of Transportation, Office of the Assistant Secretary for Research and Technology, Table 1-23: "World Motor Vehicle Production," in www.rita.dot.gov; and World Bank, "Motor Vehicles (per 1,000 people)," in http://data.worldbank.org/indicator/IS.VEH.NVEH.P3. Other figures are from Flink, *The Car Culture*, 189, and *The Automobile Age* (Cambridge, Mass.: MIT Press, 1990), 15, 300, 359; 210–31; David E. Nye, *Consuming Power: A Social History of American Energies* (Cambridge, Mass.: MIT Press, 1999), 205–7; Rae, *The Road and the Car*, 51, 94–95; and John Heitman, *The Automobile and American Life* (Jefferson, N.C.: McFarland, 2009),190.

9. Flink, *The Automobile Age*, 285–88, and *The Car Culture*, 194–98.

10. Bill Vlasic, "Bigger, Faster, More Lavish: Americans Crave S.U.V.s, and Carmakers Oblige," *New York Times* (April 12, 2017).

11. US Department of Energy, Transportation Energy Data Book, Table 2.11: Highway Usage of Gasoline and Diesel, 1973–2010 in cta.ornl.gov/data/chapter2.shtml; Ted Steinberg, *Down to Earth: Nature's Role in American History*, 2nd ed. (New York: Oxford University Press, 2009), 287.

12. Christian Pfister, "The '1950s Syndrome' and the Transition from a Slow-Going to a Rapid Loss of Sustainability," in Frank Uekoetter, ed., *The Turning Points of Environmental History* (Pittsburgh, Pa.: University of Pittsburgh Press, 2010), 90–118.

13. Thomas Zeller, *Driving Germany: The Landscape of the German Autobahn, 1930–1970* (New York: Berghahn Books, 2007), 184; Trentmann, *Empire of Things*, 301.

14. Daniel Sperling and Deborah Gordon, *Two Billion Cars: Driving Toward Sustainability* (New York: Oxford University Press, 2009), 4, 13.

15. Wanda James, *Driving from Japan: Japanese Cars in America* (Jefferson, N.C.: McFarland, 2005), 23.

16. US Department of Transportation, Office of the Assistant Secretary for Research and Technology, Table 1-23: "World Motor Vehicle Production," in www.rita.dot.gov; World Bank, "Motor Vehicles (per 1,000 people)," in http://data.worldbank.org/indicator/IS.VEH.NVEH.P3.

17. Rae, *The Road and the Car*, 62–83; Steinberg, *Down to Earth*, 210, 242–43.

18. Flink, *The Automobile Age*, 371, and *The Car Culture*, 190; Rae, *The Road and the Car*, 173–89; Nye, *Consuming Power*, 206.

19. Paul Josephson, "Technology and the Environment," in John R. McNeill and Erin Stewart Mauldin, eds., *A Companion to Global Environmental History* (Chichester, U.K.: Wiley-Blackwell, 2012), 351.

20. Jane Holtz Kay, *Asphalt Nation: How the Automobile Took Over America and How We can Take It Back* (New York: Crown, 1997), 83; Jean-Paul Rodrigue, *Geography of Transport Systems* (Thousand Oaks, Cal.: Sage, 2013). See also John Opie, *Nature's Nation: An Environmental History of the United States* (Fort Worth, Tex.: Harcourt Brace, 1998), 261–64.

21. Zeller, *Driving Germany*, 48–59; Josephson, "Technology and the Environment," 351.

22. Steinberg, *Down to Earth*, 215–17; Rae, *The Road and the Car*, 225–28.

23. Paul Robbins, *Lawn People: How Grasses, Weeds, and Chemicals Make Us Who We Are* (Philadelphia: Temple University Press, 2007), xviii, 22–29. Full disclosure: the author of these lines is one of those lawn-obsessed suburban homeowners justly derided by city-dwelling critics.

24. Kenneth T. Jackson, *Crabgrass Frontier: The Suburbanization of the United States* (New York: Oxford University Press, 1985), 246–71.

25. Rae, *The Road and the Car*, 101; Flink, *The Automobile Age*, 304–5.

26. Jackson, *Crabgrass Frontier*, 264–65; Kay, *Asphalt Nation*, 234–35; Nye, *Consuming Power*, 207.

27. Rae, *The Road and the Car*, 92–93.

28. Alan Davidson, "What's It Gonna Be, 2013?" *New York Times Magazine* (January 6, 2013), 16.

29. Rae, *The Road and the Car*, 141–43; Nye, *Consuming Power*, 222.

30. Flink, *The Automobile Age*, 375–76.

31. Daniel R. Headrick, *The Tentacles of Progress: Technology Transfer in the Age of Imperialism, 1850–1940* (New York: Oxford University Press, 1988), 55.

32. Richard Tucker, "Rubber," in McNeill and Pomeranz *The Cambridge World History*, 7 part 2: *Shared Transformations?*, 423–43.

33. Greg Grandin, *Fordlandia: The Rise and Fall of Henry Ford's Forgotten Jungle City* (New York: Metropolitan Books, 2009); Susanna Hecht and Alexander Cockburn, *The Fate of the Forest: Developers, Destroyers, and Defenders of the Amazon* (London: Verso, 1989), 85–87. See also Warren Dean, *Brazil and the Struggle for Rubber* (Cambridge: Cambridge University Press, 1987), 71–73; Paul R. Josephson, *Industrialized Nature: Brute Force Technology and the Transformation of the Natural World* (Washington, D.C.: Island Press, 2002), 134–40; and John Hemming, *Tree of Rivers: The Story of the Amazon* (London: Thames and Hudson, 2008), 265–68.

34. John R. McNeill, *Something New under the Sun: An Environmental History of the Twentieth-Century World* (New York: W. W. Norton, 2000), 298.

35. Daniel Yergin, *The Quest: Energy, Security, and the Remaking of the Modern World* (New York: Penguin, 2011), 426.

36. Yergin, *The Quest*, 231; Vaclav Smil, *Energy in World History* (Boulder, Colo.: Westview Press, 1994), 167–68.

37. On the strategic aspects of oil before World War II, see Yergin, *The Quest*, 232–33.

38. A barrel, the conventional measure of oil volume, is approximately 192 liters or 42 US gallons.

39. Yergin, *The Prize: The Epic Quest for Oil, Money, and Power* (New York: Touchstone, 1993), 792, and *The Quest*, 233–40; Pfister, "The '1950s Syndrome,'" 92–117.

40. Yergin, *The Prize*, 715, 792, and *The Quest*, 234.

41. Yergin, *The Quest*, 405, and *The Prize*, 769.

42. Yergin, *The Quest*, 241–42.

43. Yergin, *The Quest*, 229–43; quotation on 242. See also International Energy Agency, *2012 World Energy Outlook* (Paris: IEA, 2012).

44. Yergin, *The Quest*, 244–47; Smil, *Energy*, 172.

45. Yergin, *The Quest*, 254–55; Joel K. Bourne, "The Gulf of Oil: The Deep Dilemma," *National Geographic* (October 2010), 44.

46. Yergin, *The Quest*, 255–59.

47. David Biello, "What the Frack? Natural Gas from Subterranean Shale Promises U.S. Energy Independence—With Environmental Costs," *Scientific American* (March 30, 2010); Abrahm Lustgarten, "Are Fracking Wastewater Wells Poisoning the Ground Beneath Our Feet?" *Scientific American* (June 21, 2012).

48. "Seawise Giant," in www.relevantsearchscotland.co.uk (accessed July 2016) and Wikipedia (accessed June 2017.

49. "Carbon Dioxide Information Analysis Center," in cdiac.ornl.gov (accessed July 2016).

50. Jack Doyle, *Taken for a Ride: Detroit's Big Three and the Politics of Pollution* (New York: Four Walls Eight Windows, 2000), 18–19, 325–30.

51. Flink, *The Car Culture*, 173, 222, 386–87; Steinberg, *Down to Earth*, 206–8; Josephson, "Technology and the Environment," 351.

52. Kay, *Asphalt Nation*, 82–86.

53. Flink, *The Car Culture*, 223–25; Kay, *Asphalt Nation*, 80–81.

54. George A. Gonzalez, *The Politics of Air Pollution: Urban Growth, Ecological Modernization, and Symbolic Inclusion* (Albany: State University of New York Press, 2005); John R. McNeill and Peter Engelke, *The Great Acceleration: An Environmental History of the Anthropocene since 1945* (Cambridge, Mass.: Harvard University Press, 2014), 23–27.

55. Shawn William Miller, *An Environmental History of Latin America* (Cambridge: Cambridge University Press, 2007), 178–80. Rieger, "The Automobile," pp. 484–85, gives the figure of 12,500 deaths per year due to pollution in Mexico City.

56. Meera Subramanian, "Delhi's Deadly Air," *Nature* 534 no. 7606 (June 9, 2016), 166–69.

57. Joanna Burger, *Oil Spills* (New Brunswick, N.J.: Rutgers University Press, 1997), 38–61; Yergin, *The Quest*, 246–51.

58. There is a large and growing number of books (not to mention myriad articles) on the Deepwater Horizon oil spill. See in particular William R. Freudenberg and Robert Grambling, *Blowout in the Gulf: The BP Oil Spill Disaster and the Future of Energy in America* (Cambridge, Mass.: MIT Press, 2011); Joel Achenbach, *The Hole at the Bottom of the Sea: The Race to Kill the BP Oil Gusher* (New York: Simon and Schuster, 2011); Carl Safina, *A Sea in Flames: The Deepwater Horizon Oil Blowout* (New York: Crown, 2011); Bob Canvar, *Disaster on the Horizon: High Stakes, High Risks, and the Story behind the Deepwater Well Blowout* (White River Junction, Vt.: Chelsea Green, 2010); John Konrad and Tom Shroder, *Fire on the Horizon: The Untold Story of the Gulf Oil Disaster* (New York: Harper, 2011); and Antonia Juhasz, *Black Tide: The Devastating Impact of the Gulf Oil Spill* (Hoboken, N.J.: Wiley, 2011).

59. Burger, *Oil Spills*, 55.

60. Burger, *Oil Spills*, 69–73.

61. Bourne, "Gulf of Oil," 52–53.

62. Burger, *Oil Spills*, 79–89.

63. David Montgomery, *Dirt: The Erosion of Civilization* (Berkeley: University of California Press, 2007), 155, 170–74; Boris G. Rozanov, Viktor Targulian, and B. S. Orlov, "Soils," in B. L. Turner et al., *The Earth as Transformed by Human Action: Global and Regional Changes in the Biosphere over the Past 300 Years* (Cambridge: Cambridge University Press, 1990), 205.

64. Donald Worster, *The Dust Bowl: The Southern Plains in the 1930s* (New York: Oxford University Press, 1979), 69–72; John Opie, *Ogallala: Water for a Dry Land*, 2nd ed. (Lincoln: University of Nebraska Press, 1993), 39–40; W. Lockeretz, "The Lessons of the Dust Bowl," *American Scientist* 66 no. 5 (1978), 564; William E. Riebsame, "The United States Great Plains," in Turner, *The Earth as Transformed*, 561–62; Benjamin I. Cook et al., "Amplification of the North American 'Dust Bowl' Drought through Human-Induced Land Degradation," *Proceedings of the National Academy of Sciences of the USA* 106 no. 13 (2009), 4997–5001' Richard Seager and Benjamin I. Cook, "The Dust Bowl," in Peter Bobrowsky, ed., *Encyclopedia of Natural Hazards* (Dordrecht, Netherlands: Springer, 2013), 197–201.

65. Worster, *Dust Bowl*, 88–92; Opie, *Ogallala*, 95; Montgomery, *Dirt*, 146–51; Nye, *Consuming Power*, 188–91; Lockeretz, "Lessons of the Dust Bowl," 565.

66. Montgomery, *Dirt*, 148, says 40 million acres (16 million hectares) in the Midwest; Donald Worster, *Under Western Skies: Nature and History in the American West* (New York: Oxford University Press, 1992), 98–99, says 11 million acres (4.5 million hectares) in Kansas, Colorado, Nebraska, Oklahoma, and Texas.

67. Worster, *Under Western Skies*, 100, and *Dust Bowl*, 94.

68. Walter Prescott Webb, *The Great Plains* (New York: Grosset & Dunlap, 1931), 371.

69. Donald E. Green, *The Land of the Underground Rain: Irrigation on the Texas High Plains, 1910–1970* (Austin: University of Texas Press, 1973), 123; Opie, *Ogallala*, 98–99; Lockeretz, "Lessons of the Dust Bowl," 565.

70. Richard Seager and Benjamin I. Cook, "The Dust Bowl," in Peter Bobrowsky, ed., *Encyclopedia of Natural Hazards* (Dordrecht, Netherlands: Springer 2013), 197–201; Cook et al., "Amplification of the North American 'Dust Bowl'"; Lockeretz, "Lessons of the Dust Bowl," 560; Worster, *Dust Bowl*, 94, 200, 213; Opie, *Ogallala*, 98–100; Green, *Land of the Underground Rain*, 123–244; Montgomery, *Dirt*, 151–52; Ian Douglas, "Sediment Transfer and Siltation," in Turner et al., *The Earth as Transformed*, 221. The most famous novel of this period is John Steinbeck's *The Grapes of Wrath* (New York: Viking, 1939).

71. Riebsame, "The United States Great Plains," 565–71; Worster, *Dust Bowl*, 199–229; Montgomery, *Dirt*, 151–53; Lockeretz, "Lessons of the Dust Bowl," 561.

72. Worster, *Dust Bowl*, 226; see also 225–30.

73. Green, *Land of the Underground Rain*, 147; Lockeretz, "Lessons of the Dust Bowl," 568; Worster, *Dust Bowl*, 227–28; Opie, *Ogallala*, 113–16; Riebsame, "The United States Great Plains," 262–63.

74. Lockeretz, "Lessons of the Dust Bowl," 568; Worster, *Dust Bowl*, 233; Riebsame, "The United States Great Plains," 564, 571; Montgomery, *Dirt*, 173–74; Douglas, "Sediment Transfer," 228.

75. Gregory T. Cushman, *Guano and the Opening of the Pacific World* (New York: Cambridge University Press, 2013).

76. Vaclav Smil, *Enriching the Earth: Fritz Haber, Carl Bosch, and the Transformation of World Food Production* (Cambridge, Mass.: MIT Press, 2001).

77. Kenneth Pomeranz, "Advanced Agriculture," in Jerry Bentley, ed., *The Oxford Handbook of World History* (Oxford: Oxford University Press, 2011), 254–59; Giovanni Federico, *Feeding the World: An Economic History of Agriculture* (Princeton, N.J.: Princeton University Press, 2009), 89; Pfister, "The '1950s Syndrome,'" 111; Montgomery, *Dirt*, 197–200; Smil, *Energy in World History*, 182–89; Riebsame, "The United States Great Plains," 567–68.

78. Rachel Carson, *Silent Spring* (Boston: Houghton Mifflin, 1962, 2002), 17–32, 258–59.

79. Pomeranz, "Advanced Agriculture," 255; Nye, *Consuming Power*, 192; Smil, *Energy in World History*, 183.

80. Carson, *Silent Spring*, 219–76.

81. Opie, *Ogallala*, 32. See also William Ashworth, *Ogallala Blue: Water and Life on the High Plains* (New York: W. W. Norton, 2006), 17; Green, *The Land of the Underground Rain*, 145–67; and Worster, *Dust Bowl*, 234–35.

82. Opie, *Ogallala*, 124–42; Green, *The Land of the Underground Rain*, 125–48.

83. James Aucoin, "The Irrigation Revolution and its Environmental Consequences," *Environment* 21 no. 8 (October 1979),18–19; Opie, *Ogallala*, 123, 143–48; Green, *The Land of the Underground Rain*, 194; Ashworth, *Ogallala Blue,* 143–49.

84. Aucoin, "Irrigation Revolution," 19, 38–39; Ashworth, *Ogallala Blue*, 45–53.

85. According to the United States Geological Service, the Ogallala has lost 312 cubic kilometers since the 1950s; Ashworth, *Ogallala Blue*, 11, gives the figure of 120 trillion gallons, or 454 cubic kilometers.

86. Ashworth, *Ogallala Blue*, 23–27, 34–43, 153–55; Green, *The Land of the Underground Rain*, 191; Aucoin, "Irrigation Revolution," 20.

87. Green, *The Land of the Underground Rain*, 187–88; Worster, *Dust Bowl*, 234.

88. Smil, *Energy*, 190–91; see also John F. Richards, "Land Transformation," in Turner, *The Earth as Transformed*, 162–78.

89. CIMMYT, "World Wheat Facts and Trends 1998–99: Global Wheat Research in a Changing World," "World Maize Facts and Trends" (Mexico City: International Maize and Wheat Improvement Center, 1998) in libcatalog.cimmyt.org (accessed July 2016); "Soybean," in Wikipedia (accessed June 2017).

90. Figures based on Wilson Warren, *Tied to the Great Packing Machine: The Midwest and Meatpacking* (Iowa City: University of Iowa Press, 2007), table 8.2: US Retail Meat, Poultry and Fish Consumption per capita, 1910–99.

91. Alan B. Durning and Holly B. Brough, *Taking Stock: Animal Farming and the Environment* (Washington, D.C.: Worldwatch Institute, 1991) 33–35; Laurie Winn Carlson, *Cattle: An Informal Social History* (Chicago: Ivan. R. Dee, 2001), 271–73; Michael Pollan, *The Omnivore's Dilemma: A Natural History of Four Meals* (New York: Penguin, 2006), 67–84.

92. J. Webster, *Animal Welfare: Limping toward Eden: A Practical Approach to Redressing the Problem of Our Dominion over the Animals* (Oxford: Blackwell, 2005), 153; Pollan, *The Omnivore's Dilemma*, 77–78.

93. Jonathan Safran Foer, *Eating Animals* (New York: Little, Brown, 2009), 136. See also "Poultry Farming," in Wikipedia (accessed July 2016).

94. H. Herzog, *Some We Love, Some We Hate, Some We Eat: Why It's So Hard to Think Straight about Animals* (New York: HarperCollins, 2010), 167; Webster, *Animal Welfare*, 120–26.

95. Temple Grandin and Catherine Johnson, *Animals in Translation: Using the Mysteries of Autism to Decode Animal Behavior* (New York: Scribner's, 2005), 183; Peter Singer, *In Defense of Animals: The Second Wave* (Malden, Mass.: Blackwell, 2006), 176; Webster, *Animal Welfare*, 122–23, 159–68; Herzog, *Some We Love*, 167–69.

96. Marco Springmann et al., "Analysis and Valuation of the Health and Climate Change Cobenefits of Dietary Change," *Proceedings of the National Academy of Sciences of the USA* 113 no. 15 (12 April 2016), 4146–51.

97. Durning and Brough, *Taking Stock*, 18–26; Warren, *Tied*, 165–77.

98. Moises Velasquez-Manoff, *An Epidemic of Absence: A New Way of Understanding Allergies and Autoimmune Disorders* (New York: Scribner's, 2012), 180–81.

99. "List of Countries by Past and Future GDP (PPP)," in Wikipedia (accessed January 2013); figures are based on purchasing power parity, not on exchange rates.

100. Judith Shapiro, *China's Environmental Challenges* (Cambridge: Polity Press, 2012), 37–41.

101. Robert Marks, *China: Its Environment and History* (Lanham, Md.: Rowman & Littlefield, 2012), 273.

102. Judith Shapiro, *Mao's War against Nature: Politics and the Environment in Revolutionary China* (Cambridge: Cambridge University Press, 2001), 10, and *China's Environmental Challenges*, 21–23; Elizabeth Economy, *The River Runs Black: The Environmental Challenge to China's Future*, 2nd ed. (Ithaca, N.Y.: Cornell University Press, 2010), 60–68; Marks, *China*, 278–87.

103. Karl Gerth, *As China Goes, So Goes the World: How Chinese Consumers are Transforming Everything* (New York: Hill & Wang, 2010), 189–91.

104. Economy, *The River Runs Black*, 65–67; Marks, *China*, 292; David Kirby, "Ill Wind," *Discover* (April 2011), 14; Gerth, *As China Goes*, 183.

105. Marks, *China*, 267–74, 336; Economy, *The River Runs Black*, 72.

106. Chris P. Nielsen and Mun S. Ho, "Clearing the Air in China," *New York Times Sunday Review* (October 27, 2013), 4; Edward Wong, "Most Chinese Cities Fail Minimum Air Quality Standards, Study Says," *New York Times* (March 28, 2014), 8; Shapiro, *China's Environmental Challenges*, 48–49.

107. Shapiro, *China's Environmental Challenges*, 1–12; Economy, *The River Runs Black*, 74–75; Marks, *China*, 312–13; Kirby, "Ill Wind," 207–11.

108. Economy, *The River Runs Black*, 77; Marks, *China*, 314; Yergin, *The Quest*, 218; US Department of Transportation, Office of the Assistant Secretary for Research and Technology, Table 1-23: "World Motor Vehicle Production," in www.rita.dot.gov (accessed July 2016).

109. "Expressways of China," in Wikipedia (accessed February 2013).

110. Chris Buckley, "China's New Bridges: Rising High, but Buried in Debt," *New York Times* (June 11, 2017).

111. Marks, *China*, 315.

112. Marks, *China*, 315; Yergin, *The Quest*, 212, 223; Kirby, "Ill Wind," 43–53; Michael Marshall and Andy Coghlan, "China's Struggle to Clear the Air," *New Scientist* (February 9–15, 2013), 8–9.

113. Olivia Boyd, "The Birth of Chinese Environmentalism: Key Campaigns," in Sam Geall, ed., *China and the Environment: The Green Revolution* (London: Zed Books, 2013), 40–43.

114. Shapiro, *China's Environmental Challenges*, 71–72, 167–81.

Chapter 12

1. Richard B. Alley, *The Two-Mile Time Machine: Ice Cores, Abrupt Climate Change, and Our Future* (Princeton, N.J.: Princeton University Press, 2000), 118–26. See also Wolfgang Behringer, *A Cultural History of Climate*, trans. Patrick Camiller (Cambridge: Polity Press, 2010), 25–26 and 42 fig. 2.1; and John Carey, "Global Warming: Faster than Expected," *Scientific American* (November 2012), 53.

2. Alley, *The Two-Mile Time Machine*, 8–9, 126–27; quotation on 118. See also Tim Flannery, *The Weather Makers: How Man Is Changing the Climate and What It Means for Life on Earth* (New York: Grove Press, 2005), 50–53, 61, 193–95.

3. This is the thesis of William F. Ruddiman's book *Plows, Plagues, and Petroleum: How Humans Took Control of Climate* (Princeton, N.J.: Princeton University Press, 2005); see also his "The Anthropogenic Greenhouse Gas Era Began Thousands of Years Ago," *Climatic Change* 61 no. 3 (December 2003), 261–63.

4. William F. Ruddiman, *Earth's Climate, Past and Future* (New York: W. H. Freeman, 2008), 327; Jonathan Overpeck, "Arctic Environmental Change of the Last Four Centuries," *Science* 278 no. 5341 (1997), 1251–56; Behringer, *A Cultural History of Climate*, 185–89.

5. Paul J. Crutzen, "Geology of Mankind: The Anthropocene," *Nature* 415 no. 6867 (January 3, 2002).

6. John R. McNeill and Peter Engelke, *The Great Acceleration: An Environmental History of the Anthropocene since 1945* (Cambridge, Mass.: Harvard University Press, 2014), 208.

7. On the official position on the term "anthropocene," see Jan Zalasiewicz, "A History in Layers," *Scientific American* (special edition, Winter 2016), 104–11.

8. Justin Gillis, "2014 Breaks Heat Record, Challenging Global Warming Skeptics," *New York Times* (January 17, 2015), p. 1.

9. William R. L. Anderegg et al., "Expert Credibility in Climate Change," *Proceedings of the National Academy of Sciences* 107 no. 27 (June 1, 2010), 12107–9. See also Naomi Oreskes, "The Scientific Consensus on Climate," *Science* 306 no. 5702 (December 3, 2004), 1686; Elizabeth Kolbert, "Rethinking How We Think about Climate Change," *Audubon* (September–October 2014), 48; and Ruddiman, *Earth's Climate*, 341.

10. On the antecedents and creation of the IPCC, see Spencer R. Weart, *The Discovery of Global Warming* (Cambridge, Mass.: Harvard University Press, 2008), 149–59.

See also Daniel Yergin, *The Quest: Energy, Security, and the Remaking of the Modern World* (New York: Penguin, 2011), 457–59; Flannery, *The Weather Makers*, 223.

11. Weart, *The Discovery of Global Warming*, 160–92; "IPCC First Assessment Report," in Wikipedia (accessed April 2014).

12.. IPCC *Second Assessment Report, Working Group I* (1996), Preface, p. xi.

13. Yergin, *The Quest*, 484–86; "IPCC Second Assessment Report," in Wikipedia (accessed April 2014).

14. "IPCC Second Assessment Report"; see also Behringer, *A Cultural History of Climate*, 191–99; and Flannery, *The Weather Makers*, 245–46.

15. IPCC, "Summary for Policy Makers," in S. Solomon et al., eds., *Climate Change 2007: The Physical Science Basis, Contribution of Working Group I to the Fourth Assessment Report of the Inter-governmental Panel on Climate Change* (Cambridge: Cambridge University Press, 2007), 5. See also "IPCC Fourth Assessment Report," in Wikipedia (accessed April 2014), and Yergin, *The Quest*, 501–2.

16. IPCC Fifth Assessment Report 2013, Working Group 1, *Summary for Policymakers* (28 pages). The full report of Working Group 1 is 1,535 pages long. Both are at www.climatechange2013.org (accessed October 2014).

17. On the history of climate science and the discovery of global warming, see David Archer, *The Long Thaw: How Humans Are Changing the Next 100,000 Years of Earth's Climate* (Princeton, N.J.: Princeton University Press, 2009), 15–29; Weart, *The Discovery of Global Warming*, 2–8.

18. Vaclav Smil, *Energy in World History* (Boulder: Westview Press, 1994), 217–22; Alan R. Townsend and Robert W. Howarth, "Fixing the Global Nitrogen Problem," *Scientific American* (February 2010), 64–68; Ruddiman, *Earth's Climate*, 330–33.

19. On the Greenland ice-core projects, see Alley, *The Two-Mile Time Machine*, 17–79, and Behringer, *A Cultural History of Climate,* 8–19. See also Ralph F. Keeling, "Recording the Earth's Vital Signs," *Science* 319 (March 28, 2008), 1771–72, and Weart, *The Discovery of Global Warming*, 20–38.

20. McNeill and Engelke, *The Great Acceleration*, 64–69; Elizabeth Kolbert, *Field Notes from a Catastrophe: Man, Nature, and Climate Change* (New York: Bloomsbury, 2009), 43–44; Behringer, *A Cultural History of Climate*, 183–84; Henry Pollack, *A World without Ice* (New York: Penguin, 2010), 183–87; Carey, "Global Warming," 51–55; "Greenhouse Gas to Reach 3-Million-Year High," livescience.com (accessed June 6, 2013).

21. "The Enduring Technology of Coal," *Technology Review* 116 no. 3 (May–June 2013), 15; Michael Le Page and Michael Slezak, "No Sign of Emissions Letting Up as Climate Talks Begin," *New Scientist* (December 8, 2012), 11; Clive Hamilton, *Requiem for a Species: Why We Resist the Truth about Climate Change* (London: Earthscan, 2010), 5; Ruddiman, *Earth's Climate*, 328.

22. McNeill and Engelke, *The Great Acceleration*, 70–72.

23. Kerry Emanuel, "Increasing Destructiveness of Tropical Cyclones over the Past 30 Years," *Nature* 436 no. 7051 (August 4, 2005), 686–88; Archer, *Long Thaw*, 45–54; Flannery, *The Weather Makers*, 135–41, 312.

24. Collin Kelley et al., "Climate Change in the Fertile Crescent and Implications of the Recent Syrian Drought," *Proceedings of the National Academy of Sciences of the*

USA 112 no. 11 (2015), 3241–46. See also John Wendle, "Syria's Climate Refugees," *Scientific American* (March 2016), 51–55.

25. Aklo Kitoh et al., "First Super-High-Resolution Model Projections that the Ancient 'Fertile Crescent' Will Disappear in This Century," *Hydrological Research Letters* 2 (2008), 1–4.

26. McNeill and Engelke, *The Great Acceleration*, 69–70.

27. Mark Carey, *In the Shadow of Melting Glaciers: Climate Change and Andean Society* (New York: Oxford University Press, 2010), 147; Steven Mithen, *Thirst: Water and Power in the Ancient World* (Cambridge, Mass.: Harvard University Press, 2012), 280.

28. Michael E. Mann, Raymond S. Bradley, and Malcolm K. Hughes, "Global-Scale Temperature Patterns and Climate Forcing over the Past Six Centuries," *Nature* 392 (April 23, 1998), 779–87; Michael Wines, "Climate Change Threatens to Strip the Identity of Glacier National Park," *New York Times* (November 23, 2014), 20, 26; Robert Marks, *China: Its Environment and History* (Lanham, Md.: Rowman & Littlefield, 2012), 317; Pascal Acot, *Histoire du climat* (Paris: Perrin, 2003), 260; Behringer, *A Cultural History of Climate*, 190–98; Pollack, *A World without Ice*, 198–201, 116–17; Judith Shapiro, *China's Environmental Challenges* (Cambridge: Polity Press, 2012), 47–48.

29. Julienne Stroeve et al., "Arctic Sea Ice Decline: Faster than Forecast," *Geophysical Research Letters* 34 (May 1, 2007), no. L09501; Pollack, *A World without Ice*, 118–26, 206–209, 224–25; Steger, *Deep Future*, 140.

30. Jill Jäger and Roger G. Barry, "Climate," in B. L. Turner et al. eds., *The Earth as Transformed by Human Action* (Cambridge: Cambridge University Press, 1990), 335–41; Flannery, *The Weather Makers*, 123–34, 145–47, 194; Carey, "Global Warming," 54; Pollack, *A World without Ice*, 127, 220; Steger, *Deep Future*, 122–25.

31. Ruddiman, *Earth's Climate*, 344–48; Le Page and Slezak, "No Sign of Emissions Letting Up"; Behringer, *A Cultural History of Climate*, 193–96; Alley, *The Two-Mile Time Machine*, 172–73; Hamilton, *Requiem for a Species*, 6–7; Carey, "Global Warming," 52; "IPCC," in Wikipedia (accessed June 2013); Pollack, *A World without Ice*, 239.

32. David Victor and Charles Kennel, "Climate Policy: Ditch the 2°C Warming Goal," *Nature* 514 (2 October 2014), 30–31.

33. IPCC Fifth Assessment Report 2013, *Summary for Policymakers*.

34. Brad Plumer, "Assessing the Economic Bite from Rising Temperatures," *New York Times* (June 30, 2017), 20.

35. Marks, *China*, 317.

36. Estimates differ. According to IPCC, by the end of the twenty-first century, sea levels could rise by over 3 feet; the US Army Corps of Engineers estimates that it could rise by 5 feet; the National Oceanic and Atmospheric Administration estimates the rise as up to 6.5 feet. See Elizabeth Kolbert, "The Siege of Miami," *New Yorker* (December 21 and 28, 2015).

37. Albert Ammerman and Charles E. McClennen, "Saving Venice," *Science* 289 (2000), 1301–2; Callum Roberts, *The Ocean of Life: The Fate of Man and the Sea* (New York: Penguin, 2012), 93–97; Steger, *Deep Future*, 129–37.

38. Gardiner Harris, "As Seas Rise, Millions Cling to Borrowed Time and Dying Land," *New York Times* (March 29, 2014), p. 10.

39. David Rind et al., "Potential Evotranspiration and the Likelihood of Future Drought," *Journal of Geophysical Research* 95 (June 20, 1990), 9,983–10,004; Ruddiman, *Earth's Climate*, 350–56, and *Plows, Plagues, and Petroleum*, 167–81; Behringer, *A Cultural History of Climate*, 198–99; Acot, *Histoire du climat*, 261–62; Kolbert, *Field Notes from a Catastrophe*, 110–11; Joe Barnett and Neil Adger, "Climate Dangers and Atoll Countries," *Climatic Change* 61 (2003), 321–37; Flannery, *The Weather Makers*, ch. 32; Pollack, *A World without Ice*, 214–19, 233.

40. Ruddiman, *Earth's Climate*, 346–57, and *Plows, Plagues, and Petroleum*, 151–63; Alley, *The Two-Mile Time Machine*, 172.

41. IPCC Fifth Assessment Report 2013, *Summary for Policymakers*.

42. Steger, *Deep Future*, 50–71.

43. Pollack, *A World without Ice*, 258.

44. Robert DeConto and David Pollard, "Contribution of Antarctica to Past and Future Sea-Level Rise," *Nature* 531 (30 March 2016), 591–97; Archer, *The Long Thaw*, 141–45.

45. Stefan Rahmstorf et al., "Exceptional Twentieth-Century Slowdown in Atlantic Ocean Overturning Circulation," *Nature Climate Change* 5 (2015), 475–80, also in doi: 10.1038/nclimate2554.

46. Alley, *The Two-Mile Time Machine*, 148–50, 183–84 (quotation on p. 169).

47. IPCC Fifth Assessment Report 2013, *Summary for Policymakers;* Kolbert, *Field Notes from a Catastrophe*, 128; Flannery, *The Weather Makers*, 61, 190–95; Ruddiman, *Earth's Climate*, 356; Steger, *Deep Future*, 19–20.

48. Ted Schuur, "The Permafrost Prediction," *Scientific American* 315 no. 6 (December 2016), 56–62; Sarah E. Chadburn et al., "An Observation-Based Constraint on Permafrost Loss as a Function of Global Warming," *Nature Climate Change* (online April 10, 2017); K. M. Walter, S. A. Zimov, et al., "Methane Bubbling from Siberian Thaw Lakes as a Positive Feedback to Climate Warming," *Nature* 443 (September 7, 2006), 71–75; Archer, *The Long Thaw*, 131–36; Flannery, *The Weather Makers*, 199–201; Carey, "Global Warming," 52–54.

49. Flannery, *The Weather Makers*, 196–99; John Hemming, *Tree of Rivers: The Story of the Amazon* (London: Thames and Hudson, 2008), 322–24.

50. For an overview of these issues, see Anthony Giddens, *The Politics of Climate Change* (Cambridge: Polity Press, 2009).

51. Different sources give somewhat different numbers: "Earth Summit," in Wikipedia (accessed April 2014) gives 172 government and 108 heads of state or of government; Behringer, *A Cultural History of Climate*, 192, puts the numbers at 178 countries; Yergin, *Quest*, 472, mentions 160 heads of state, government, and international organizations. On the international politics of global warming, see McNeill and Engelke, *The Great Acceleration*, 76–82. See also Ramachandra Guha, *Environmentalism: A Global History* (New York: Longman, 2000), 141.

52. Quoted in Benjamin Kline, *First Along the River: A Brief History of the U.S. Environmental Movement*, 3rd ed. (Lanham, Md.: Rowman & Littlefield, 2007), 135. See also Guha, *Environmentalism*, 141–43.

53. Yergin, *Quest*, 468–73; "United Nations Framework Convention on Climate Change," in Wikipedia (accessed July 2013); Flannery, *The Weather Makers*, 223, 243; Kline, *First Along the River*, 110–13; Hamilton, *Requiem for a Species*, 98.

54. William K. Stevens, "Greenhouse Gas Issue: Haggling over Fairness," *New York Times* (November 30, 1997); Yergin, *Quest*, 485–95, 512–13; "Kyoto Protocol," in Wikipedia (accessed April 2014).

55. Flannery, *The Weather Makers*, 243; Behringer, *A Cultural History of Climate*, 192–95; Acot, *Histoire du climat*, 256–58; Kolbert, *Field Notes from a Catastrophe*, 15–72, 197; Hamilton, *Requiem for a Species*, 98; "Earth Summit" and "United Nations Framework Convention on Climate Change," in Wikipedia (accessed July 2013).

56. Flannery, *The Weather Makers*, 223–31; Kolbert, *Field Notes from a Catastrophe*, 197; Yergin, *Quest*, 489–99; Kline, *First Along the River*, 135–36, 150–72; Behringer, *A Cultural History of Climate*, 192–95; Stevens, "Greenhouse Gas Issue."

57. "2009 United Nations Climate Change Conference" (also known as the "Copenhagen Summit") in Wikipedia (accessed July 2013); Marks, *China*, 316; Yergin, *Quest*, 515.

58. "2010 United Nations Climate Change Conference" (also known as the "Cancún Summit") in Wikipedia (accessed July 2013).

59. Steven Lee Myers and Nicholas Kulish, "Growing Clamor about Inequities of Climate Crisis," *New York Times* (November 12, 2013).

60. Joe Romm, "Record First: Global CO2 Emissions Went Flat in 2014 While the Economy Grew" (March 13, 2015) in thinkprogress.org/climate/2015/03/13/3633362/iea-co2-emissions-decouple-growth (accessed October 2015); Sandy Dechert, "How Big a Deal Is Economy-Energy CO2 Decoupling" (March 15, 2015) in cleantechnica.com/2015/03/15/big-deal-economy-energy-co2-decoupling (accessed October 2015).

61. Hamilton, *Requiem for a Species*, 14.

62. S. Fred Singer and Dennis T. Avery's *Unstoppable Global Warming: Every 1,500 Years* (Lanham, Md.: Rowman & Littlefield, 2007).

63. Christopher C. Horner, *Red Hot Lies: How Global Warming Alarmists Use Threats, Fraud, and Deception to Keep You Misinformed* (Washington, D.C.: Regnery, 2008), and *Power Grab: How Obama's Green Policies Will Steal Your Freedom and Bankrupt America* (Washington, D.C.: Regnery, 2010); Brian Sussman, *Eco-Tyranny: How the Left's Green Agenda Will Dismantle America* (Washington, D.C.: WND Books, 2012); and Christopher Booker, *The Real Global Warming Disaster: Is the Obsession with 'Climate Change' Turning Out to Be the Most Costly Scientific Blunder in History?* (London: Continuum, 2010).

64. Joshua B. Howe, *Behind the Curve: Science and the Politics of Global Warming* (Seattle: University of Washington Press, 2004), 118–46, 170–96; Naomi Oreskes and Erik M. Conway, *Merchants of Doubt: How a Handful of Scientists Obscured the Truth on Issues from Tobacco Smoke to Global Warming* (London: Bloomsbury Press, 2011), 183–215.

65. Shawn Lawrence Otto, "America's Science Problem," *Scientific American* (November 2012), 65; Oreskes, "The Scientific Consensus," 1686; "Climate of Distrust" (editorial), *Nature* 436 (July 7, 2005), 1; Flannery, *The Weather Makers*, 241.

66. Parker, *Global Crisis*, 687.

67. Mann, Bradley, and Hughes, "Global-Scale Temperature Patterns," 779–87; Michael E. Mann et al., "Northern Hemisphere Temperatures during the Past Millennium: Interferences, Uncertainties, and Limitations," *Geophysical Research Letters* 26 no. 6 (March 15, 1999), 759–62; Pollack, *A World without Ice*, 255–57.

68. A. W. Monford, *The Hockey Stick Illusion: Climatologists and the Corruption of Science* (London: Stacey International, 2010); Michael E. Mann, *The Hockey Stick and the Climate Wars: Dispatches from the Front Lines* (New York: Columbia University Press, 2012); Dale Jamieson, *Reason in a Dark Time: Why the Struggle against Climate Change Failed and What It Means for Our Future* (Oxford: Oxford University Press, 2014), 61–62; "Climate of Distrust."

69. Robin Lloyd, "Why Are Americans so Ill-Informed about Climate Change?" *Scientific American* (February 23, 2011), online at www.scientificamerican.com/article.cfm?id=why-are-american-so-ill; Otto, "America's Science Problem."

70. Ruddiman, *Plows, Plagues, and Petroleum*, 189.

71. Jamieson, *Reason in a Dark Time*, 81–96; Kolbert, "Rethinking," 46–48; Oreskes and Conway, *Merchants of Doubt*, 186–90, 213.

72. Patrick Michaels and Robert C. Balling Jr., *Climate of Extremes: Global Warming Science They Don't Want You to Know* (Washington, D.C.: Cato Institute, 2009), 7; see also Michaels's *Meltdown: The Predictable Distortion of Global Warming by Scientists, Politicians, and the Media* (Washington, D.C.: Cato Institute, 2004).

73. Flannery, *The Weather Makers*, 240–48; quotation on page 240. See also "Climate Skeptic Group Works to Reverse Renewable Energy Mandates," *Washington Post* (November 24, 2012); Hamilton, *Requiem for a Species*, 100–106.

74. Pew Research Center for People and the Press, "Fewer Americans See Solid Evidence of Global Warming," www.people-press.org/2009/10/22/fewer-americans-see-solid-evidence-of-global-warming (released October 22, 2009); Pollack, *A World without Ice*, 152; Hamilton, *Requiem for a Species*, 120–23.

75. Clive Hamilton, *Earthmasters: The Dawn of the Age of Climate Engineering* (New Haven, Conn.: Yale University Press, 2013), 86.

76. Hamilton, *Requiem for a Species*, 34.

77. Kari Marie Norgaard, *Living in Denial: Climate Change, Emotions, and Everyday Life* (Cambridge, Mass.: MIT Press, 2011), 2.

78. Alison Kopicki, "Americans More Worried about 'Warming' than 'Climate Change,'" *New York Times* (June 15, 2014).

79. Behringer, *A Cultural History of Climate*, 216.

80. Varun Sivaram, "The Global Warming Wild Card: Energy Decisions that India Makes in the Next Few Years Could Profoundly Affect How Hot the Planet Becomes This Century," *Scientific American* 316 no. 5 (May 2017), 48–53.

81. Hiroko Tabuchi, "As Beijing Joins Climate Fight, Chinese Companies Build Coal Plants," *New York Times* (July 2, 2017), 10; Michael Forsythe, "China Plans a Big Increase in Spending on Renewable Energy," *New York Times* (January 6, 2017), A6.

82. Ruddiman, *Plows, Plagues, and Petroleum*, 82–83.

83. This is the point made by Craig Simons in *The Devouring Dragon: How China's Rise Threatens Our Natural World* (New York: St. Martin's, 2013).

Chapter 13

1. Jean-Baptiste Lamarck, *Philosophie zoologique*, quoted in Callum Roberts, *The Ocean of Life: The Fate of Man and the Sea* (New York: Penguin, 2012), 80.

2. Thomas Henry Huxley, "Inaugural Address of the Fishery Conference," *Fisheries Exhibition Literature* 4 (1884), 18, quoted in Brian Fagan, *Fish on Friday: Feasting, Fasting and the Discovery of the New World* (New York: Basic Books, 2006), 235.

3. Marcel Hérubel, *Sea Fisheries: Their Treasures and Toilers* (London: T. Fisher Unwin, 1912), quoted in Roberts, *Ocean of Life*, 243.

4. Rachel Carson, *The Sea around Us* (first published in 1950) (New York: Oxford University Press, 1989), 14–15.

5. Figures vary; see Millennium Ecosystem Assessment, *Ecosystems and Human Well-Being: Synthesis* (Washington, D.C.: Island Press, 2005), 8; Ray Hilborn, "Marine Biota," in B. L. Turner et al., eds., *The Earth as Transformed by Human Action: Global and Regional Changes in the Biosphere over the Past 300 Years* (Cambridge: Cambridge University Press, 1990), 371, 380–83; Paul Greenberg, "Time for a Sea Change," *National Geographic* (October 10, 2010), 81, and *Four Fish: The Future of the Last Wild Food* (New York: Penguin, 2010), 136; and Roberts, *Ocean of Life*, 50–51, 243.

6. John F. Richards, *The Unending Frontier: An Environmental History of the Early Modern World* (Berkeley: University of California Press, 2003), 574–85; Richard Ellis, *Men and Whales* (New York: Knopf, 1991), 40–46.

7. Mark Cioc, *The Game of Conservation: International Treaties to Protect the World's Migratory Animals* (Athens: Ohio University Press, 2009), 107–16.

8. Ellis, *Men and Whales*, 47–66; Richards, *Unending Frontier*, 589–610; Cioc, *Game of Conservation*, 106.

9. Dauril Alden, "Yankee Sperm Whalers in Brazilian Waters, and the Decline of the Portuguese Whale Fishery (1773–1801)," *The Americas* 20 no. 3 (January 1964), 267–72; Shawn William Miller, *An Environmental History of Latin America* (Cambridge: Cambridge University Press, 2007), 99–101.

10. Alden, "Yankee Sperm Whalers," 273–88; Miller, *Environmental History*, 100; Richards, *Unending Frontier*, 576.

11. Richard Ellis, *The Empty Ocean: Plundering the World's Marine Life* (Washington, D.C.: Island Press, 2003), 242; Hilborn, "Marine Biota," 375.

12. Cioc, *Game of Conservation*, 104–6, 120–24.

13. Johan N. Tønnessen and A. O. Johnsen, *The History of Modern Whaling*, trans. R. I. Christophersen (Berkeley: University of California Press, 1982), 292–95; Ellis, *Men and Whales*, 345, 357–61.

14. Tønnessen and Johnsen, *History of Modern Whaling*, 472–75; Ellis, *Men and Whales*, 346–87; William M. Tsutsui, "Landscapes in the Dark Valley: Toward an Environmental History of Wartime Japan," in Richard P. Tucker and Edmund

Russell, eds., *Natural Enemy, Natural Ally: Toward an Environmental History of War* (Corvallis: Oregon University Press, 2004), 200–207.

15. Ellis, *The Empty Ocean*, 248–49.

16. Cioc, *Game of Conservation*, 132–45.

17. Gordon Jackson, *The British Whaling Trade* (St. John, Newfoundland: International Maritime Economic History Association, 2005), 216–31; quotation on p. 229.

18. William M. Tsutsui and Timo Vuorisalo, "Japanese Imperialism and Marine Resources," in Simo Laakkonen, Richard P. Tucker, and Timo Vuorisalo, eds., *The Long Shadows: A Global Environmental History of the Second World War* (Corvallis: Oregon State University Press, 2017), 269; Ellis, *Empty Ocean*, 246–51, and *Men and Whales*, 406–8, 499.

19. Daniel Cressey, "World's Whaling Slaughter Tallied," *Nature* 519 (12 March 2015), 140–41.

20. "List of cetacean species," in Wikipedia (accessed June 2018).

21. Tim Flannery, *The Weather Makers: How Man Is Changing the Climate and What It Means for Life on Earth* (New York: Atlantic Monthly, 2005), 96–98; Henry N. Pollack, *A World without Ice* (New York: Avery, 2009), 211; Edward O. Wilson, *The Diversity of Life* (Cambridge, Mass.: Harvard University Press, 1992), 270–71.

22. Mark Kurlansky, *Cod: Biography of a Fish that Changed the World* (New York: Penguin, 1998), 13, 17, 34–49; Greenberg, *Four Fish*, 137–41.

23. Fagan, *Fish on Friday*, 62–64, 226–27, 242, 259–60; Richards, *Unending Frontier*, 547–49; Kurlansky, *Cod*, 19–29, 51; Ellis, *Empty Ocean*, 59–60.

24. W. Jeffrey Bolster, *The Mortal Sea: Fishing in the Atlantic in the Age of Sail* (Cambridge, Mass.: Harvard University Press, 2012), 31–34.

25. Ellis, *Men and Whales*, 47; Richards, *Unending Frontier*, 552–59; Fagan, *Fish on Friday*, 226–50, 265–68.

26. Roberts, *Unnatural History of the Sea* (Washington, D.C.: Island Press, 2007), 131–53, 314; Kurlanksy, *Cod*, 131, 152.

27. Paul R. Ehrlich and Anne H. Ehrlich, *Extinction: The Causes and Consequences of the Disappearance of Species* (New York: Random House, 1981), 107.

28. Roberts, *Ocean of Life*, 43–44, 229–31, and *Unnatural History of the Sea*, 44, 187–95, 314; Charles Clover, *The End of the Line: How Overfishing Is Changing the World and What We Eat* (New York: New Press, 2006), 97–103; Hilborn, "Marine Biota," 379; Kurlansky, *Cod*, 132–58, 209.

29. "Bagehot: The Parable of the Clyde," *Economist* (August 31, 2013), 50; Roberts, *Ocean of Life*, 53–55.

30. Paul R. Josephson, "When Stalin Learned to Fish: Natural Resources, Technology, and Industry under Socialism," in Jeffrey M. Diefendorf and Kurk Dorsey, eds., *City, Country, Empire: Landscapes in Environmental History* (Pittsburgh, Pa.: University of Pittsburgh Press, 2005), 162–92; Paul Greenberg and Boris Worm, "When Humans Declared War on Fish," *New York Times* (May 10, 2015), 4.

31. Graeme Wynn, "Foreword: This is More Difficult than We Thought," in Dean Bavington, *Managed Annihilation: An Unnatural History of the Newfoundland Cod Collapse* (Vancouver: University of British Columbia Press, 2010), xii–xiii;

Graeme Wynn, *Canada and Arctic North America: An Environmental History* (Santa Barbara: ABC-CLIO, 2007), 355–56; Kurlansky, *Cod*, 138–41; Roberts, *Unnatural History*, 188–89, 326–27; Ellis, *Empty Ocean*, 67; Hilborn, "Marine," 383.

32. Millennium Ecosystem Assessment, 12; Bavington, *Managed Annihilation*, 17; Wynn, *Canada*, 357.

33. Kurlansky, *Cod*, 153–69; Roberts, *Unnatural History*, 190; Ellis, *Empty Ocean*, 68; Wynn, *Canada*, 357; Greenberg and Worm, "When Humans Declared War on Fish."

34. Kurlansky, *Cod*, 171–81, 221–23; Ellis, *Empty Ocean*, 69; Clover, *End of the Line*, 112; Bavington, *Managed Annihilation*, 31–32; Greenberg, *Four Fish*, 129.

35. Ellis, *Empty Ocean*, 68–72; Clover, *End of the Line*, 114; Bavington, *Managed Annihilation*, 1–2; Greenberg, *Four Fish*, 129, 142–45, 151; Kurlansky, *Cod*, 3–4, 186–88; Wynn, *Canada*, 357–58.

36. Carmel Finley, *All the Fish in the Sea: Maximum Sustainable Yield and the Failure of Fisheries Management* (Chicago: University of Chicago Press, 2001), 2–3, 155, 165; Arthur F. McEvoy, *The Fisherman's Problem: Ecology and Law in the California Fisheries, 1850–1980* (Cambridge: Cambridge University Press, 1990), 6; Clover, *End of the Line*, 105–10; Wynn, "Foreword," xix–xxi.

37. Finley, *All the Fish in the Sea*, 7; see also Kurlansky, *Cod*, 185, and Clover, *End of the Line*, 114.

38. Andrew J. Pershing et al., "Slow Adaptation in the Face of Rapid Warming Leads to Collapse of Gulf of Maine Cod," *Science* (online) (October 29, 2015).

39. Clover, *End of the Line*, 110; McEvoy, *The Fisherman's Problem*, 6–7; Roberts, *Unnatural History*, 320–21.

40. Greenberg, *Four Fish*, 130, 147; Kurlansky, *Cod*, 9, 202; Roberts, *Unnatural History*, 199–213, 323–25.

41. Kurlansky, *Cod*, 198–200; Clover, *End of the Line*, 136–39; Roberts, *Ocean of Life*, 49–56, 249–51; Pascal Lorance et al., "Habitat, Behaviour and Colour Patterns or Orange Roughy *Hoplostethus atlanticus* in the Bay of Biscay," *Journal of the Marine Biological Association of the UK* 82 (2002), 321–31.

42. Andrew Jacobs, "China's Appetite Pushes Fish Stocks to Brink: Overfishing by Massive Fleet Exacts a Toll on Oceans Worldwide," *New York Times* (April 30, 2017).

43. Joseph E. Taylor III, *Making Salmon: An Environmental History of the Northwest Fisheries Crisis* (Seattle: Washington University Press, 1999), 5–38.

44. Jim Lichatowich, *Salmon without Rivers: A History of the Pacific Salmon Crisis* (Washington, D.C.: Island Press, 1999), 85–90; David R. Montgomery, *King of Fish: The Thousand-Year Run of Salmon* (Boulder, Colo.: Westview Press, 2003), 174, 229; Richard White, *The Organic Machine* (New York: Hill & Wang, 1995), 91–92; Ellis, *Empty Ocean*, 80–81; Taylor, *Making Salmon*, 41–66.

45. On the dams on the Columbia River and its tributaries, see White, *The Organic Machine*, 59–88.

46. Lichatowich, *Salmon without Rivers*, 202–17; Montgomery, *King of Fish*, 180, 230; Ellis, *Empty Ocean*, 81; Richard White, *The Organic Machine*, 98–102.

47. Montgomery, *King of Fish*, 146–47, 230; Lichatowich, *Salmon without Rivers*, xiii; Taylor, *Making Salmon*, 3–4.

48. Taylor, *Making Salmon*, 69–132.

49. Lichatowich, *Salmon without Rivers*, 115–17, 124–30, 207–15; Montgomery, *King of Fish*, 150–70; Taylor, *Making Salmon*, 10.

50. Robert J. Behnke, *Trout and Salmon of North America* (New York: Free Press, 2002); "Alaska salmon fishery," in Wikipedia (accessed September 2013).

51. Richard C. Hoffmann, *An Environmental History of Medieval Europe* (Cambridge: Cambridge University Press, 2014), 272–76, and "Economic Development and Aquatic Ecosystems in Medieval Europe," *American Historical Review* 101 no. 3 (June 1996), 631–69.

52. "Aquaculture," in Wikipedia (accessed July 2017).

53. K. Heen et al., *Salmon Aquaculture* (New York: Halsted Press, 1993); Robert Stickney, *Encyclopedia of Aquaculture* (New York: Wiley, 2000); Roberts, *Ocean of Life*, 244–49, 258–59; Clover, *End of the Line*, 301–3; Lester R. Brown, "Fish Farming May Soon Overtake Cattle Ranching as a Food Source," *Earth Policy Institute* (October 3, 2000) in www.earth-policy.org/plan_b_updates/2000/alert9

54. Lichatowich, *Salmon without Rivers*, 211; Montgomery, *King of Fish*, 171–75; Ellis, *Empty Ocean*, 83–91; Roberts, *Ocean of Life*, 253–56; "Aquaculture of salmon," in Wikipedia (accessed September 2013).

55. Timothy D. Jickels, Roy Carpenter, and Peter Liss, "Marine Environments," in Turner et al., *The Earth as Transformed*, 313–34; Roberts, *Ocean of Life*, 264–65.

56. R. J. Diaz and R. Rosenberg, "Spreading Dead Zones and Consequences for Marine Ecosystems," *Science* 321 no. 5891 (August 15, 2008), 926–29; Roberts, *Ocean of Life*, 120–30, 257.

57. C. J. Moore, "Synthetic Polymers in the Marine Environment: A Rapidly Increasing Long-Term Threat," *Environmental Research* 103 no. 2 (2008), 131–39.

58. Personal observation.

59. Charles G. Moore and Cassandra Phillips, *Plastic Ocean: How a Sea Captain's Chance Discovery Launched a Determined Quest to Save the Oceans* (New York: Penguin, 2011); Donovan Hohn, *Moby-Duck: The True Story of 28,800 Bath Toys Lost at Sea, and the Beachcombers, Oceanographers, Environmentalists, and Fools—Including the Author—Who Went in Search of Them* (New York: Viking, 2011); Alan Weisman, *The World without Us* (New York: St. Martin's Press, 2007), 142–59; Patricia Newman, *Plastic, Ahoy! Investigating the Great Pacific Garbage Patch* (Minneapolis: Millbrook Press, 2014).

60. Terry P. Hughes et al., "Global Warming and Recurrent Mass Bleaching of Corals," *Nature* 543 (March 16, 2017), 373–77; Michelle Innis, "Climate-Related Death of Coral around World Alarms Scientists," *New York Times* (April 10, 2016); Elizabeth Kolbert, *The Sixth Extinction: An Unnatural History* (New York: Henry Holt, 2014), 138–41, 161–62; Flannery, *Weather Makers*, 104–13; Roberts, *Ocean of Life*, 2, 85–86, 108; "Corals and Coral Reefs," in *Smithsonian Museum of Natural History Ocean Portal* (ocean.si.edu/coral-and-coral-reefs) (accessed September 2013).

61. Orrin H. Pilkey and Rob Young, *The Rising Sea* (Washington, D.C.: Island Press, 2009), 32–39, 49–51, 66–79.

62. Pilkey and Young, *The Rising Sea*, 132–35, 142–57.

63. Elizabeth Kolbert, "The Darkening Sea," *New Yorker* (November 20, 2006), 65–77, and *The Sixth Extinction*, 114–23; Robert E. Service, "Rising Acidity Brings an Ocean of Trouble," *Science* 337 no. 6091 (July 12, 2012), 146–48; Scott C. Doney, "The Dangers of Ocean Acidification," *Scientific American* 294 no. 3 (March 2006), 58–65; "The Ocean in a High CO_2 World," *Eos, Transactions, American Geophysical Union* 85 no. 37 (2004), 351–53.

64. Boris Worm et al., "Impacts of Biodiversity Loss on Ocean Ecosystem Services," *Science* 314 (November 3, 2006), 787–90. See also Douglas J. McCauley et al., "Marine Defaunation: Animal Loss in the Global Ocean," *Science* 347 no. 6219 (January 16, 2015).

Chapter 14

1. Joel R. Greenberg, *A Feathered River across the Sky: The Passenger Pigeon's Flight to Extinction* (New York: Bloomsbury Press, 2014); Tim Flannery, *The Eternal Frontier: An Ecological History of North America and Its Peoples* (New York: Grove Press, 2001), 312–15.

2. John H. Lawton and Robert M. May, eds., *Extinction Rates* (New York: Oxford University Press, 1995), 3–6, 43; David M. Raup, *Extinction: Bad Genes or Bad Luck?* (New York: W. W. Norton, 1991), 3, 108, and "Diversity Crises in the Geological Past," in Edward O. Wilson and Frances M. Peter, eds., *Biodiversity* (Washington: National Academy Press, 1988), 52–54; Richard Leakey and Roger Lewin, *The Sixth Extinction: Patterns of Life and the Future of Humankind* (New York: Doubleday, 1995) 39, 46, 232.

3. Mark V. Barrow Jr., *Nature's Ghosts: Confronting Extinction from the Age of Jefferson to the Age of Ecology* (Chicago: University of Chicago Press, 2009), ch. 1.

4. Savvy readers will no doubt draw some parallels with the contrasting histories of revolutionary France and of reforming Britain in the lifetimes of Cuvier and Lyell.

5. Among the leaders of this paradigm shift were David Raup and Jack Sepkowski; see their "Periodicity of Extinctions in the Geological Past," *Proceedings of the National Academy of Sciences of the United States of America* 81 (1984), 801–4, and "Periodic Extinction of Families and Genera," *Science* 231 (1986), 833–36. See also Norman MacLeod, *The Great Extinctions: What Causes Them and How They Shape Life* (London: Firefly Books, 2013), 187–88.

6. There are several accounts of this discovery; the most readable is Walter Alvarez, *T-Rex and the Crater of Doom* (Princeton, N.J.: Princeton University Press, 1997).

7. Peter Ward and Joe Kirschvink, *A New History of Life: The Radical New Discoveries about the Origins and Evolution of Life on Earth* (New York: Bloomsbury Press, 2015), ch. 16; quotation on p. 306.

8. Ward and Kirschvink argued that there were actually nine mass extinctions before the present, four of them before the Late-Ordovician mentioned here. See *A New History of Life*, 329–30.

9. Michael J. Benton, *When Life Nearly Died: The Greatest Mass Extinction of All Time* (London: Thames and Hudson, 2003), 9–15, 262–83; Ward and Kirschvink, *A New History of Life*, ch. 12.

10. Raup, "Diversity Crises," 52, and *Extinction*, 5; Edward O. Wilson, *The Diversity of Life* (Cambridge, Mass.: Harvard University Press, 1992), 29–31; Leakey and Lewin, *The Sixth Extinction*, 44–45; Elizabeth Kolbert, "The Sixth Extinction?" *New Yorker* 85 no. 15 (May 25, 2009).

11. MacLeod, *The Great Extinctions*, 190–91; Wilson, *Diversity of Life*, 31–32, 330; Leakey and Lewin, *The Sixth Extinction*, 48–54.

12. Peter Raven, foreword in Paul Harrison and Fred Pierce, *AAAS Atlas of Population and Environment* (Berkeley: University of California Press, 2000). For other, similar opinions by eminent scientists, see Leakey and Lewin, *The Sixth Extinction*, 235; Stuart L. Pimm, G. J. Russell, J. L. Gittleman, and T. M. Brooks, "The Future of Biodiversity," *Science* 269 (July 21, 1995), 347–50; Stuart L. Pimm and Clinton Jenkins, "Sustaining the Variety of Life," *Scientific American* (September 2005), 66–75; Ross D. E. MacPhee and Clare Fleming, "*Requiem Aeternam*: The Last Five Hundred Years of Mammalian Species Extinctions," in Ross D. E. MacPhee, ed., *Extinctions in Near Time: Causes, Contexts and Consequences* (New York: Kluwer Academic, 1999), 333–72; Colin J. Bibby, "Recent Past and Future Extinctions in Birds," in Lawton and May, *Extinction Rates,* 98; Jan Schipper et al., "The Status of the World's Land and Marine Mammals: Diversity, Threat, and Knowledge," *Science* 322 (October 10, 2008), 225–30; Paul R. Ehrlich and Ann H. Ehrlich, *Extinction: The Causes and Consequences of Disappearance of Species* (New York: Random House, 1981), 214–21; Wilson, *The Diversity of Life*, 243–44 and 274–80; and Norman Myers, *The Sinking Ark: A New Look at the Problem of Disappearing Species* (Oxford: Pergamon Press, 1979), 31.

13. Paul S. Martin and David W. Steadman, "Prehistoric Extinctions on Islands and Continents," in MacPhee, *Extinctions in Near Time*, 17–19. See also Wilson, *Diversity of Life*, 246–49.

14. Anthony D. Barnosky et al., "Has the Earth's Sixth Mass Extinction Already Arrived?" *Nature* 471 no. 7336 (March 3, 2011), 51–57.

15. Richard H. Grove, *Green Imperialism: Colonial Expansion, Tropical Island Edens and the Origins of Environmentalism, 1600–1860* (Cambridge: Cambridge University Press, 1995), 44.

16. David Quammen, *The Song of the Dodo: Island Biogeography in an Age of Extinctions* (New York: Scribner's, 1996), 261–63.

17. Elizabeth Kolbert, *The Sixth Extinction: An Unnatural History* (New York: Henry Holt, 2014), 56–58.

18. Bibby, "Recent Past and Future Extinctions in Birds," 107; MacLeod, *The Great Extinctions*, 177; Robert L. Peters and Thomas E. Lovejoy, "Terrestrial Fauna," in B. L. Turner II et al., eds., *The Earth Transformed by Human Action: Global and Regional Changes in the Biosphere over the Past 300 Years* (Cambridge: Cambridge University Press, 1990), 353.

19. Terry Glavin, *The Sixth Extinction: Journey among the Lost and Left Behind* (New York: St. Martin's Press, 2006), 1; MacLeod, *Great Extinction*, 184–85; Schipper et al., "Status of the World's Land and Marine Mammals."

20. Leakey and Lewin, *The Sixth Extinction*, 233.

21. Andrew R. Blaustein and Andy Dobson, "Extinction: A Message from the Frogs," *Nature* 439 (January 12, 2006), 143–44; Peters and Lovejoy, "Terrestrial Fauna," 354; Kolbert, *Sixth Extinction*, 4–18.

22. John W. Fitzpatrick, executive director, Cornell University Laboratory of Ornithology, "Saving Our Birds," *New York Times Sunday Review* (August 31, 2014).

23. David S. Wilcove, *The Condor's Shadow: The Loss and Recovery of Wildlife in America* (New York: W. H. Freeman, 1999), 230–33.

24. Barrow, *Nature's Ghosts*, 349–52.

25. Barrow, *Nature's Ghosts*, 352–59.

26. John R. Platt, "Poachers Drive Javan Rhino to Extinction in Vietnam," *Scientific American* (October 25, 2011), 43–45; Kolbert, *Sixth Extinction*, 222, 254–55.

27. Glavin, *Sixth Extinction*, 202; Paul R. Ehrlich, "The Scale of Human Enterprise and Biodiversity Loss," in Lawton and May, *Extinction Rates*, 216.

28. Wilson, *Diversity of Life*, 256; Leakey and Lewin, *The Sixth Extinction*, 234–35; Peters and Lovejoy, "Terrestrial Fauna," 353.

29. Jared Diamond, "Quaternary Megafaunal Extinctions: Variations on a Theme by Paganini," *Journal of Archaeological Science* 16 no. 2 (March 1989), 169.

30. Robert Marks, *China: Its Environment and History* (Lanham, Md.: Rowman & Littlefield, 2012), 297–98.

31. Ehrlich and Ehrlich, *Extinction*, 117–28.

32. John R. McNeill and Peter Engelke, *The Great Acceleration: An Environmental History of the Anthropocene since 1945* (Cambridge, Mass.: Harvard University Press, 2016), 89.

33. Kolbert, *Sixth Extinction*, 151–52.

34. Edward O. Wilson, "Threats to Biodiversity," *Scientific American* (261 no. 3 (September 1989), 108–16; Kolbert, *Sixth Extinction*, 183.

35. Wilson, "Threats to Biodiversity"; Kolbert, *Sixth Extinction*, 167–81.

36. Tim M. Blackburn et al., "Avian Extinction and Mammalian Introductions on Oceanic Islands," *Science* 305 (2004), 1955–58; Donald Grayson, "The Archeological Record of Human Impact on Animals," *Journal of World Prehistory* 15 no. 1 (2001), 1, 17–31; Martin and Steadman, "Prehistoric Extinctions," 26.

37. Kolbert, *Sixth Extinction*, 105–6.

38. Richard N. Holdaway, "Introduced Predators and Avifaunal Extinction in New Zealand," in MacPhee, ed., *Extinctions in Near Time*, 189–238; Alfred Crosby, *Ecological Imperialism: The Biological Expansion of Europe, 900–1900* (Cambridge: Cambridge University Press, 1986), 220–22; Atholl Anderson, *Prodigious Birds: Moa and Moa Hunting in Prehistoric New Zealand* (Cambridge: Cambridge University Press, 1989); Tim Flannery, *The Future Eaters: An Ecological History of the Australasian Lands and People* (Chatswood, NSW: Reed, 1994), 195–96, 243–46; Grayson, "Archaeological Record," 8–11.

39. Peter Tyson, *The Eighth Continent: Life, Death, and Discovery in the Lost World of Madagascar* (New York: William Morrow, 2000), 127–44; David A. Burney, "Rates, Patterns, and Processes of Landscape Transformation and Extinction in Madagascar," in MacPhee, ed., *Extinctions in Near Time*, 145–64; Sharon Levy, *Once and Future*

Giants: What Ice Age Extinctions Tell Us about the Fate of Earth's Largest Animals (New York: Oxford University Press, 2011), 20–25.

40. Tim Flannery, *Here on Earth: A Natural History of the Planet* (New York: Atlantic Monthly Press, 2010), 90, and *The Eternal Frontier*, 191; Franz J. Broswimmer, *Ecocide: A Short History of the Mass Extinction of Species* (London: Pluto, 2002), 26.

41. Stuart L. Pimm, Michael P. Moultan, and Lenora J. Justice, "Bird Extinctions in the Central Pacific," in Lawton and May, *Extinction Rates*, 75–87; Tim M. Blackburn and Kevin J. Gaston, "Biological Invasions and the Loss of Birds on Islands," in Dov E. Sax, John Stachowicz, and Steven J. Gaines, eds., *Species Invasions: Insights into Ecology, Evolution, and Biogeography* (Sunderland, Mass.: Sinauer Associates, 2005), 85–110; John R. McNeill, "Of Rats and Men"; Peters and Lovejoy, "Terrestrial Fauna," 357–58; Wilson, *Diversity of Life*, 244–46; Bibby, "Recent Past and Future Extinctions in Birds," 106; Alexander H. Harcourt, *Human Biogeography* (Berkeley: University of California Press, 2012), 240.

42. Kolbert, *Sixth Extinction*, 203.

43. Blackburn and Gaston, "Biological Invasions."

44. Yvonne Baskin, *A Plague of Rats and Rubbervines: The Growing Threat of Species Invasions* (Washington, D.C.: Island Press, 2002), 100–102; Quammen, *Song of the Dodo*, 321–37; Bibby, "Recent Past and Future Extinctions in Birds," 106; Kolbert, *Sixth Extinction*, 203.

45. Jared Diamond, N. P. Ashmole, and P. E. Purves, "The Present, Past and Future of Human-Caused Extinction," *Philosophical Transactions of the Royal Society of London. Series B: Biological Sciences* 325 (November 6, 1989), 469–77; R. M. Pringle, "The Origins of the Nile Perch in Lake Victoria," *BioScience* 55 (2005), 780–87; Dirk Verschuren et al., "History and Timing of Human Impact on Lake Victoria, East Africa," *Proceedings of the Royal Society B* 269 no. 1488 (February 7, 2002), 289–94; Tijs Goldschmidt, *Darwin's Dreampond: Drama in Lake Victoria*, trans. Sherry Marx-Macdonald (Cambridge, Mass.: MIT Press, 1996).

46. Dan Egan, *The Death and Life of the Great Lakes* (New York: W. W. Norton, 2017), 38–46, 65–70, 109–24, 179–81.

47. Brian Coman, *Tooth & Nail: The Story of the Rabbit in Australia* (Melbourne: Text, 1999).

48. Baskin, *A Plague of Rats and Rubbervines*, 44–46.

49. Wallace Stegner, *American Places* (1981), quoted in Baskin, *A Plague of Rats and Rubbervines*, 43.

50. Harcourt, *Human Biogeography*, 241.

51. "Kudzu," "Water hyacinth," "Burmese python," and "Red imported fire ant," in National Invasive Species Information Center (invasivespeciesinfo.gov) (accessed July 2017).

52. Kolbert, *Sixth Extinction*, 204; "Chestnut blight," "Dutch elm disease," and "Hemlock wooly adelgid," in National Invasive Species Information Center (invasivespeciesinfo.gov) (accessed July 2017).

53. *Arctic Climate Impact Assessment* (Cambridge: Cambridge University Press, 2005), passim; Susan Joy Hassol, *Impacts of a Warming Arctic: Arctic Climate Impact*

Assessment (Cambridge: Cambridge University Press, 2004), 10, 22–43, 78–81; Julienne Stroeve et al., "Arctic Sea Ice Decline: Faster than Forecast," *Geophysical Research Letters* 34 (May 1, 2007), no. L09501; Henry Pollack, *A World without Ice* (New York: Penguin, 2010), 118–26, 206–9, 224–25.

54. Hassol, *Impacts of a Warming Arctic*, 10, 45–47, 58–59, 68–73; Curt Stager, *Deep Future: The Next 100,000 Years of Life on Earth* (New York: Thomas Dunne Books, 2011), 146; Flannery, *Weather Makers*, 99–102; Pollack, *World without Ice*, 210.

55. Glavin, *Sixth Extinction*, 2, 195; Glavin, 196, draws an interesting parallel with the extinction of languages, one of which is lost every two weeks.

56. Camille Parmesan and Gary Yohe, "A Globally Coherent Fingerprint of Climate Change Impacts across Natural Systems," *Nature* 421 (January 2, 2003), 37–42, and Camille Parmesan et al., "Poleward Shift in Geographical Ranges of Butterfly Species Associated with Global Warming," *Nature* 399 (June 10, 1999), 579–84. See also Tim Flannery, *The Weather Makers: How Man Is Changing the Climate and What It Means for Life on Earth* (New York: Atlantic Monthly Press, 2005), 89–91, and Gian-Reto Walther, "Ecological Responses to Recent Climate Change," *Nature* 416 (March 28, 2002), 389–95.

57. Callum Roberts, *The Unnatural History of the Sea* (Washington, D.C.: Island Press, 2007), 82–89.

58. Kolbert, *Sixth Extinction*, 159; Wilson, "Threats to Biodiversity."

59. Katherine Bagley, "Climate Change Mix-Up," *Audubon* 115 no. 6 (November–December 2013), 42–47.

60. Anna M. Whitehouse, "Tusklessness in Elephant Population of the Addo Elephant National Park, South Africa," *Journal of the Zoological Society of London* 257 (2002), 249–54; H. Jachmann, P. M. S. Berry, and H. Imae, "Tusklessness in African Elephants: A Future Trend," *African Journal of Ecology* 33 no. 3 (1995), 230–35.

61. Stephen R. Palumbi, *Evolution Explosion: How Humans Cause Rapid Evolutionary Change* (New York: W. W. Norton, 2001), 153–55, 168–86.

62. Etienne Benson, *Wired Wilderness: Technologies of Tracking and the Making of Modern Wildlife* (Baltimore: Johns Hopkins University Press, 2010); Daniel Duane, "The Unnatural Kingdom: If Technology Helps Us Save the Wilderness, Will the Wilderness Still Be Wild?" *New York Times Sunday Review* (March 13, 2016).

63. Kolbert, *Sixth Extinction*, 263.

64. Jane Goodall, *Hope for Animals and Their World: How Endangered Species Are Being Rescued from the Brink* (New York: Grand Central, 2009), 140–44; Kolbert, *Sixth Extinction*, 219–21.

65. Michael Soulé, "Conservation Tactics for a Constant Crisis," *Science* 253 no. 5021 (August 16, 1991), 749–50; Kolbert, *Sixth Extinction*, 260.

66. David S. Wilcove, *The Condor's Shadow: The Loss and Recovery of Wildlife in America* (New York: W. H. Freeman, 1999), 239–40; Noel F. R. Snyder and Helen Snyder, *The California Condor: A Saga of Natural History and Conservation* (San Diego, Cal.: Academic Press, 2000); Goodall, *Hope for Animals*, 27–36.

67. Jon Mooallem, "Rescue Flight," *New York Times Magazine* (February 22, 2009), 30–35, and *Wild Ones: A Sometimes Dismaying, Wildly Reassuring Story about Looking at*

People Looking at Animals in America (New York: Penguin, 2013), 195–282; Goodall, *Hope for Animals*, 105–20.

68. Daniel Glick, "Back from the Brink," *Smithsonian Magazine* (September 2005); Sarah Zielinski, "What Price Do We Put on an Endangered Bird?" *Smithsonian Magazine* (April 2011).

69. Felicity Barringer, "Swim to Sea? These Salmon Are Catching a Lift," *New York Times* (April 19, 2014), 8.

70. Felicity Barringer, "Taking Up Arms Where Birds Feast on Buffet of Salmon," *New York Times* (August 16, 2014), 11, 14.

71. Alfred W. Crosby, *Ecological Imperialism: The Biological Expansion of Europe, 900–1900* (Cambridge: Cambridge University Press, 1986), ch. 10.

72. Elizabeth Kolbert, "The Big Kill: New Zealand's Invasive Mammal Species," *New Yorker* (December 22 and 29, 2014), 120–29; Abigail Tucker, *The Lion in the Living Room: How House Cats Tamed Us and Took Over the World* (New York: Simon and Schuster, 2016), 70–76.

73. J. Michael Scott et al., "Conservation-Reliant Species and the Future of Conservation," *Conservation Letters* 3 no. 2 (April 2010), 91–97.

Chapter 15

1. Adam Rome, *The Genius of Earth Day: How a 1970 Teach-In Unexpectedly Made the First Green Generation* (New York: Hill and Wang, 2013).

2. Tim Flannery, *Here on Earth: A Natural History of the Planet* (New York: Atlantic Monthly Press, 2010), 101–3.

3. Tim Flannery, *The Future Eaters: An Ecological History of the Australasian Lands and People* (New York: Grove Press, 2002), 288–89.

4. John R. McNeill, "Of Rats and Man: A Synoptic Environmental History of the Island Pacific," *Journal of World History* 5 no. 2 (Fall 1994), 308–9.

5. William M. Denevan, "Pre-European Impacts on Neotropical Lowland Environments," in Thomas T. Veblen, ed., *The Physical Geography of South America* (Oxford: Oxford University Press, 2007), 275.

6. Clarence J. Glacken, *Traces on the Rhodian Shore: Nature and Culture in Western Thought from Ancient Times to the End of the Eighteenth Century* (Berkeley: University of California Press, 1967), 148–49.

7. Yi-fu Tuan, "Discrepancies between Environmental Attitudes and Behaviour: Examples from Europe and China," *Canadian Geographer* 12 no. 3 (1968), 94.

8. Mark Elvin, "3,000 Years of Unsustainable Growth: China's Environment from Archaic Times to the Present," *East Asian History* 6 (1993), 7–46.

9. Robert B. Marks, *China: Its Environment and History* (Lanham, Md.: Rowman & Littlefield, 2012), 190.

10. Shawn William Miller, *An Environmental History of Latin America* (Cambridge: Cambridge University Press, 2007), 196.

11. Giovanni Battista Stefinlongo, *Pali e palificazioni della laguna di Venezia* (Sottomarina di Chioggia: Il Leggio, 1994), 40; Frederic Chapin Lane, *Venice: A Maritime Republic* (Baltimore: Johns Hopkins University Press, 1973), 384, and *Venetian Ships and Shipbuilding of the Renaissance* (Baltimore: Johns Hopkins University Press, 1973), 220–28.

12. Conrad D. Totman, *The Green Archipelago: Forestry in Preindustrial Japan* (Berkeley: University of California Press, 1989), 57–68, 88–93, 271–72.

13. Richard H. Grove, *Ecology, Climate and Empire: Colonialism and Global Environmental History, 1400–1940* (Knapwell, U.K.: White Horse Press, 1997), 53–66.

14. Richard H. Grove, *Green Imperialism: Colonial Expansion, Tropical Island Edens and the Origins of Environmentalism, 1600–1860* (Cambridge: Cambridge University Press, 1995), ch. 5, and "Origins of Western Environmentalism," *Scientific American* 267 no. 1 (1992), 42–47.

15. Glacken, *Traces*, 130; Grove, "Origins," 44–46, and *Green Imperialism*, 29–31.

16. Ramachandra Guha, *Environmentalism: A Global History* (New York: Longman, 2000), 26–30; José Augusto Padua, "Environmentalism in Brazil: A Historical Perspective," in John R. McNeill and Erin Stewart Mauldin, eds., *A Companion to Global Environmental History* (Chichester: Wiley-Blackwell, 2012), 460–61; Miller, *Environmental History*, 197–201.

17. Stephen Mosley, *Environment in World History* (New York: Routledge, 2010), 29.

18. Guha, *Environmentalism*, 45–46; Grove, "Origins," 46–47.

19. Mark Cioc, *The Game of Conservation: International Treaties to Protect the World's Migratory Animals* (Athens: Ohio University Press, 2009), 13.

20. David Lowenthal, "Awareness of Human Impacts: Changing Attitudes and Emphases," in B. L. Turner et al., eds., *The Earth as Transformed by Human Action: Global and Regional Changes in the Biosphere over the Past 300 Years* (Cambridge: Cambridge University Press, 1990), 121–35; Guha, *Environmentalism*, 27–32.

21. George Perkins Marsh, *Man and Nature; or, Physical Geography as Modified by Human Action* (New York: Charles Scribner, 1864), 32.

22. Guha, *Environmentalism*, 5–20.

23. Henry David Thoreau, *Walden; or, Life in the Woods* (Boston: Ticknor and Fields, 1854).

24. John Muir, *A Thousand Mile Walk to the Gulf* (Boston: Houghton Mifflin, 1916), 207–8.

25. Benjamin Kline, *First Along the River: A Brief History of the U.S. Environmental Movement*, 3rd ed. (Lanham, Md.: Rowman & Littlefield, 2007), 47–57; John Opie, *Nature's Nation: An Environmental History of the United States* (Fort Worth: Harcourt Brace, 1998), 386–88; Lowenthal, "Awareness of Human Impacts," 129.

26. Aldo Leopold, *Game Management* (New York: Scribner's, 1933).

27. Guha, *Environmentalism*, 55.

28. Aldo Leopold, *A Sand County Almanac, and Sketches Here and There* (New York: Oxford University Press, 1949), 224–25.

29. Kline, *First Along the River*, 47–60; quotation on p. 54.

30. Kline, *First Along the River*, 55.

31. Robert W. Righter, *The Battle over Hetch Hetchy: America's Most Controversial Dam and the Birth of Modern Environmentalism* (New York: Oxford University Press, 2005).

32. Donald Worster, *Dust Bowl: The Southern Plains in the 1930s* (New York: Oxford University Press, 1979), 198–203; Lowenthal, "Awareness of Human Impacts," 123–24.

33. William B. Wheeler and Michael J. McDonald, *TVA and the Tellico Dam, 1936–1979: A Bureaucratic Crisis in Post-Industrial America* (Knoxville: University of Tennessee Press, 1986).

34. Opie, *Nature's Nation*, 377–78.

35. Paul S. Sutter, *Driven Wild: How the Fight against Automobiles Launched the Modern Wilderness Movement* (Seattle: University of Washington Press, 2002), passim; Opie, *Nature's Nation*, 377–401.

36. John R. McNeill and Peter Engelke, *The Great Acceleration: An Environmental History of the Anthropocene since 1945* (Cambridge, Mass.: Harvard University Press, 2016), 186.

37. Rachel Carson, *Silent Spring* (Greenwich, Conn.: Fawcett, 1962).

38. Carson, *Silent Spring*, 8 and 15.

39. Carson, *Silent Spring*, passim. See also Kirkpatrick Sale, *Green Revolution: The American Environmental Movement, 1962–1992* (New York: Hill & Wang, 1993), 3–7; Opie, *Nature's Nation*, 414; Guha, *Environmentalism*, 69–73; and Lowenthal, "Awareness of Human Impacts," 129.

40. Lois Marie Gibbs, *Love Canal and the Birth of the Environmental Health Movement* (Washington, D.C.: Island Press, 2010); Richard Newman, *Love Canal: A Toxic History from Colonial Times to the Present* (New York: Oxford University Press, 2016).

41. J. Samuel Walker, *Three Mile Island: A Nuclear Crisis in Historical Perspective* (Berkeley: University of California Press, 2004).

42. Christopher Sellers, *Crabgrass Crucible: Suburban Nature and the Rise of Environmentalism in Twentieth-Century America* (Chapel Hill: University of North Carolina Press, 2012), 267–70; McNeill and Engelke, *The Great Acceleration*, 188; Opie, *Nature's Nation*, 393–94; Sale, *Green Revolution*, 16, 34–35.

43. Anna Bramwell, *Ecology in the Twentieth Century: A History* (New Haven, Conn.: Yale University Press, 1989), 214–16; McNeill and Engelke, *The Great Acceleration*, 193; Sale, *Green Revolution*, 20–21, 32–33, 60–65, 89; Kline, *First Along the River*, 77–90, 109.

44. Kline, *First Along the River*, 109–18; Sale, *Green Revolution*, 257–60; Guha, *Environmentalism*, 87–88.

45. Caroline Merchant, *Earthcare: Women and the Environment* (New York: Routledge, 1995), 139–66.

46. Council on Environmental Quality, *Environmental Quality: The Twenty-First Annual Report of the Council on Environmental Quality together with the President's Message to Congress* (Washington: US Government Printing Office, 1990), 271 table 9.

47. Opie, *Nature's Nation*, 447–51; Sale, *Green Revolution*, 49–77; Kline, *First Along the River*, 101–34, 155–68; Lowenthal, "Awareness of Human Impacts," 131.

48.. Raymond H. Dominick, *The Environmental Movement in Germany: Prophets and Pioneers, 1871–1917* (Bloomington: Indiana University Press, 1992), 43–66, 215.

49. Simo Laakkonen, "Environmental Policies of the Third Reich," in Simo Laakkonen, Richard P. Tucker, and Timo Vuorisalo, eds., *The Long Shadows: A Global Environmental History of the Second World War* (Corvallis: Oregon State University Press, 2017), 55–74; Jonathan Olsen, "How Green Were the Nazis? Nature, Environment, and Nation in the Third Reich (review)," *Technology and Culture* 48 no. 1 (January 2007), 207–8; Dominick, *The Environmental Movement in Germany*, 81–115.

50. Frank Uekoetter, *The Age of Smoke: Environmental Policy in Germany and the United States, 1880–1970* (Pittsburgh: University of Pittsburgh Press, 2009), 260–66; McNeill and Engelke, *The Great Acceleration*, 185; Dominick, *The Environmental Movement in Germany*, 148–58, 169–79, 191–92, 215–16.

51. Dominick, *The Environmental Movement in Germany*, 160–68, 218–26; Sale, *Green Revolution*, 43–44.

52. Uekoetter, *The Age of Smoke*, 260–62.

53. Stephen Brain, "The Environmental History of the Soviet Union," in McNeill and Mauldin, *A Companion to Global Environmental History*, 225–28.

54. Murray Feshbach and Alfred Friendly Jr., *Ecocide in the USSR: Health and Nature under Siege* (New York: Basic Books, 1992), 43–45; Douglas R. Weiner, *Models of Nature: Ecology, Conservation and Cultural Revolution in Soviet Russia* (Bloomington: Indiana University Press, 1988), 2–4, 149, 178–89, 229–35; Brain, "Environmental History," 228–35.

55. Douglas R. Weiner, *A Little Corner of Freedom: Russian Nature Protection from Stalin to Gorbachev* (Berkeley: University of California Press, 1999).

56. Feshbach and Friendly, *Ecocide in the USSR*, 229–40; McNeill and Engelke, *The Great Acceleration*, 197.

57. Glenn E. Curtis, *Russia: A Country Study* (Washington, D.C.: Federal Research Division, Library of Congress, 1998), 146.

58. Feshbach and Friendly, *Ecocide in the USSR*, 45, 108, 247–58; Zhores A. Medvedev, *The Legacy of Chernobyl* (Oxford: Basil Blackwell, 1990).

59. Brain, "Environmental History," 236.

60. Bao Maohong, "Environmentalism and Environmental Movements in China since 1949," in McNeill and Mauldin, *A Companion to Global Environmental History*, 480–83; Marks, *China*, 327–28.

61. Elizabeth Economy, *The River Runs Black: The Environmental Challenge to China's Future* (Ithaca, N.Y.: Cornell University Press, 1990), 99–102; Bao, "Environmentalism and Environmental Movements," 484–85; Marks, *China*, 322–23.

62. Economy, *The River Runs Black*, 224–26; Marks, *China*, 317–18.

63. Marks, *China*, 296–97.

64. Marks, *China*, 294–95; Economy, *The River Runs Black*, 67.

65. Judith Shapiro, *China's Environmental Challenges* (Cambridge, U.K.: Polity Press, 2012), 57–77; Bao, "Environmentalism and Environmental Movements," 484–86; Shapiro Economy, *The River Runs Black*, 96–115; Marks, *China*, 295, 322–23.

66. R. Edward Grumbine, *Where the Dragon Meets the Angry River: Nature and Power in the People's Republic of China* (Washington, D.C.: Island Press, 2010); Olivia Boyd, "The Birth of Chinese Environmentalism: Key Campaigns," in Sam Geall, ed., *China and the Environment: The Green Revolution* (London: Zed Books, 2013), 45–53, 63–66; Bao, "Environmentalism and Environmental Movements," 474–88 (quotation on p. 487); Shapiro, *China's Environmental Challenges*, 103–33; Marks, *China*, 319–24.

67. Shapiro, *China's Environmental Challenges*, 62–63, 71–72, 169.

68. McNeill and Engelke, *The Great Acceleration*, 190.

69. Joan Martínez-Alier, "The Environmentalism of the Poor: Its Origins and Spread," in McNeill and Mauldin, *A Companion to Global Environmental History*, 513–29; Guha, *Environmentalism*, 98–108.

70. Quoted in Guha, *Environmentalism*, 22.

71. Padam Nepal, *Environmental Movements in India: Politics of Dynamism and Transformations* (Delhi: Authorspress, 2009), 101–2.

72. Ramachandra Guha, *The Unquiet Woods: Ecological Change and Peasant Resistance in the Himalayas* (Berkeley: University of California Press, 1990), 48–55, 153–79, and *Environmentalism*, 110–16; Martínez-Alier, "The Environmentalism of the Poor," 516–21; McNeill and Engelke, *The Great Acceleration*, 192.

73. Madhav Gadgil and Ramachandra Guha, "Ecology and Equity," ch. 3 in *The Use and Abuse of Nature* (New Delhi: Oxford University Press, 2000); Ramachandra Guha and Joan Martínez-Alier, *Varieties of Environmentalism: Essays North and South* (London: Earthscan, 1997), 4–11; Nepal, *Environmental Movements in India*, 103–14.

74. N. Patrick Peritore, *Third World Environmentalism* (Gainesville: University of Florida Press, 1999), 65–74.

75. Miller, *An Environmental History of Latin America*, 205–15.

76. Padua, "Environmentalism in Brazil," 455–65.

77. Padua, "Environmentalism in Brazil," 459–66; Miller, *An Environmental History of Latin America*, 209–12.

78. Susanna Hecht and Alexander Cockburn, *The Fate of the Forest: Developers, Destroyers and Defenders of the Amazon* (London: Verso, 1989), 174–83; Padua, "Environmentalism in Brazil," 467–68. Quotation in Guha, *Environmentalism*, 116–17.

79. John Hemming, *Tree of Rivers: The Story of the Amazon* (London: Thames and Hudson, 2008), 298–321; Padua, "Environmentalism in Brazil," 469–70.

80. Sale, *Green Revolution*, 42.

81. Economy, *The River Runs Black*, 97–98. See also Bao, "Environmentalism and Environmental Movements," 474–80.

82. Miller, *An Environmental History of Latin America*, 206.

83. Opie, *Nature's Nation*, 480–81.

84. Opie, *Nature's Nation*, 469, 481–82.

85. Edward O. Wilson, *The Future of Life* (New York: Alfred A. Knopf, 2002), 184–85; Daniel Yergin, *The Quest: Energy, Security, and the Remaking of the Modern World* (New York: Penguin, 2011), 459–60; Kline, *First Along the River*, 105.

86. Yergin, *The Quest*, 468–73; Kline, *First Along the River*, 110–13; Opie, *Nature's Nation*, 482–84; Guha, *Environmentalism*, 141–43.

87. Edward O. Wilson, "Afterword," in Carson, *Silent Spring* (50th anniversary edition, Boston: Houghton Mifflin, 2012), 362.

88. Wilson, *The Future of Life*, 185–87.

89. Sale, *Green Revolution*, 94.

90. Chris Buckley, "China Burns Much More Coal than Reported, Complicating Climate Talks," *New York Times* (November 4, 2015).

91. Marks, *China*, 293.

Epilogue

1. Respectively: Curt Stager, *Deep Future: The Next 100,000 Years of Life on Earth* (New York: Thomas Dunne Books, 2011), 241–42; Dave Foreman, *Rewilding North America: A Vision for Conservation in the 21st Century* (Washington, D.C.: Island Press, 2004), 230; John F. Richards, *The Unending Frontier: An Environmental History of the Early Modern World* (Berkeley: University of California Press, 2001), 622; and Richard Fortey, *Life: A Natural History of the First Few Billion Years of Life on Earth* (New York: Random House, 1998), 322.

2. William C. Clark, "Managing Planet Earth," *Scientific American* 261 no. 3 (September 1989), 41–54.

3. Edward O. Wilson, *The Future of Life* (New York: Alfred A. Knopf, 2002), 23.

4. Anthony D. Barnosky, *Dodging Extinction: Power, Food, Money, and the Future of Life on Earth* (Berkeley: University of California Press, 2014), 46–51.

5. Edward O. Wilson, *Half-Earth: Our Planet's Fight for Life* (New York: W. W. Norton, 2016), 167–83, and "The Global Solution to Extinction," *New York Times* (March 13, 2016); Tony Hiss, "The Wildest Idea on Earth," *Smithsonian* (September 2014), 68–78.

6. Curtis H. Freece et al., "Second Chance for the Plains Bison," *Biological Conservation* 136 (April 2007), 175–84; Eric W. Sanderson et al., "The Ecological Future of the North American Bison: Conceiving Long-Term, Large-Scale Conservation of Wildlife," *Conservation Biology* 22 no. 2 (April 2008), 254.

7. Eli Kintisch, "Born to Rewild," *Science* 350 no. 6265 (December 4, 2015), 1148–51.

8. Suzanne Daley, "From Untended Farmland, Reserve Tries to Recreate Wilderness from Long Ago," *New York Times* (June 14, 2014), 8.

9. Josh Donlan et al., "Rewilding North America," *Nature* 436 (August 18, 2005), 913–14, and "Pleistocene Rewilding: An Optimistic Agenda for 21st Century Conservation," *American Naturalist* 168 no. 5 (November 2006), 660–81; Foreman, *Rewilding North America*; Paul Martin, *Twilight of the Mammoths: Ice Age Extinctions and the Rewilding of America* (Berkeley: University of California Press, 2005), 200–16.

10. Wilson, *Half-Earth*, 167.

11. Beth Alison Shapiro, *How to Clone a Mammoth: The Science of De-Extinction* (Princeton, N.J.: Princeton University Press, 2015), chs. 4–8; Barnosky, *Dodging Extinction*, 133–42.

12. Malia Wollan and Spencer Lowell, "Arks of the Apocalypse," *New York Times Magazine* (July 16, 2017), 35–47; M. R. O'Connor, *Resurrection Science: Conservation, De-extinction, and the Precarious Future of Wild Things* (New York: St. Martin's Press, 2015), 132–40.

13. Carl Zimmer, "Bringing Them Back to Life," *National Geographic* (April 2013); "Capra pyrenaica pyrenaica," in Wikipedia (Spanish) (accessed December 2014).

14. Erik Stokstad, "Bringing Back the Aurochs," *Science* 350 no. 6265 (December 4, 2015), 1144–47.

15. O'Connor, *Resurrection Science*, 188–204; Barry Yeoman, "From Billions to None," *Audubon* (May–June 2014), 28–33; "The Great Passenger Pigeon Comeback," longnow.org/revive (accessed December 2014).

16. Nathaniel Rich, "The Mammoth Cometh," *New York Times Magazine* (February 27, 2014); Shapiro, *How to Clone a Mammoth*, ch. 4.

17. Shapiro, *How to Clone a Mammoth*, ch. 8.

18. Michael Specter, "The Climate Fixers," *New Yorker* (May 14, 2012), 100.

19. Mark Z. Jacobson and Mark A. Delucchi, "A Path to Sustainable Energy by 2030," *Scientific American* 301 no. 5 (November 2009), 58–65.

20. For an optimistic appraisal of renewable energy potential, see Tim Flannery, *The Weather Makers: How Man Is Changing the Climate and What It Means for Life on Earth* (New York: Grove Press, 2005), chs. 27–31.

21. Vaclav Smil, "The Long Slow Rise of Solar and Wind," *Scientific American* 310 no. 1 (January 2014), 52–57.

22. Paul J. Crutzen, "Albedo Enhancement by Stratospheric Sulfur Injection: A Contribution to Resolve a Policy Dilemma?" *Climatic Change* 77 no. 3–4 (August 2006), 211–20.

23. Clive Hamilton, *Earthmasters: The Dawn of the Age of Climate Engineering* (New Haven, Conn.: Yale University Press, 2013), 174.

24. Hamilton, *Earthmasters*, 18, 107. For other, more accessible works on this subject, see Jeff Goodell, *How to Cool the Planet: Geoengineering and the Audacious Effort to Fix Earth's Climate* (Boston: Houghton Mifflin, 2010), and, more succinctly, Flannery, *The Weather Makers*, 249–57, and Henry Pollack, *A World without Ice* (New York: Penguin, 2010), ch. 16.

25. James Rodger Fleming, *Fixing the Sky: The Checkered History of Weather and Climate Control* (New York: Columbia University Press, 2010), 2.

26. Specter, "The Climate Fixers," 100.

27. James Rodger Fleming, *Fixing the Sky: The Checkered History of Weather and Climate Control* (New York: Columbia University Press, 2010), ch. 7; Ross N. Hoffman, "Controlling the Global Weather," *Bulletin of the American Meteorological Society* 83 no. 2 (February 2002), 241.

28. Clive Hamilton, *Requiem for a Species: Why We Resist the Truth about Climate Change* (London: Earthscan, 2010), 183–86, and *Earthmasters*, 85–93, 120–34.

29. Hamilton, *Earthmasters*, 2.

30. Hamilton, *Earthmasters*, 25–50; Pollack, *A World without Ice*, 266, 280–82.

31. Callum Roberts, *The Ocean of Life: The Fate of Man and the Sea* (New York: Penguin, 2012), 282–85; Stager, *Deep Future*, 116–17; Hamilton, *Earthmasters*, 36–45.

32. Specter, "The Climate Fixers," 99; Pollack, *A World without Ice*, 265–66; Hamilton, *Earthmasters*, 52–57.

33. Erin O'Donnell, "Buffering the Sun: David Keith and the Question of Climate Engineering," *Harvard Magazine* (July–August 2013), 26; David Rotman, "A Cheap and Easy Plan to Stop Global Warming," *MIT Technology Review* 116 no. 2 (March–April 2013), 52–59.

34. O'Donnell, "Buffering the Sun"; Rotman, "A Cheap and Easy Plan"; Specter, "The Climate Fixers."

35. Crutzen, "Albedo Enhancement," 217; Rotman, "A Cheap and Easy Plan"; Hamilton, *Earthmasters*, 60–71.

36. Eduardo Porter, "To Curb Global Warming, Science Fiction May Become Fact," *New York Times* (April 5, 2017); Jon Gertner, "Pandora's Umbrella, *New York Times Magazine* (April 23, 2017), 58–63.

Index

For the benefit of digital users, indexed terms that span two pages (e.g., 52–53) may, on occasion, appear on only one of those pages.

CPSIA information can be obtained
at www.ICGtesting.com
Printed in the USA
BVHW032152300821
615673BV00006B/234